# Science, Technology, Innovation, and Development in the Arab Countries

# Science, Technology, Innovation, and Development in the Arab Countries

Omar F. Bizri

ELSEVIER

ACADEMIC PRESS
An imprint of Elsevier

Academic Press is an imprint of Elsevier
125 London Wall, London EC2Y 5AS, United Kingdom
525 B Street, Suite 1800, San Diego, CA 92101-4495, United States
50 Hampshire Street, 5th Floor, Cambridge, MA 02139, United States
The Boulevard, Langford Lane, Kidlington, Oxford OX5 1GB, United Kingdom



**Library of Congress Cataloging-in-Publication Data**
A catalog record for this book is available from the Library of Congress

**British Library Cataloguing-in-Publication Data**
A catalogue record for this book is available from the British Library

ISBN: 978-0-12-812577-9

For information on all Academic Press publications visit our website at https://www.elsevier.com/books-and-journals



*Publisher:* Candice Janco
*Acquisition Editor:* J. Scott Bentley
*Editorial Project Manager:* Susan Ikeda
*Production Project Manager:* Punithavathy Govindaradjane
*Cover Designer:* Victoria Pearson

Typeset by TNQ Books and Journals

*This book is dedicated firstly to all the women and girls who bore the brunt of social injustice, political turmoil and armed conflict that has afflicted the Arab countries for many years past. May their children and grandchildren enjoy a better future.*

*On a more personal level, I would secondly like to dedicate this book also to my late grandmothers and my mother who taught me so much.*

*Thirdly, this book is dedicated to my wife Nimat for her unflinching support through many difficult times.*

*Fourthly, I dedicate this book to our two daughters, Maria and May and to their daughters Leia, Lila, Maya, and Yasmina. May they help bring about a better future for all.*

# Contents

# List of Figures

# List of Tables

# List of Frames

# Preface

As these lines are being typed, armed conflicts, leading to destruction of life and property with droves of refugees fleeing their towns and villages, continue to be daily events in several parts of the Arab world. Meanwhile, several Arab countries are gripped by political, religious, and sectarian tensions that hinder economic activity and threaten to explode into open strife. All this is taking place in a world that is undergoing political, economical, and technological developments that are transforming established power relationships across the globe toward unpredictable ends. In many ways, the turmoil that now engulfs many parts of the Arab world should come as no surprise. For many years past, statistics and opinion surveys spoke volumes of desperate Arab youth faced with limited job opportunities and inadequate social services, even in countries with adequate natural resources and among the first to gain independence as well as establish parliamentary systems of government.

In all countries of the Arab world, power has essentially been the prerogative of small groups of men, who either seized power through military coups, or who merely headed families to which departing colonialists handed absolute rule. Rentier approaches, adopted by both groups of rulers in overall governance and in running their countries' affairs, in particular, resulted in distorted relationships between rulers and rules that permeated all walks of life. Being connected to ruling elites is an essential prerequisite for a life of privilege as well as the chance for amassing untold wealth, sometimes amounting to many billions of dollars, often spirited abroad to collect interest in foreign bank accounts or to be invested in projects with returns that their compatriots never see.

Uses and abuses of religious and ultranationalist authoritarian doctrines to maintain the status quo and quell freedom of speech had their impact for quite sometime. They were soon to be swamped, however, by dashed expectations, constantly nurtured by many factors including more or less open access to international media in an increasingly connected world, while failed attempts at equitable, inclusive, and sustainable socioeconomic development became the rule rather than an exception.

This book adopts the view that many, if not most, problems faced by the Arab countries now emanate from rentier approaches, allowing poor access to, and implementation of, knowledge leading to faltering development on the one hand, and increasing poverty and inequality on the other. In particular, access to and implementation of knowledge, rooted in science, technology, and innovation

(STI) aimed at attaining essential development objectives, have repeatedly been thwarted by inadequate policies and inefficient planning. Based on this premise, this book is essentially an attempt at analyzing the Arab countries' STI performance against a background of their particular economic conditions and stated priorities.

This book is organized in six chapters. An introductory chapter sets the scene with historical notes, laying down the basis for alternative STI futures with novel policies, enhancing government contributions while wider responsibilities are shared by a host of new actors including civil society institutions, networks of scientists and engineers, and local community actors.

Chapter 2 is devoted to a review of major economic trends in each of the Arab countries, in which estimated and forecasted economic growth patterns expressed in terms of gross domestic product are outlined, unemployment and poverty statistics are reviewed, and challenges briefly discussed. Chapter 3 takes a look at STI systems in the Arab countries, with focus on STI governance, policy directives, as well as institutional landscapes. Because the Arab countries are neither socially nor economically homogeneous, both of these chapters consider their economic performance as well as STI development paths individually, while attempting, whenever possible, to draw on similarities and complementarities that may constitute the basis for future STI-based collaboration.

Chapter 4 provides a detailed and up-to-date account of STI output, in terms of quantity and quality, with emphasis on research publications produced, and patents registered and granted by the Arab countries over the past two decades or so. Special attention is awarded to analyzing the quality of research publications through comparative analysis of citation indicators.

Chapter 5 attempts an evaluation of efforts made by the Arab countries toward promoting innovation with reference to access to requisite technologies and related inputs, on the one hand, and legislative and regulatory arrangements aimed at encouraging enterprises, on the other. A special section in this chapter is dedicated to analyzing recent statistics on access to information and communication technologies. Access to the Internet, in general, and by educational institutions, in particular, thereby impacting overall socioeconomic development, is also considered.

Orientations and objectives of future policies aimed at technology acquisition, dissemination, and effective utilization in innovative enterprises are the subject of the final chapter in this book, which addresses challenges facing the Arab countries with emphasis on governance, education, unemployment, poverty, and inequality. This chapter ends with a set of policy guidelines aimed at radical changes in the way STI capacity building is to be approached, in view of past failings and future challenges and with emphasis on opportunities posed by new partnerships and contributions by novel actors at local community levels.

Work on this book took the best part of 1 year. However, it rests on an experience of over 20 years of intimate engagement in pure and applied scientific research in the region coupled to an additional decade and a half of work with

United Nations organizations and regional foundations with special focus on technology for development. Despite some of the sentiments expressed in this preface and in the book's dedication as well as what might be interpreted as a political stance toward rentier systems of governance, this book is essentially technical with no other aspiration than laying down the ground for a better understanding of past failures for the sake of a better future.

This book was not intended to tell a happy tale. To do so would have run counter to daily observations all round for many past decades. Despite some bright sparks here and there, prospects for the evolution of robust STI capabilities in the entire Arab countries along more promising trajectories in the near future are effectively doomed. Misrule by rentier systems, in control to varying degrees all over the region, by heirs to centuries of feudalism and distorted value systems, on the one hand, or fake nationalistic ideologies that regularly betray the human values they pledged to uphold, on the other, has constantly thwarted the creation of dynamic and forward-looking STI systems.

On the other hand, the fact that contradictions created by these rentier systems continue to defy all notions of socioeconomic progress will eventually render them futile and ephemeral. At the end of day, hope must rest on the yearning of every woman and every man for a better life and a better future for their children and grandchildren. As the history of the region has shown many times over, such aspirations will always prevail.

# Acknowledgments

I have many people to thank for their help, both direct and indirect, in making this book a reality. Thanks must go first to trendsetters on the path toward capacity building in science and technology. In particular, I would like to pay special gratitude to the late Dr. A. Watheq Chahid and the late Dr. Usama Khouli as well as their, and my, dear friend Dr. Antoine Zahlan.

Next, I am indebted to many friends and colleagues who facilitated access to seminal reports used in enriched parts of the book with respect to individual Arab countries. In particular, my thanks go to Ms. Rita Wehbe, Ms. Hania Dimassi, and Ms. Jana El Baba. Thanks must also go to colleagues who contributed materials on STI activities in the Arab countries that I was able to quote in this book. I am particularly grateful to Dr. Abdelkader Djeflat at the University of Lille in France, Dr. Mohamed Mrayati at the United nations Development Program (UNDP), Saudi Arabia, as well as Dr. Mouin Hamze, Secretary General of the National Centre for Scientific Research in Lebanon and Dr. Zeki Fattah, Ex-Chief of the Economic Development Division at the United Nations Economic and Social Mission for Western Asia (ESCWA).

While working on this book, I had many discussions with friends, colleagues, and experts, who shared their insights and helped me crystallize several ideas. In particular, I would like to thank Mr. Mohammad Bizri, Dr. Bashir Mounajed, and Dr. Ramzi Salamé.

My very special thanks are also due to Ms. Susan Ikeda and Dr. Scott Bentley at Elsevier for their constant help while working on this book over the past year and to Ms. Punithavathy Govindaradjane for her patient and expert assistance in proofreading manuscripts, rendering this book a reality.

Naturally, none of the persons mentioned above must bear any blame for whatever misplaced sentiments and critical references that may be found in the following pages. All such blame must rest with the author.

Finally, I am most indebted to my wife Nimat for her patient support and for the opportunities that her work, in aid of educating Syrian refugee children, offered us both for meeting so many inspiring young, and not so young, women and men; instilling enormous hopes for a much brighter future for the Arab countries and their peoples.

# Abbreviations and Acronyms

| | |
|---|---|
| AC | Article counts |
| ACSAD | Arab Center for the Study of Dry and Arid Zones |
| ACT | Africa City for Technology |
| ALECSO | Arab League Educational, Cultural and Scientific Organization |
| ALI | Association of Lebanese Industrialists |
| ARC | The Agricultural Research Corporation |
| ASEZ | Aqaba Special Economic Zone |
| ASRT | Academy of Scientific Research and Technology (Cairo) |
| ASRT | Academy of Scientific Research and Technology |
| ASTI | Agricultural Science and Technology Indicators |
| ATP | Advanced Technology Programme |
| AUB | American University of Beirut |
| AUC | American University of Cairo |
| BSc | Bachelor of Science |
| BA | Bachelor's degree |
| BCSR | Bahrain Centre for Studies and Research |
| BDP | Berytech Digital Park |
| CDS | Centre for Development Studies |
| CGIAR | Consultative Group on Agricultural Research |
| CIA | Central Intelligence Agency |
| CII | Composite Innovation Index |
| CNRST | The National Centre for Scientific and Technological Research |
| DGSRDT | The General Directorate for Scientific Research and Technological Development |
| DIA | Dutch Incubation Association |
| DRD | Department of Research and Development |
| EBRD | European Bank for Reconstruction and Development |
| EC | European Commission |
| EIU | Economist Intelligence Unit |
| EJADA | Euro Jordanian Enterprise Development Action |
| EoDB | Ease of doing business |
| EPO | European Patent Office |
| ESPRIT | European Strategic Programme for Research and Development in Information Technology |
| ESTIO | Egyptian Science, Technology and Innovation Observatory |
| ETC | ESCWA Technology Centre |
| EU | European Union |
| EUREKA | Framework for European Technological Cooperation |
| FAC | Fractional article counts |
| FAO | Food and Agriculture Organization of the United Nations |
| FC | Fractional count or fractional article count |
| FDI | Foreign direct investment |

| | |
|---|---|
| FP7 | Seventh framework cooperation programme |
| FTA | Free Trade Agreement |
| FTE | Full-time equivalent |
| FWP | Framework Programme |
| GCC | Gulf Cooperation Council |
| GCI | Global Competitiveness Index |
| GCR | Global Competitiveness Report |
| GDP | Gross domestic product |
| GNP | Gross national product |
| GTZ | German Agency for Technical Cooperation |
| H and SS | Humanities and Social Sciences |
| HAST | Hassan II Academy of Science and Technology |
| HCDP | Human Capacity Development Program |
| HCSR | Higher Council for Scientific Research |
| HCSR | Higher Commission for Scientific Research |
| HCST | Higher Council for Science and Technology |
| HDIPI | Health, Development, Information and Policy Institute |
| IAS | Iraqi Academy of Sciences |
| ICARDA | International Centre for Agricultural Research in Arid Areas |
| ICCI | Iraqi Commission for Computers and Informatics |
| ICT | Information and communications technology |
| IDB | Islamic Development Bank |
| IDRC | International Development Research Centre |
| IER | Innovation Efficiency Ratio |
| IFPRI | International Food Policy Research Institute |
| I-Hub | The Innovation Hub |
| IIC | The Industrial Innovation Centre |
| ILO | International Labour Office |
| IMF | International Monetary Fund |
| INSEAD | Institut Européen d'Administration des Affaires |
| INTILAC | National Program for Technological Specialized Incubators |
| IPC | International Patent Classification |
| IPR | Intellectual Property Rights |
| IRCC | Industrial Research and Consultancy Centre |
| IRESEN | Research Institute for Solar Energy and New Energies |
| IRI | Industrial Research Institute |
| IRPA | Intensification of Research for Priority Areas |
| ISNAR | International Service for National Agricultural Research |
| ISO | International Organization for Standardization |
| IT | Information Technology |
| ITU | International Telecommunication Union |
| JCE's | Joint Centres of Excellence |
| JD | Jordanian Dinar |
| JEDCO | Jordan Enterprise Development Corporation |
| KACST | King Abdulaziz City for Science and Technology |
| KAUST | King Abdullah University for Science and Technology |
| KD | Kuwait Dinar |
| KFAED | Kuwait Fund for Arab Economic Development |
| KFAS | Kuwait Foundation for the Advancement of Science |
| KFFG | Kuwait's Fund for Future Generations |
| KFUPM | King Fahd University for Petroleum and Minerals |
| KISR | Kuwait Institute for Scientific Research |
| KISSR | Kurdistan Institution for Strategic Studies and Scientific Research |

| | |
|---|---|
| KRG | Kurdistan Region's Government |
| KRI | Kurdistan Region of Iraq |
| KRRP | Kuwait Research Review Panel |
| KSU | King Saud University |
| LAN | Local Area Network |
| LARI | Lebanese Agricultural Research Institute |
| LASRT | Libyan Authority for Scientific Research and Technology |
| LE | Egyptian Pound |
| LRRC | Livestock Resources Research Corporation |
| LS | Syrian Pound |
| MSc | Master of Science |
| MA | Master's degree |
| MARD | Moroccan Association for Research and Development |
| MBRF | Mohammed bin Rashid Foundation |
| MENA | Middle East and North Africa |
| MHESR | Ministry of Higher Education and Scientific Research |
| MNC | Multinational company |
| MoHESR | Ministry of Higher Education and Scientific Research |
| MoST | Ministry of Science and Technology |
| MSME | Micro, small, and medium enterprise |
| NA | Not available |
| NARC | National Agricultural Research Centre |
| NCI | National Innovation Centre |
| NCR | National Centre for Research |
| NCSR | National Council for Scientific Research |
| NCTR | The Nile Center for Technology Research |
| NGO | Nongovernmental organization |
| NIC | Newly industrialized country |
| NIS | National Information System |
| NIST | National Institute of Standards and Technology |
| NPRP | National Priorities Research Program |
| NR | Nature Research |
| NRIE | National Research Institute for Energy |
| NRIFS | National Research Institute for Food Science |
| NSI | National System of Innovation |
| NSST | National Science and Technology System |
| NSTIP | National Science, Technology and Innovation Policy |
| NSTRP | National Scientific and Technological Requirements and Potential |
| OECD | Organisation for Economic Cooperation and Development |
| PALAST | Palestine Academy for Science and Technology |
| PC | Personal computer |
| PCT | Patent Convention Treaty |
| PhD | Doctoral degree |
| PICTI | Palestinian ICT Incubator |
| PIRD | Prime d'investissement en Recherche et Developpement |
| PITA | Palestinian IT Association |
| PMV | Plan for Green Morocco |
| PNM | National Mobilizing Programmes |
| ppp | Purchasing power parity |
| PSI | Palestinian Standards Institute |
| QEWC | Qatar Electricity and Water Company |
| QNRF | Qatar National Research Fund |

| | |
|---|---|
| QSTec | Qatar Solar Technologies |
| QSTP | Qatar Science and Technology Park |
| R&D | Research and development |
| RIAL | Rapid Innovation Action Learning |
| RSS | Royal Scientific Society |
| RTD | Research and Technological Development |
| RTDI | Research, Technology Development and Innovation |
| S&T | Science and Technology |
| SABIC | Saudi Basic Industries Corporation |
| SAEC | Sudan Atomic Energy Commission |
| SAS | The Sudan Academy of Sciences |
| SBIR | Small Business Innovation Research |
| SEBC | Syrian Enterprise and Business Centre |
| SERST | Secretariat of State for Scientific Research and Technology |
| SESAME | Synchrotron-Light for Experimental Science and Applications in the Middle East |
| SMART | Small Firms Merit Award for Research and Technology |
| SME | Small- and medium-sized enterprises |
| SPRU | Science and Technology Policy Research Unit |
| SRC | Scientific Research Council |
| SRTA City | The City for Scientific Research and Technology Applications |
| STD | Syrian Trust for Development |
| STI | Science, Technology and Innovation |
| STIP | Science, Technology and Innovation Policy |
| SVDMC | The Smart Villages Development and Management Company |
| TOKTEN | Transfer of Knowledge Through Expatriate Nationals |
| TRC | The Research Council |
| TRI | Technological Readiness Index |
| TRIPS | Trade-Related Aspects of Intellectual Property Rights |
| UAEU | United Arab Emirates University |
| UCASTI | University College for Applied Science's Technology Incubator |
| UN-ACSTD | United Nations Advisory Committee on Science and Technology for Development |
| UNCSTD | United Nations Commission on Science and Technology for Development |
| UNCTAD | United Nations Conference on Trade and Development |
| UNDP | United Nations Development Programme |
| UNESCO | United Nations Education, Science and Culture Organization |
| UNESCWA | United Nations Economic and Social Commission for Western Asia |
| UNIDO | United Nations Industrial Development Organization |
| USAID | United States Agency for International Development |
| USPTO | United States Patent and Trademark Office |
| WAN | Wide Area Network |
| WEF | World Economic Forum |
| WESI | Water and Environment Studies Institute |
| WFC | Weighted fractional count |
| WIPO | World Intellectual Property Organization |
| WTO | World Trade Organization |
| ZCST | Zewail City of Science and Technology |

Chapter 1

# Science, Technology, and Innovation: Paths for Alternative Futures

## Chapter Outline



## 1.1 INTRODUCTION

Political turmoil and armed conflict have engulfed many parts of the Arab region for several years, claiming hundreds of thousands of lives, destroying capital assets worth hundreds of billions and turning millions of its inhabitants into refugees. The human and economic costs of the conflict are still on the rise with no end yet in sight.

Analysis of the immediate causes of the conflict tends to blame a wide range of factors. Some even invoke complex conspiracies by outside powers, eager to loot the region's wealth and spread their hegemony. However, although plots and conspiracies may have been the bread and butter of human history, what is glaringly obvious is a total lack of immunity possessed by large segments of Arab societies against them, whether imaginary or real. Devoid of

Science, Technology, Innovation, and Development in the Arab Countries
http://dx.doi.org/10.1016/B978-0-12-812577-9.00001-5

such immunity—normally engendered by good governance, solid educational systems, plentiful employment opportunities, leading to uniform distribution of incomes and wealth—even worse fates are possible.

More plausibly, the roots of disasters that unfolded in many Arab countries are firmly rooted in persistent failure to acquire, adapt, and implement knowledge for inclusive and sustainable socioeconomic development. On the other hand, acquiring, adapting, and implementing such knowledge would have to rest squarely on the above pillars; namely good governance, quality education as well as transparent economic and social transactions.[1] In order to escape this chicken-and-egg conundrum, it is essential to encourage evolutionary processes in politics, economics as well as in scientific and technological knowledge that subvert vicious cycles, installing more benign ones in their place. For example, governance, both good and bad, requires that modern telecommunication capabilities are acquired and disseminated. Once available, such capabilities may be used to enhance the quality of education as well as access useful information on assembling and delivery of a variety of goods and services, driving a multitude of innovative enterprises while at the same time enhancing people's abilities to share crucial everyday information and organize to counter bad governance.

It is also likely that events of the current decade might convince policy makers to adopt alternative courses, opening wider possibilities for access to and implementation of knowledge, particularly knowledge linked to science, technology, and innovation (STI) capabilities in a manner that benefits all segments of society.

The current race by several Arab countries to set up their own knowledge economies is at least partly in response to some such notions. However, given goodwill and best intentions all round, trajectories toward establishing truly functional knowledge economies and societies are complicated by the fact that rentier practices have taken firm hold during the decades of misrule, inherited feudal values, and distorted legal and regulatory frameworks, thus engendering corruption and cronyism and negatively impacting public–private collaboration. Additionally, crucial coordination between concerned national STI institutions, which are at any rate ill-equipped, inadequately funded, and often mismanaged, with the aim of tackling essential development problems is absent or at best incidental.

## 1.2 A HISTORICAL PERSPECTIVE

Since their independence from colonial rule, toward the middle of the 20th century, development efforts in the Arab countries have mostly been directed toward institution and infrastructure building. Attempts were made as early as the 1950s to establish science councils. Scientific research centers came into being in many countries in the 1970s. Some of the countries even established ministries of state for technology, as well as ministries for telecommunications and technology. Higher commissions for scientific research were set up by several Arab countries under the auspices of the Council of Ministers. Many of these initiatives certainly did make their mark on the national STI landscape; however, for the most part changes they were able to bring about were mostly cosmetic and palliative, rather than fundamental.

Reviews of scientific research and technological development activity undertaken in some of the worst hit Arab countries compared with other countries in the region and around the globe indicate meager progress over the past years for which statistics are available, even prior to the onset of hostilities they witnessed several years later. With regard to innovation and innovative activities undertaken for the benefit of enterprise and job creation, several reports issued by international bodies, notably, the World Economic Forum, also indicate that countries such as Syria, Iraq, Libya, and Yemen have, for a number of years, occupied a less than satisfactory position even in comparison with other developing countries.

This situation may not be entirely blamed on the lack of will—by concerned government institutions as well as the STI community—to develop and implement viable STI initiatives. Indeed, as the following chapters will indicate, several attempts were made at relevant institution building by all Arab countries, particularly during the latter three decades of the past century and toward the beginning of the present century. Thus, many Arab countries, including Egypt, Syria, Iraq, Lebanon, Jordan, Tunisia as well as Kuwait and Saudi Arabia, among the Arab Gulf countries, established research and development facilities in the 1970s. This led to research output, at least some of which eventually translated into significant gains for national development. An example of such developments is provided by the near self-sufficiency in agricultural produce achieved by Syria during much of the 1980s and 1990s. Success attained by this country owed a great deal to endogenous agricultural scientific research and technological development aided by international and regional research centers,[2] as well as the creation of agricultural research stations by concerned government authorities. Nevertheless, several factors[3] conspired to limit and redirect research and development efforts in this country toward less, even nonproductive domains. At any rate, national institutions involved in scientific, technological, and innovation activities suffered the same if not a higher degree of undue political intervention and corruption that many other institutions in the country had to cope with, thereby curtailing their independence and abilities to contribute to national development.

## 1.3 TOWARD A BETTER FUTURE FOR SCIENCE, TECHNOLOGY, AND INNOVATION IN THE ARAB COUNTRIES

Capabilities in STI have assumed critical importance for all aspects of socioeconomic activity in the developed as well as developing countries. The Arab countries are no exception. As indicated later, particularly in Chapters 2 and 3, building knowledge economies aided by a host of advanced technologies appears to be a prime preoccupation for many in the Arab world. In this regard, and in order that future efforts lead to optimal outcomes, it is essential to consider the following points:

- Was STI capacity building progressing satisfactorily prior to the onset of the current turmoil?
- How might STI capacity building proceed in the shadow of rentier practices, attendant corruption and cronyism, widespread unemployment, and poverty?
- What kind of innovation systems and policies are needed in the immediate and near futures?
- Who else, beside governments, might facilitate STI capacity building?

Subsequent paragraphs in this chapter provide introductory comments while detailed consideration of the above questions is taken up in ensuing chapters.

### 1.3.1 STI Capacity Building Was Already Suffering Before the Current Turmoil

It is also essential to note at the very outset that the roots to current problems in the region, with their heavy impact on STI capacity building, date back to decades if not centuries ago, imprinted by a series of historical events and deeply embedded in cultural systems that enshrine rentier practices. See Frame 1.1.

Even before the Arab uprisings, which resulted in stagnation on many fronts including innovation in many aspects of socioeconomic life, it was patently evident that innovative potential inherent in these countries and their populations was far from being fully realized. Thus, in comparison with several other developing countries around the world and within their immediate vicinity, several Arab countries could claim:

- reasonable education scores by regional standards;
- a number of higher educational institutions, with some pockets of excellence;
- considerable endowments in human resources, including multilingual engineers and other professionals;
- highly skilled diaspora, particularly in Europe, Africa, and the United States;
- reasonable IT infrastructures and an IT-savvy younger generation;
- a pool of entrepreneurs with keen business sense and ambitions;
- access to large markets in the region and overseas.

Despite the above, and apart from isolated but notable success stories, actual innovation performance linked to R & D activity has been rather disappointing for reasons that include:

- limited linkages between enterprising business communities and university research centers;
- inadequate networking between home research and innovation institutions with their counterparts abroad;

**Frame 1.1 Arab Socioeconomic Development Problems are Deeply Rooted in the Region's History**

As indicated in the concluding chapter, rentier practices in the Arab countries date far before the days of hydrocarbon wealth. Indeed, they were firmly rooted in the region's tribal past, to be later consecrated by feudal and colonial regimes.

Only a little more than seven decades ago, almost all the Arab countries were held by European powers under direct colonial rule, mandate or protection arrangements. For several centuries before World War I, they had mostly been either under direct Ottoman rule or owed allegiance to the High Port. Colonial powers handed over the rule of the countries they occupied over to heads of important tribes, as was the case in the Gulf Arab countries or to nationalist leaders, as was the case in countries of the Levant and North Africa. However, these leaders were soon overthrown by coup d'états staged by army officers establishing state hegemony over all aspects of socioeconomic activity. A well-known academic and commentator remarked, the "exercise of power and public authority in the Arab World has always been defined by small groups of men—usually anchored in military establishments—who seized and sat in the seats of power."[a]

With the creation of the State of Israel, ensuing wars and constant tensions, Arab regimes had what they considered perfect pretexts for assuming absolute power. All opposition had to be stifled, in order that nothing might interfere with the ongoing struggle for the liberation of Palestine. Only a semblance of representative democracy was allowed whenever absolutely necessary and convenient, while heads of state and their privileged cronies could operate unhindered, amassing huge wealth that was often spirited away into safe havens and investment projects abroad.

Naturally, creating effective capabilities in quality education, STI became a much lower priority in the shadow of rentier systems and attendant corruption and cronyism.

---

[a]"Persistent Roots of Arab Weakness and Relinquished Sovereignty" Rami G. Khouri; April 10, 2016.

- a minute proportion of university research leading to patents or effective applications;
- virtual lack of research activity within the private sector.

With regard to innovation among existing enterprises, on the other hand, shortcomings included:

- lack of serious physical or virtual clusters in the manufacturing sectors;
- a small proportion of products and services with embedded novel technological content is exported into the region and overseas;
- only few enterprises were plugged into global value chains.

### 1.3.2 Failed Policies and Wasted Resources Could Have the Effect of Subverting the Very Systems That Initiated Them

Whatever guises it takes, political, religious, or sectarian, popular upheaval witnessed during the past few years by several Arab countries, is but one manifestation of the impasse that Arab governments have reached on many fronts, with vicious cycles feeding upon one another, and with the adoption, dissemination, and utilization of knowledge, for equitable socioeconomic development relegated lowest priority.

In the case of Syria's uprising, for example, and notwithstanding the need for more research to elucidate its root causes, and the role of various parties in its evolution, it is generally acknowledged that inadequate implementation of irrigation technologies coupled to a succession of dry winters, in Syria's Jazira province, led to catastrophic crop failures. Internal migrations of farm laborers and their families toward neighboring urban centers temporarily quelled the impact before informal social safety nets gave way, placing several urban centers under unprecedented social and economic pressures and ultimately providing extremist groups with eager recruits.[4]

In effect, many of the challenges that now confront the Arab countries, whether socioeconomic or environmental, are shared with numerous other developing, and even some developed countries. However, nowhere are the socioeconomic and environmental burdens more acute than they are in many of the Arab countries. Income inequality, unemployment, on the one hand, and environmental crises such as soil erosion, creeping desertification, water scarcity, on the other, stand at the very roots of unrest that threatens to continue unabated in the absence of radical remedies.

Moreover, although developed countries and quite a few developing ones have built research capabilities and made evident progress in mastering or adapting technologies required for tackling such threats, most Arab countries are still largely bereft of such means. In effect, whatever capacity building that took place in STI has, for the most part, been dominated by supply-side approaches as well as narrow interests dictated by rentier and corrupt practices. Indeed, such practices have become so entrenched that it would take many years of hard work to reverse their negative impacts on all aspects of life, including higher education, scientific research, technological development innovation, and entrepreneurship.

The ability of resource-rich Arab countries to deal with many of the challenges that confront them today, whether socioeconomic or environmental, is essentially due to their hydrocarbon wealth. However, most of today's challenges are expected to continue further into the future, outlasting hydrocarbon resources. Furthermore, it is likely that many such problems will not go away by simply throwing more cash at them.

Success in the acquisition, adaptation, and dissemination of science and technology inputs and their effective utilization in innovative practices across the board must rest on sound government policies, based on shared values by all segments of society, effective institutions, and infrastructures, as well as the ability to implement strategies guided by reasoned analyses rather than prejudiced arguments and corrupt practices. Attaining such success also presupposes capabilities for taking part in long-term cooperative ventures within and beyond national borders with focus on technology acquisition, dissemination, and effective utilization targeting equitable socioeconomic development.

### 1.3.3 Cash Making Innovations May Be Good but Others Are Also Needed

While innovation and entrepreneurship are generally acknowledged as drivers of national economic success and social development throughout the world,[5] current conditions that prevail in most of the Arab countries render it essential to extend the goals of innovation initiatives, whether by enterprises, government departments, or NGOs, to include social development as well as commercial returns.

Innovation policy directives promoted by many international organizations, and overzealously echoed by a variety of incubation and enterprise support institutions in the region, are mostly designed to benefit educated urban elites, the so-called "creative class," through ventures also directed toward urban elites. In the meantime, prevailing conditions in the Arab countries call for innovation on several fronts, targeting and benefiting other segments of society and targeting a variety of both traditional and novel production and services sectors.

Thus, the bulk of policy directives and subsequent initiatives are aimed at quick-win solutions based on entrepreneurial ventures that cater for fortunate urban populations with little, if any, regard for the needs of often impoverished and marginalized rural populations. Goals including the development and dissemination of improved agricultural practices, better educational aids as well as wider access to a variety of social and health services have been largely sidelined in practice despite being mentioned in various planning documents.

To attain long-lasting benefits, including job creation and social development, innovation may not be confined to ventures targeting creation of new computer games, the fashion industry, and attractive gadgets for interior decoration. Left to their own devices, however, most innovative entrepreneurs would have to respond to demand for such products while more urgent needs lie elsewhere, e.g., in the agro-food and artisanal village industries, for better education and health services in urban slums.

In several instances, incentives of various kinds may be necessary to guide the bulk of innovative entrepreneurs toward such ends. In most cases, however, there is need to marshal efforts by a variety of other actors, government authorities, NGOs as well as aid agencies, toward well-designed ventures with evident medium- and long-term social rather than more immediate financial rewards.

### 1.3.4 Government Contributions to STI Capacity Building Is Essential, but Contributions to Many Others Will Be Crucial

To a very large extent, actors on the STI landscape in most Arab countries tend to be central government institutions. Exceptions arise in the Gulf Arab countries as well as a few countries in North Africa and the Levant, where private enterprises, investment firms, and multinational companies are involved in enterprise incubation facilities, designed to tackle a variety of innovative applications, with some considerable focus on applications of information and communication technologies (ICTs).

On the other hand, a wide range of nongovernmental and civil society institutions, including charities, professional associations, and federations of commerce and industry as well as local government institutions are hardly ever involved in setting up and supporting STI facilities.

Clearly, for optimal effects, in terms of enhanced funding as well as relevance to actual needs, arrangements should be made for much greater involvement by novel nongovernmental actors. Civil society institutions, trade and industry federations as well as professional associations, including networks of scientists and engineers should help create more dynamic STI capabilities through linking up to the large Arab STI diaspora and providing their expertise to schemes coordinated by both central as well as local authorities, reaching out with STI inputs where these inputs are most needed.

Above all, future initiatives aimed at building STI capacity would need to capitalize on the Arab youths' potential in building an environment conducive to scientific discovery, communication, and advocacy as well as explore the viability of programs aimed at "brain circulation" rather than "brain drain" and greater mobility of academics and researchers within the region and with partners overseas.

For many of the above modalities to be realized in harmony, two prerequisites are of the essence:

- a candid evaluation of past initiatives and outcomes;
- national and regional charters that guarantee best practices and ethical standards in research, technology acquisition, and development.

## 1.4 CONCLUDING REMARKS

Effective capacity building in all the countries of the Arab world calls for a radical paradigm shift emphasizing the needs of much wider segments of society and discarding rentier practices and the ill effects they invariably engender on many fronts.

Focus must be maintained at all times on creating conditions for inclusive and sustainable development, with immediate goals including raising educational standards, building robust research and development capabilities that target national priorities and providing conducive environments for innovative enterprises to service a broad range of national needs.

Reform plans need to be implemented for ensuring that public, private as well as civil society institutions are able to operate efficiently within regulatory frameworks that provide incentives while maintaining as much focus on equitable local community development. To do so, the need has never been more acute for close monitoring and evaluation schemes as well as open and transparent reporting channels.

The importance of enhancing allocations to STI activities is now widely acknowledged as a prerequisite for socioeconomic development. That STI inputs will open up opportunities for diversified, and hence more robust, national economies, less dependent on natural and nonrenewable resources, is also acknowledged by official policies and pronouncements by government officials. However, it may be confidently stated that with very few exceptions, limited concrete action has been taken by many Arab governments on the basis of such knowledge.

In fairness, several Arab governments in the region, including Syria, Iraq and Egypt, did launch initiatives back in the 1960s and 1970s with a view to strengthening scientific and technological capabilities and with economic development as an ultimate objective. However, in most cases such initiatives faltered or failed to provide their full benefits because of mismanagement, unstable or inconsistent financing and support. In numerous cases, failures of such initiatives were essentially due to managers being selected on account of their loyalty to the ruler or ruling party rather than their professional qualifications.

The Arab countries today occupy a region of the world that formed the cradle of human civilization. This fact is often overshadowed by the extent of the challenges that these countries and the region at large faced in recent years and are expected to endure for several years to come. Nevertheless, as many would argue, the region where the Arab countries are situated today has witnessed much better days as well as disasters of comparable proportions, then managed to collect the pieces and rebuild its future. Although history is not bound to repeat itself, hope remains with the people of the region to conquer forces of division and strife, political, religious, and sectarian, rebuild their ruined homes once more, and establish better futures for themselves.

Science, technology and innovation will provide precious tools for overcoming the numerous hurdles that the Arab countries are bound to face toward inclusive and sustainable development. Although some of these countries are confronted by insurmountable odds, hope must rest always in the ingenuity and resilience of common humanity.

## ENDNOTES

1. A recent report by the United Nations Economic and Social Commission for Western Asia states that although the immediate causes of the conflict in Syria, for example, may be debatable, mounting evidence points to years of short-sighted social and economic development policies, inadequate responses to new global and regional realities as essential causes for popular resentment. See "National Agenda for the Future of Syria." A Project Document by the United Nations Economic and Social Commission for Western Asia (UN-ESCWA), June 20, 2015.
2. Namely the International Center for Agricultural Research in Arid Areas (ICARDA) and the Arab Center for the Study of Dry and Arid Zones (ACSAD). See Section 3.2.5 in Chapter 3.

3. Including sanctions imposed by the United States and Europe on the import of certain materials and equipment items into Syria.
4. Similar stories may undoubtedly be told about each and every other Arab country, whether or not they are suffering the conditions now witnessed by Iraq, Libya, Syria, and Yemen.
5. See, for example, "Innovation Policies for Development" by M Goedhuys et al., United Nations University and Maastricht University. Published as Chapter 3 of the Global Innovation Index 2015 Report. Downloaded in March 2017 from http://www.wipo.int/edocs/pubdocs/en/wipo_pub_gii_2015-chapter3.pdf.

Chapter 2

# The Arab Countries' Economies and Development Policies

**Chapter Outline**




**Science, Technology, Innovation, and Development in the Arab Countries**
**http://dx.doi.org/10.1016/B978-0-12-812577-9.00002-7**

## 2.1 INTRODUCTION

The combined population of the Arab countries,[1] was estimated in 2014 at over 392 million or 5.3% of the world's population. Inhabitants of the Arab countries are one of the world's most youthful populations. Under normal conditions, this would surely count as a blessing. However, for Arab governments with dysfunctional institutions, limited, or no social services and obsolete educational systems, their young populations have turned from an onerous burden into a grave menace.

The combined gross domestic product (GDP) of the Arab countries was estimated in 2014 at approximately US$ 6 trillion,[2] or 5.8% of the world's combined GDP in the same year. However, as Table A.2.1 indicates both GDP and population figures vary widely from one Arab country to another. Thus, although annual GDP per capita reached an estimated US$ 174,000 in Qatar in 2011, it fell down to between US$ 600 and 450, in Somalia. Inequalities persist and may even be more stark and damaging within countries; between regions and social groups.

Economic growth throughout the Arab region declined during the past few years due to the global economic slowdown and falling oil prices. On the other hand, political instability and outright armed conflict produced even more disastrous humanitarian as well as economic consequences in countries such as Iraq, Libya, Syria, and Yemen.

Events that took place toward the beginning of the present decade in Tunisia, followed by even more violent unrest in several other Arab countries have been the subject of analytical studies by journalists and scholars. A few notions may be distilled from the plethora of articles on the subject:

- Protests were motivated by enhanced social and economic expectations as well as wider awareness of the situation people found their countries in, in terms of cronyism, corruption, mismanagement, relative to countries around the world.
- In the absence of possibilities to engage in legitimate political action, and with political parties either completely subservient to or in close alliance with the ruling elite, many young people fell prey to religious movements verging toward extremist ideologies and deformed interpretations of Islam.
- Inequality of income distribution, perhaps more than abject poverty, played an important part in the series of uprisings starting with Tunisia and then spreading to its immediate and more distant neighbors.
- Advances in communication technologies played an important part in empowering protestors but failed to allow further gains as authorities all over took measures to curb or undermine their use.

The following sections present brief summaries of the economies of the Arab countries within five groups, essentially on geographical basis, but also reflecting the extent of shared natural resources as well as social and cultural characteristics. A close look will be taken at how the Arab countries' economies evolved over the past few years, sectoral contributions to their economic growth, and challenges they face, with some emphasis on employment creation and poverty reduction. Future national visions and economic development plans formulated by some Arab countries will also be briefly reviewed in preparation for more detailed consideration of their STI policies and related capacity building efforts in later chapters.

## 2.2 THE LEVANT GROUP OF ARAB COUNTRIES

All the countries in the Levant region, with the possible, albeit partial exception of Jordan, witnessed intense political strife over the past few years. At the time of writing these lines, extensive military operations continue in both Iraq and Syria, although normal socioeconomic activity, including education, research, development, enterprise creation, in Palestine is routinely hampered by Israeli occupation. Lebanon in which armed conflict has been limited to areas close to its borders with Syria had more than its fair share of it during its civil war in the 1970s and 1980s, whose trauma are continually revived by political and sectarian maneuvers. By contrast, Jordan, which also witnessed armed strife in the early 1970s, has since enjoyed relative stability, coupled to both positive and negative consequences, of political and economic changes in its neighborhood.

In 2014, the Levant Arab countries were home to around 19% of the Arab countries' population and produced 13% of these countries' domestic product. The following paragraphs will examine economic development in the Levant Arab countries.

### 2.2.1 Iraq

Iraq's mainly state-controlled economy is dominated by its oil wealth. The oil sector provides in excess of 90% of government revenues and over 80% of foreign exchange earnings.

In addition to nearly four turbulent decades in the past, Iraq has recently had to confront several crises, including the rise of the fanatical Islamic State (IS) and its complete takeover of wide swathes of the country including one of the country's principal cities, Mosul. Simultaneously, an oil price shock resulted in decreased government investment curtailing its ability to employ Iraqi youth and provide contracts to private enterprises for much needed infrastructure rebuilding and repair.

The above crises have served to expose a number of inherent vulnerabilities, including sectarian divisions, generally outdated institutions, and decision-making processes as well as widespread corruption,[3] within and outside government circles.

Fig. 2.1 indicates that contributions to GDP by the country's industrial and services sectors are comparable, falling within 45%–50%, holding out good prospects for future nonoil economic activity. However, agriculture contributes only around 6% to the country's GDP, far below the country's potential, while employing over 20% of Iraq's total workforce, which was estimated in 2010 at around 9 million, indicating strong possibilities for massive productivity improvements. By contrast, employment in the industrial sector falls far below its contribution to the national economy. This is essentially because the oil sector employs only around 12% of the national workforce.[4]

### 2.2.1.1 Future Economic Growth

Iraq's macroeconomic outlook is subject to internal developments as well as the evolution of the regional and global environments. On the national front, fighting the IS insurgency, sectarian infighting, and rampant corruption take center stage. Resolving regional instabilities is a medium- to long-term proposition, although a low rate of global economic growth and excess global oil supplies would certainly exert downward pressures on the economy's growth, in both the short and medium terms. It is such eventualities that lead to predictions of lower growth, which would also be expected to curtail reform efforts on many fronts.

Fig. 2.2 indicates that the growth of nonoil economic activity over the period 2017–2021 is expected to assume comparable rates to oil-based economic growth. Overall, the real GDP growth is likely to fall within the range

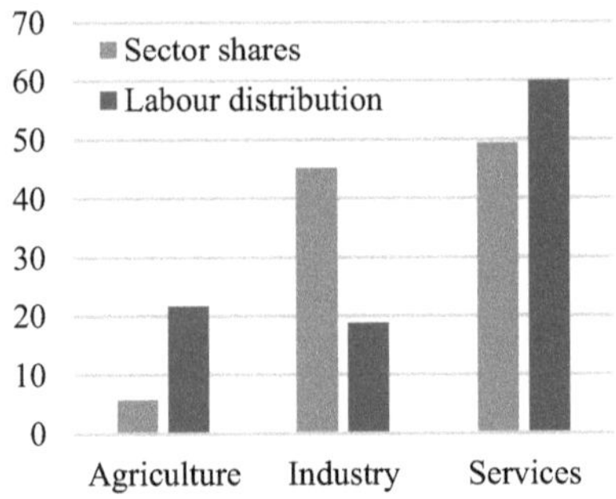


**FIGURE 2.1** Sectoral shares in GDP and labor distributions across sectors in Iraq (%).

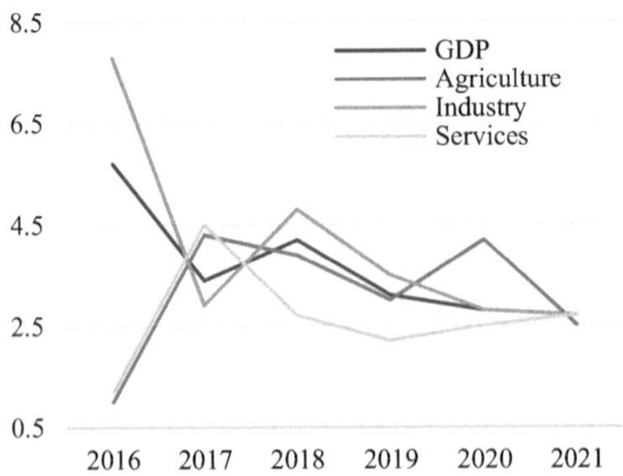


**FIGURE 2.2** Estimated and forecast economic growth in Iraq (%).

3.4%–4% during 2017–2018. This forecast is based on planned development of oilfields in southern Iraq and the nonoil economy regaining strength, particularly as the agricultural sector returns to normal, opening up fresh possibilities for science, technology, and innovation inputs. Yet, in the absence of radically new approaches, future economic activity in Iraq will continue to rely heavily upon oil exports, with national oil output expected to average around 4.6 m b/d in the near and medium futures, as pipeline and export terminal constraints curtail major expansions in oil exports.

Nevertheless, as indicated by Fig. 2.2, agriculture is not expected to grow over the coming 4–5 years beyond the values it has achieved in the past due to intrinsic weaknesses, compounded by both internal and external migration of rural populations.

### 2.2.1.2 Unemployment and Poverty in Iraq

Unemployment remains an acute problem throughout the country. While in 2012 unemployment figures stood at around 12%, more recently the proportion of the unemployed was reported in excess of 19%.[5] Total unemployment among youth, which is defined as the percentage of both males and females aged between 15 and 24 within the national workforce, is estimated by International Labor Organization (ILO) at 34.6%.[6]

In the past, attempts at improving the national employment profile, Iraq's government acted in a manner that is typical of many developing countries in the region. Thus, more Iraqi graduates and others with lesser qualifications were employed by the public sector during the past decade, thereby swelling the ranks of public sector employees from 1.1 million in 2004 up to around 3.5 million in 2014.[7,8]

If properly managed, the current inability by the bloated public sector to absorb more employees could turn out to be a blessing in disguise, encouraging the creation of private enterprises. However, for the enterprise sector to thrive there is a need for radical reforms to facilitate efforts by both Iraqi citizens and foreign enterprises starting new business ventures.

Prevalent corruption, outmoded infrastructures, inadequate essential services, shortage of skilled labor, and obsolete regulatory arrangements will continue to conspire against such developments, although slow progress is being made toward enacting laws and creating institutional capabilities for the design, implementation, and follow-up economic policies.

Iraq was reported in 2008 with 25% of its population below the poverty line.[9] With current conflict and security problems forcing many citizens to leave their home, this figure may have only increased since then.

In summary, events over the past four decades militated against sound socioeconomic development in Iraq. More recently, armed conflict at home and in neighboring Syria, combined to falling oil prices, failed attempts at unified control of oil revenues[10], as well as rampant corruption have reduced the federal government's revenues with negative impact on public spending on a wide range of services, including health and education.

Preoccupation with the overall security situation and the need to recapture territory taken over by the IS have stalled policy reforms included in the past development plans,[11] some of which were meant to reform higher education, scientific research, and technology development. For the near term, sustained improvements in overall living standards and public services remain dependent upon a rebound in global oil prices.

For the longer run, sustainable gains in socioeconomic terms will require radical reforms, which are expected to remain hostage to the current tug-of-war situation between various sectarian and political factions. Without such reforms, attempts aimed at improving the economic climate, particularly through science, technology and innovation inputs, will most likely remain limited in extent and impact.

### 2.2.2 Jordan

Jordan's economy is among the smallest of its Middle Eastern neighbors.[12] In addition to inadequate water supplies, it possesses only modest mineral resources in the form of oil shale and phosphate rock. Oil shale deposits are being evaluated,[13] although phosphate rock is mined and exported mainly as a raw material. Only a limited proportion of extracted phosphate ore is processed locally.[14]

Jordan's economy suffered several setbacks over the past few years, both directly and indirectly due to armed conflicts in Syria and Iraq, as well as rising tension elsewhere in the region, including

- attacks in Sinai on natural gas pipeline supplying Jordan with fuel for electricity generation and other uses;[15]
- an unprecedented influx of refugees;
- disrupted trade routes due mainly to armed conflict in both Syria and Iraq;
- lower rates of foreign investments;
- declining tourism.

Contributions by the main economic sectors to the national economy as well as their corresponding shares of the labor force, which was estimated at

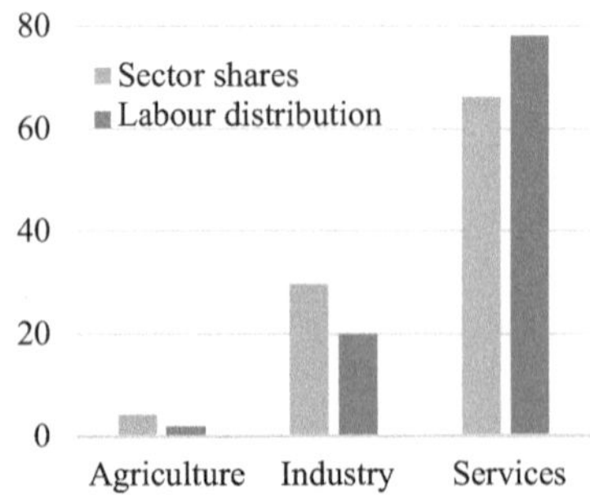


**FIGURE 2.3** Jordan's sectoral contributions to GDP and labor distribution across sectors (%).

a total of 2.2 million in 2016, are presented in Fig. 2.3, which indicates the predominant role of the services sector, compared to a miniscule contribution by the agricultural sector as well as an apparently efficient utilization of labor resources by this latter sector.

Major challenges facing Jordan's socioeconomic development include

- high unemployment, see the section on unemployment and poverty below;
- dependency on grants and remittances from Gulf economies;
- continued pressure on natural resources;
- a chronically high poverty rate, with population below the poverty line estimated in 2002 at over 14%;
- budget and current account deficits and increasing government debts.

At any rate, the global economic slowdown and regional turmoil made for slower growth during the past 6 years. Thus, in the following 4 years of relative recovery, partially aided by the availability of cheap refugee labor, Jordan's economy slowed down in 2015. Real GDP growth was estimated at 2.4% in 2015,[16] compared to 3.1% in 2014.

Fig. 2.4 indicates that GDP growth is expected to improve to reach 2.8% in 2018, although industrial growth is estimated at 3.7% in the same year.[17]

Lower commodity, raw material and food product prices in 2008 helped boost consumption. Nevertheless, the economy experienced declines starting in 2009, which have been linked to reduction in remittances by Jordan's immigrant workforce in the GCC countries. Decreased exports due to the global economic crisis have also had a pronounced negative impact.

### *2.2.2.1 Government Economic Policies*

Policies enacted by Jordan's government during the first decade of the century were advertised as aimed at revitalizing the national economy, including attempts at expanding foreign trade and privatization of state-owned companies, which resulted in attracting foreign investment, contributing to an average annual economic growth of 8% during 2004–2008. Jordan's privatization drive

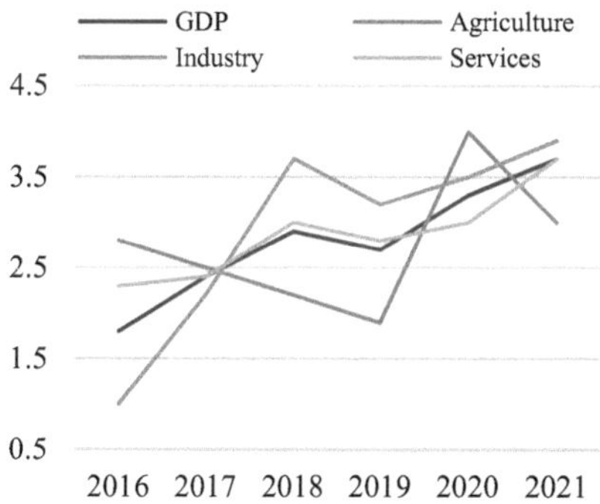


**FIGURE 2.4** Jordan's estimated and forecast economic growth (%).

is reported as having moved to private ownership facilities in the transportation, communication, water, electricity, education, and health care sectors, leaving government ownership of facilities in the education and health sectors at ~60% in 2010.

Additionally, Jordan embarked on structural reforms in education, health, during the past decade. Social protection systems and subsidies were overhauled, creating opportunities for public–private partnerships in infrastructure building. Further measures are said to be in the pipeline with the aim of improving the investment climate and the business environment.[18]

Although growth-enhancing regulatory reforms may be expected to reduce the economy's sensitivity to remittances and external shocks, it would also be essential to renovate a host of sectoral policies with emphasis on acquiring proven technology inputs coupled to educational and retraining reforms.

A number of steps were taken to streamline taxing, e.g., temporary laws were passed to overhaul the tax system, including the introduction of a 14% flat tax for nearly all corporations and a simpler, two-tier income tax for individuals.[19] Income generated by individual farmers will continue to be tax exempt. However, agribusinesses will enjoy full exemption on the first JD75,000 in income, but would be required to pay 14% on income generated beyond that threshold.[20] Of particular significance to building capacity in science, technology, and innovation is the fact that the new law does away with taxes covering universities, scientific research, and vocational training.

#### *2.2.2.2 Creating Employment in Times of Crisis*

Jordan's economy has always struggled to provide employment for the national workforce even without the added burden of the competition by its nearly 1.4 million Syrian refugees, willing to work unofficially, and often for lower pay. Government attempts to remedy the situation, through programs designed to integrate Syrian refugees in occupations that also benefit nationals. However, despite international financial support, only isolated successes have thus far been reported.

Jordan's Department of Statistics (DOS) is reported as indicating that unemployment in Jordan rose to around 15.8% in the third quarter of 2016, an increase of more than 2% from late 2015.

However, unofficial estimates put overall unemployment even higher at between 25% and 30% of the labor force.[21] The slow pace of growth in industrial activities, including construction and mining, which employ men for the most part is blamed for declining male unemployment, whereas female unemployment rates among Jordan's working population was reported by DOS as having stabilized at around 25%. This is explained by service sectors such as financial services, healthcare, and education, with traditionally higher female participation, witnessing steady expansion.

Despite government efforts at reforming educational programs, also enjoying international aid, unemployment among graduates is quite high at around 27%, among men and exceedingly high at nearly 80% among women with university degrees. While educational reforms will understandably take a long time to show concrete results, an additional factor that is quoted for low employment among young graduates is difficulty in accessing financial aid for small enterprises, essentially due to high borrowing costs, suppressing opportunities for enterprise, and subsequent job creation.

As measured by the percentage of the population living on an income below US$ 3.1 per day,[22] poverty in Jordan is reported as having declined from levels of around 22% in the late 1980s down to around 12% in 1997 and then further to reach 2% in 2010.[23] More recent figures on poverty rates are unavailable and neither are there any figures indicating the extent of rural poverty. It may be noted that such figures should understandably be handled with some caution. At any rate, the country's Gini coefficient, measuring overall inequality was reported in 2010 as relatively low at 0.33.

In conclusion, government economic policies are generally guided by the need to shield segments of its population from the impact of the economic downturn, the armed conflict in neighboring Iraq and Syria as well as rising tensions throughout the region. Attempts are being made to encourage greater private sector involvement in the economy, including promotion of public–private partnerships. However, the likelihood is that things will proceed rather slowly on this front, considering risk aversion by private businesses and the financial sector.[24]

## 2.2.3 Lebanon

Lebanon's economy is essentially service oriented with banking and tourism as its principal growth sectors; see Fig. 2.5. Remittances by its sizeable diaspora make up a significant proportion of the country's GDP.[25]

The civil war during 1975–90 dealt a severe blow to the country's economic infrastructure, drastically reducing national output, and wrecking its established status as a Middle Eastern banking hub. Following the end of the civil war

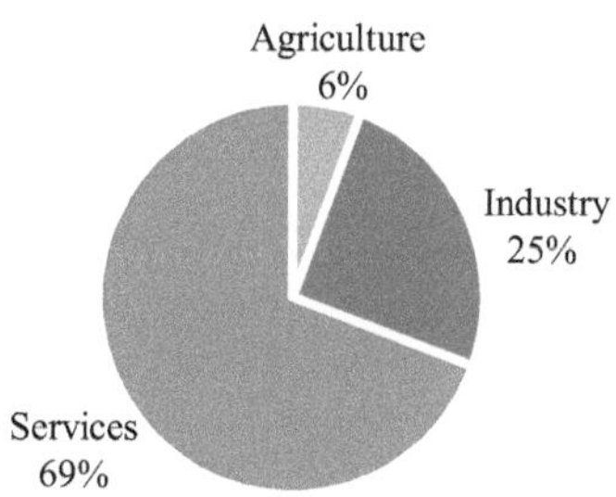


**FIGURE 2.5** Sectoral contributions to GDP in Lebanon (%).

Lebanon had to borrow heavily in order to rebuild its physical and financial infrastructure. Chronic fiscal shortfalls have tended to increase the debt-to-GDP ratio, which was estimated in 2016 as the fourth highest in the world, at 161.5% of GDP. However, most of Lebanon's debts are domestic, i.e., owed to Lebanese banks.

Although Lebanon has always been a free-market economy with a tradition of openness toward foreign investment, its domestic climate suffers a good deal of bureaucratic hurdles and arbitrary licensing decisions, often reinforced by sectarian considerations. Corrupt practices making use of complex customs procedures, high tariffs and archaic legislation further curtail business operations.

### 2.2.3.1 Economic Growth in Lebanon

In 2016 Lebanon's growth rate was estimated at a modest 1.0% with growth in the agricultural, industrial and service sectors estimated in the same year to fall around 2.0%, 1.4% and 0.8%, respectively.[26] Future patterns of growth for GDP as well as the main sectors of the economy were forecast to follow similar patterns, as indicated in Fig. 2.6, with growth in services showing higher rates in 2021, at 3.5% compared to growth in agriculture and industry at 3.0 and 2.9, respectively.

Weak economic growth, undoubtedly linked to turmoil in the region and particularly to the conflict in neighboring Syria, have placed limits on tax revenues, thereby curtailing the government's ability to invest in infrastructure improvements, including water and electricity supplies.

The discovery of reportedly sizeable hydrocarbon resources off Lebanon's coast is expected to have a considerable impact on the country's economy. However, delays due to internal politics, falling hydrocarbon prices, and regional tensions are expected to extensively delay exploration and exports for quite a few years ahead.

Surveys undertaken in preparing Lebanon's Science, Technology and Innovation Policy document indicated that Lebanon's industrial sector faced

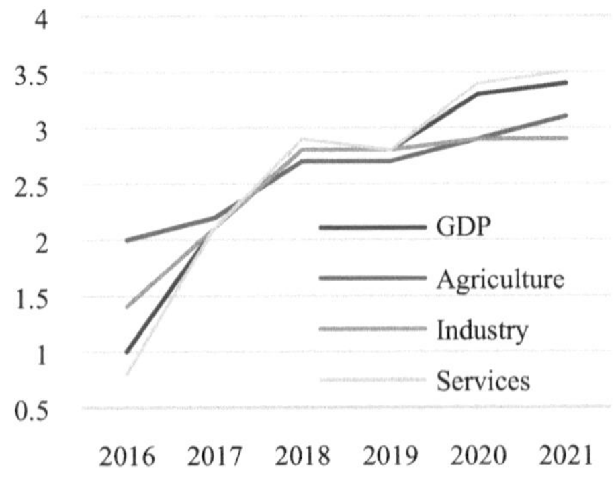


**FIGURE 2.6** Lebanon's estimated and forecast GDP and sectoral growth (%).

a number of internal and external challenges listed in Frame 2.1 below. Several of these challenges may be overcome or reduced to manageable proportions through well-designed science and technology inputs and related interventions.

### *2.2.3.2 Unemployment and Inequality in Lebanon*

With more than 1.1 million registered Syrian refugees in 2014, labor supply far outweighs demand, reportedly pushing Lebanese into unemployment. However, Lebanon's high unemployment level is nothing new. Thus, figures published by the World Bank indicate that total unemployment, that is, the percentage of the unemployed relative to the country's entire labor force oscillated between 7% and 8% for most of the past 30 years, whereas it stood at 6.4 in 2014. On the other hand, youth unemployment has recently been reported at 24.5%.[27] Additionally, the poverty head count ratio to the total population relative to the national poverty line was estimated at 28.6% in 2004.[28]

**Frame 2.1 Internal and External Challenges Facing Lebanon's Industrial Sector**

Internal challenges outlined by the STIP document include:

- Small-size of industrial firms;
- Labor-intensive production and meager investments;
- Limited diversity in production;
- Predominance of family businesses, characterized by monopolized decision making;
- Limited industrial investments;
- High production costs, including cost of raw material and energy;
- Generally poor quality of products, limiting their ability to penetrate foreign markets;
- High transaction costs;
- High energy intensity;
- Limited resources at the Ministry of Industry for proper management and support of the industrial sector and looking after its research needs;
- Limited control in formulation and application of industrial standards and norms.

On the other hand, external challenges identified by the STIP document include:

- Poor competitive ability on the world market, due to high production costs and inadequate conformity to quality standards;
- High competition for foreign direct investment (FDI), especially in the absence of a conducive FDI environment in Lebanon;
- Increased number of regulations, resolutions, standards and codes that the industry has to abide by since the signing of bilateral and multilateral trade agreements with various Arab countries and the EU and in acceding to the World Trade Organization.

The need for policies that ensure more equitable distribution of wealth through improved entrepreneurship and job creation activity is evident to most policy makers and practitioners but continually hindered by an archaic system based on sectarian divisions. Thus, Lebanon with a Gini coefficient estimated at 85.5, occupies a position among the six countries on the top of the world's wealth inequality league.[29] Ensuring lasting political stability and stemming the country's notorious brain drain may not be attained under conditions that perpetuate inequality.

## 2.2.4 Palestine

Israeli policies, coupled to political tensions between the West Bank and Gaza, Israeli and Egyptian restrictions of movement of goods and people in and out of the Gaza strip as well as settler activities in the West Bank, continue to disrupt many aspects of national development, with serious results for Palestinian industrial and business activity.[30] Since 2007, Israel's border control measures have become more obstructive toward Gaza in particular, severely limiting export activities, which constitutes the main source of income for many businesses, resulting in high unemployment and widespread poverty as well as curtailing private sector activities.[31] Egyptian policies aimed at dismantling smuggling of fuel, construction material, and consumer goods, have also played an important part in exacerbating shortages.

### *2.2.4.1 Sources of Government Revenues*

Palestine continues to rely on donor aid in order to pay salaries as well as engage in infrastructure building. Donor aid enabled government-funded services and fuelled consumption-driven growth from 2007 to 2012.

Table 2.1 outlines contributions by agriculture, industry, and services toward Palestine's GDP, indicating the small percentage contributed by

**TABLE 2.1 Contributions by Main Economic Sectors to Palestine's GDP**[a]

| GDP Composition by Sectors | Agriculture | Industry | Services |
|---|---|---|---|
| Sector shares (%) | 3.5 (3.3) | 25.2 (21.7) | 71.7 (61.9) |
| Labor distribution by sectors (%) | 11.5 (5.2) | 34.4 (10) | 54.1 (84.8) |
| Labor force 828,000 (1.157 million)[b] | | | |

[a]*Figures for Gaza between brackets.*
[b]*Data on GDP and GDP per Capita include Gaza.*
From CIA factsheet on the West Bank. Downloaded in January 2017 from: https://www.cia.gov/library/publications/the-world-factbook/geos/we.html and CIA factsheet on Gaza. Downloaded in January 2017 from: https://www.cia.gov/library/publications/the-world-factbook/geos/gz.html.

agriculture and the relatively large contribution by services. The contribution by the agricultural sector in the West Bank is comparable to that of the Gaza strip. However, the contribution by the agriculture sector in the West Bank is generated by 11.5% of its labor force, whereas that of the Gaza strip is produced by around 5% of its total labor force. This clearly raises issues of efficient use of human resources with implications for training and technology acquisition.

A similar situation arises in reference to the industry sector. Thus, 34% of the entire labor force engaged in industrial activity in the West Bank contributed around 25% to GDP in 2014. Corresponding figures for the Gaza strip were estimated at around 22% of GDP and 10% of the total labor force. Naturally, such differences are closely linked to the nature of industrial activities in both locations, as well as restrictions placed by Israel's occupation of the West Bank and its blockade of the Gaza strip. At any rate, future socioeconomic development and STI capacity building measure would need to investigate the range of industrial activities undertaken in the West Bank and the Gaza strip with view to addressing efficient utilization of human resources.

Donor support took a significant dip in recent years, with both direct and indirect deleterious effects on unemployment, with overall rates estimated in 2016at 27%; 18% in the West Bank and 42% in Gaza, with youth unemployment at around 58%. In 2012, estimates of populations living below the poverty line in the West Bank and Gaza were listed as 18% and 30%, respectively. And, although nearly 80% of Gaza's residents receive some form of aid, poverty rates are very high.

### 2.2.4.2 GDP Growth

Fig. 2.7 [32] compares the variation of the combined GDP per capita for the West Bank and Gaza to averages for the Arab countries taken together and the world at large, since 2001, highlighting the disparity that separates the Palestinian

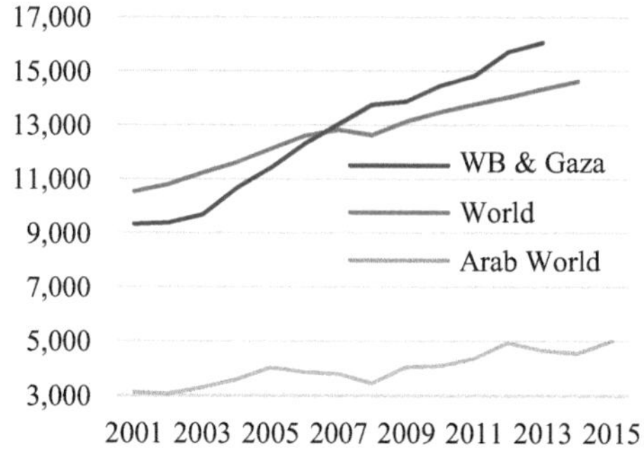


**FIGURE 2.7** Growth of GDP per capita, in the West Bank and Gaza, the Arab countries and the World since 2001 (US$).

territories from the rest of the world as well as its immediate surroundings, within the Arab region. It is not only that GDP in the Palestinian territories is typically around one-third of those for the Arab countries taken together and for the World at large, additionally, growth rates have been considerably different, with GDP growth in both the West Bank and Gaza, throughout the period covered by this figure a mere 21% of that for the Arab countries and 42% of that for the entire world, during the same period.

Fig. 2.8,[33] on the other hand provides a more detailed view of measured, estimated, and forecast GDP growth in the West Bank and the Gaza strip during the period 2011–18. This figure clearly illustrates the sharp decline in GDP growth due to the war on Gaza in 2014. In addition to the dead, the wounded and the human suffering, the 2014 war on Gaza, caused economic damages estimated at US$ 1.7 billion. The effects of this war will persist for years to come as reconstruction is hindered by Israel's restrictions on building materials entering the Gaza strip, on the one hand, and as donor contributions are delayed, on the other.

Although considerable construction work has been undertaken, however, the Gazan economy is not expected to regain its pre-2014 war levels any time soon. In addition to its slow recovery, Gaza suffers severe disruptive shortages including fuel for electric power, clean water, and at times medical supplies. These shortages are unlikely to find remedies in the near future, barring an easing of restrictions by both the Israeli and Egyptian governments as well as improved relations between Palestinian political leaders.

Fig. 2.8 also indicates the modest growth attained immediately following the Gaza war, as well as the level of growth expected in 2017 and 2018, which is in no way comparable to levels attained toward the beginning of the period covered by this figure.

Israel's blockade of the Gaza strip and occupation of the West Bank will continue to restrict economic activity and employment opportunities. On the assumption that violence does not flare up uncontrollably, growth is

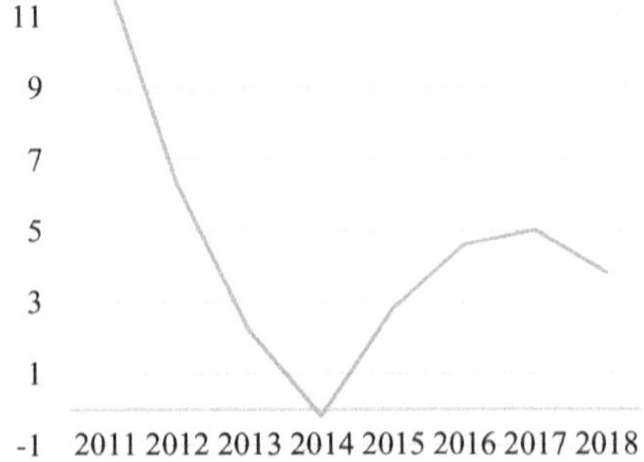


**FIGURE 2.8** Actual, estimated and forecast real GDP growth in the West bank and Gaza (%).

expected to reach 5% in 2017, with the services sector witnessing particular growth in the West Bank. Recovery is also likely in Gaza during 2017–18 but at a much slower pace due to Israel's restrictions on rebuilding and movement of people and goods. Palestine's real GDP growth is forecast to fall to around 3.8%, in 2018, which is well below its potential as reconstruction in Gaza slows down. However, even such levels are contingent upon a relatively stable political environment, revenue transfers by Israeli authorities and donor support.[34]

#### *2.2.4.3 Economic Activity Under Occupation*

Improving the state of the economy as well as social welfare in Gaza and the West Bank is contingent upon an environment that allows sustainable, private sector investment, financial support from the international community as well as Arab country donors. Although Israel's border restrictions continue to drive off potential private investments, lower oil prices and current conflicts in the region have resulted in reduced donations by the Arab countries. Additionally, global economic and political developments render increased international support less likely.

With limited chances of substantive progress in the peace process and with many parts of the West Bank and Gaza under siege, questions for Palestine's policy makers should include consideration of how investment in scientific research, technology development, and innovation initiatives might reduce or temper down the impact of Israel's blockade of Gaza and the occupation in the West Bank. Issues in need of priority attention and subsequent action should include access to clean water, alternative sources of electric power, for which technology inputs are widely available. Investment aimed at accessing such inputs would certainly require building local capabilities in relevant research and technical services in order to ensure proper adaptation to local conditions, and for subsequent tangible and cost-effective responses to materialize. In addition to improved living conditions, technology inputs for alternative energy and cleaner water supplies should create conditions for wider investment opportunities for businesses in both Gaza and the West Bank.

### 2.2.5 Syria

Syria is going through an economic and humanitarian crisis of enormous proportions. The country's economy declined by at least 62% between 2010 and 2014 due to armed conflict that started in 2011. The effects of international sanctions, widespread damage to infrastructure, diminished consumption and production activity, and dwindling foreign exchange reserves have resulted in enormous budget deficits. Devaluation of the Syrian pound has also resulted in falling household purchasing power and widespread poverty in urban centers as well as rural areas.

### *2.2.5.1 Syria's Refugee Communities*

Conditions for Syrian families that took refuge in neighboring countries have gone from bad to worse, with most totally dependent upon handouts or work performed for extremely low wages by both their male and female members, often children and adolescents, particularly as agricultural workers in rural locations. International assistance has been only partly helpful, as the number of people in need within Syria has increased beyond 12 million inhabitants and the number of Syrian refugees in neighboring countries increased to more than 4 million. A large number of Syrians have also been internally displaced as a result of the fighting as well as deals reached between the authorities and rebel groups. Although some of these communities have managed to receive limited assistance, including rudimentary educational and health services the vast majority are reported to lack many basic needs.

### *2.2.5.2 GDP Growth and Sectoral Contributions*

Table 2.2 indicates that real GDP growth rates experienced massive declines since 2010. However, assuming that armed conflict would continue to lose its intensity, and as limited reconstruction activity commences, modest GDP growth has been forecast from 2017 onward.[36] Given such conditions, expectations are that GDP growth will turn to positive values in the range of around 3.0% in 2017. However, several factors might conspire to thwart or hinder positive economic growth. These include depleted public finances, possible failure to recapture oil facilities and related infrastructures, insufficient external financial support, compounded by limited availability of qualified labor.

**TABLE 2.2 Basic Data on Syria's Economy**[35]

| | 2010 | 2013 | 2015 |
|---|---|---|---|
| Gross domestic product (GDP; PPP) US$ billion[1] | 97.5 | 61.9 | 55.8 |
| GDP/Capita US$ thousand | NA | NA | 2900[1] |
| GDP real growth rate | −30.9[1] | −36.5[2] | -9.9[3] |
| GDP composition by sectors: | Agriculture | Industry | Services |
| Sector shares (%) | 19.5 | 19 | 61.5[3] |
| Labor distribution by sectors (%) | 17 | 16 | 67[4] |
| Total labor force: 3.37 million[3] | | | |
| GDP at official exchange rate 2015 estimated US$ 24.6 billion.[2] 3. 2016 estimate. 4. 2008 estimate | | | |

1. Data in 2015 US dollars. 2. 2014 estimate. 3. 2013 estimate.

Prior to 2011, Syria had taken steps to liberalize its economic policies and attempted to encourage growth by cutting lending interest rates, licensing private banking, adopting multiple exchange rates, and setting up a Stock Exchange. Nevertheless, its economy remained regulated, with special constraints on foreign trade. Declining oil production,[37] high unemployment rates, rising budget deficits, pressure on water supplies, rapid population growth, industrial expansion, and water pollution are factors that are often blamed for preconflict stagnation and have even been reported as indirect causes of the uprising that eventually deteriorated into open-ended armed conflict.

As indicated in Table 2.2 industrial production in Syria accounted for around 20% of the country's GDP in 2010 and utilized 19% of its work force, with industrial activity principally based on oil and gas extraction, storage, transportation and export, textiles, food processing, beverages, tobacco, phosphate rock mining, cement, oil seeds crushing, as well as assembly of a limited range of electrical equipment. Due to events of the past few years, the industrial sector was estimated in 2016 as having suffered a recession of around 2.4%.

A comparable proportion of the national work force worked in agriculture, with main products including wheat, barley, cotton, lentils, chickpeas, olives, sugar beets; beef, mutton, eggs, poultry, and dairy products.

Fig. 2.9 presents a view of contributions in terms of value added to GDP by industry, agriculture, and the services, during 1985–2007. This figure clearly

**FIGURE 2.9** Value added due to industry, agriculture and services in Syria during 1985–2007 (% of GDP).

indicates that value added by agriculture experienced more or less steady decline since the early nineties, relative to that added by industry and services activities. However, the evolution of the industrial sector's contribution appears to have taken an upward track in the mid-nineties, whereas that of the services sector took a steady upward path toward the end of that decade. Detailed study of the causes behind such trends as well as the implications they hold for future socioeconomic development would be extremely valuable in formulating sectoral as well as national STI policies.

#### *2.2.5.3 Poverty and Unemployment in Syria*

Poverty, mainly in the rural areas, is known to have played an important factor in the protest movement that preceded the armed conflict. It is also known to have created impetus for many young men, particularly in northern and north eastern districts of the country to join armed rebel groups.[38]

Syria ranked 202nd worldwide, in terms of the level of unemployment, which was estimated, in 2015 and 2016, at around 50%. In 2014, the percentage of Syria's population living below the poverty line was estimated at around 83%. Figures on poverty in Syria were unavailable prior to the onset of its armed conflict. They are presently estimated to exceed 50% at home with higher proportions of impoverished Syrians in refugee camps spread over three neighboring countries, namely Jordan, Lebanon, and Turkey.

In conclusion, it would seem that nothing short of a complete and utter rethinking of national development policies will enable Syria to gain a semblance of its even most recent past, despite the many shortcomings that this past held for all to see. Criticism is often made of steps made toward liberalization of Syria's economy, a few years before the outbreak of protests and ensuing violence. It is to be granted that further analysis of such steps and the manner in which liberalization measures were enacted is essential. However, even more serious ills may have long been festering, in bureaucratic and corrupt government machinery and in the manner with which criticism was systematically stifled. More serious maladies may have been due to selection criteria for those in positions that impact development efforts, resting almost entirely upon loyalty to the ruling party and powerful elites, with other far more relevant qualifications for discharging their duties trailing very far behind.

Once all the dust has settled down, the suffering endured by this country must be channeled into efforts aimed at rebuilding the country and its institutions, ensuring good governance through a wide range of partnerships, including local community bodies and NGOs in addition to government authorities adopting highest possible levels of transparency and accountability.

## 2.3 ARAB COUNTRIES OF THE NILE VALLEY; EGYPT AND SUDAN

Egypt and its southern neighbor, Sudan, are endowed with natural resources including the fertile plains of the Nile valley as well as significant hydrocarbon resources. Additionally, for many decades, Egypt was able to derive enormous benefits from its geographic position at the cross-road of international trade that is still dependent upon the Suez Canal.

With a combined population of around 129 million, representing over one-third of the total population of all the Arab countries, Egypt and Sudan possess vast pools of human capital. Although traditionally averse to migrating for work abroad, many Egyptian and Sudanese workers and professionals now work abroad, particularly in the Gulf Arab countries, with their remittances contributing to economic and social stability at home.[39]

The combined GDP of Egypt and Sudan amounted to just over US$ 1 trillion in 2014, representing only 18% of the Arab total. However, both countries face serious challenges on several fronts, not least of which is their growing and youthful populations, with even more serious threats expected as Ethiopia's projects designed to dam and utilize a considerable proportion of Nile waters take their final shape.

### 2.3.1 Egypt

Egypt has a diverse middle-income economy that is well developed in comparison to many of the developing countries in North Africa and the Middle East. As indicated by Fig. 2.10, agriculture contributes to around 10% of Egypt's GDP although nearly 30% of the country's labor force is involved in this sector's activities. The country's main agricultural products include cotton, rice, corn, wheat, beans, fruits, vegetables, as well as a variety of farm animals. At around 36% and 53%, respectively, industry and services contribute higher percentages to Egypt's GDP.

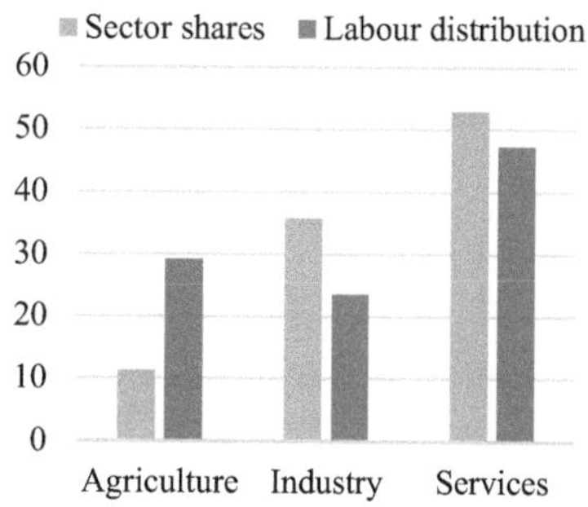


**FIGURE 2.10** Egypt's sectoral GDP contributions and labor distribution across sectors (%).

Egypt's recent loan from the International Monetary Fund (IMF), of US$ 12 billion, should ease the implementation of socially sensitive structural reforms, including the introduction of a value-added tax of 13%. The first installment of US$ 2.75 billion was made available to Egypt's government with future installments linked to conditions posed by IMF, demanding that further reforms that Egypt had meant to implement in the past are enacted. The slow pace of such reforms is generally due to fears that they might drive inflation upward as well as aggravate unemployment, on the one hand, and to pressure exerted by dominant beneficiaries from current practices, both in government circles as well as the business sector, on the other.[40]

In 2016, Egypt's government adopted a plan requiring that 20% of bank loans are made to small and medium enterprises (SMEs). However, it is unlikely that unregistered enterprises, which constitute a considerable proportion of SMEs, would be eligible as recipients. Indeed, it is unlikely that a sufficient number of viable SMEs are in operation and in need of absorbing such a magnitude of loans. On the other hand, enterprises in general are heavily constrained by red tape and the need to bribe corrupt officials.

#### *2.3.1.1 Economic Growth in Egypt*

Egypt's economy had been growing at annual rates of 4%–5% a year during much of the past decade. However, following the events of 2011, the country's GDP growth dropped down to levels below 2% per annum, starting on an upward trend in 2014 to reach levels slightly beyond 4% in 2015.[41]

Traditionally, Egypt's economic growth has been driven by the expansion of industrial and services activities, with the country's principal industries, including textiles, food processing, tourism, chemicals, pharmaceuticals, hydrocarbons, construction, cement, metals, light manufactures. In particular, tourism a leading source of national income, witnessed severe declines over the past 5 years.

Industrial activity in Egypt is currently exposed to a multitude of hardships, including soaring costs of production inputs as well as subdued demand due to rising inflation, which was reported at 23.3% in December 2016.[42,43] The growth rate of industrial production was estimated in 2016 at 0.6%. Clearly, difficulties encountered by this sector would render it difficult for industrial enterprises to grow and acquire productivity enhancing technologies. However, as shown in Fig. 2.11, which presents estimated and forecast growth rates for Egypt's GDP as well as its principal economic sectors, growth of Egypt's industrial sector is expected to rise to levels approaching 3.5% over the coming few years and to approach 5% by 2020. A somewhat stable growth is forecast for the country's agriculture, at rates of around 3%, whereas the services sector is forecast to grow at rates of around 5.5% by 2021. Fig. 2.11 also forecasts rather buoyant near-term futures for the services sector, expecting it to grow at rates approaching 6% by 2021.

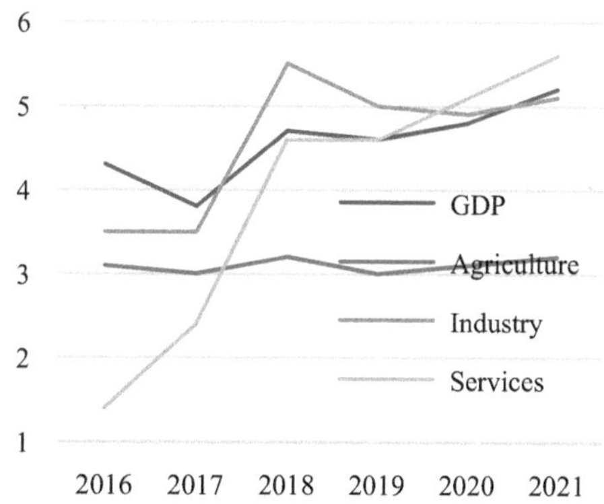


**FIGURE 2.11** Estimated and forecast growth rates for Egypt's GDP and main sectors (%).

### *2.3.1.2 Unemployment and Poverty in Egypt*

Egypt's unemployment rate was estimated at 12.5% in 2016, an increase of between 2.5% and 3.5% from its level before 2011. Highest rates of unemployment have traditionally been encountered among youth and women. Thus, youth unemployment was reported at around 25%, even before the events of 2011. However, estimates for 2014 point to youth unemployment rates of around 42%.[44] The country's youthful population is essentially a product of the high fertility and population growth rates that prevailed in the late 1980s and early 1990s. The former was above 4.3 per woman till 1992, whereas the latter stood above 2.5% between 1980 and 1989. Naturally, at 2.1% per annum, Egypt's present population growth rate is still rather high, placing added pressures upon the country's present and future infrastructure and services.

Another shortcoming of public policies in Egypt is manifested in the country's inability to train its young citizens. Thus, while university education is free, the quality of higher education and training institutions is poorly matched to actual needs. Medical and engineering graduates might be an exception, however, a large proportion of these graduates end up working in the Gulf region rather than in their home country.

Official data indicate that 28% of the population lived below the poverty line in 2015. However, it is estimated that poverty rates rise to levels as high as 60% in rural Upper Egypt.

### *2.3.1.3 Egypt's Economic Recovery*

Short-term risks due to security challenges, and the social implications of floating the Egyptian pound in November 2016, will continue to pose threats to economic growth and social welfare in the midst of other, more endemic obstacles that faced Egypt's economic recovery in the past, including red tape, cronyism, and corruption coupled to a high population growth rate. Further notes on factors that appear to be complicating Egypt's economic recovery are presented in Frame 2.2.

**Frame 2.2 Factors Complicating Egypt's Economic Recovery**

Both external and internal factors complicate Egypt's economic recovery following the widespread unrest witnessed in 2011. The former include political turmoil in the region as well as sluggish economic recovery in Europe and the world at large.

Internal factors, on the other hand, include the country's investment climate, suffering red tape, and corruption added to the fact that economic policy is seldom implemented with due regard to tangible benefits.

An example of this tendency is provided by various mega-projects, such as the expansion of the Suez Canal, which seems to have done very little to enhance economic returns. Indeed, for a variety of reasons the Suez Canal revenues actually fell since the expansion project was completed in August 2016.

Further complications arise due to prevalent opacity regarding information about the national economy, particularly where the army's activities are concerned. Much of the armed forces' budget and accounts are considered state secrets, allowing corruption and cronyism to flourish unabated.

An example of the kind of corruption that pervades many walks of life in Egypt is provided by the manner with which wheat imports, storage, and final utilization in making bread, a staple food item in Egypt are handled. Egypt one of the world's largest wheat importers, purchases this commodity on the international markets to be used in making heavily subsidized bread. Home-grown wheat is also purchased by government at higher prices in order to encourage local farmers. However, corrupt practices intervene, first with farmers mixing foreign wheat with their own and selling to government authorities, making high profits, thereby. Second, officials in charge of government's wheat silos regularly exaggerate the amount of wheat they have in store and pocket a portion of the allocated subsidies. Third, even the use of new technology to stem corruption seems to have failed, as a smart-card system meant to track bread purchases in Egypt was hacked, allowing some bakers to make illicit profits from subsidized flour.

The country's future socioeconomic development will also be affected by developments in Egypt's private sector. Thus, the country's private sector has traditionally been dominated by a relatively small number of large firms with strong connections to high ranking officials and army officers, enabling them to employ monopolistic practices, thereby creating rapid returns. In the face red tape and corrupt bureaucracy many SMEs are forced to operate within the informal sector, where they are unable to expand, create new jobs, or access novel competitive technologies.[45] Naturally, small unregistered enterprises do not pay taxes, on the one hand, whereas on the other hand they find it very difficult to borrow money in order to innovate. An estimated 18 million businesses in Egypt are unmonitored and untaxed by the government, leading to estimates of the informal economy of around two-thirds the size of the formal one.

In addition to links that senior army officers maintain to civilian enterprises, the Egyptian army is called upon to implement many government projects, exaggerating its already inflated role in the national economy. Thus, the army operates many factories and businesses of no relevance to defense needs and is thought to control as much as 40% of the country's economy.[46]

## 2.3.2 Sudan

With the exception of a few years of tranquility, Sudan has experienced relentless socio-political upheaval and armed conflict ever since its independence.

For much of its recent history, Sudan was consistently classified by various international organizations among the least developed and highly indebted economies. However, since the late 1990s, due to a series of macroeconomic reform policies, Sudan registered considerable improvement on most macroeconomic indicators, with rapid increases in per capita incomes, turning it from a low income into a lower medium income economy. However, Sudan is still classified by the United Nations Development Programme (UNDP) among countries with low values of the Human Development Index.[47]

### *2.3.2.1 Sudan's Oil Wealth*

Sudan began to export crude oil in 1999, recording its first trade surplus in 1999 and attaining an average rate of GDP growth of around 7.5% during 2000–08. However, with the secession of its oil rich Southern provinces in 2011 Sudan lost three-quarters of its oil production, which accounted for over half of government revenues and 95% of exports.

As shown in Fig. 2.12, oil rent accounted for around 27% of GDP in 2008, but fell to 5% by 2013. However, as indicated in Fig. 2.13, the country's oil production is estimated to rise up to around 163,000 b/d by the year 2021, that is almost 40% higher than its 2016 production level. However, it would still fall far below pre-2011 levels.

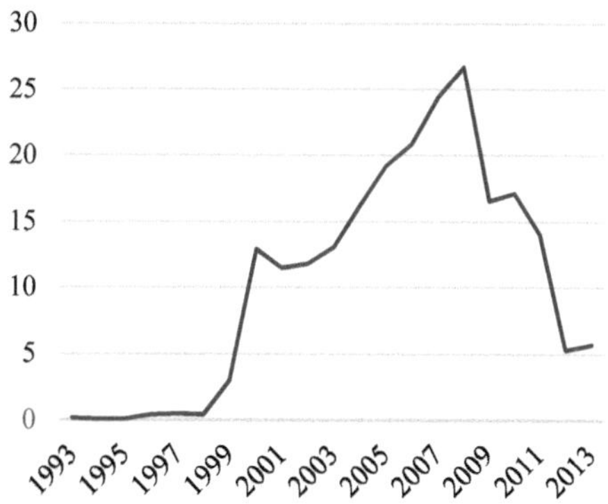


**FIGURE 2.12** Variation of Sudan' s oil rent as percentage of GDP (1993–2013).

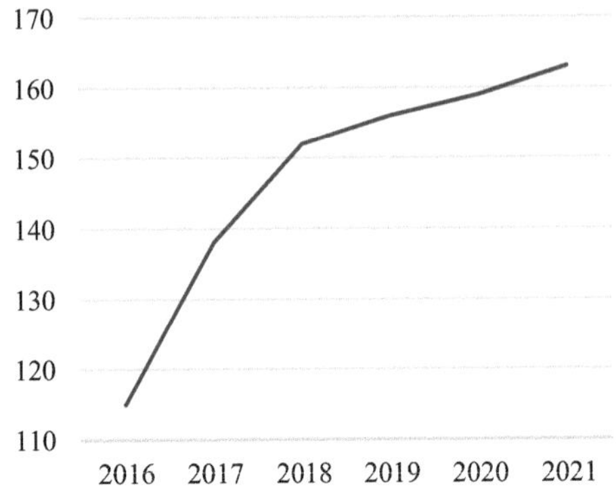


**FIGURE 2.13** Estimated and forecast oil production figures for Sudan during 2016–21 ('000 b/d).

The national economy continues to suffer the shock waves of Southern Sudan's secession to date. Huge macroeconomic and fiscal challenges ensued, manifested most clearly in reduced economic growth, double-digit consumer price inflation, which together with increased fuel prices, triggered violent protests in September 2013. Toward the end of 2014, Sudan's external debt stood at $ 43.6 billion in nominal terms, 85% of which in arrears.[48]

Sudan's government has been attempting to shore up its economy and make up for the loss of foreign exchange earnings, whereas still subjected to strict US sanctions.[49] Additionally, ongoing conflicts in at least three Sudanese provinces,[50] in addition to continuing unrest in South Sudan, which caused an estimated 100,000 refugees to seek sanctuary in Sudan in October 2014,[51] place further pressures upon the country's resources. Lack of basic infrastructure throughout, together with overreliance on subsistence agriculture, has conspired to keep nearly 47% of the population at or below the poverty line.[52]

As a response to these challenges, Sudan's government devalued its currency in June 2012, and adopted austerity measures, including gradual dismantling of fuel subsidies. Government policies have also emphasized development of nonoil sources of revenues, while austerity measures were introduced to reduce government spending. Additionally, attempts are being made to enhance gold mining capabilities. Nevertheless, the country's economy continued to struggle with high inflation, amounting to 47% on annual bases in November 2012, but subsequently going down to 18% in 2015.

### *2.3.2.2 GDP Growth and Sectoral Contributions*

Fig. 2.14 presents a view of estimated and forecast GDP growth, as well as growth in agriculture, industry, and services over the years 2016–21, indicating projected GDP growth in excess of 2.5% till 2019, and approaching 4% by 2020. Industrial growth is also expected to climb considerable to exceed 5% in 2020. Agriculture and services are expected to grow at rates between 3% and 4% during 2017–21, creating serious possibilities for expansion in related

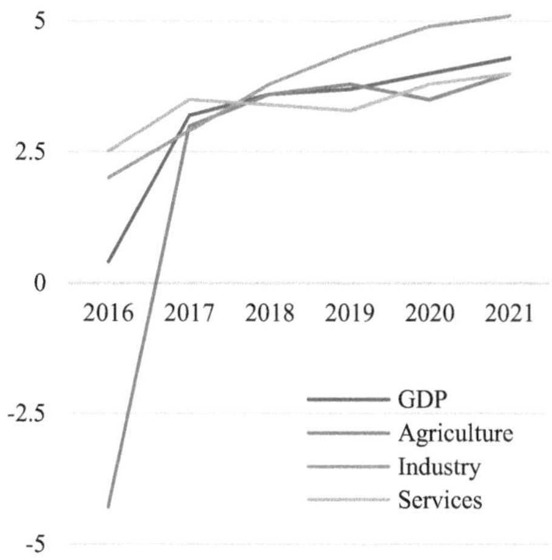


**FIGURE 2.14** Estimated and forecast growth rates of GDP and contributions by sectors to GDP in Sudan (2016–2021) (%).

research, technology acquisition, adaptation as well as human capital formation at a variety of levels.

Contributions to GDP by the agriculture, industry, and services sectors in Sudan were estimated at around 28%, 21%, and 52% in 2007. However, percentages of Sudan's total work force engaged in these sectors were estimated at 80%, 7%, and 13%, respectively, indicating tremendous scope for productivity enhancement, within the agricultural sector, in particular, normally attained through specific S&T inputs, as well as through education and training and extension services.

Some encouraging signs may be read into Sudan's Interim Poverty Reduction Strategy Paper and its 5-year Program for Economic Reforms approved by its parliament in December 2014. Thus, both documents recognize the importance of greater attention to agriculture and animal husbandry as means of diversification and export promotion, which could have profound socioeconomic impacts since, as indicated above, the agriculture and animal husbandry sector employs 80% of the work force.[53] Given the country's fertile land and water availability, this sector could play a crucial role in Sudan's economic diversification. Although the agriculture and animal husbandry sector presently contributes ~35%–40% of GDP, it could contribute considerably more, if allocated sufficient investments allowing it access to novel STI inputs, and guaranteed improved governance it could make considerably greater contributions to the country's medium-term macroeconomic stability.

### *2.3.2.3 Unemployment and Poverty*

As shown by Fig. 2.15, World Bank figures for total youth unemployment during 1991–2014, hovered in the range 22%–24%, during 1991–2014. However, for females, this figure rose to around 27%, in 2008.[54]

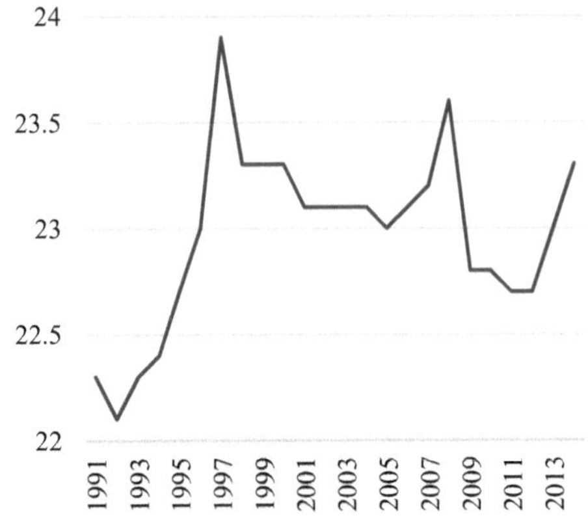


**FIGURE 2.15** Youth unemployment in Sudan during 1991–2014 (% of total labor force aged 15–24; modeled International Labor Organization estimate).

Poverty is prevalent all over the country due to factors including conflicts that have been raging for decades as well as policies that marginalized agricultural activity, as oil became the main source of the country's source of income alternative sources of energy. In 2009, World Bank estimate of the overall poverty headcount ratio at national poverty lines fell at around 46.5%. In rural areas, the estimates rose even higher, to 57.6%, according to the same criteria.

In essence, many of Sudan's development problems are due to the absence of public institutions capable of mediating power and wealth sharing between the center and the periphery. Policies aimed at mending this particular situation should help overcome economic fragility as well as political unrest that has repeatedly led to armed conflict. Events over the past two decades clearly indicate the need for policies that allow optimal utilization of oil wealth, and indeed wealth generated from any other natural resource, to promote human capital development, economic growth, and diversification, particularly within impoverished rural communities.

In Sudan's case, not only was the relatively short-lived oil wealth used unwisely; in addition, the country's government took paths that aggravated political instability and did little to stem corruption in its ranks, thereby generating a climate of economic uncertainty that compromised social development, economic growth, and employment creation. Coupled to other longstanding constraints, such as the country's chronically inadequate infrastructural arrangements, particularly with regard to electricity and transport facilities, poor access to finance and ill-conceived taxation policies, the present situation faced by the country was a foregone conclusion.

Another lesson provided by Sudan's experience over the past two decades relates to the need for fair distribution of resources between the center and the periphery. Since, in the absence of fair distribution of resources, ensuring that benefits are evenly shared, and thereby engendering proactive commitment on the part of the widest range of population groups, governance failures are generally magnified and the credibility, indeed legitimacy of government agencies is severely diminished. In turn, this will lead to insufficient incentives for further investment and participatory ventures, feeding back into even more inadequate policies and implementation programs.

## 2.4 THE ARAB COUNTRIES OF NORTH AFRICA

Algeria, Mauritania,[55] Morocco, Tunisia, and Libya make up nearly 24% of the Arab countries' total population but are responsible for around 17% of their total GDP. With the exception of Libya, all were under French rule till the 1950s and 1960s. Italian domination of Libya, on the other hand, extended between 1910 and 1947.

Among these five countries, only Mauritania ranks with the group of low human development countries. Tunisia, Libya, and Algeria are classified among high human development countries, whereas Morocco is ranked as a medium human development country.

### 2.4.1 Algeria

Algeria's Government remains the primary actor on the economic development front despite all reforms designed to grant meaningful roles to other actors, including enterprise and even some civil society institutions.

#### *2.4.1.1 Algeria's Hydrocarbon Wealth*

With the 10th largest volume of natural gas reserves in the world, and ranking 16th in oil reserves, Algeria's economy may be expected to remain largely dominated by hydrocarbon revenues, which are entirely controlled by State institutions. Indeed, the country's hydrocarbon wealth constitutes the backbone of the economy, accounting for around 60% of budget revenues, 30% of GDP, and over 95% of export earnings.[56]

Historically, hydrocarbon exports enabled Algeria to maintain macroeconomic stability and accumulate a sizeable reserve of foreign currency as well as means for stabilizing the national budget. The fact that Algeria's external debt is extremely low, at around 2% of GDP, is another positive sign regarding future economic growth and social development. Declining oil prices, however, forced Algeria's government to deplete its reserves at a rather high rate in order to maintain various social spending programs.

Although high public spending has helped Algeria's economy in recent years, constraints are expected to arise in the near future, due to reduced oil output and curtailed consumer spending. This is expected to reduce overall growth in 2017, to just 1.6%. It is just likely, however, that as higher rates of growth are expected beyond 2017, helped by forecast higher oil prices, the government might simply revert to older strategies, which would most likely dampen attempts at diversifying the economy, thereby postponing, rather than eliminating causes for future unrest.

### 2.4.1.2 Economic Diversification and Growth

Recent attempts at economic diversification have mostly relied upon foreign investment projects, such as a planned automotive assembly plant as well as pharmaceutical manufacturing facilities. Should ventures in these two, as well as other domains, prove effective, the nonoil sector might then be expected to grow at a somewhat robust pace. Nevertheless, it is still unlikely that

- limited investment projects will improve the situation drastically;
- the public sector's restrictive operating environment will end abruptly;
- domestic private investment growth will take off as rapidly and as widely as it should due to inadequacies in local private sector financing.

This has prompted the expectation that nonoil growth might remain well below potential levels, which, coupled to limited government consumption would confine growth to levels of around 2.4% in the near future (see Fig. 2.16).[57] Incidentally, causes for the low growth in estimated and forecast public consumption, indicated in this figure, are essentially linked to the government's need to dedicate resources to spending on salaries and essential services.

Therefore, in the shadow of current policies, the hydrocarbons sector will continue to dictate growth patterns in the medium term. Additionally, further investments in the sector are likely to be delayed due to licensing rounds, and it will be constrained by subdued oil prices, a cumbersome bureaucratic framework and concerns about security risks. On the other hand, natural gas production will rise in 2017–21, as several new projects commence production. In fact,

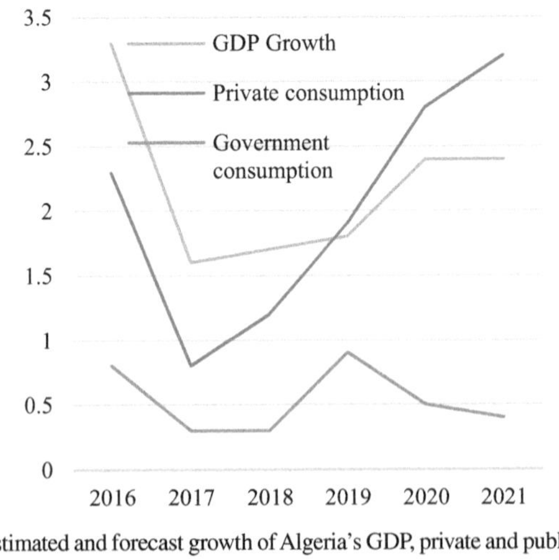


**FIGURE 2.16** Estimated and forecast growth of Algeria's GDP, private and public consumption (%).

part of the enhanced growth forecast in the medium term is due to this factor. However, low energy prices are expected to hinder implementation of many projects on the drawing board, before the beginning of the next decade, thereby placing an upper limit on GDP growth.

#### 2.4.1.3 Sectoral Contributions and Growth Patterns

Algeria's agricultural sector contributed over 13% to the country's GDP. With contributions to GDP at around 39% and 48%, respectively,[58] the industry and services sectors, on the other hand, employ the majority of the country's active labor force. Actual, estimated, and forecast growth rates of the agriculture, industry, and services sectors are presented in Fig. 2.17,[59] which indicates rather low growth rates, particularly for the latter two sectors in years leading toward the end of the present decade. Although these growth rates must at least be partly due to global factors, there is little doubt that national policies could contribute to higher growth rates. Thus, investment, both public and private in the construction sector might at the same time, help resolve the country's housing problem as well as provide opportunities for creating new enterprises and associated job openings.

#### 2.4.1.4 Foreign Direct Investment

Foreign direct investment inflows improved considerably following legislation on investment coupled to limited economic reforms undertaken in the late 1990s. The hydrocarbon sector attracted a substantial portion of FDI inflows during 1999–2003. Investments originating from other Arab countries, notably the United Arab Emirates, Kuwait, and Egypt, also witnessed considerable growth during the past two decades. The telecom, construction materials, real estate, pharmaceuticals, and the automotive sectors were principal beneficiaries

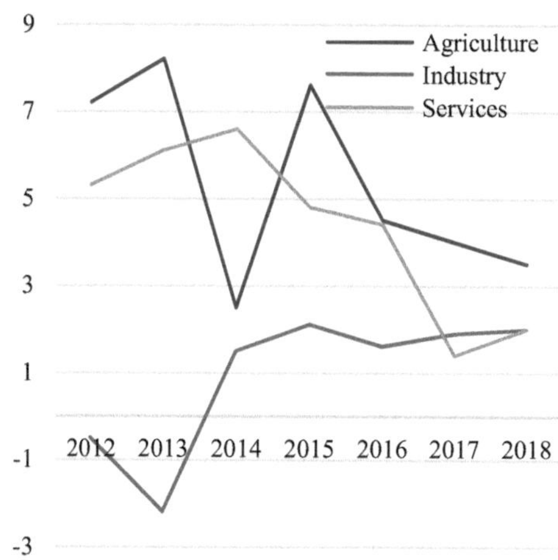


**FIGURE 2.17** Estimated and forecast sectoral growth in Algeria (2012–18).

with a total of around US$ 4.5 billion during 2001–07.[60] However, as shown in Fig. 2.18, having peaked around 2009–10, FDI inflows appear to have declined considerably down to around 0.7% of GDP in 2013–14.

### *2.4.1.5 Unemployment in Algeria*

Available estimates place overall unemployment in 2016 Algeria between 12.4% and 10.4% of the total work force, which is estimated during the same year at about 12 million. As Fig. 2.19 indicates, short-term unemployment is forecast as increasing by at least 0.8% between 2016 and 2018.

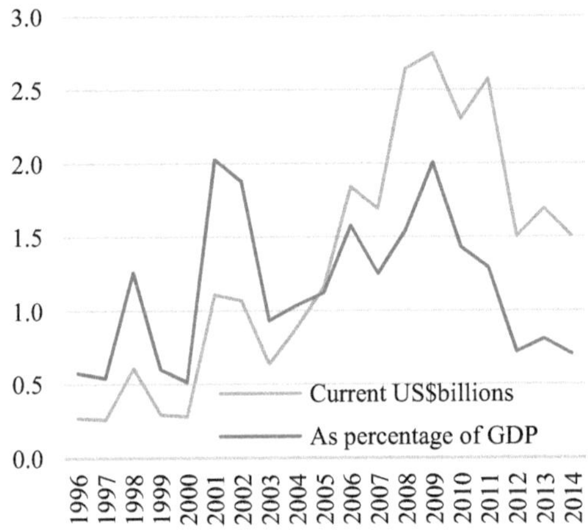


**FIGURE 2.18** Algeria's net foreign direct investment inflows (1996–2014).

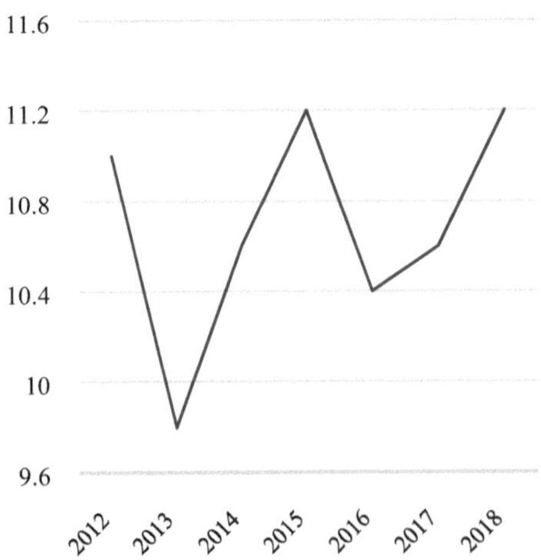


**FIGURE 2.19** Actual, estimated and forecast unemployment in Algeria (recorded averages %).

A recent report[61] indicates that a campaign launched in August 2012 by the ministries of trade and the interior to eradicate unlicensed trade had limited results, with the state closing over 1000 informal marketplaces by the end of 2016, or around 75% of the total. In parallel, government assigned the equivalent of US$ 109 million toward the creation of nearly several hundred new marketplaces, within which space could be rented out to registered traders. Significant results were attained as a result of the initiative with a little over 40% of the 50,000 traders previously working in the informal sector have brought their businesses to the new legal marketplaces. However, only 230 out of a total of 635 planned markets are fully operational as yet.

No fresh estimates are available for the percentage of Algeria's population living below the poverty line. However, available figures for 2006 place this percentage at 23%.[62]

### *2.4.1.6 Future Challenges to Algeria's Development*

For many years now, Algeria has faced three key challenges all closely linked to central government policies:

- utilizing its hydrocarbon wealth to ensure long-lasting benefits of its population;
- creating growth and employment in a diversified economy;
- providing better public services.

Concerning the first challenge, Algeria has largely failed to develop non-hydrocarbon industries, despite relative availability of financial resources that would have enabled it to promote a variety of such industries. This is essentially due to emphasis on sluggish state-driven growth, with the result that youth unemployment rates remain high and the country continues to suffer housing shortages.

With regard to the second and third, challenges, restrictive economic policies, centered on public enterprise activity with limited encouragement for private enterprise development, exacted a rather heavy price in the recent past. Thus, in order to pacify protesters early in 2011, the Algerian Government had to offer more than $ 23 billion in public grants and retroactive salary and benefit increases.

Algeria's economy suffered several setbacks over the past two decades, and is certain to meet even more serious challenges in the near future. Small-scale protests, over taxes, housing shortages, unemployment, and cost of living have erupted regularly over the past two decades. Persistent high youth unemployment, the overbearing role of the state, limited political expression and perceptions of corruption are likely to constitute bases for wider social and political unrest, particularly as government is forced to raise taxes and reduce subsidies, whereas unable to absorb large-scale public-sector recruitment.

Radical policy revisions are needed on many fronts. Although diversifying the economy away from its high levels of reliance on hydrocarbon exports, ought to remain long-term policy priorities, short- and medium-term policy priorities would have to include bolstering private sector development, attracting foreign investment and creating job opportunities for Algeria's youth.

With particular reference to STI policies, there is need for overhauling existing STI-related plans with emphasis on education and vocational training with enterprise creation playing a much larger role than hitherto in economic diversification. Important roles may be played in all of these domains by well-designed civil society interventions, which scored many successes in the country's past.

## 2.4.2 Morocco

Morocco was a heavily indebted country in the 1980s. Austerity measures and promarket reforms have been credited with positive outcomes, including economic stability, steady growth at low inflation rates, and a gradual fall in unemployment. Morocco's present economic policies appear committed to building a diverse, open, market-oriented economy, essentially based on:

- capitalizing on the country's European neighborhood;
- exploiting its newly forged links to other countries in Africa;
- the fact that Moroccan labor may still be employed at much lower costs;
- increased decentralization, including granting greater powers to regional authorities;
- maintaining strong links to the GCC countries, particularly Saudi Arabia.

### *2.4.2.1 Morocco's Economic Growth*

In 2006, Morocco signed a bilateral Free Trade Agreement with the US and an Advanced Status agreement with the EU in 2008. Additionally, with an export dominated outlook in full swing Morocco embarked upon a series of infrastructure projects including wide ranging improvements in its ports, transportation, and industrial infrastructures to position itself as an industrial and business hub, with new port and free trade zone near Tangier, going some way toward improving Morocco's competitiveness. Additionally, in order to reduce outlays that burdened the country's budget and current account, Morocco eliminated in 2014 subsidies for automotive and heating fuels, dramatically. However, subsidies on cooking fuel and certain staple food products were retained. Based on continued reduction in consumption subsidies as well as "strong revenue performance," Morocco is expected by the World Bank to be able to stabilize debt owed by its central government at around 64% of GDP.[63]

Agriculture, tourism, aerospace, automotive, phosphates, textiles, apparel, and electronics subcomponents constitute mainstay of Morocco's economy. Agricultural production suffers frequent swings and tourism is also highly dependent on external demand, which is very much influenced by security concerns. By contrast, mining and industrial production constitute more stable sectors. On the energy front, Morocco's championing of electrical power production from renewable sources should allow coverage of around 50% of its energy requirements by 2030.[64,65]

Morocco's economy was reported as having grown at a considerably lower rate in 2016 than it did over the past two decades. A large contraction in agricultural production has been pointed out as the cause for this slow down, whereas growth in other sectors was maintained at around 2.5% (see Fig. 2.20). Astute monetary policy and falls in international commodity prices are given by a recent World Bank overview, as reasons for this relatively steady growth in the services and industrial sectors, enjoyed by Morocco and reflected in the country's GDP.

A forecast by the Economist Intelligence Unit indicates that real GDP is likely to grow by 4% in 2017, from an estimated 1.5% in 2016, with agricultural production expected to returns to normal following its sharp decline in during 2015–16.[66] This decline clearly reflects the sensitivity of Morocco's economy to the performance of its agricultural sector, which employs nearly 40% of its work force, as indicated in Fig. 2.21.

Nevertheless, Morocco is expected to be able to enhance its economic growth rate while maintaining macroeconomic stability in the medium-term. This forecast is based on expected strong performance by the country's recent industrial ventures, based essentially on FDI projects, initiated toward the present decade, in automobiles, aeronautics, and electronics as well as expansion of Moroccan companies' activities into West Africa (see Fig. 2.22).

On the other hand, macroeconomic stability as well as widespread benefits across population tranches would require macroeconomic policies aimed at expanding structural reforms, with particular emphasis on enhancing productivity. Based on such policies it should be possible to reduce youth unemployment, increase female participation in the work force, and further diminish poverty and inequalities.[67]

### *2.4.2.2 Poverty and Unemployment*

Growth in Morocco's per-capita income during recent years is contributed to drastically reducing abject poverty as well as significantly reducing poverty overall.[68] The above World Bank's overview states that the poverty rate declined from around 9% in 2007 to 4.2% in 2014. However, according to the same source nearly 19% of the country's rural population still live in poverty or are vulnerable.[69,70]

However, despite these achievements, the country still suffers high unemployment, poverty, and illiteracy rates, particularly in its rural areas. The overall unemployment rate in both 2015 and 2016 was estimated at around 10% (see Fig. 2.23). At around 47% in the second quarter of 2016, the employment

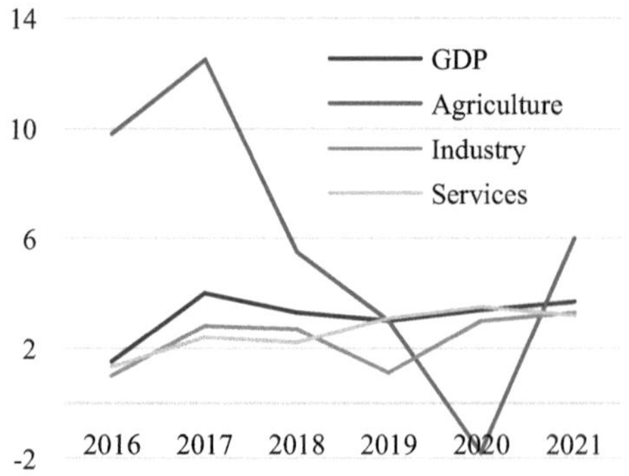


**FIGURE 2.20** Marocco's GDP and sectoral growth (%).

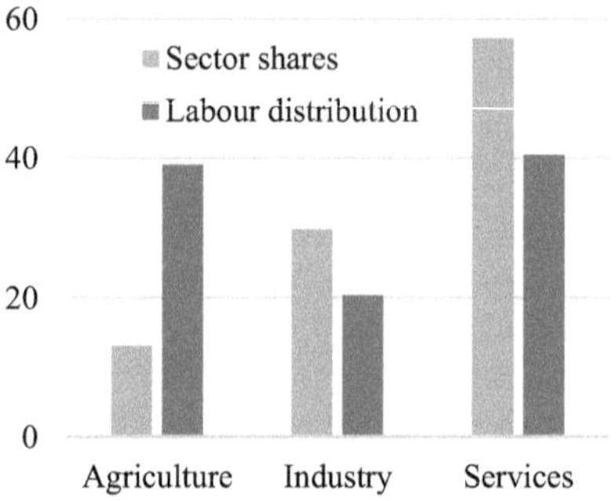


**FIGURE 2.21** Sectoral GDP contributions and distribution of Morocco's labour force.

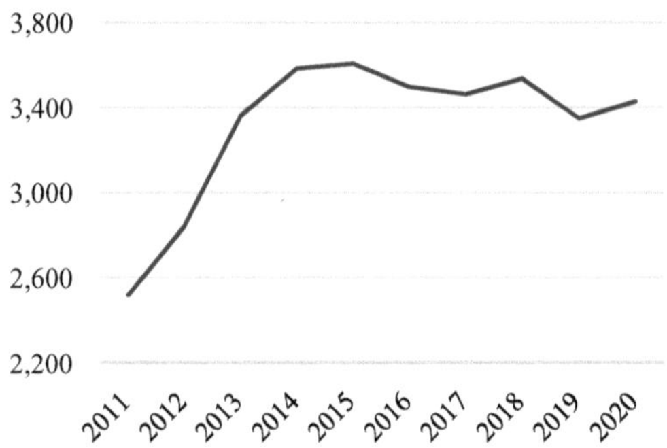


**FIGURE 2.22** Estimated and forecast volume of foreign direct investments hosted by Morocco (US$ millions).

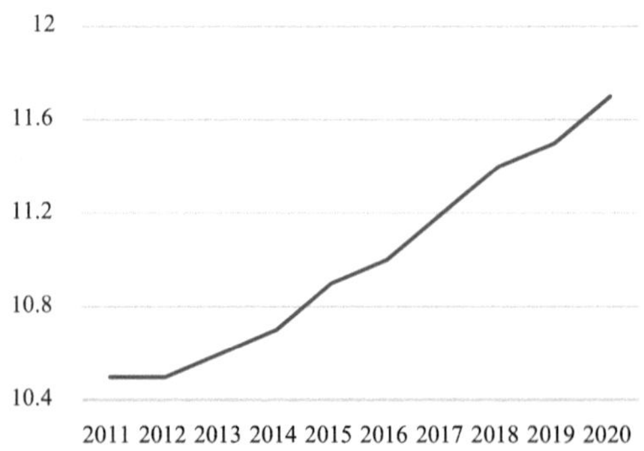


**FIGURE 2.23** Estimated and forecast unemployment in Morocco; 2011–20 (millions).

rate, as reported in the above-mentioned World Bank's report, is quite low. The same report indicates that new employment opportunity being created are generally informal and somewhat precarious. Additionally, whereas overall unemployment has remained broadly stable at around 9% in recent years, the rate among urban youth is much higher, increasing to around 39% by June 2016. An estimated that around 15% of Moroccans lived below the national poverty line in 2007.

Unemployment and generally low per capita incomes will tend to limit internal demand for high-value goods produced by recently established high-technology industrial facilities. On the other hand, targeting the export market with such products would further expose the country to external factors, in a manner that is reminiscent of the situation with regard to the tourism sector.

#### *2.4.2.3 Future Challenges*

Although Morocco's economy is forecast to grow on the strength of recent economic reform measures, foreign investments as well as possible hydrocarbon resources,[71] challenges will persist for future socioeconomic development, with relatively high unemployment and unequal income distribution. Challenges will be especially serious for populations engaged in agriculture activity, which suffers frequent instability due to seasonal factors and output volatility as well as inadequate infrastructures. Thus, nearly half of the rural population does not have access to paved roads.[72]

The Ministry of Agriculture and Fisheries' agricultural development strategy "Plan for Green Morocco (PMV)" included channeling investments totaling around US$ 15 billion into the country's agricultural sector over the 2008–20 period. The plan's objectives included stabilizing and diversifying output,[73] whereas embracing a shift away from government protection toward market-oriented policies.[74] It goes without saying that policy shifts of this nature are quite likely to backfire if undertaken too rapidly and particularly if deprived of support for vulnerable and impoverished rural communities. Thus, although PMV has claimed successes in introducing drip irrigation techniques, with their use rising by around 20%, since the plan was launched, an estimated 80% of the country's 4 million farmers toil on plots of less than 5 ha are too poor to access such techniques. Many cannot even afford fertilizers or even access to water from wells.[75] On the other hand, export-oriented large-scale farming concerns, largely by a small number of wealthy Moroccans, including members of the royal family, attract the largest proportion of capital investment.[76]

### 2.4.3 Tunisia

For several decades in the past, Tunisia's diverse, market-oriented economy was held out as a model for other countries in Africa and the Middle East. The free trade agreement with the European Union combined with progressive dismantling of tariff barriers created considerable opportunities for a young Tunisian industry, in total contrast to protectionist orientations that prevailed till the late eighties.[77] The country's pre-2011 economic strategies focused on exports, foreign investment, and tourism, which became central to the country's economic health.

### 2.4.3.1 Economic Growth

Investments in education and infrastructure, coupled to the above strategies, resulted in decades of annual GDP growth in the range of 4%–5%, with commensurate improvements in living standards. However, cronyism and corruption that marked the reign of the country's pre-2011 ruling class had negative impacts on social development. Rising unemployment, particularly among the country's university graduates ultimately contributed to the government's overthrow, with subsequent events sharply affecting tourism and further investment activities.

Although Tunisia's political transition made steady progress up to 2014, with a new constitution adopted and both parliamentary and presidential elections successfully held, successive terrorist attacks in 2015, coupled to strikes in the phosphate sector, considerably slowed down growth.[78] Thus, GDP growth was estimated at around 1% in 2015 and then only due to an enhanced performance by the agricultural sector, particularly in olive production, by around 9%. On the other hand, most other sectors of the economy either contracted or stagnated.

Strikes in phosphate mining regions and continuing declines in oil and gas production caused the chemical industries to contract by around 5%, whereas oil refining suffered an even more massive decline, estimated at around 18%.

Fig. 2.24 presents a view of changes experienced by Tunisia's foreign direct investment inflows, as a percentage of GDP, between 2007 and 2015, indicating a general decline from figures in the range of 5% in 2008 down to

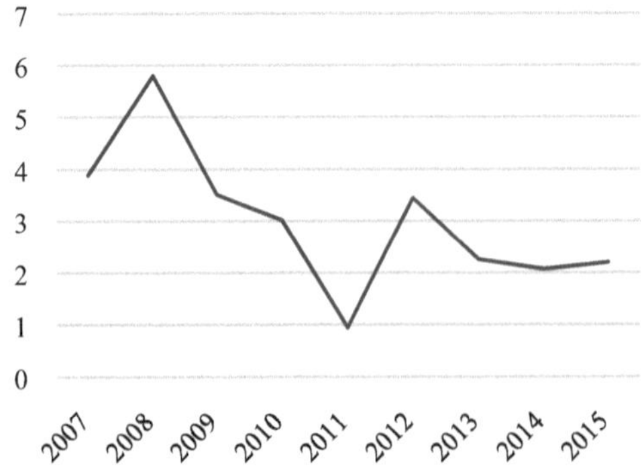


**FIGURE 2.24** Tunisia's net foreign direct investment inflows as a percentage of GDP.

around 2% in year 2014, before going up by 9% in 2015.[79] This decline is at least partly due to the political turmoil witnessed by Tunisia since 2011 and subsequent deterioration in the country's security situation. It is noteworthy that FDI took a downward trend somewhat earlier in 2008, following its record value in 2006.[80]

Despite of the many difficulties faced by the country, modest economic successes particularly, over the past 3 years, may be underlined with overall growth estimated at 1% in 2016, and inflation rate down to 4.5% in 2016 from 4.9% in 2014.

### *2.4.3.2 Sectoral Contributions*

Estimates made in 2014 indicated that a large percentage, around 62%, of Tunisia's GDP, was generated by Services, whereas industry and agriculture contributed just over 28% and 10%, respectively. The country's labor force, which was estimated at over 4 million in 2016 is distributed among economic sectors somewhat in step with their GDP contributions, as indicated in Fig. 2.25.

Fig. 2.26 presents a view of Tunisia's measured, estimated, and forecast GDP growth as well as growth in its agricultural, industrial and services sectors

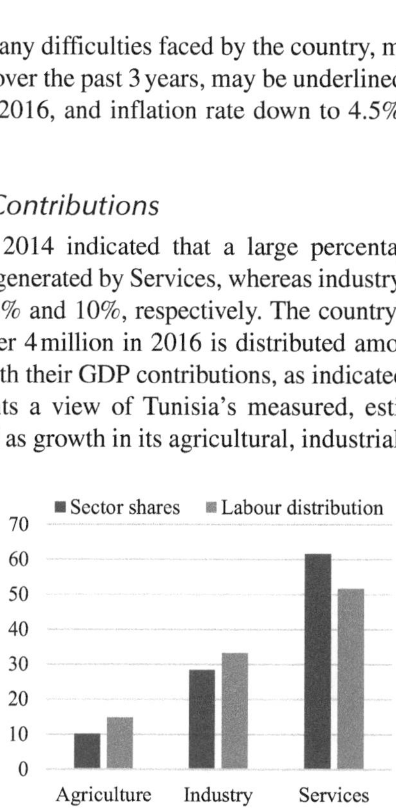


**FIGURE 2.25** Sectoral shares in Tunisia's GDP and labor distribution across sectors (%).

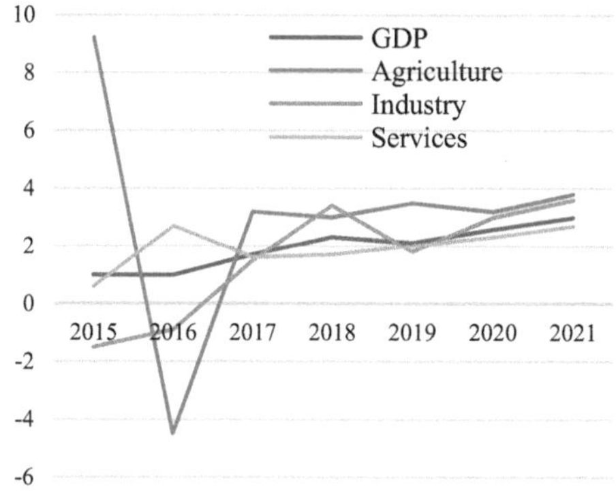


**FIGURE 2.26** Tunisia's estimated and forecast GDP growth.

over the period 2015–21, based on assumptions concerning global growth patterns, regional conditions as well as relative political stability and continued freedom from major security breaches at home.[81] Based on such assumptions, forecasts indicate that all three main sectors will grow at rates between 2% and 4%, during the period 2017–21.

The main contributions to Tunisia's exports come from sectors such as textiles and garments, food products, chemicals, and phosphate mining. Nearly 80% of the country's exports go to the European Union, Tunisia's main economic partner.

#### *2.4.3.3 Unemployment and Poverty*

Unemployment in Tunisia fell to 15.3%, from figures approaching 17% in 2011. However, it remains above pre-2011 levels of around 13%, and is reported as exceeding 30% in the phosphate mining belt of the southwest.

No official statistics are available on poverty in Tunisia following 2010. The World Bank's projections suggest that, based on the 2011 PPP US$3.1 poverty line, poverty declined significantly from 7.6% in 2013 to 7.1% in 2015, reflecting the capacity of the social protection system to safeguard a certain level of consumption and some control over inflation. On the other hand, setbacks in the mining and tourism sectors could adversely affect incidence of poverty particularly in rural areas, but also in more exposed urban areas.

#### *2.4.3.4 Future Challenges*

Having inspired a series of Arab uprisings, essentially on the basis of massive unemployment and rising public expectations, Tunisia is still experiencing difficulties in moving toward a democratic regime. Social and political tensions that dominated Tunisia since, together with the threat of terrorist attacks, have had immense impacts upon the country's economic performance negatively impacting social development. Prospects for a rapid recovery of Tunisia's economy are constrained by limited growth in Europe, Tunisia's principal market and mainstay of its tourist industry, the global economic situation as well as difficulties in reaching consensus on a variety of reform and development measures within fragile internal politics. The hydrocarbons and manufacturing sectors are also experiencing difficulties amid low investment and lagging investor confidence. With around 650,000 unemployed, including over 200,000 with postsecondary education, job creation is presently a prime economic and social priority.

Several challenges that face the country necessitate improvement of the competitiveness of its strategic sectors. In the short run, the free trade agreement with the EU might be expected to continue producing some positive outcomes, by instigating improved management, organization, and enhanced productivity.[82] In the long run, however, there will be a greater need for stimulating technological innovation through instilling capacity learning at the enterprise level as well as promoting greater collaboration, both formal and informal between enterprises, on the one hand, and universities, laboratories, and national as well as foreign research organizations, on the other.

Past policies enacted as part of successive national 5-year plans emphasized moves toward a knowledge-economy, as well as creating a cadre of efficient civil servants, and an able business community, even within the highly competitive sectors such as textiles and clothing.[83] Other assets that Tunisia might build upon include established industrial clusters in various areas. Finally, the country's government is also known to adopt relatively transparent procedures with data on national accounts, sectoral and regional data with ample information on various aspects of education, unemployment, poverty, and gender disparity, facilitating policy revisions and subsequent planning at national as well as institutional levels.

Reassuring businesses and investors balancing the budget and reducing deficits as well as lowering unemployment will remain high on the list of national priorities for Tunisia's governments.

## 2.4.4 Libya

Almost entirely dependent on hydrocarbon exports, Libya's economy has struggled over the past few years as the country plunged into chaos and world oil prices dropped to less than 50% of their value. Armed conflict that erupted in 2015 for control of the country's major oil terminals precipitated a deeper economic decline. Thus, in May 2017, Libya's oil output stood at around one-half of its average pre-Revolution highs of 1.6 million barrels per day.[84]

In 2016, it was estimated that Libya's GDP was down to around 50% of its pre-revolution levels. Budget revenues and proceeds from hydrocarbon exports reached their lowest levels on record due to depressed oil production and prices. Net foreign reserves are being rapidly depleted in order to remedy shortfalls.[85]

Nevertheless, it appears that the government continued to pay salaries to most of Libya's workforce as well as fund fuel and food subsidies, causing an estimated budget deficit of around 49% of GDP.

### *2.4.4.1 Sectoral Development*

The private sector was estimated as accounting for a mere 5% of Libya's GDP, with 95% of this sector made up of SMEs, operating mostly within the informal economy. Due to a variety of factors, including regulatory arrangements that govern the business sector as well as high levels of corruption Libya's SMEs are known to have performed poorly in relation to generating employment opportunities, in terms of productivity, competitiveness, and generating added value.[86]

Fig. 2.27 indicates that Libya's economy is largely dependent upon its industry and services sectors, with agriculture estimated as contributing a mere 1.9% to the country's GDP in 2016. The fact that this sector employs 19% of the national work force, estimated in 2016 at around 1.2 million, indicates plenty of room for efficiency enhancements.[87]

Negative growth has dominated all sectors as well as national GDP and is expected to continue to do so at least till the onset of the 2020s, when high growth rates in GDP as well as the economy's three sectors are forecast assuming a close end to hostilities (see Fig. 2.28[88,89]).

### 2.4.4.2 Unemployment Poverty and Welfare

Deteriorating security and political chaos reigned following 2011, halting a number of development projects, some involving significant STI inputs. Real incomes have plunged over the past few years, exacerbated by extremely high inflation. The conflict has hit certain segments of Libya's population quite hard, with around 1.3 million people having no access to life-saving healthcare services. Living conditions, in general, have witnessed serious reversals, as access to clean drinking water became more difficult in several parts of the country, and as armed conflict caused inhabitants of troubled regions to migrate into safer locations. Widespread power outages are common in the country's cities, mainly due to fuel shortages, curtailing serious educational activities, let alone scientific research, and technology development.

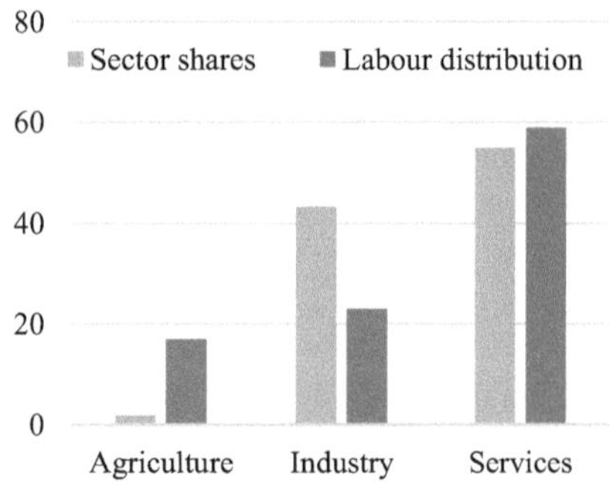


**FIGURE 2.27** Sectoral contributions to GDP and labor distribution across sectors (%).

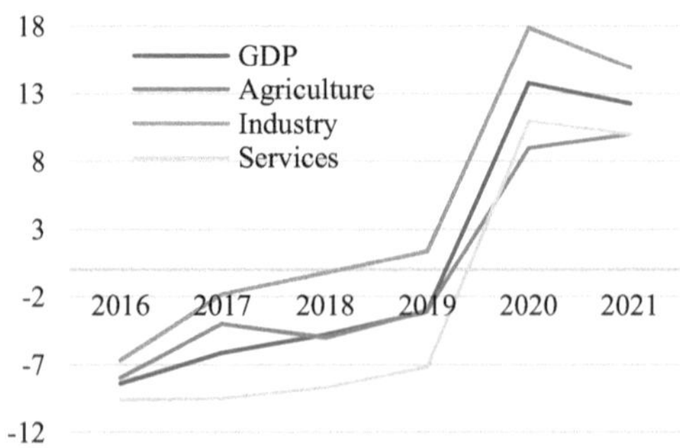


**FIGURE 2.28** Estimated and forecast growth rates of Libya's GDP and main sectors of the economy (2016–21).

Estimated at around 43% in 2010, youth unemployment was already quite high even before the countrywide turmoil.[90] It was estimated at nearly 49% in 2014. Total unemployment, as a percentage of the total work force, was also quite high, estimated at 18.6% in 2010 and then at 19.2% in 2014. No figures on poverty or income inequality are available for Libya from customary sources.

### *2.4.4.3 Policy Challenges*

Immediate challenges facing Libya's policy makers revolve around managing fiscal spending pressures, whereas restoring and improving basic public services, and providing a wide range of job opportunities.[91] Developing the framework and requisite institutional entities for a more diversified market-based economy, broadening the economic base beyond the oil and gas sector are certainly longer-term goals for the time being.[92] Both sets of challenges would greatly benefit from creating institutions dedicated to promoting and undertaking scientific research, technology acquisition, adaptation and development and the creation of innovative enterprise and employment creation schemes in urban centers as well as rural locations with agenda that priorities inclusive and sustainable socioeconomic development.

## 2.4.5 Mauritania

Mauritania is a West African Arab country whose population was estimated in 2014 at around 5 million inhabitants, 40.3% of whom were below 14 years of age. The country has considerable mineral wealth, including iron ore, gold, copper, gypsum, and phosphate deposits.[93] Additionally, Mauritania possesses modest oil reserves and considerable offshore gas deposits, which have not yet created large revenues for its economy. As indicated in Fig. 2.29 Mauritania's GDP grew from levels close to US$ 5.6 billion in 1990 to exceed US$ 14.8 billion by 2014.[94]

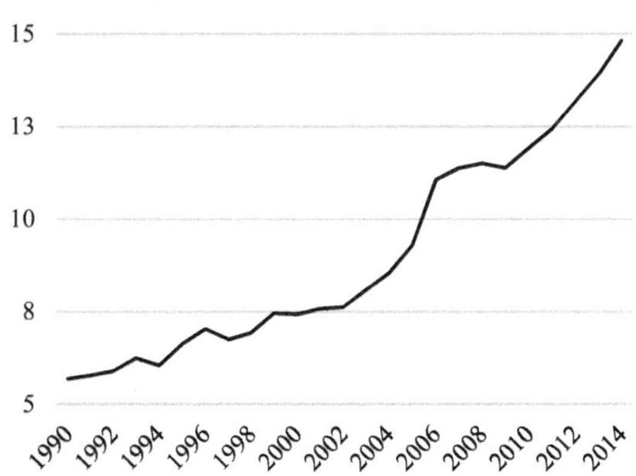


**FIGURE 2.29** GDP growth during 1990–2014 (US$ billions PPP at constant 2011 international $).

### 2.4.5.1 Recent GDP Growth

Current pressures on the country's economy are due to the collapse of iron ore and oil prices in 2014. A drop of over 10% in mineral production took place in 2014. This, coupled to slower economic activity during 2015 led real GDP growth to fall down to 3% by the end of that year, from 6.4% in 2014. As indicated in Fig. 2.30,[95] the prolonged plunge in global prices of metals and oil in particular, has severely reduced contributions the country's GDP, exposing serious shortcomings in Mauritania's development strategies, with their overreliance on mining as well as the country's emerging hydrocarbons sector.[96] Responses based on expansion of the nonextractive sector will undoubtedly require extensive STI inputs.

Previous GDP growth was largely driven by foreign investment in the mining as well as the oil sector. Hence, domestic sectors that rode the back of the mining sector, e.g., utilities, transport, and telecommunications slowed down along with the mining sector. Despite sizeable public investment spending, the construction sector was also affected, diminishing imports of construction materials by 3.8%. The national currency was devalued by 5% compared to 2014. Core inflation more than doubled to 2.6% in 2015 as compared to 1.2% in 2014. However, headline inflation remained close to 0.5% due to a 1.1% decline in food prices.

### 2.4.5.2 Sectoral Contributions to Mauritania's Economy

Fig. 2.31 provides a summary view of value added by the industry, manufacturing and agriculture sectors, over the period 2000–14, as a percentage of GDP, indicating negative change over that period for manufacturing as well as agriculture, whereas industry's contribution grew considerably, but rather unsteadily. Fig. 2.31 clearly implies limited inputs designed to improve the productivity of at least the former two areas of economic activity, namely manufacturing and agriculture, whereas growth observed in industry is most likely due to contributions by the oil industry, which is run by multinationals.

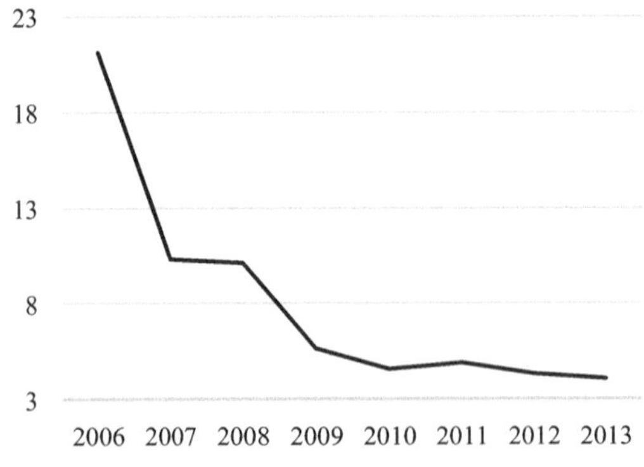


**FIGURE 2.30** Variation of Oil rent as percentage of GDP during 2006–13.

On average, the mining sector and fisheries accounted for nearly 80% of total exports between 1990 and 2000. Apart from crude oil entering the picture around 2006, Mauritania's economy has not witnessed any recent diversification. More recently, exports became even more heavily skewed toward mineral resources, with iron representing over 50% of total exports.[97]

Unlike many other Arab countries, half of Mauritania's population is dependent upon agriculture and livestock for their livelihood.[98] However, as indicated in Fig. 2.32, agriculture contributes only around 24% to the country's GDP, whereas it employs 50% of the country's work force. That only 2% of the country's labor force, which was estimated at around 1.4 million in 2016, works in the industrial sector, whereas this sector contributes nearly 35% to the country's GDP, may be explained by the fact that this sector is essentially dominated by oil and minerals extraction and export, activities, conducted by multinational firms.

### *2.4.5.3 Unemployment in Mauritania*

Policy challenges facing Mauritania include the need to improve employment in its various regions, considering its geographic expanse and the fact that

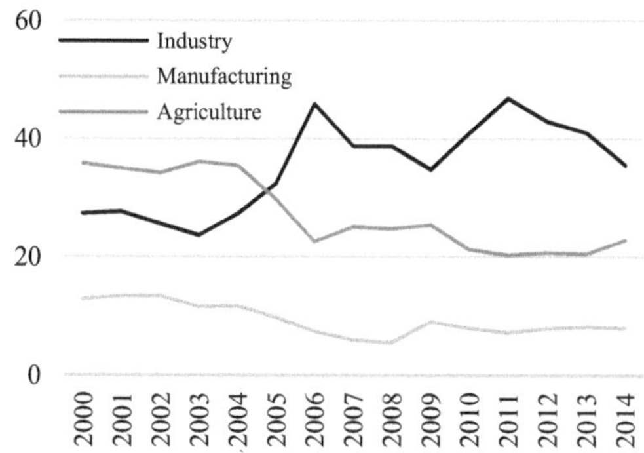


**FIGURE 2.31** Value added as percentage of GDP by industry, manufacturing and agriculture (2000–14)

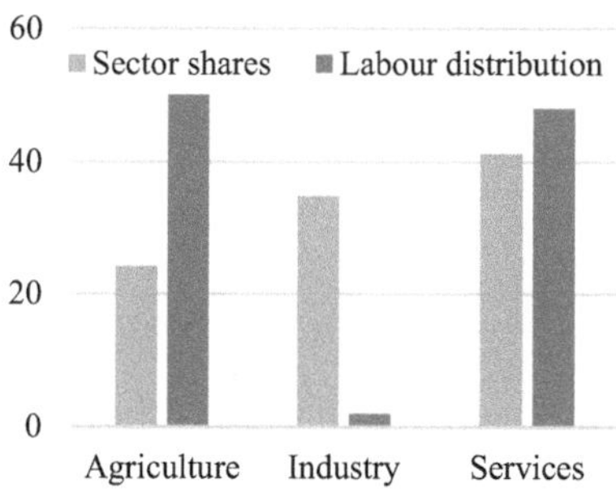


**FIGURE 2.32** Mauritania's GDP composition and labor distribution across sectors (%).

economic growth and social development are still concentrated in the country's capital. Additionally, whereas Mauritania's government has been attempting to tackle major infrastructure weaknesses, financial constraints may slow down progress over the medium term, especially with the difficulties encountered by traditional European and GCC donors.[99]

Unemployment levels in Mauritania were reported at about 31% in 2014. An overview published by the World Bank[100] put the poverty rate at 42% in 2008, a reduction by around 5% from 2004.[101] Rural poverty is far more acute, with a rate of around 59.4%.[102] Urban poverty according to the same indicator is far lower, at around 22.3%.

#### 2.4.5.4 *Obstacles to Poverty Reduction*

Principal obstacles to poverty reduction in Mauritania are low productivity of the rural economy, inadequate governance, which is largely responsible for inefficient use of natural resources revenues, with the poor quality of public services, laying additional limits on competitiveness, as well as vulnerability to external factors, including global commodity prices. Low literacy rates would also need to be tackled if future employment and poverty reduction policies are to have any chance of success (see Table 2.3).

### 2.4.6 Policy Challenges and Future Prospects

Mauritania's dependence on extracted minerals exposes its economy to fluctuations in the commodity markets. The mining industry contributed around 29% of total revenue in 2013.[103] The nation's coastal waters are among the richest fishing areas in the world, and fishing accounts for about 25% of budget revenues. However, overexploitation by foreign fishing fleets poses serious threats to the sustainability of this key source of revenue.

**TABLE 2.3** Youth and Adult Literacy Rates for Males and Females in Mauritania in 2000 and 2007 (Percentage of the Entire Population)

| | 2000 | | | 2007 | | |
|---|---|---|---|---|---|---|
| Age Groups | Female | Male | Both Sexes | Female | Male | Both Sexes |
| Youth (15–24 years) | 55.5 | 67.6 | 61.3 | 47.7 | 66.4 | 56.1 |
| Adult literacy (15+ years) | 43.4 | 59.5 | 51.2 | 35.3 | 57.4 | 45.5 |

From World Bank's Development Statistics. Databank accessed in February 2017.

In the agricultural and fisheries sectors, which employ a large proportion of the labor force, there is significant potential for revenue improvement and reinvestment. Thus, policies aimed at further processing of raw output with view to exporting processed products to neighboring and developed country markets as well as enhanced overall productivity should enable higher earnings than has hitherto been possible.

Both the mining and oil industries have naturally tended to form economic, technological, and management enclaves with limited linkages to the broader economy. However, despite being highly capital intensive and isolationist in nature, both industries stand to benefit from well-designed and managed links to their national surroundings. Improving overall STI capabilities could enhance these sectors' contribution to the country's socioeconomic development, through enabling closer linkages to domestic support and service enterprises. Furthermore, integrating STI policies within national socioeconomic development policies should allow reorientation of R&D activities toward improving extraction and processing activities, with positive impacts on the overall economic and environmental fronts.

Mauritania's economy faces natural, political, and security risks, including recurring droughts, dependence on foreign aid and investment, as well as political turmoil in its immediate vicinity.[104] Government policies over the past two decades attempted to tackle most of Mauritania's challenges with varying degrees of success. Major challenges remain with regard to infrastructure building, economic diversification coupled to private and financial sectors' development, as well as improved public administration and management capabilities. In particular, future social development will depend upon success in reducing corruption, poverty, and inequality;[105] priorities closely linked to progress in human capital formation and STI capacity building, within an equitable regulatory environment.

Mauritania's sustained and equitable development is also hampered by the small size of its economic base, extremely limited diversity of economic activities, inadequate physical infrastructure as well as legal frameworks and policy regimes.[106] Additional key challenges are posed by the need to reduce inequality and tackle disparities in wealth redistribution, in the face of widespread corruption. Some success has recently been reported with regard to improved tax collection, which may significantly impact social services. Thus, a 2016 World Bank report[107] estimated that between 2012 and 2014, Mauritania increased the amount it collected in taxes by nearly 50% through reforms to improve public resources management.

Mauritania's future economic policy is expected to focus on a number of priority issues, including

- maintaining macroeconomic stability while directing government investment toward expanding the productive sector;
- reducing poverty and tackling inequality may also be expected to constitute policy priorities for decision makers in Mauritania.

Policy directives aimed at achieving the above goals would greatly benefit from specific STI inputs. Hence, the need to marry such directives with selected strands of the country's STI policies.

In conclusion, future development policies may only achieve tangible results if aimed at the acquisition of technologies specifically targeting enhanced productivity and competitiveness. Improved education and technical training facilities as well as reduced vulnerability to climate change should also play a central role in the country's future development.

Good governance is essential for all of the above policies to succeed, particularly where the mining sector is concerned and in relation to the supervision of state enterprises. Good governance should also play a key role in allocating funds and implementing specific initiatives aimed at improving social services, particularly those directed at maternal health, infant and child mortality, and combating hunger among impoverished populations.

## 2.5 ARAB COUNTRIES OF THE GULF

With the exception of Yemen the remaining six Arab countries of the Arabian Peninsula, namely Bahrain, Kuwait, Oman, Qatar, Saudi Arabia, and the United Arab Emirates are all members of the Gulf Cooperation Council (GCC). These six countries are home to only 13% of the entire population of the Arab world, they command 50% of their collective GDP. All GCC countries were associated one way or another with British colonial exploits during the past. However, due to its strategic location, the region that is now covered by the Sultanate of Oman was attacked and occupied by the Portuguese as early as the first decades of the 16th century.

All six GCC countries are heavily dependent on their hydrocarbon resources as well as foreign labor at various levels, considerably swelling their ranks of inhabitants.

Although Yemen is not a fellow GCC member, its political and economic development has always been closely linked to that of its neighbors. A large contingent of Yemeni workers and professionals work and live in neighboring countries. The ongoing catastrophe that befell their country is more likely to render most Yemeni's more rather than less in need of support by GCC members.

### 2.5.1 Bahrain

Bahrain is the smallest economy among the GCC countries. Oil makes up 86% of the country's budget revenues, despite attempts at economic diversification. Efforts aimed at attracting multinational firms for this purpose have not been successful due to severe political unrest in 2011.

Tourism is reported as having played an important part in the country's economic recovery during 2012–15. Nevertheless, several factors, including low oil prices, have resulted in Bahrain's economy experiencing a sizeable deficit of almost 14% of GDP in 2016. With few overseas assets and with gross national savings reported in 2016 at around 20% of GDP, down from 32% in 2014, the country has limited options for covering such shortfalls.[108,109]

Steps were taken during 2014 to enhance government revenues, including higher tobacco and alcohol taxes and government services fees, including primary healthcare. Additionally, petrol price increases by up to 60% are expected to result in annual savings of around US$ 148 million. Gradual reductions are projected for 2019, in electricity, water, diesel, and kerosene subsidies.[110] In attempting to limit social unrest, measures aimed at reducing subsidies are expected to be designed in such a manner so that expatriates, rather than nationals would bear their brunt. Proposals have also been made with regard to privatization of several state-owned businesses are also being considered to help curb the deficit.

### *2.5.1.1 GDP Growth and Sectoral Developments*

As indicated in Fig. 2.33, the agricultural sector makes an insignificant contribution to GDP. The services sector is predominant, contributing around 66% to GDP and an almost equal percentage of the work force engaged in operating this sector.

Fig. 2.34 presents forecast GDP growth rates extending till 2021 based on a number of international and national assumptions. This figure indicates that Bahrain's economy grew by just under 3.4% in 2016, but was expected to go down to around 2% in 2017, essentially due to lack reduced growth of the nonoil sector as well as continued difficulties within the oil sector.[111]

Additionally, the government's reduced spending coupled to the regional liquidity crunch and domestic politics are expected to impact both private consumption and investment. As indicated in Fig. 2.34, some limited recovery is forecast for the near future. Thus, in addition to real-estate and infrastructure construction projects, the completion of key industrial projects is expected to boost economic performance in the later stages of the forecast period.[112]

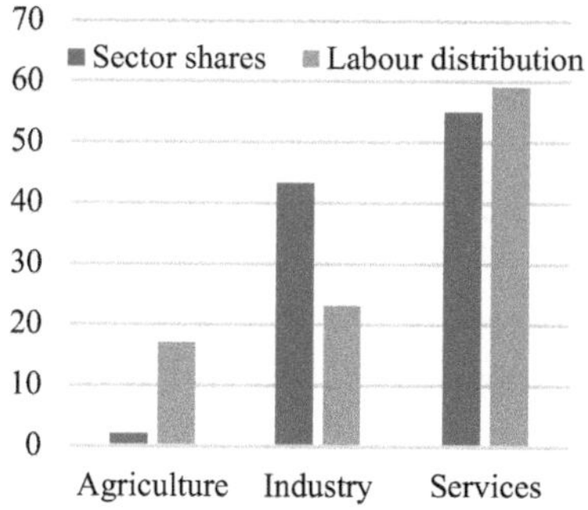


**FIGURE 2.33** Bahrain's GDP composition and labor distribution across sectors (%).

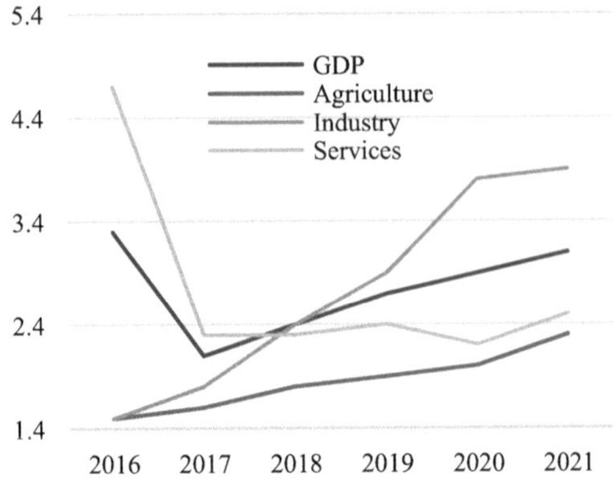


**FIGURE 2.34** Estimated and forecast growth of Bahrain's GDP and main sectors of its economy (%).

### *2.5.1.2 Unemployment and Poverty in Bahrain*

Unemployment fell to 3.1% in September 2015, from 3.8% toward the end of 2014. However, total youth unemployment, modeled on ILO estimates, is around 10.9% of total labor force aged between 15 and 24.[113]

Figures on poverty in Bahrain are unavailable from customary sources.[114] Bahrainis belonging to the Shia sect tend to command lower incomes in general than their Sunni compatriots, thereby aggravating sectarian tensions. Nevertheless, it is unlikely that poverty is widespread among nationals. On the other hand, immigrant workers who constitute around 44% of the country's labor force, and, together with their families, make up 50% of Bahrain's residents generally commanding much lower resources.

### *2.5.1.3 The Bahrain Economic Vision 2030*

Bahrain was one of the first GCC countries to lay down a long-term development plan, in 2008. The country's Economic Vision 2030 focused on reducing its economy's dependence on oil, through diversification, with tourism, business services, logistics, and manufacturing as alternatives.

Increased private sector participation in the economy is one of the main pillars of Bahrain's Vision, with government taking on a supervisory role. This, in addition to encouraging foreign direct investment, the Vision's authors hoped, would help create high-wage jobs for Bahraini's giving rise to a thriving middle class.

Furthermore, the Economic Vision 2030 seeks to encourage innovation and entrepreneurship by ensuring a conducive business environment, allowing the country to improve its status within the global value chain. Thus, knowledge-based firms were to be encouraged, whereas the financial sector would continue to perform a critical role in the economy.

A central component of the Vision's implementation is "competition through higher productivity." Thus, government would offer incentives to reward enhanced productivity within the private sector, whereas its markets would remain open, challenging home enterprises to adopt best practices.

With particular regard to enhancing the productivity of Bahrainis and local companies, monitoring progress in the Vision's implementation would rely mainly on indicators, such as improved GDP per capita, Bahrainis employment in high-paying positions, as well as the total number of Bahraini citizens in gainful employment, in addition to FDI levels.

Bahrain's Vision document acknowledges at the outset that education is critical for its realization, stating that the current education system did not provide requisite skills. Thus, one of the principal objectives of the Vision 2030 plan is set as improving the quality of education, ensuring that more Bahrainis are gainfully employed and bringing down the number of expatriates to a minimum. This, the government would realize through strategies aimed at improving performance and standards within learning institutions, some of which would target methods of assessing students and teachers, as well as enhancing critical thinking and creativity.[115] Empowering teachers through enhanced training, improved performance management as well as enhancing the attractiveness of teaching professions were among tools to be used in arriving at the Vision's objectives. Additionally, government would adopt measures aimed at encouraging higher focus on research and development at institutions of higher learning, in order to facilitate creation of a knowledge-based economy.

One of the major problems that realization of Bahrain's Economic Vision 2030 is very likely to meet is the lack of clear outlines as to how its goals might be achieved. The lack of short- and medium-term and milestones and benchmarks, with focus on objectives more than two decades away is a particular weakness that could have been avoided at the time the Vision and its implementation plans were drafted. Another serious, though very common, deficiency is the lack of numerical indicators, even relating to goals that should be within reach, such as student-to-teacher ratio within various levels of the educational process. Lack of such indicators is also encountered in the Vision's treatment of the quality of the country's healthcare sector.

Perhaps more crucial is the fact that limited measures are embedded in order to invest the Vision as well as its implementation modalities with the dynamism needed in dealing with changes that are bound to be met in over two decades of its lifetime. Thus, the country's priorities are likely to change during the period set for realizing the Vision's objectives, due to both regional as well as global factors. This is particularly relevant because a great deal of those objectives relate to the country's regional and global roles. Plans might have been adopted, whereby such a long-term vision is implemented in phases, allowing for revisions and adjustments as the country, its institutions as well as its citizens and foreign workers evolved from one phase to the next. The fact that the Vision is implemented by a Bahrain Economic Development Board, and that changes might be introduced as needed, could be quoted against such arguments. However, embedding dynamism as plainly as possible into the Vision and its implementation modalities, whereas not diluting any of its elements or goals would have been a far better option.

#### *2.5.1.4 Bahrain's Current and Future Challenges*

Several challenges face Bahrain in attempting to address its current development problems. Long-term challenges are mainly due to the need for boosting the country's attractiveness as a regional and international business hub with focus on manufacturing, finance, and tourism. Other challenges, mostly in the medium- and short-terms, concern matching revenue constraints with the need for maintaining state subsidies and supporting a bloated public sector that encourages masked unemployment, in order to stem public unrest.

Given the existing social tensions, and in light of government plans to cut subsidies, the economy remains vulnerable to civil unrest as well as regional tensions, for the foreseeable future. Tourism and financial services activity might suffer some stagnation as a result of the overall slowdown and increased military spending in the region. Although some might consider such conditions a good cause for austerity and reduced spending on STI capacity building, it is likely that carefully crafted government policies aimed at accelerated STI capacity building could in fact lead to employment and enterprise creation as well as resolving several endemic problems, regarding, for example

- water scarcity through modern water treatment technologies suitable for distributed applications techniques;
- high cost of food imports through home production using novel agricultural techniques;
- wider access to lifelong learning and related educational services, using information and communication technologies (ICTs) to reach wider segments of the population.

### 2.5.2 Kuwait

Kuwait is a small country whose open oil-rich economy is supported by crude oil reserves of over 100 billion barrels—around 6% of the world's reserves. Revenues created by its oil sales accounts for over half of GDP, 94% of export revenues and 90% of government income.[116]

Kuwait is one of the few countries in the world to set up a "Fund for Future Generations," annually allocating around 10% of government revenues to this fund. Hence, its economy was cushioned against the impact of lower oil prices.

### 2.5.2.1 GDP and Sectoral Growth

The country's public sector accounts for more than 70% of the country's GDP and employs more than 85% of the national workforce.[117] Agriculture in Kuwait, along with most other GCC countries, makes minimal contributions to the country's GDP, whereas industry and services contribute the most, with the latter's contribution at 40% falling considerably behind the former's, at nearly 60%.

Fig. 2.35 presents a view of attained and forecast annual growth in all three sectors, between 2016 and 2021. Fig. 2.35 indicates expectations of GDP growth in the vicinity of 3% during 2018–21, whereas growth of between 3% and 4% is forecast for all three sectors around 2018. Interestingly highest rates are forecast for the miniscule agricultural sector, which might eventually approach 5% by 2021. Industry and services, on the other hand are forecast to grow at around 3% in 2021.[118]

Construction, within Kuwait's industrial sector, witnessed considerable development in recent decades. Infrastructure and residential construction projects are likely to remain at the focus for the government as well as private sector concerns, with substantial investment planned in transport and housing projects, whose STI requirements are likely to be mostly traditional.

Kuwait's budget went into deficit in 2015 for the first time during the past 15 years. Attempts to reduce this deficit through decreasing subsidies met with limited success. Nevertheless, growth was resilient in 2015 as rising output in the nonoil sector compensated for the oil sector's contraction. Ongoing and projected infrastructure projects are expected to provide further support for growth in the coming few years.

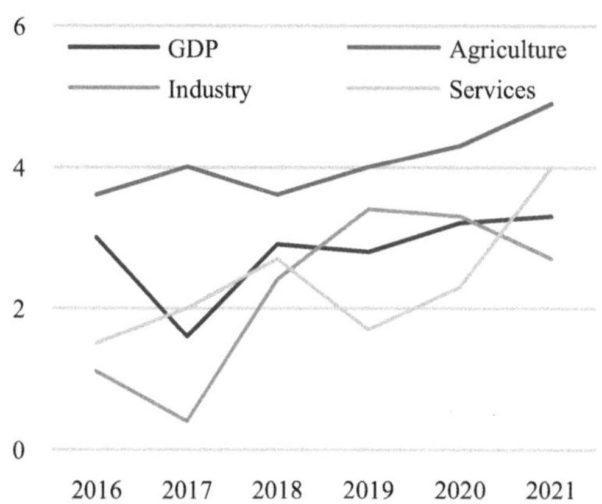


**FIGURE 2.35** Estimated and forecast growth of Kuwait's GDP and main sectors of its economy (%).

### 2.5.2.2 Economic Challenges and Development Priorities[119]

Official announcements and national development plans over the past two decades have emphasized the need for diversification of Kuwait's economy, whereas at the same time acknowledging oil as the backbone of the country's economy and that it will continue to play an important role in driving growth in the foreseeable future. However, only limited progress has been reported by successive governments with regard implementing the long-term economic development plan, which was initially adopted in 2010.

Key economic challenges for Kuwait include the need to ensure fiscal sustainability, improve management of wealth derived from the country's hydrocarbon resources, and achieve tangible progress on the economic diversification front.

With particular reference to diversification, past efforts have tended to fail due several factors including friction between the National Assembly and the executive branch, which is quoted as contributing to stalled ventures aimed at economic reform in general. Thus, although $ 104 billion were earmarked by government for economic diversification, attracting investment, and boosting private sector participation in the economy, several projects did not materialize, mainly due to internal political differences. Additional factors include an oversized public sector that tends to crowd out private employment of Kuwaiti nationals and the lack of an overall climate that is conducive to sustainable entrepreneurial activity.

### 2.5.2.3 Unemployment and Poverty

Unemployment in Kuwait has been reported at around 3% in two consecutive years, 2015 and 2016. Total unemployment among youth as a percentage of the total labor force aged between 15 and 24 was estimated at around 14.6% by national authorities.

Poverty figures are not available from customary sources.[120] A recent World Bank report indicates that poverty was not an issue for Kuwaiti citizens. However, human rights groups as well as the regular press have often reported on the inadequate conditions endured by expatriate workers and domestic servants as well as stateless persons residing in Kuwait, who continue to live in the country without access to work or social services.[121]

### 2.5.2.4 Kuwait's Development Challenges

Improving the quality of life for all Kuwaitis, creating the right climate for enterprise and employment creation and safeguarding Kuwait's environmental integrity—considered crucial for the country's present and future prosperity—are at the center of the country's declared priorities. The latter issue necessitates development and widespread utilization of renewable energy resources, allowing added value to be realized through sound processing of hydrocarbon resources thus saved, with due regard to environmental implications.

Kuwait's geographical location and its accumulated wealth are considered essential basis for positioning the country as a trading hub and financial center.[122] A wide range of scientific and technological and innovation capabilities are acknowledged by a variety of national planning documents as essential basis for attaining these objectives. Several reports have also emphasized the need to identify ways and means that would allow the country to play a more effective role in STI capacity building in collaboration with its GCC partners. However, as indicated in sections dedicated to reviewing recent STI capacity building efforts, little has been achieved in these respects.

## 2.5.3 Oman

Oman is a middle-income country and one of the Arab Gulf's smaller economies. Compared to its neighbors, Oman is only a modest oil producer and its oil resources are known to be dwindling, prompting adoption of enhanced oil recovery techniques.[123] The fact that oil resources presently constitute around 84% of government's revenue, together with the massive drop in global oil prices, has therefore been causes for deep concern.[124]

Although the country continues to enjoy remarkable stability compared to some of its close neighbors, popular sentiments that surfaced between 2011 and 2013 revealed a certain degree of political and social frustration.

### *2.5.3.1 Current Economic Situation and Future Plans*

Reforms instituted in 2015, included doubling gas prices for industry, as well as over 30% increase in diesel and petrol prices, in addition to measures aimed at limiting wage increases.

The budget has come under increased pressures for several consecutive years, due to falling oil prices, as the government decided to abandon planned cuts in social welfare benefits in recognition of what such measures might lead to—in view of unrest elsewhere in the Middle East and North Africa. As a result, Oman suffered a budget deficit of around $ 6.5 billion, nearly 11% of GDP in 2015. Thus, in 2016, Oman's government had to borrow US$ 1 billion from a consortium of banks and issue sovereign bonds amounting to US$ 2.5 billion. Reports indicate that Oman's government would still need to use US$ 1.5 billion of its general reserves as well as borrow between US$ 5 and 10 billion on the international market.[125]

Oman has put in place diversification, industrialization, and privatization plans as far back as 1996 (see Frame 2.3). The country's long-term objective is to reduce oil's contribution to GDP down to around 9% by 2020, from its present levels of around 46%, with tourism as one of the key ingredients in its development strategy.

Nevertheless, fiscal consolidation and diversification are expected to continue as principal tools of the country's economic policy. With regard to

**Frame 2.3 Recent National Development Plans**[126]

**Oman's Vision 2020**

A national Vision 2020 document[127] was formulated with essential objectives, including the development of human resources and other capabilities to keep up with technical development, as well as managing change on the domestic and global arena, in seeking "a sustainable and diversified economy."[128]

The Vision document targeted the creation of a stable economic climate as well as developing a private sector capable of optimizing the use of human capital and natural resources, whereas safeguarding environmental integrity. More specifically, the Vision document recognized the need for favorable conditions for optimal exploitation of the natural resources and reduced disparities among regions and income groups, ensuring "that all citizens benefit from the development process."[129] Along related lines, the Vision document aimed to use oil and gas revenues to achieve economic diversification, whereas government performed its role "towards the provision of health and education social services, training citizens and developing their skills, and pursuing such policies aimed at promoting decent standard of living."[130]

**Oman's 9th Five-Year Plan**

This is the last of the series within plans designed to implement Vision 2020. The 9th five-year plan covered the years 2014–20 and was designed to

- achieve an average annual GDP growth rate of 3%;
- reduce the contribution of oil in GDP at current prices from44%[131] to 26%;
- maintain the inflation rate within safe levels at an average of 2.9%;
- focus on private sector and activate the public private partnerships PPPs;
- create job opportunities with emphasis on SMEs.

The private sector constituted the backbone of this plan, with support through either public private partnerships or providing access to facilities. The plan targets investments amounting to around US$ 107 billion, 52% of which would be sought from private investments, whereas the rest would come from the public sector.[132]

**Oman's Vision 2040**

Oman appears to be actively seeking to formulate its Vision 2040, with a special committee set up in 2013 for the purpose.[133] Available reports indicate that this Vision will build upon the outcomes of Vision 2020, whereas taking a closer look at sectoral issues and that community outreach will constitute a significant component of activities advocated by Vision 2040. Promoting the business sector and tourism also appears to have received considerable attention as well.

the latter, Oman is adopting an economic diversification program designed along lines that Malaysia implemented in order to transform its own economy, with particular support directed toward logistics, manufacturing, and tourism, which would require extensive improvements in various STI infrastructures, with particular regard to ICTs and a host of "green technologies."

### 2.5.3.2 Recent and Forecast Economic Growth

Annual GDP growth went down to around 1.8% in 2016 from levels close to 3% in the previous 2 years. As indicated by Fig. 2.36, economic growth figures below 3% are expected in the near future.[134] This is essentially due to lower oil sector investments coupled to lower public spending, whereas nonhydrocarbon GDP growth is estimated to go down to around 4% from higher previous levels as public spending declined impacting investment and consumption.

Growth of Oman's services sector underwent a catastrophic decline earlier in the decade but is expected to grow at levels of between 2% and 3% during 2018–21. The agricultural sector is expected to grow at higher rates as a percentage of GDP than the industry and services sectors throughout the period covered by Fig. 2.36, approaching 3.5% by 2021, although GDP growth as well as growth in the services and industry sectors is expected to approach 3% by 2021.

Past legal and institutional arrangements aimed at FDI promotion over the past decade do not seem to have borne fruit as indicated by Fig. 2.37,[135] exhibiting a somewhat steep trend from values of between 5% and 8% of the country's GDP down to around 1% during the first 4 years of the current decade.

### 2.5.3.3 Unemployment in Oman

Overall unemployment is estimated at around 7.2% of the total work force[136] However, youth unemployment assumed considerably higher rates, averaging

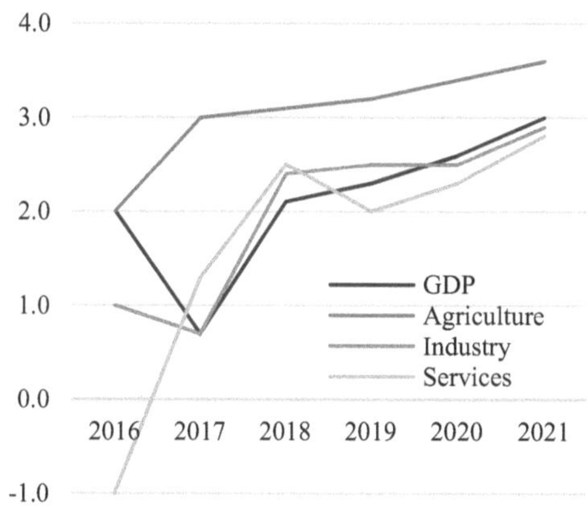


**FIGURE 2.36** Estimated and forecast GDP and sectoral growth in Oman as a percentage of GDP (2016–21)

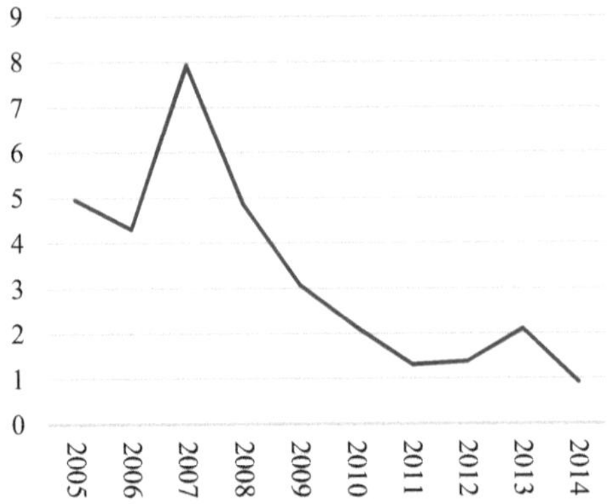


**FIGURE 2.37** Oman's foreign direct investment net inflows as percentage of GDP.

around 19.7% during 1991–2014, with very little variation from 1 year to the next. This is a serious problem because almost 40% of the country's population is less than 25 years old. Oman's government has adopted plans aimed at creating more employment opportunities however, little information is available concerning their outcome.

No statistics are available on poverty from customary sources. However, it is highly likely that poverty is widespread among immigrant worker communities from the subcontinent, estimated to constitute around 30% of the country's total labor force, and around 290,000 residents.[137]

### *2.5.3.4 Future Development Prospects*

Attempts at prioritizing economic diversification by promoting agriculture, tourism, the gas sector and nonoil industries have met with mixed success. Although significant consolidation efforts have led to some fiscal savings, deficits maintained high levels, and are in need of invasive measures, which the country can ill-afford under the present and foreseeable, global as well as regional conditions.

As means of gradual enhancement of national growth, traditional advice by analysts schooled in classical economics advocate pro-business reforms, including 100% foreign ownership of enterprises in Oman, implementation of legal and regulatory arrangements that are more conducive FDI as well as the liberalization of key sectors. Limited, if any mention is made, however, of well-funded countrywide initiatives aimed at national enterprise and employment creation, based on novel technology inputs as well as improved education and life-long learning, as alternatives or subsidiary means of enhancing diversification and growth. Greater attention ought to be awarded to such initiatives with emphasis on local STI capacity building in collaboration between local research institutions, donor countries and international organizations and central as well as local authorities and civil society institutions.

## 2.5.4 Qatar

Qatar's economy witnessed remarkable prosperity, with continued high real GDP growth, particularly since the start of the 21st century. Although lower oil prices prevailing since August 2014 have had their effects on the country's economy, Qatar was the only GCC member that avoided a budget deficit in 2015.[138] However, a $ 12.8 billion deficit, 6% of GDP was projected for 2016.[139]

Qatar's economy is largely driven by its oil and gas resources, which account for around 92% of export earnings and 56% of government revenues. With proven oil reserves in excess of 25 billion barrels and proven reserves of natural gas exceeding 25 trillion cubic meters, or about 13% of the world's total reserves, this country's reliance on its hydrocarbon resources might be expected to go on for many years to come.[140] Recent growth in manufacturing, construction, and financial services has lifted nonoil contributions to around 50% of the country's nominal GDP.[141]

### 2.5.4.1 *Qatar's Vision 2030*

Qatar's 2030 Vision document was issued in 2008, following ratification of the country's constitution. The document recognizes that Qatar's abundant wealth created unpredicted opportunities as well as formidable challenges, including:

- modernization while preserving traditional values;
- responding to the present generation's needs as well as safeguarding those of future generations;
- managed growth and uncontrolled expansion;
- the size and the quality of the expatriate workforce in relation of selected development paths;
- economic growth, social development and safeguarding the environment.

In responding to the above challenges, Qatar's Vision document underscores the need to allocate returns from the country's hydrocarbon wealth prudently and efficiently in order to avoid stresses posed by economic overheating and imbalances that accompany uncontrolled expansion.[142]

A good deal of concern is expressed by the Vision document regarding the consequences of recruiting expatriate workers, and their impact on the national identity. Additionally, the document warns that the pursuit of a rapidly growing, diversified, and technologically sophisticated economy must never be allowed to sacrifice environmental integrity for economic gains.

Guiding principles of Qatar's National Vision emphasize promoting justice, moral, and religious values and traditions, and the need to guarantee security, stability, and equal opportunities. Qatar's National Vision is proclaimed as resting upon four pillars, namely human, social, economic and environmental development. In outlining its stance toward the First Pillar, the Vision document recognizes that oil and gas resources will eventually run out and stresses the need to build a viable human capital that is capable of meeting future challenges. Establishing advanced educational and health systems, as well as increasing the effective participation of Qataris in the workforce, is considered central components of the human development pillar. In this context, the Vision document looks toward:

- becoming a prime center for scientific research and intellectual activity in the region;
- developing an integrated system for health care, managed according to world-class standards.

Further STI-related objectives outlined by Qatar's Vision 2030 document are addressed in the following chapter, whereas succeeding paragraphs will take a look at selected production activities, in agriculture and industry including water desalination, with close links to technology acquisition and adaptation in Qatar.

### 2.5.4.2 *Sectoral Developments*

Industries that are central to the country's economy include extraction of oil and gas, natural gas liquefaction, crude oil refining, production of ammonia, fertilizers, production of some basic petrochemicals, steel reinforcing bars, cement, and commercial ship repair.

### 2.5.4.3 *Industrial Production*

Industrial production growth was estimated at about 1.5% in 2016. This estimate is based on plans for a number of industrial projects that were either recently completed or were scheduled to be completed in Qatar. One of these projects is the Al Sejeel Petrochemicals complex at Ras Laffan. A joint venture by the Qatar Petrochemical Company and Qatar Petroleum, which is scheduled to go on stream in 2018. Frame 2.4 presents information regarding this project with emphasis on its output, construction management, and technology sources. However, although construction management is to be conducted by well-known international firms, available literature does not mention possibilities for joint activities, linking Qatari research and technology development institutions and technology providers with view to future adaptation and development of the technologies to be utilized. The fact that the project will result in creation of only 1500 jobs is also noteworthy.

Acquiring higher returns on the basis of hydrocarbon wealth is an essential issue for Qatar, as well as its other GCC neighbors. Moving toward production of specialty, rather than basic commodity petrochemicals, should allow higher returns as well as create a wider range of employment opportunities. However, little is mentioned about such moves in recent planning documents. Indeed, a recent report states that "Qatar's chemical and petrochemical industry is considered by many to be at a standstill." It further goes on to indicate that future expectations for this industry hinge on how Qatar Petroleum decides to move the sector forward.[143]

**Frame 2.4 Al Sejeel Petrochemicals Project in Ras Laffan, Qatar**

The Al Sejeel project is part of Qatar's efforts to increase the petrochemicals sector's output to 23 million metric tons per annum (MMTA) by 2020, from its present level of around 10 MMTA, as well as further diversify its economy (see Fig. 2.38).

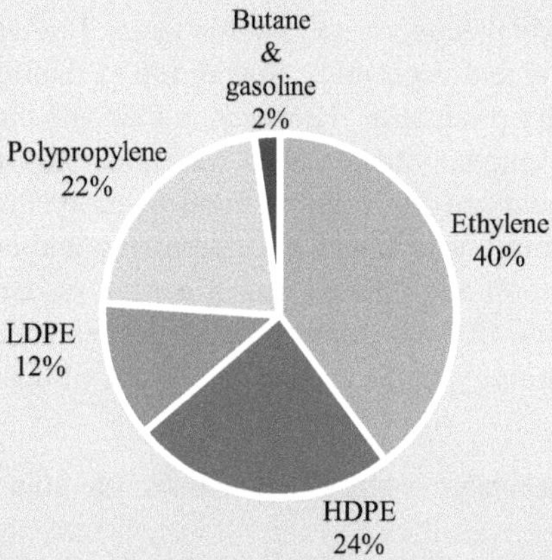


**FIGURE 2.38** Output of the Sejeel petrochemicals complex at Ras Laffan (Total of 3.5 MMTA)

The project will use a large-scale mixed-feed steam cracker, with a mixed feedstock of ethane, butane and naphtha from natural gas projects located in the north of Qatar in order to produce around 3.5 MMTA of ethylene, high- and low-density polyethylene, polypropylene, butane and gasoline. See the adjoining pie chart.

The polypropylene production facility is designed to produce homopolymers, random copolymers and impact copolymers, destined for use in packaging, durable goods, automotive parts, nonwovens, and fibers. Polyethylene, on the other hand, is for use in packaging mainly.

Technology license contracts for the project were signed in June 2012 with Univation Technologies providing the polyethylene technology and Union Carbide Chemicals & Plastics Technology, a subsidiary of The Dow Chemical Company, providing the polypropylene technology.

Bechtel is providing project management services for the construction of the complex, as well as initial services for the front-end engineering design phase.

It would seem, therefore, that decision makers in Qatar are confident of the fundamental strength of the country's position, based on its hydrocarbon wealth, the returns it is creating as well as the investment capabilities they have been able to build over the years, and do not feel the need for radical changes in their industrial strategies.

Strategies aimed at investing in the manufacture of specialty chemicals as well as downstream products based on such products, e.g., automotive parts, furniture items, etc., through partnering with foreign technology owners appear to have been considered by officials and investors. However, the tendency to create such facilities abroad in partnership with technology owners, e.g., in the Far East and Europe, would miss out on a key objective of such a strategy, namely job and subsidiary enterprise creation in Qatar.

### *2.5.4.4 Agriculture and Food Production*

Agriculture is a marginal economic activity in Qatar, limited to production of a narrow range of fruits and vegetables, in addition to fishing as well as limited beef, poultry, and dairy production. However, in line with the country's National Vision 2030, which emphasized the need for enhancing self-reliance in food production, Qatar launched its National Food Security Programme (QNFSP) in 2008. The Programme was aimed at modernizing and boosting agricultural production, conservation and optimal use of natural resources, through utilizing green technologies. QNFSP emphasizes vertical expansion of food production using high technology as far as possible. Lines of action proposed by the Programme include:

- utilization of renewable energy with focus on solar energy for water desalination;
- extensive water management initiatives including distribution of desalinated water for agricultural use and for recharging aquifers;
- intensifying agricultural production through novel inputs, e.g., drip irrigation and hydroponics;
- promotion of modern food processing methods with accent on food safety and nutritional health through the establishment of an Agro-industrial Park.

Clearly, massive planning efforts and considerable investments would be required in order to attain viable outcomes based on the above lines of action. Revised policies, legislative and regulatory arrangements are needed to start with, along with implementation of research and extension strategies, as well as intensive training for operators in various agricultural subsectors. This seems to have caused QNFSP to confine activities to coordination of the efforts of concerned organs in the food production chain, as well as the development of an overarching national policy for sustainable agricultural development.[144]

On the other hand, press reports over the past few years have indicated a certain level commercial activity in alternative agricultural technologies, with some focus on hydroponic production of a rather narrow range of vegetables. No indications are available as to the proportion of domestic demand covered by such novel techniques, however. Judging by the above-mentioned press reports it is safe to assume that commercialization of hydroponic agriculture is in its early beginnings in Qatar, and possibly in other Arab Gulf countries.[145]

### *2.5.4.5 Water Desalination*

Qatar's reliance on desalinated water is almost total, with negligible amounts of natural water resources that are in any case rapidly dwindling. In 2015, Qatar was reported as having water supplies sufficient for only 48 h according to prevailing consumption levels, exposing it to extreme hardship in case of interruptions. Reservoir projects slated to come into operation in 2017–18 were supposed to render reserves sufficient for at least 1 week.

Recent seawater desalination projects planned by Qatar included the construction of a facility announced in 2015 by Qatar Electricity and Water Company (QEWC). The US$ 3.15 billion facility, which should go on stream in the southern part of the country in 2018, is being constructed at a time when Qatar is witnessing unprecedented growth in demand for both water and energy.[146] The plant is designed, constructed and will be operated on the basis of a joint venture agreement according to which Qatar Petroleum, Qatar Foundation, Mitsubishi Corporation, and Tokyo Electric Power Company are taking part. Power and water produced by this facility will be purchased by the Qatar General Water and Electricity Corporation (Kahramaa) for 25 years. This facility is designed to produce around 620,000 $m^3$ of potable water per day.[147] Qatar has also been reported as building a reverse osmosis water desalination plan in 2015, with QEWC reported as intending to develop a US$ 500 million plant which was reported to produce 136,000 $m^3$ daily by the third-quarter of 2016.[148] Considering environmental as well as likely economic benefits presented by reverse osmosis, it would seem that efforts at adopting this technology were considerably delayed.

Other water desalination and storage projects have been reported including five mega-reservoirs expected to be completed by mid-2018. Thus, in March 2015, contracts were awarded for new water desalination plants and mega-reservoir facilities, including a US$ 4.67 billion contract dedicated to building water desalination and storage facilities.

### *2.5.4.6 Unemployment and Poverty in Qatar*

With regard to unemployment and poverty levels, it is well to remember that oil and gas resources have rendered Qatar the world's highest per-capita income country, with unemployment figures estimated at 0.7% and 0.4% in 2016 and 2015, respectively. No figures are available from reliable sources on poverty headcounts during recent years, which are in any case expected to be rather low and possibly confined to expatriate workers.

### *2.5.4.7 Future Prospects*

Qatar possesses both vast nonrenewable resources, in the form of oil and gas, as well as limited renewable land and marine resources. Its growing population coupled to expanding urban and industrial activity is likely to place enormous burdens on its future sustainable development.

Reliable information on the outcome of implementing recommendations included in Vision documents and development plans is scarce. However, it is likely that government spending aimed at enhancing Qatar's image, such as hosting costly international sports events, might constrain spending on more pressing long-term improvements in education and STI capacity building, particularly in view of the current global economic climate.

Comprehensive plans aimed at integrating environmental, economic and social aspects of development goals and enlightened strategies for their attainment will be essential for tackling Qatar's longer-term challenges. Innovative industrial development and urban designs, promoting research and technology development in selected niche sectors could go a long way toward ensuring sustainable socioeconomic development as well as job creation.

Astute designs for such plans coupled to aggressive implementation strategies will be essential for success, considering the currently depressed hydrocarbon markets and regional tensions as well as tough competition from both regional and international actors.

## 2.5.5 Saudi Arabia

The Saudi economy is largely oil-based and strong government control is exercised over major economic activities. The country possesses about 16% of the world's proven petroleum reserves and ranks among the largest exporters of petroleum, performing a leading role on the international oil market. The oil sector is the source of around 87% of national budget revenues, constitutes 42% of GDP, and 90% of export earnings.[149]

### *2.5.5.1 GDP Growth and Sectoral Contributions*

Saudi Arabia's GDP growth suffered a considerable decline, by almost one-third from 3.5 down to 1.2%, between 2015 and 2016.[150] This decline, which is linked to reduced hydrocarbon rents, is likely to have had repercussions on planned expansion in development expenditures, including education and research activity. Increased defense spending, from a rather high level of almost 8% of GDP in 2010 up to 12.6% in 2015, could further constrain additional development spending.[151]

Fig. 2.39 summarizes information on sectoral contributions to the country's GDP as well the distribution of the country's working population across sectors, indicating the minimal role played by agriculture in Saudi economy as well as the small proportion of the working force engaged by the industrial sector. This is to be expected given that diversification of industry's contribution to GDP is still limited.

Fig. 2.40 presents a view of estimated and forecast GDP growth economic performance parameters for Saudi Arabia, during 2011–21. Fig. 2.40 underscores the sharp declines suffered by the country's GDP growth as well as contribution it received from various sectors of economic activity during the period

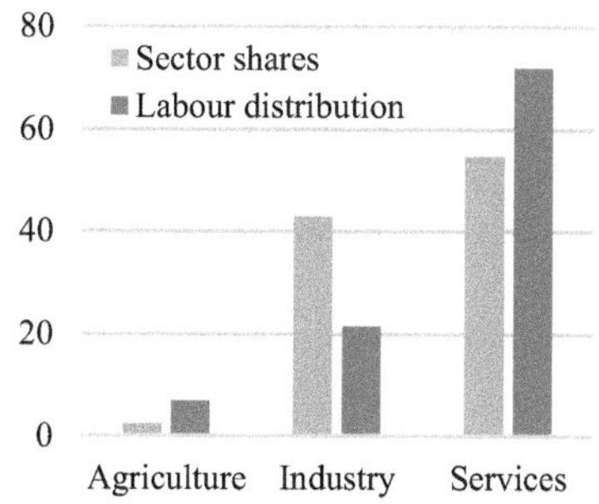


**FIGURE 2.39** Sectoral contributions to GDP in Saudi Arabia and labor distribution across main sectors (%).

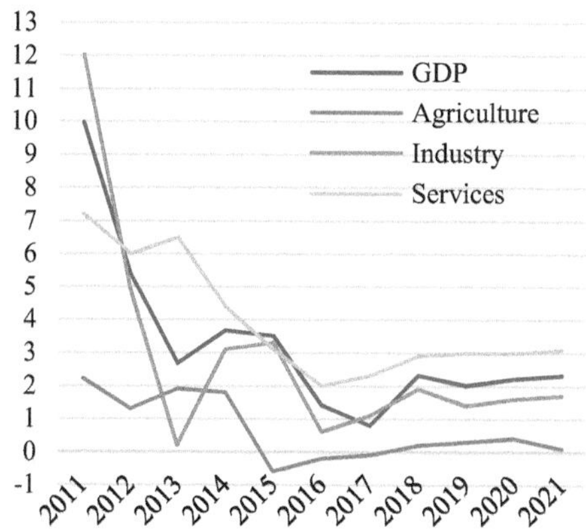


**FIGURE 2.40** Estimated and forecast GDP and sectoral growth in Saudi Arabia's (%).

2012–14. Although the shares of both the industrial and services sectors subsequently maintained positive growth during 2014–16, albeit at declining rates, agriculture's share contracted by 0.6% in 2015. Such changes will naturally have a significant impact on the intensity of technology acquisition and implementation across these sectors, as well as nationally, in relation to associated educational and training activities.

### 2.5.5.2 Saudi Arabia's Budget Deficit

Both regional and international developments, including the reduction of oil prices and enhanced defense purchases, due to the country's involvement in neighboring Yemen and elsewhere in the region, are responsible for a budget deficit of around 13% of GDP announced in 2015. A deficit of around US$ 87 billion was also expected in 2016. Although such deficits might be financed by bond sales and reserve withdrawals, the Saudi government nevertheless announced its intention to cut capital spending in 2016. Other steps being discussed with view to cutting deficits include the introduction of a value-added tax and the reduction of subsidies on electricity, water, and petroleum products. Additionally, it has been announced that shares of its state-owned petroleum company, ARAMCO would be publicly listed with view to increasing revenues and outside investment.[152]

Anecdotal evidence points to the possibility that spending cuts caused by budget deficits might negatively impact STI expenditures. Clearly, there is a need for further research regarding such claims before they may be accepted as basis for established trends.

### 2.5.5.3 *Saudi Arabia's Vision 2030*

Reform of Saudi Arabia's economy has been a recurring theme in national development plans over the past decades. Economic diversification, privatization and providing wider employment opportunities for Saudi nationals were central to almost all past development plans. Attempts at economic reform gained greater prominence over the past few years due to both internal and external economic as well as security challenges. More recently, the concept of creating a knowledge-based economy has featured more prominently in planning documents and official pronouncements, including Saudi Arabia's Vision 2030, which was made public in 2016, with close focus on reducing dependence on oil and economic diversification.[153] Improvements in health, education, infrastructure, recreation, and tourism feature prominently with specific goals to be achieved by 2020 and 2030. Table 2.4 presents a list of actions with associated goals in various domains, including the country's economic performance, global competitiveness, and its attractiveness as a location for foreign direct investment.

Saudi Arabia's Vision 2030 is based upon three principal themes: building a vibrant society, a thriving economy and an ambitious population. The Vision 2030 document places a good deal of emphasis on adherence to traditional values while seeking moderate interpretation of religious teaching is highlighted.

#### 2.5.5.3.1 STI Capacity Building for Realising Vision 2030

The Vision document outlines measures to be taken and goals to be met, several of which are linked more or less closely to scientific, technological, and innovative capacity building. Ties to specific aspects of the national science, technology, and innovation system are most evident in relation to reducing dependence on hydrocarbon resources and diversifying the national economy as well as upgrading the performance of service sectors including among others: health,[157] education,[158] ICT infrastructure, as well as a variety of cultural and recreational activities, including tourism.

Actions listed in Table 2.4 would greatly benefit from radical upgrading of the country's technological capabilities, calling for acquisition of new and

**TABLE 2.4** Numerical Goals of Saudi Arabia's Vision 2030

**Economic performance**

By 2030:

- Move from current position as 19th largest world economy into the top 15 economies;[154]
- Rise from current 25th rank on the Global Competitiveness Index to one of the top 10 countries worldwide;
- Raise global ranking in the Logistics Performance Index from 49 to 25;
- Increase Public Investment Fund's assets, from US$ 160billion to over US$ 1.8trillion;
- Enhance FDI from 3.8% to the international level of 5.7% of GDP;
- Increase the private sector's contribution to GDP from 40% to 65%;
- Enhance share of nonoil exports in GDP from 16% to 50%;
- Increase nonoil revenue sixfold from about $ 43.5billion annually to $ 267billion;
- Enhance household savings from 6% to 10% of total household income.

**Education**

By 2020: train half a million government employees through distance learning. By 2030: at least five Saudi universities among the top 200 universities in international rankings.

**Health and welfare**

By 2030: increase the average life expectancy from 74 to 80years.

**Urban development and home ownership**

By 2020: increase percentage of home owners from 47% to 52%. By 2030: three Saudi cities among top-ranked 100 cities in the world.

**Energy** By 2030: build up the renewable energy sector to provide 9.5GW of renewable energy.

**Industry**

By 2020: enhance the contribution to GDP by the mining industry sector to US$ 26billion, creating 90,000 job opportunities.

By 2030:

- increase local activities in the oil and gas sectors from 40% to 75%;
- lower the rate of unemployment from 11.6% to 7%;
- localize over 50% of military equipment spending.

**Employment and SME creation**

By 2020: add 1 million employment opportunities to the country's retail sector, bringing total employed to 2.5million.[155] By 2030:

- increase SME contribution to GDP from 20% to 35%;
- encourage financial institutions to allocate up to 20% of overall funding to SMEs;
- increase women's participation in the workforce from 22% to 30%.

**Government**

By 2030: enhance Saudi Arabia international ranking on the Government Effectiveness Index, from 80 to 20.

**Information and communications technology (ICT)**

By 2030:

- raise the country's ranking on the E-Government Survey Index from its current 36th to one of the top five countries worldwide;
- ICT coverage to exceed 90% housing in densely populated cities and 66% in other urban zones;[156]

By 2020 increase the contribution of modern trade and e-commerce to 80% of the retail sector.

**Defence** By 2030: localize over 50% of military equipment spending by 2030.

competitive technology inputs as well as improved ICT infrastructures. This is particularly the case with regard to attaining objectives such as

- enhancing the contribution of the private sector up to 65% from its present level of around 40%; and
- increasing the share of nonoil exports to GDP from 16% to 50% and increasing nonoil revenue.

Additionally, if supported by adequate strategies, implementation plans, regulatory arrangements and monitoring modalities, it should be possible to secure a variety of such inputs both through FDI,[159] including improved technology acquisition.

Implementation of the Vision's activities would necessarily require considerable investments. Notably, the document includes provisions for enhancing assets of the Public Investment Fund to be increased by a factor of more than 11, from US$ 160 billion to over US$ 1.8 trillion. However, it is likely that a considerable proportion of resources available to this fund will be dedicated to traditional sovereign investments, both within the country and overseas.

#### 2.5.5.3.2 Diversification and Privatization According to Vision 2030

Past efforts at diversification were principally confined within sectors such as power generation, telecommunications, natural gas exploration, and petrochemicals. More recently, however, the Saudi government seems to be targeting an expanding role by the private sector in the healthcare, education and tourism industries.

Privatization and private enterprise development have been features of past plans and pronouncements. They are also taken up as a prominent feature of the Vision 2030 document. Thus, even in relation to healthcare the document states that Saudi Arabia would seek to improve healthcare services through the introduction of "corporatization into the sector by transferring the responsibility for health care provision to a network of public companies that compete both against each other and against the private sector." This, the document states should provide citizens with the highest quality of health care, whereas government would focus on legislative, regulatory, and supervisory tasks.

The fact that diversification and creating employment opportunities are closely linked to imparting educational and technical skills to the country's large youth population is a crucial aspect that needs to be urgently tackled through educational reforms. However, no concrete initiatives in this direction are clearly articulated by the Vision 2030 document. Thus, although other initiatives aimed at educational reform might be in the pipeline, this document poses only two numerical goals related to education. The first, to be attained by 2020, requires that the country train half a million government employees through distance learning, although the second stipulates that at least five Saudi universities be ranked among the 200 best worldwide. This, together with the limitations that might be imposed on public spending, might foretell rather limited outcomes regarding educational reforms.

#### 2.5.5.3.3 Unemployment and Social Development

Tackling unemployment is a critical element of Vision 2030. Goals set by the Vision document include cutting unemployment among Saudis men down to 7% and increasing the percentage of women in the workforce from 22% to 30% by 2030. Over 6 million foreign workers are currently employed by various Saudi businesses, operating mainly in the oil and services sectors. On the other hand, unemployment in Saudi Arabia was reported in the third-quarter of 2016 at a 4-year high level of over 12%.[160,161,162] Additionally, unemployment among women and youth in general has been on the rise.

Table 2.5 lists two sets of actions and associated numerical goals aimed at social and cultural development set for attainment by 2020 and 2030, respectively. Goals aimed at upgrading the role of nonprofit organizations in providing cultural and social services will inevitably require wider utilization of ICTs and associated tools, including a variety mobile applications as well as networking arrangements.[163]

#### 2.5.5.3.4 Implementing Saudi Arabia's Vision 2030

Various ministries acting in coordination and collaboration are allocated responsibilities for detailed design and implementation of Saudi Arabia's Vision initiatives. Nevertheless, it is highly likely that the Vision's realization might hit

**TABLE 2.5 Vision 2030 Actions and Goals Related to Nonprofit Sectors, Social and Cultural Settings**

**By 2020**

- increase the number of amateur clubs providing a variety of cultural and entertainment activities by more that 450;
- more than one-third of Saudi nonprofit organizations' projects should have measurable and deep social impact.

**By 2030**

- raise Saudi Arabia's ranking as measured by the Social Capital index from 26 to 10;
- raise the nonprofit sector's contribution to GDP from less than 1% to 5%;
- increase household spending on culture and entertainment within Saudi Arabia by up to 6%;
- ensure support by 1 million volunteers per year, compared to the current figure of around 11,000.

several snags, essentially due to the fact that activities relating to the design, implementation, monitoring, and reporting have been assigned to a single institutional entity, namely the Council of Economic and Development Affairs. Although reports indicate that a National Centre for Performance Management has been entrusted with transparent reporting on outcomes of the Vision's implementation, it is unlikely that this Centre might effectively overcome entrenched hierarchical structures.

The fact that the Vision's goals span across a wide range of sectors, involving a multitude of issues on which decision will fall to a variety of institutional entities may present considerable obstacles, particularly where reasonable measures of coordination are needed.

With particular reference to some of the Vision's STI-related goals, a special section in Chapter 3 will examine steps taken to implement the country's national Science, Technology and Innovation Plan (NSTIP), which was in fact approved in 2002. Thus, although NSTIP predates the vision and since one of its essential goals is to transform Saudi Arabia into a knowledge-based economy by 2025–29, it is highly likely that a good deal of interaction would take place between programs designed for the Vision's implementation and those designed for NSTIP realization.

Although it is likely that several goals of Vision 2030 may well be realized, skeptics point out that the Vision's extensive scope, its focus on economic outcomes with limited or no reference to political reform, and the level of challenge it poses for deep-rooted norms and achievement criteria are factors which might impede its implementation. These and other factors should clearly be taken into account in years to come.

## 2.5.6 The United Arab Emirates

The Emirates' economy is the Arab countries' second largest, and evidently the most open economy, in the region with a high per capita income and a considerable trade surplus. The Emirates is one of the few resource-rich countries to have effectively managed considerable diversification of its economy; reducing the portion of GDP based on oil and gas output down to 25%.[164]

Real estate and infrastructure developments during the recent past have contributed to making the country a global tourist destination.[165] As a result, contributions by the tourism sector to the Emirates' GDP increased considerably over the past decade.

### *2.5.6.1 Economic Growth in the Emirates*

The UAE possess over one-third of the region's oil and gas industries, chemical industries, energy, water and garbage recycling projects, with exports including petroleum and petrochemicals; fishing, aluminum, cement, fertilizers, commercial ship repair, construction materials, handicrafts, textiles.

The UAE government responded to lower oil prices by reduced social spending.[166] However, over the years, it did allocate massive resources to job creation and infrastructure expansion as well as private enterprise promotion. The country's free trade zones, which offer 100% foreign ownership and zero taxes, have helped attract foreign investors from all over the globe.

Nevertheless, long-term challenges facing the UAE economy include continuing dependence on oil, rising inflation as well as its largely expatriate workforce, estimated at around 85% of its inhabitants.

Fig. 2.41 indicates that real GDP growth, at constant market prices witnessed two reversals during 2013–16, essentially due to depressed oil prices—and possibly due to the Emirates' involvement in the Yemeni war as well. However, GDP growth was expected to rebound in the near future, due to enhanced oil and nonoil revenues. Growth in the industrial and services sectors is also expected to recover somewhat following sharp declines, particularly in the latter sector during 2013–15.[167] Additionally, continued demand in established free-zone developments is expected to sustain nonoil growth, whereas real estate activities are expected to remain somewhat resilient.

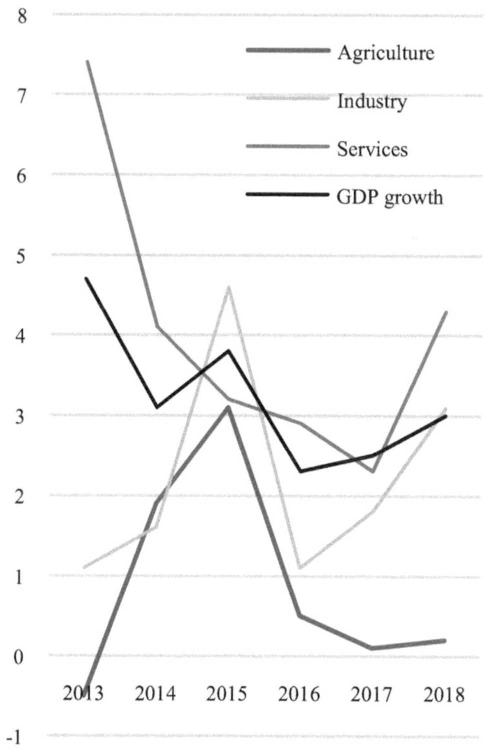


**FIGURE 2.41** Actual, estimated and forecast GDP and sectoral growth in the United Arab Emirates.

### 2.5.6.2 Unemployment and Poverty

Unemployment in the UAE was estimated in 2014 at 3.6%, relative to the total labor force. Higher rates of around 17% and 8% were recorded for females and males within the age group 15–24.[168] An earlier figure for overall unemployment is quoted at 2.4 for the year 2001.[169] No recent figures are available on poverty in the UAE.[170] In fact, the UAE is yet to develop integrated capabilities for comprehensive measurement and analysis of household welfare, with each Emirate at present running its own statistical agency. A federal statistical bureau was established as recently as 2009. However, survey methodologies and statistical agenda for Emirates-wide welfare assessment have not been accomplished as yet.[171]

### 2.5.6.3 The United Arab Emirates' Vision 2021

The UAE's Vision 2021 document was launched in 2010 by the Emirates' Government, setting key objectives for the country's sustainable socioeconomic development and inaugurating moves toward a diversified and knowledge-based economy. Implementing Vision 2021 is supervised by the Ministry of Cabinet Affairs and the Future (see Fig. 2.42).

The Vision 2021 document further highlights the creation of more job opportunities for nationals. The document further states that the Emirates' future as a "diversified and flexible knowledge-based economy will be powered by skilled Emiratis and strengthened by world-class talent to ensure long-term prosperity

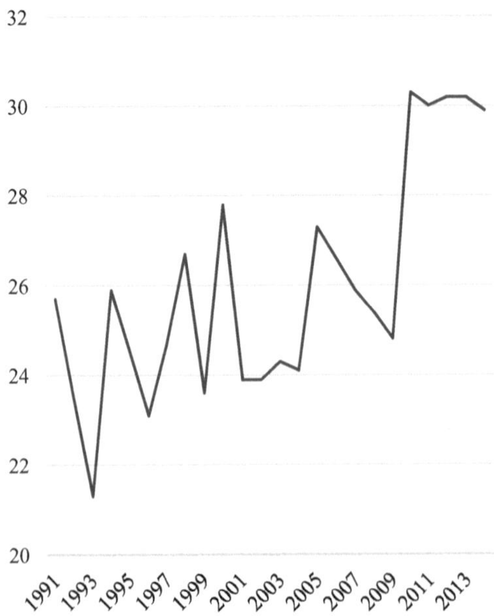


**FIGURE 2.42** Youth unemployment in Yemen (% of total labor force aged 15–24).

for the UAE." Improved education and increased private sector employment are also key elements of this document. Thus, the Vision 2021 document emphasises "equal opportunity and access to first-rate education" allowing UAE citizens to "achieve their true potential, contributing positively to society."[172]

Innovation, research, science, and technology are also underscored by the Vision 2021 document, which goes on to state that capabilities in these domains constitute "pillars of a knowledge-based, highly productive and competitive economy, driven by entrepreneurs in a business-friendly environment where public and private sectors form effective partnerships."

The UAE Vision 2021 document further emphasizes the need to preserve the environment, targeting the status of a "leader of the green revolution," the document pledges to combat the impact of "climate change on its habitat and ecosystems" in the interest of an environmentally sustainable world. This too would require considerable investment in STI inputs to ensure implementation of environmentally sound technologies in its service and manufacturing activities as well as monitoring the impact of measures taken to safeguard the environment.

### *2.5.6.4 Abu Dhabi's Economic Vision 2030*[173]

Abu Dhabi's Economic Vision 2030, which was issued in 2007, emphasized a set of macroeconomic goals including GDP growth averaging at 7% through to 2015, but down to 6%, thereafter, as well as nonoil GDP growth at a higher rate than that sought for the oil sector, with the aim of reaching parity by 2028. Economic stability is claimed as a prime objective.

Areas targeted as focus for sustained attention include human capital development, research and development, infrastructure development, transport as well as ICT. Additionally, this Vision document stresses the need for encouraging financial markets' further development so as to service various "economic sectors, industries, and projects."

A wide range of industries occupy central attention with the aim of further diversification of Abu Dhabi's economy, including petrochemicals, pharmaceuticals, biotechnology, healthcare equipment and services, metals, aviation, aerospace and defense industries, tourism, telecommunication, transportation, trade and logistics and the media.

As indicated above, education is accorded a good deal of attention by Abu Dhabi's Vision 2030 document. Higher education and research are highlighted, with particular emphasis on:

- hosting campuses of "renowned foreign universities and academic institutions;"
- attracting some of the world's best academics and specialists by providing attractive living environments and world-class research facilities and endowments;
- encouraging public–private partnerships at the primary and secondary level in order to improve public education;
- developing vocational and technical training base.

Benchmarks adopted as guidelines for monitoring progress toward the Vision's realization focus on overall economic indicators. Thus, as indicated in Fig. 2.43, the Vision 2030 document seeks to increase the Emirate's real GDP by up to fivefold, from US$ 78 billion to around US$ 416 billion. Additionally, as indicated in Fig. 2.44, Abu Dhabi's Vision document targets increasing nonoil contributions to real GDP from around 41% in 2007 up to 64% in 2030.

However, no numerical benchmarks are provided for objectives concerning developments targeted within various other sectors. Thus, despite repeated mention of education and research and the importance accorded to building superior educational research facilities no figures are quoted by the Vision document with regard to levels of spending or prospective outcomes.

Additionally, it would seem that detailed sectoral analyses are yet to be performed in pursuit of at least some of this Vision's goals. Thus, to take the Vision document's interest in pharmaceuticals, biotechnology and the life sciences as an example, factors for including activity within these domains among the Vision document's priorities are linked to:

- global demographic changes, and the aging populations of Europe and North America and increased "spending power in emerging markets;"
- the country's increasing popularity as a medical tourism destination, on account of relatively low costs, compared to developed countries in Europe, North America, and Oceania.
- the UAE pharmaceuticals market was expected to grow from US$ 3 in 2015 to US$ $ 5.7 billion by 2020.[174]

Moreover, the Vision document merely requires that activities in these domains produce an estimated US$ 700 billion in returns. The Vision document further considers that the development "of a vibrant and successful pharmaceuticals and life sciences sector" would benefit from the burgeoning education sector and the presence of international institutions.

The Vision document, nevertheless, acknowledges pharmaceuticals, biotechnology, and life sciences were still at an embryonic state of development in Abu Dhabi, and that significant resources were needed to finance R&D in

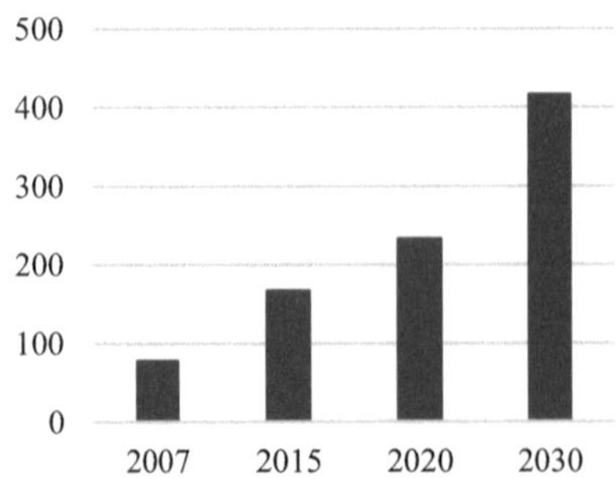


**FIGURE 2.43** Abu Dhabi's projected real GDP growth (2005 US$ billions).

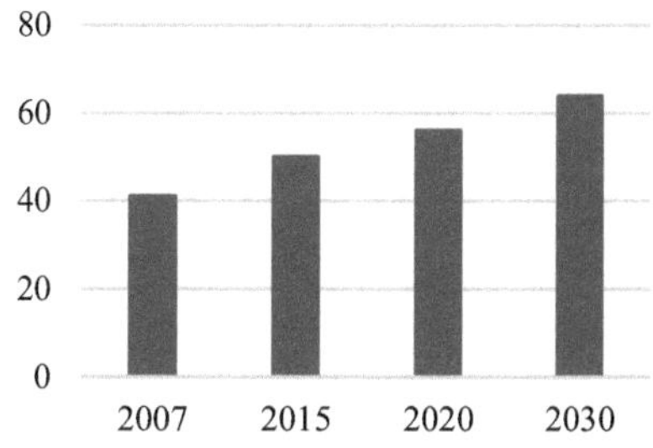


**FIGURE 2.44** Abu Dhabi's projected non-oil contributions as percentage of real GDP (2005 US$).

these domains and that additional measures would also be required to attract the world's most successful actors. Moreover, the Vision document indicates the need for the United Arab Emirates to enhance intellectual property rights legislation, revise international trade agreements, establish reliable drug testing and approval systems as well as develop modalities to further attract investment, marketing, and distribution capabilities.

Should only considerations such as the above be taken as basis for including pharmaceuticals, biotechnology, and life sciences as strategic priorities then it is more than likely that the UAE would serve as just another offshore facility for international producers. Although welcome on commercial ground, such developments would require the design and enforcement of elaborate measures by government authorities in order to ensure adherence with international norms governing clinical trials and other health related issues, in such a sensitive domain of industrial activity. On the other hand, in order to attain optimal benefits in terms of providing employment opportunities for the Emirate's youth, there is need to establish far closer links between industrial and investment concerns, on the one hand, and educational and training institutions, on the other.

### *2.5.6.5 Dubai's Industrial Strategy 2030*

Dubai's industrial sector contributes between 11% and 14% to this Emirate's GDP, placing it in third position after trade and logistics. This sector is reported as having demonstrated solid performance despite the global financial crisis and regional problems. Dubai has 18 industrial areas, nearly half of which are located within free zones.

Dubai's Industrial Strategy 2030 was launched in June 2016 with the overriding objective of elevating Dubai into a "global platform for knowledge-based, sustainable and innovation-focused businesses." The strategy identified the following key pillars as foundations for Dubai's industrial future:

- increasing total output and value-addition of the manufacturing sector;
- enhancing depth of knowledge and innovation;
- making Dubai a preferred manufacturing platform for global businesses;

- promoting environmentally friendly and energy-efficient manufacturing;
- becoming a global center for the Islamic products.

Industries singled out for special attention included: aerospace, maritime, aluminum and fabricated metals, pharmaceuticals and medical equipment, machinery and equipment as well as food and beverages. Furthermore, the Strategy identified 75 initiatives to transform Dubai into a global platform for knowledge and innovation based industries.

An interesting feature of Dubai's industrial strategy is its projected increase of research and development spending by around US$ 190 million. Additionally, the strategy was projected as generating an estimated US$ 43 billion, in additional contributions to Dubai's GDP and an estimated 27,000 employment opportunities for specially qualified workers by 2030.

#### 2.5.6.6 Future Prospects

The Emirates' economic development is indebted to an extremely centralized but, often informed decision-making processes, coupled to significant government involvement in the economy that made it possible to coordinate considerable State, foreign investments, and, to a large extent, local private sector investments.

Although a centralized approach,[175] such as that currently implemented in the Emirates, in general, and particularly in Dubai, might rightly be subject to criticism on grounds of democracy and human rights, it has nevertheless succeeded in creating a functioning economic system with advanced infrastructures as well as credible future promise.

In other words, the Emirates must be considered as a very special case with regard to the economic development model it adopted and appears to be consolidating. Although enjoying considerable successes on account of such approaches it lies prone to the vagaries of global changes. To improve its chances of sustainable economic growth and social development the strongest of links are needed between sectors of economic activity, on the one hand, and educational institutions, on the other.

However, complications due to regional tensions as well as international development around the Arab region are more likely to delay rather than accelerate moves toward further socioeconomic reforms. This is likely to be particularly the case should such reforms be unrelated or weakly linked to business and investment prospects.

### 2.5.7 Yemen

Although these lines are being typed, Yemen, a country that was referred to in Roman times as Arabia Felix, or "happy Arabia," is going through one of the worst periods of its long history, with armed conflict, famine conditions and outbreaks of contagious diseases.

A low-income country, with endemic socioeconomic challenges, including a high population growth rate, high unemployment and declining water resources, Yemen has always struggled to stabilize its economy. And for many years in the recent past, malpractices perpetrated both by government figures and tribal authorities resulted in chronic poverty and underdevelopment. Difficulties in access to basic services such as education, electricity, clean water and healthcare were often experienced in many parts of the country. The current conflict has certainly exacerbated such difficulties.

With various provinces currently gripped by high intensity armed conflict a great deal of economic activity has come to a standstill. Yemen's exports have been halted and the national currency's exchange rate has fallen driving up inflation. As a result of military operations, damage to infrastructure has been quite severe in many locations.[176]

Perhaps more importantly, access to food and medical supplies has been severely curtailed, with even some of the established welfare funds terminating their operations.

### *2.5.7.1 Economic Trends*

Oil and gas accounted for roughly 25% of GDP and 65% of government revenue prior to the conflict. Yemen's economic growth started receding several years before the present crisis broke out, first in the form of public protests in 2011 and later in 2015 when Saudi Arabia aided by several other GCC countries intervened militarily. Thus, as Fig. 2.45, which presents annual GRP rates of growth between 1991 and 2009,[177] indicates, the rate of the country's GDP growth had been declining at an overall rate of 0.16% per annum, albeit with sharp oscillations from 1 year to the next. GDP growth took a steep plunge in 2011, and assumed negative values estimated at −0.2%, −2.8% and −17% in 2014, 2015, and 2016, respectively. Additionally, as indicated by Fig. 2.46, percentage contributions to Yemen's GDP by rent on natural resources—mainly oil, with natural gas contributing significantly only by 2010—declined at an appreciable rate during the period 2003–13.

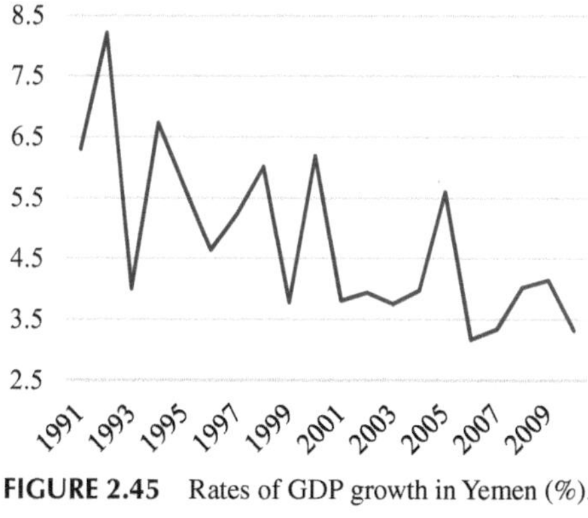


**FIGURE 2.45** Rates of GDP growth in Yemen (%).

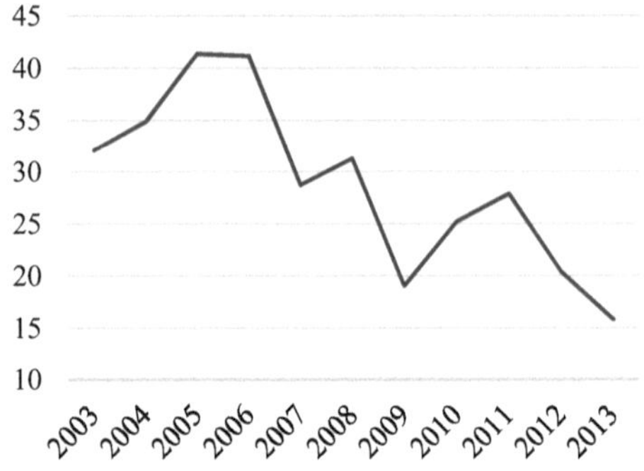


**FIGURE 2.46** Total natural resources rent (% of GDP).

The country had to deal with annual budget shortfalls even before the conflict, however. Additionally, attempts were made to diversify Yemen's economy through reform programs in support of nonoil sectors of the economy and foreign investment were met with little success. Civil strife followed by outright military conflict put a stop to all economic reform efforts. Furthermore, Yemen's Central Bank reserves were reduced to around $ 1.5 billion, from preconflict levels of more than $ 5 billion. The private sector has also been severely hit by the conflict, with businesses laying off substantial numbers of their employees.

### *2.5.7.2 Outlook for Future Economic Development*

Figs. 2.47 and 2.48 present estimated and forecast percentages for GDP growth as well as growth of sectoral contributions to GDP, during 2016–21. Forecasts presented in these figures are based on the assumption that hostilities will abate and relative calm will be restored around the middle of 2017. Therefore, all figures appear to indicate positive, following several years of

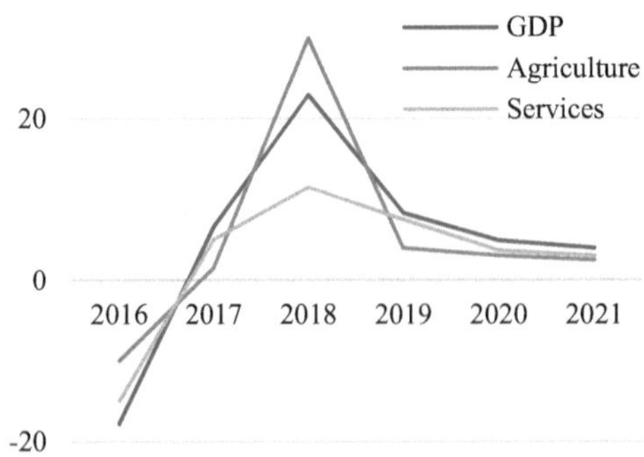


**FIGURE 2.47** Estimated and forecast growth rates of GDP and the agricultural and services sectors in Yemen: 2016–21 (Percentage of GDP).

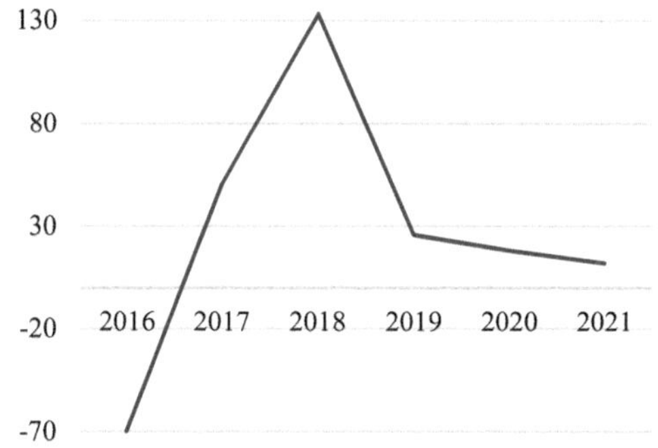


**FIGURE 2.48** Estimated and forecast growth rates of the industrial sector in Yemen; 2016–21 (Percentage of GDP).

negative, growth as of 2017. However, forecasts also indicate that this growth will be tempered by:

- the diversion of whatever resources are in fact available to restarting oil and gas facilities as quickly as possible in order to generate revenues; and
- since donor disbursements have recently slowed down,[178] and while the GCC countries are expected to continue their support once hostilities are over, they too might suffer shortages for a number of reasons.

At any rate, structural economic reforms, on the other hand, are likely to remain on the back burner. This should not, however, prevent planners from developing improved policy regimes for radically dealing with main socioeconomic development issues that bedevilled the country for many years in the past, emphasizing local community based initiatives with civil society participation. Considerable international assistance will also be needed by Yemen during and following the conflict in order to provide humanitarian aid and stabilize the economy.

### *2.5.7.3 Unemployment and Poverty in Yemen*

Yemen's youth unemployment, in particular, has always been high, ranging between 22% and 28% for much of the 1990s and 2000s, before rising to even higher levels more recently, where it oscillated around the 30% mark during 2010–14.

Figures for overall unemployment, on the other hand, went down from levels of around 35% of the total work force down to around 27% in 2014.[179]

An estimate of the percentage of the population below poverty line was put at 54% in 2014. This estimate has most likely been exceeded, as conditions assumed disaster proportions in many more parts of the country. Thus, in 2016, UN estimates suggested that over 80% of the population in need of humanitarian assistance, and around one-third of the population, or almost 8 million inhabitants suffering severe food insecurity with more than 3 million internally displaced.

### 2.5.7.4 *Prospects for the Future*

Decades of mismanagement and corruption, coupled to high population growth rates, failed education systems, depleting natural resources and half-hearted attempts at creating alternative sources for national income, resulted in the catastrophic conditions witnessed by Yemen today.

Unemployment, chronic poverty, and poor health conditions throughout the country should be handled as top priorities by future relief and socioeconomic development policies, with specific STI based inputs, designed of provide essential public services; electric power, clean water, health services as well as building national capacity for the adaptation, dissemination and maintenance of such inputs.

Yemen will remain heavily dependent upon foreign aid for both direct budgetary support and humanitarian assistance. The Gulf countries and the international community are expected to provide further aid once peace is reached. In the long term, there can be no alternative to policies enabling national capacity building based on sustainable and inclusive socioeconomic development, emphasizing local community needs. The road ahead will undoubtedly be long and arduous. However, that alternatives would be infinitely harder to handle, has been made obvious time again over the past few years.

## 2.5.8 Socioeconomic Development in the Arab Gulf Countries

Although the Arab Gulf countries' hydrocarbon wealth has enabled them to take significant leaps in terms of overall socioeconomic development, with particular emphasis on infrastructures and institution building, it has also resulted in some serious challenges. Thus, not only are hydrocarbon resources exhaustible, additionally their prices have recently been particularly unstable, rendering related rents uncertain.

The fact that some of the Gulf Arab countries have now moved up on world human development rankings might be taken to signify weaknesses in ranking systems rather than concrete outcomes in terms of inclusive and sustainable human development.

In general, it would be true to say that the Arab Gulf countries represent the closest incarnation of the fabled Dutch disease. Symptoms of this disease naturally vary from one country to another. However, for the most part they include structural imbalances in labor markets, growing unemployment among nationals coupled to a culture of dependence on public sector occupations, limited incentives for youth and women to partake in economic and social activities. Additionally, several Arab Gulf countries lack adequate policy framework to regulate public and private sector activities with underdeveloped transparency and accountability for the most part.

Vision and long-terms strategy documents have been drawn up by several Arab Gulf countries, with emphasis on building or approaching knowledge economies. For the most part, however, these documents seem to approach glaring economic and social problems using modalities known to have had limited if any tangible positive outcomes. Many of the Vision and Strategy documents reviewed in preparing this chapter lacked numerical goals, or were exceedingly selective in stating such goals. Additionally, almost all such documents had little to say about implementation plans as well as monitoring and evaluation modalities. In short, they may not be considered valid tools for facing the challenges created by the depletion of hydrocarbon resources or their price instability, and may well fall short on many scores, including diversification, local technological capacity building, and labor market restructuring.

Finally, it is extremely unlikely that the GCC countries might achieve extended socioeconomic development, whereas Yemen, suffers endemic underdevelopment and abject poverty. Inclusive regional development is not favored only on moral grounds, it is even more essential for practical reasons.

## 2.6 ECONOMIES OF THE ARAB EAST AFRICAN COUNTRIES; COMOROS, DJIBOUTI AND SOMALIA

Comoros, Djibouti and Somalia are the least endowed of all the Arab countries, with their population representing 0.3% of the Arab countries' total population, whereas their GDP amounts to a mere 0.13% of the Arab countries' combined GDP.

Somalia gained its independence in 1960 and was followed by Comoros in 1975 then Djibouti in 1977. All three countries proceeded to join the League of Arab States. Somalia leading the way in 1974, followed by Djibouti in 1977 and Comoros in 1993. Each one of these countries has had its full share of political problems almost since their independence, from the British and Italians in the case of Somalia and from French colonial rule in Djibouti and Comoros. The country most affected and still suffering frequent strife and armed conflict, with evident impact on socioeconomic development has been Somalia. Nevertheless, indications abound as to the role that mismanagement and corruption have played in retarding socioeconomic development in all three countries. As a result, all three countries rank close to the bottom of world rankings on a variety of international human and economic development indicators.

Although well-designed STI initiatives should enable these countries to create better conditions for their citizens as well as viable futures for forthcoming generations, it is seldom that such initiatives were designed or launched. A variety of reasons are given for this, including scarce resources as frequent adverse climatic conditions forced governments to divert resources toward humanitarian relief operations.

The following paragraphs present a summary review of economic conditions in these countries, followed by a brief look at arrangements dedicated to promoting science, technology and innovation, which take place at their institutions of higher learning, mediated, for the most part by donor agencies and international aid organizations.

### 2.6.1 Comoros

Comoros is one of the world's poorest countries, lacking natural resources and with a relatively high birth rate coupled to low levels of educational achievement. Thus, in 2014, around 40.3% of its nearly 800,000 population was below 14 years of age, whereas the primary schooling completion rate for both sexes was estimated at 73.6% in 2012. Inadequate transportation links between and within the three islands that constitute its territories present further obstacles to its development.

Comoros has suffered endemic political disputes that hindered several development initiatives during its recent past. The constitutional reform, adopted in 2009, allowing the country's islands some autonomy, whereas reaffirming the country's unity and of the role of the federal government, is generally considered as a positive development, because it helped reduce internal tensions, allowing development projects to move ahead, albeit rather slowly.

The Comoros economy is small, undiversified and highly dependent on imports, foreign grants, and technical assistance. Remittances from around 200,000 of its citizens working abroad contribute about 25% to the country's GDP, whereas agriculture, fishing, hunting, and forestry account for 50% of GDP, employing a large proportion of the workforce, which was estimated at around 250,000 in 2013. Nevertheless, Comoros is far from self-sufficient, as far as food production is concerned; importing nearly 70% of its food requirements. Export income relies on three main crops; vanilla, cloves, and ylang-ylang, however, export earnings in the past have occasionally been disrupted by climatic conditions. As indicated in Fig. 2.49, at around 12% of GDP, industry's contribution to the national economy is quite low.

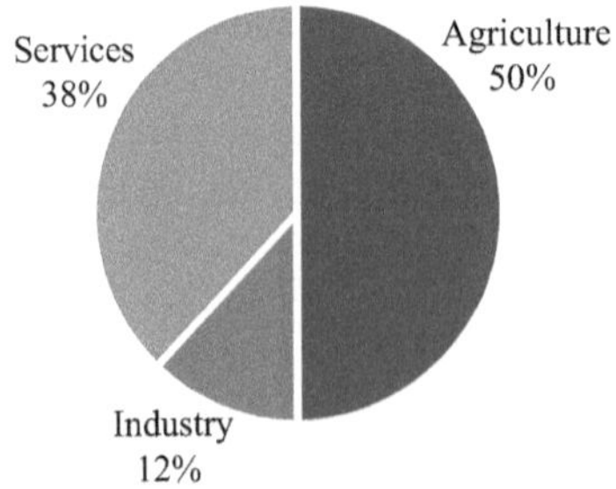


**FIGURE 2.49** Sectoral GDP composition in Comoros (%).

Real GDP growth was down by almost 50% in 2016 to around 1% from levels it held in 2014, but is expected to gradually increase, approaching 1.7% by 2018.[180] Thus, although electric power shortages, deficient road maintenance and inadequate telecommunications have constrained economic activity for many years, developments on all these three fronts should enable significant growth in services and industry. On the other hand, agricultural growth is expected to be limited due to the country's limited fertile land areas. Obviously, the above predictions are based on the assumption that political stability will be maintained and progress will be made toward improving the business environment.

At any rate, steps taken to improve education and technical training, health services as well as attempts made to diversify exports and promote tourism are expected to have positive impact on overall socioeconomic development in the medium, rather than the short term.

#### *2.6.1.1 Social Development, Unemployment and Poverty*

The Comoros, occupied 159th position among 188 countries on the United Nations Development Programme's Human Development Index in 2015, with efforts on combating hunger and malnutrition in need of far greater government attention. Thus in 2012, almost 32% of children under the age of five, particularly in rural areas, suffered chronic malnutrition and stunted growth.[181] Infant mortality was also relatively high, at a rate of 36 per 1000 births for children under the age of one and 15 per 1000 births for children under the age of five, as recorded in 2012.

At an estimated 6.5% in 2013, unemployment in Comoros is rather low, by regional standards. However, a 2014 household survey estimated that 42.4% of the population, or around 320,000 people had per capita monthly consumption levels below the national poverty line and that nearly 18% of the population lived below the international poverty line of US$ 1.9.[182,183]

### 2.6.2 Djibouti

Djibouti is a small country of around 900,000 inhabitants on the red sea. Nearly 75% inhabitants live in the capital city, whereas the remaining inhabitants are mostly nomadic herders.

Djibouti's economy is highly dependent upon service activities linked to its capital's location and its deep-water Red Sea port as well as foreign aid, which is vital for supporting its balance of payments and development investments. Imports, exports, and reexports constitute 70% of Djibouti's container terminal's activities, with reexports dominated by coffee output from its land-locked neighbor, Ethiopia.

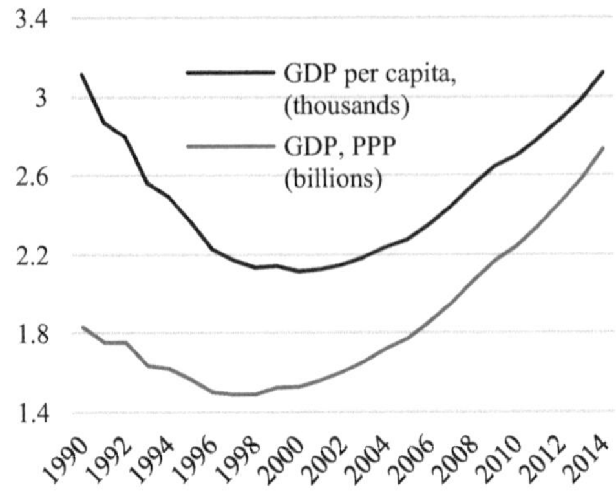


**FIGURE 2.50** Djibouti's total and per capita GDP (PPP at constant 2011 international $).

As indicated by Fig. 2.50, which covers a rather long span between 1990 and 2014, Djibouti's GDP enjoyed reasonable growth rates during the past few years. GDP growth was estimated at around 6.7% in 2016 and is expected to rise slightly up to an annual average of 7% in 2017–18, driven by infrastructure investment and trade with neighboring Ethiopia. Activities by the construction sector will notably include building of the US$ 4 billion gas pipeline between Ethiopia and Djibouti. Additionally, an estimated sum of US$ 1.6 billion is scheduled for investment in an oil pipeline between the two countries and two international airports are already under construction at a cost of US$ 600 m, due to come into service in 2018.

The agricultural sector contributes around 3% to the country's GDP. Low rainfalls and less than 4% of the country's landmass suitable for cultivation place serious limits upon agricultural development. In fact, most food products need to be imported. This, together with Djibouti's reliance on diesel-generated electricity exposes consumers to global price changes.

In line with government's development plans that emphasize support for infrastructure development with transportation and energy production as priorities, construction of an export processing zone for handling shipments from China into Africa was launched early in 2017, and a 300-mw solar plant is being built, with an initial phase set for completion before 2018. The intention has been expressed by Djibouti's government to generate 100% of its power needs from renewable sources.[184]

#### *2.6.2.1 Unemployment, Poverty and Humanitarian Relief in Djibouti*

Unemployment was estimated at 60% of the labor force, and 80% among youth in 2014. In 2015 poverty was estimated in 2015 at around 23% of the entire population.[185,186]

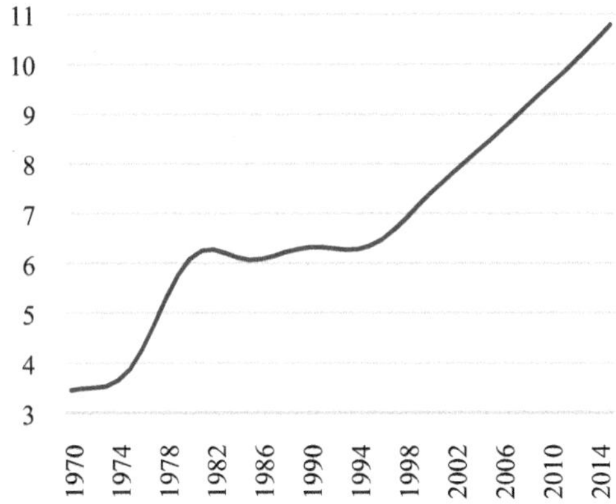


**FIGURE 2.51** Growth of Somalia's population; 1970–2014 (millions).

Drought that persisted for the ninth consecutive year has enhanced vulnerability of the unemployed and impoverished, with the situation further worsening due to the influx of Yemenis fleeing the war into the Djibouti and with El Niño weather having displaced Ethiopian families in the first quarter of 2016. With child malnutrition rife in many parts of the country, The United Nations International Children's Emergency Fund (UNICEF) reported a funding gap of about 90% in all sectors of emergency responses it was managing in the country.[187]

## 2.6.3 Somalia

Somalia is a small East African country that has been beset by political turmoil and armed conflict for almost 25 years. In 2015, its population was estimated of around 11 million inhabitants, over 46% of whom are less than 14 years of age (see Fig. 2.51). The country is going through a process of state and peace building, with the aid of several donor countries.[188]

Despite significant improvements since the devastating famine during 2011–12, malnutrition and child mortality remain high, particularly in the country's largest and most populous central/southern zone, which has been plagued by instability over the past two decades. A food security and nutrition assessment conducted by UNICEF in 2013 found that 870,000 people across Somalia "remained food insecure, while an additional 2.3 million people remain highly vulnerable to shocks and were on the margins of food insecurity."[189]

Much of the country's economic infrastructure is in disarray, following so many years of conflict. However, based on a provisional constitution, the Federal Government of Somalia took power in 2012, reaching an agreement, with international agencies in 2013. This agreement provided an organizing framework or the delivery of assistance to Somalia according to a set of national priorities.[190]

Somalia's GDP was estimated at around US$ 4.4 billion in 2014. Later estimates predicted that it would rebound reaching US$ 6.2 billion in 2016, with per capita GDP at US$ 450, the lowest on record throughout the world. Consumption accounts for most of the country's GDP, whereas investment represented only 8% of GDP in 2015.

Somalia suffers a huge trade deficit, with imports, financed mainly through remittances and international aid, accounting for more than two-thirds of GDP, whereas the ratio of exports to GDP was estimated at around 14% in 2015.

Somalia's largely informal economy is based on agriculture and animal husbandry—which account for about 40% of GDP and more than 50% of export earnings—remittances from its expatriates across the globe as well as the telecommunications sector. Its external debt, mostly in arrears, was estimated at 93% of GDP in 2014.

Economic activity was estimated to have grown by 3.7% in 2014 due to developments in the agriculture, construction and telecommunications sectors.[191] A recent attempt at creating a small industrial sector, dedicated to processing agricultural products is reported as having failed and general economic growth has yet to expand outside the country's capital, with security concerns dominating all business activity.

Public expenditure by the Somali government has increased considerably, from around $35 million in 2012 to more than US$135 million in 2015, due to improvements in revenue collection. Thus, domestic revenue grew by 36% from US$84.3 million in 2014 to US$114.3 million in 2015. This increase is due mainly to improved collection of trade taxes. Total revenue as a ratio of GDP accounted for 2.8% of GDP, however.[192] As such, revenue collected by government is still insufficient to allow delivery of basic services, especially since the administrative and security sectors account for more than 85% of total spending, whereas economic and social services sectors account for only 10% of total expenditures.

#### *2.6.3.1 Unemployment and Poverty in Somalia*

Nomadic and semipastoralists make up a large portion of Somalia's population and poverty is rampant with nearly 52% of the population living below the poverty line. Income differences are reported between and within urban settings, with poor populations approximating 60% in Mogadishu and more than 40% in other urban locations. Highest poverty rates pertain in settlements established for internally displaced persons. Remittances help limit even more widespread poverty.[193]

### 2.6.4 Challenges Facing Future Development

A great deal is yet to be achieved with regard to future development goals in Comoros, Djibouti, and Somalia. Science, technology and innovation inputs are playing, and will play an even more essential role in approaching such goals.

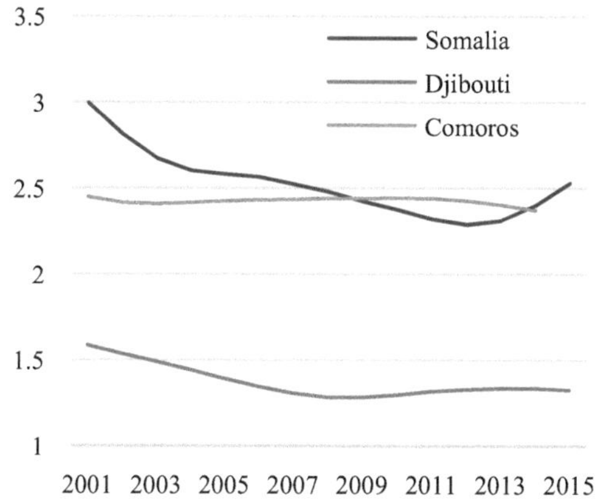


**FIGURE 2.52** Population growth rates (%).

Political instability and credibility of government institutions pose serious challenges, throughout East Africa, rendering it the continent's lowest recipient of FDI in 2009, with the exception of Djibouti. Thus, due to its strategic location and port facilities, FDI flowed most abundantly into Djibouti, amounting to around 19.6% of GDP in that year.

On account of its relatively small population and low population growth rates, see Fig. 2.52, heightened interest in its strategic location, as well as steps it has taken toward reduced dependence on fossil fuels, Djibouti is likely to forge ahead sooner than either Comoros or Somalia. Nevertheless all three countries will need to formulate and adopt sustainable development policies, ensuring that investments on all fronts will have widespread benefits, with particular regard to wider dissemination of education and health services and humanitarian relief.

## 2.7 CONCLUSIONS

Available statistics indicate that for two decades since 1990, the Arab countries collective GDP values increased at comparable rates to the world's average (see Fig. 2.53). However, GDP of the Arab countries grew at an appreciably lower rate than the world's average since 2010. This is undoubtedly due to factors such as

- lower oil prices affecting the collective GDP of the Arab countries;
- declining economic activity, due to regional turmoil in general to armed conflict in countries such as Libya, Syria, Yemen, and Iraq, as well as sporadic attacks in both Egypt and Tunisia.

GDP values and associated resources are unevenly distributed among the Arab countries. Thus, as indicated in Table 2.6, the GCC countries produce 50%

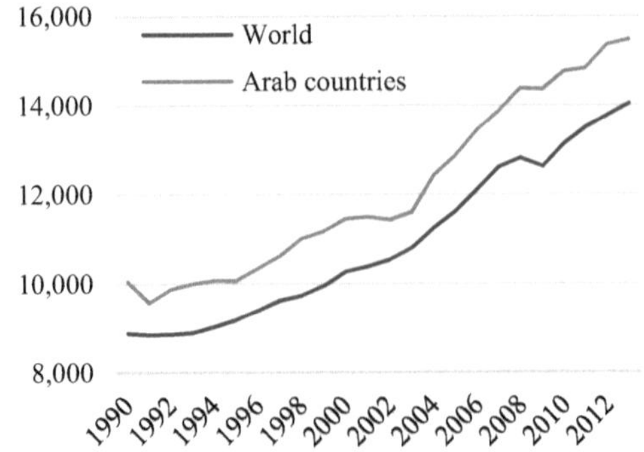


**FIGURE 2.53** Growth of GDP per capita in the Arab countries and the World 1990–2013.

**TABLE 2.6 Distribution of Populations and Annual GDP Across Groups as Percentages of Totals for All the Arab Countries**

| | Population (% of Total) | GDP (% of Total) |
|---|---|---|
| The Levant | 18.8 | 12.6 |
| Egypt and Sudan | 33.4 | 18.1 |
| North Africa | 24.3 | 17.5 |
| GCC countries | 13.3 | 50 |
| East Africa and Yemen | 53.2 | 1.7 |

of the Arab countries total GDP while home to only 13% of their collective population. Unequal distribution of wealth is also evident within the Arab countries, with values of the Gini index beyond 70% in half the 22 Arab countries, and close to or exceeding 80% in seven countries, namely Egypt, Lebanon, Libya, Morocco, Oman, Saudi Arabia, and Comoros.

The uneven distribution of populations and economic resources in the Arab countries is further complicated by higher dependency ratios in the less endowed countries. Indeed, as indicated in Fig. 2.54 the percentage of Arab country populations below the age of 14 is considerably higher than the world's average.

The picture is further complicated as one views assessments of corruption based on opinion surveys performed by the World Economic Forum (WEF). Thus, the Forum's Global Competitiveness Report for the years 2016–17 indicates that corruption was ranked among most problematic factors affecting business operations in almost all the Arab countries, with countries such

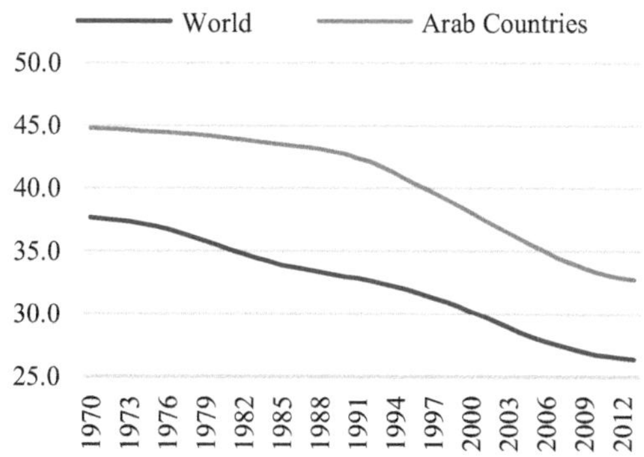


**FIGURE 2.54** Percentages of World's and Arab country populations below the age of 14.

as Mauritania, Lebanon, Algeria, Kuwait, Morocco, and Tunisia on top of the list. A similarly disappointing situation is observed with regard to responses concerning favoritism in decisions of government officials, based on the same WEF survey, where out of 188 countries surveyed, six Arab countries, namely Lebanon, Kuwait, Algeria, Tunisia, Mauritania, Morocco, ranked close to or below 50th position.

To a large extent, efforts aimed at diversification in countries such as Kuwait, Oman, Bahrain, Qatar and Saudi Arabia have not produced their desired impact. Many explanations are put forward for such failures. The most convincing of which are the least tractable, since they are closely linked to core rentier development and management models impacting every aspect of social and economic life, particularly in the resource rich Arab countries of the Gulf region.

In many countries around the region, and not just in the Arab Gulf countries most economic activity is tied to systems of rent collection and distribution by political authorities. This appears to place severe limits upon the expansion of economic activity in any way that threatens delicately balanced status quos. Thus, with the need for emerging industrial enterprises to pay rent in various forms and guises to powerful figures in authority, several manufacturing industries, in particular, have to face additional challenges due to competing imported products, on the one hand, as well as, lack of trained personnel at various levels, on the other.

It is to be granted that rentierism may not be legislated out with a pen stroke. However, more widespread realization of its negative impacts coupled to persistent efforts at building alternative paradigms, through policies as well as direct action, e.g., by supporting novel industrial structures and investment landscapes with associated STI facilities should eventually relax its current grip on all socioeconomic development throughout the Arab countries. The road may be hard and long, but no other feasible routes in sight.

## ANNEX TABLE

**TABLE A.2.1** Population, Surface Area, and GDP Data for the Arab Countries

| | Population (Millions) | Surface Areas (km²) | GDP, PPP (Millions Constant 2011 intl. $) | GDP per Capita, PPP (Constant 2011 intl. $) |
|---|---|---|---|---|
| Algeria | 38.934 | 2,381,740 | 527,202 | 14,193 |
| Bahrain | 1.314 | 770 | 59,119 | 45,500 |
| Comoros | 0.788 | 2235 | 1102 | 1364 |
| Djibouti | 0.888 | 23,200 | 2734 | 3120 |
| Egypt | 89.580 | 1,001,450 | 899,898 | 10,046 |
| Iraq | 34.812 | 435,240 | 500,073 | 14,365 |
| Jordan | 6.607 | 89,320 | 75,956 | 11,496 |
| Kuwait | 3.892 | 17,820 | 272,866 | 70,107 |
| Lebanon | 4.547 | 10,450 | 75,746 | 16,659 |
| Libya | 6.259 | 1,759,540 | 93,134 | 14,880 |
| Mauritania | 3.516 | 1,030,700 | 14,814 | 3912 |
| Morocco | 34.378 | 446,550 | 257,399 | 7365 |
| Oman | 4.236 | 309,500 | 156,118 | 36,855 |
| Palestine (West Bank and Gaza) | 4.422 | 6020 | 22,200 | 4715 |
| Qatar | 2.172 | 11,610 | 291,453 | 134,182 |
| Saudi Arabia | 30.887 | 2,149,690 | 1,530,030 | 49,537 |
| Somalia | 10.787 | 637,657 | 4186 | 450 |
| Sudan | 39.350 | 1,879,358 | 152,767 | 3882 |
| Syria[2015] | 22.158 | 185,180 | 55,800 | 2900 |
| Tunisia | 10.996 | 163,610 | 119,971 | 10,910 |
| United Arab Emirates | 9.086 | 83,600 | 586,625 | 64,563 |
| Yemen | 26.184 | 527,970 | 93,536 | 3663 |

From World Bank tables.

## ENDNOTES

1. The Arab countries are defined here as those countries who founded or later sought to become members of the League of Arab States.
2. GDP values in this chapter are based on World Bank's estimates for purchasing power parity (PPP) 2011 constant 2011 international dollars.
3. In its Corruption Perceptions Index, issued late in January 2016, the Germany-based non-governmental organization "Transparency International," ranked Iraq 166th among 176 countries. The Economist Intelligence Unit's Iraq Country Report. "Fiscal policy outlook," dated February 22, 2017. EIU Report generated March 22nd 2017.
4. "Reforms in MENA Countries: Are We Missing Science in Iraq's Reform Agenda?" Posted by S Raouf, Ex Deputy Minister at Iraq's Ministry of Science and Technology on a website "Science technology and innovation for Iraq & Arab development." Downloaded in May 2017 from: https://stiiraqdev.wordpress.com/2017/01/24/reforms-in-mena-countries-are-we-missing-science-in-iraqs-reform-agenda/.
5. See endnote 4.
6. Figure reported by the World Bank within its online World Development Indicators, last updated in 17th February 2016. Data downloaded in March 2016.
7. See endnote 4.
8. At more than 10% of the population, this is around three times the world's average.
9. A somewhat smaller figure, 22.4%, was reported by the World Bank's Development Indicators database for the year 2006, under the heading "Poverty headcount ratio at national poverty lines as percentage of population".
10. Failed attempts at reaching agreement between Iraq's federal government and the autonomous Kurdistan Regional Government (KRG) with regard control of exports from northern oil fields are reported as having deprived the Baghdad central government of sizeable resources. See the EIU Country Report generated on 22nd March, 2017 and downloaded in March 2017 from: www.eiu.com.
11. Several such reforms were described in the recent National Development Plan for the 5-year period 2013–17. A copy of this plan was downloaded in May 2017. Downloaded in May 2017 from: http://www.theiqdteamconnection.com/iraq-news/national-development-plan-2013-2017-iraq-ministry-of-planning.
12. Jordan is classified by the World Bank as lower-middle income country.
13. The Jordan Oil Shale Company (JOSCO), a wholly owned subsidiary of Royal Dutch Shell plc, registered in Jordan in 2009, was granted a large concession agreement to explore oil shale deposits, committing to handing land back to the government along with oil shale data it gathered through exploration operations.
14. Processing of phosphate rock in Jordan is carried out with the purpose of manufacturing di-ammonium phosphate fertilizer, phosphoric and sulfuric acids as well as small quantities of aluminum fluoride. The following quantities, in metric tons, of these four products were manufactured in 2014: DAP fertilizer: 590, phosphoric acid: 292, sulfuric acid: 932, aluminum fluoride: 9. Information downloaded from the Jordan Phosphate Mines Company PLC, in May 2017, from: http://www.jpmc.com.jo/echobusv3.0/SystemAssets/e0c3e564-f9c6-4824-992f-c810490482a2.pdf.
15. This brought to the forefront the urgent need for diversifying the country's energy supplies. Toward this end, Jordan secured several contracts for liquefied natural gas and is presently exploring nuclear power generation, exploitation of abundant oil shale reserves and renewable technologies, as well as making use of Israel's offshore gas.
16. World Bank's Overview dated October 2016. Retrieved in February 2017 from: http://www.worldbank.org/en/country/jordan/overview.
17. Economist Intelligence Unit's March 2017 Country Report for Jordan. Downloaded in March 2017 from: www.eiu.com.
18. See endnote 16.

19. The law includes tax exemption for households with income of up to JD24,000 (US$ 33,850). A small rise in rates for income above the JD24,000 level is introduced. Industrial, trade and tourism companies will pay a 14% tax. Tax on financial and telecommunications companies were reduced to 30% from previous rates at 35%.
20. Previously, the agricultural sector was entirely exempt from tax.
21. See endnote 17.
22. In purchasing power parity 2011 dollars.
23. World Bank's Development Indicators. Data generated in February 2016. Accessed in March 2016 from: http://data.worldbank.org/data-catalog/world-development-indicators.
24. The Aqaba Special Economic Zone (ASEZ), which was inaugurated in 2001, provides an example of government initiatives aimed at offering tax benefits and minimal regulatory restrictions for private sector led economic growth. However, the ASEZ website does not include a great deal of information on firms that have since joined ASEZ. A master plan is supposed to have been finalized with view to town development and construction of residential and touristic facilities. However, there are no indications that specific activities industrial activities have taken place apart from the "Aqaba Development Corporation" which was established in 2004 and a laboratory having been set up to conduct physical, chemical and microbiological tests. The Aqaba Special Economic Zone's website. Information retrieved in February 2017, at: http://www.aqabazone.com/.
25. The World Bank's Database consulted in June 2017 indicates a percentage of almost 16%.
26. See endnote 17.
27. See endnote 23.
28. See endnote 23.
29. "Global Wealth Databook 2014;" Research Institute of Credit Suisse AG. Report published in October 2014. Downloaded in June 2017 from: https://publications.credit-suisse.com/tasks/render/file/?fileID=5521F296-D460-2B88-081889DB12817E02.
30. In particular, 2015 witnessed a series of measures by Israeli authorities, including security restrictions combined to Israel's withholding taxes and other fees collected on behalf of the Palestinian Authority's, causing the PA to suspend salary payments to employees, with broader effects on business activity.
31. Access to Gaza remains highly controlled. Additionally, around 60% of the West Bank; under the designation Area (C) is inaccessible to Palestinians.
32. See endnote 23.
33. See endnote 17.
34. See endnote 17.
35. CIA factsheet on Syria. Downloaded in January 2017 from: https://www.cia.gov/library/publications/the-world-factbook/geos/sy.html.
36. See endnote 17.
37. Oil output is reported by the Economist Intelligence Unit in its March 2017 Country Report as barely reaching 32,000 barrels/day (b/d), compared with 385,000 b/d in 2011.
38. The section titled "Concluding remarks" in Chapter 3 discussed at some length policy inadequacies that led to poverty and widespread unemployment in parts of rural Syria.
39. Egypt, for example, was among the top five recipients of expatriate remittances with nearly US$ 14 billion. Egypt's "country profile" from the "OECD Science, Technology and Industry Outlook 2012," published online by "Innovation Policy Platform." Profile accessed in March 2017 at: https://www.innovationpolicyplatform.org/content/egypt.
40. Parliament has held up reforms targeting Egypt's bloated civil service amid fears that firing civil servants would create widespread unrest.
41. The World Bank reported annual rates of GDP growth of around 4% in 2014/15 and 2015/16, up from an average of only 2% during the period 2010/11–2013/14. Furthermore, the World Bank expected that the country's GDP would grow at rates averaging around 4.6% annually during the years 2017–21. These rates are consistent with data presented in Fig. 2.11. "Egypt Economic Outlook;" published by Focus-Economics, in February 7, 2017. Downloaded in March 2017 from: http://www.focus-economics.com/countries/egypt.

42. News Report "Egypt's economy: outlook and challenges in the eyes of international financial institutions;" Mohammad Samir, January 21st 2017. Viewed in March 2017 on: http://www.dailynewsegypt.com/2017/01/21/egypts-economy-outlook-challenges-eyes-international-financial-institutions/.
43. "Egypt's economy, State of denial;" The Economist, 6th August, 2016. Article downloaded in March 2017 from: http://www.economist.com/news/middle-east-and-africa/21703393-egypt-has-squandered-billions-dollars-aid-more-way-it.
44. See endnote 43.
45. Thus, Egypt ranked 131st on the World Bank's ease-of-doing-business, with investors starting a new project required to obtain permits from around 80 official bodies. A promised "one-stop shop" representing all official bodies has not yet materialized.
46. Reports published by the regular media as well as business publications contracts worth billions of dollars are awarded annually to construction, manufacturing and other companies established by Egypt's armed forces, reinforcing their already controversial role in the country's civilian economic affairs. See for example "The Military Crowds Out Civilian Business in Egypt;" by A Morsy. Published by the Carnegie Endowment for International Peace in June 2014. Accessed in March 2017 at: http://carnegieendowment.org/2014/06/24/military-crowds-out-civilian-business-in-egypt-pub-55996.
47. UNDP report "The Human Development Report; Human development for everyone." eISBN: 978-92-1-060036-1. Report downloaded in May 2017 from: http://hdr.undp.org/sites/default/files/2016_human_development_report.pdf.
48. Sudan is eligible for debt relief under the Highly-Indebted Poor Countries Initiative. However, in order to take advantage of this initiative it would have to reach an understanding with its main creditors in association with South Sudan.
49. Damage to Sudan's economy was especially difficult since the secession came during a period of high oil prices and due to fact that interruption of oil production in South Sudan due to armed conflict in 2012 caused the loss of oil transit fees.
50. Namely, Southern Kordofan, Darfur, and the Blue Nile.
51. The number of refugees is reported as swelling on daily basis since the outbreak of violence in Juba in 2016. See the overview of Sudan's economic and social development by the World Bank. Downloaded in January 2017 from: http://www.worldbank.org/en/country/sudan/overview.
52. Based on an estimate dating back to 2009.
53. The country is the world's largest producer and exporter of gum Arabic, estimated at between 75% and 80% of the world's total output.
54. Youth unemployment figures quoted here are estimates of the percentages of the unemployed within the age bracket 15–24, modeled by the World Bank on ILO estimates. When worked out based on national estimates these figures assume slightly lower values at around 23 for total youth unemployment and 26% for female youth unemployment.
55. Mauritania is usually classified among countries of West Africa. However, socioeconomic development patterns have tended to bring it closer to its Arab neighbors in North Africa.
56. CIA factsheet for Algeria. Downloaded in January 2017 from: https://www.cia.gov/library/publications/the-world-factbook/geos/ag.html.
57. See endnote 17.
58. See endnote 56.
59. See endnote 17.
60. Abdelkader DJEFLAT, "Building Knowledge Economies for job creation, increased competitiveness, and balanced development: Individual country overviews with data on STI systems in the Arab countries. Report produced for the High Level International Conference, Organized by the Tunisian Government and ISESCO, In collaboration with the World Bank, Carthage, December 1–3, 2009.
61. See endnote 17.
62. See endnote 56.
63. World Bank's Overview of country performance, dated September 2016. Downloaded in January 2017 from: http://www.worldbank.org/en/country/morocco/overview.

64. The German engineering firm Siemens is reported as having invested $109 million into the Moroccan energy market, building a plant for the manufacture of giant long wind turbine blades in the country.
65. The goal of producing 42% of the country's installed capacity from renewable sources by the year 2020 is also mentioned in some references.
66. The Economist Intelligence Unit further forecasts that the slowdown of growth witnessed by China, combined with Europe's unsettled economic situation, will limit overall GDP growth attained by Morocco to around 3% in 2018 and 2019 before going up to around 3.6% in 2020.
67. The above-mentioned World Bank Overview expects that full implementation of comprehensive reforms following parliamentary elections toward the end of 2016 would lead to regained growth at around 4% over the medium-term, with inflation kept at around 2%.
68. See endnote 63.
69. See endnote 63.
70. The World Bank's database does not include figures for the year 2014 for the indicator "Poverty headcount ratio at national poverty lines as percentage of population. Figures included for this indicator for the years 1998, 2000 and 2007 are reported as 16.3%, 15.3% and 8.9%, respectively.
71. As reported by the Economist Intelligence Unit, considerable natural gas resources were verified in August 2016, in Eastern Morocco. Initial estimates indicate that they might contain several trillion cubic feet of natural gas.
72. EIU Country Forecast, published in May 2016.
73. Essentially away from cereals in favor of more lucrative crops such as olives and fruits.
74. A report published in May 2016 by the Economist Intelligence Unit indicated that over 60% of PMV objectives had been achieved; adding that Morocco's government estimated that around US$ 9 billion would be needed for the 2014–20 period.
75. See endnote 72.
76. See endnote 72.
77. "Innovation in Tunisia: Empirical Analysis for Industrial Sector;" Moez El Elj, Laboratory of Economics and Industrial Management. Tunisia Polytechnic School and High School of Management, Tunisia. Paper published by the *Journal of Innovation Economics & Management*, 1/2012 (n°9), p. 183–197. Downloaded in April 2017 from: https://www.cairn.info/revue-journal-of-innovation-economics-2012-1-page-183.htm.
78. The tourism and phosphate mining sectors make up around 15% of the country's GDP.
79. Data accessed in April 2017 at: http://data.worldbank.org/indicator/BX.KLT.DINV.WD.GD.ZS?locations=TN.
80. As indicated by World Bank data, FDI in Tunisia reached its highest level in 2006, when it exceeded 9.4% of GDP.
81. Data presented in Fig. 2.26 is based on measured, estimated and forecast growth figures published by the Economist Intelligence Unit Country Report, dated March 13th, 2017 including Tunisia's economic outlook for the period 2017–2021. www.eiu.com.
82. See endnote 77.
83. "Transforming Arab Economies: Traveling the Knowledge and Innovation Road." Centre for European Integration, the European Development Bank and the World Bank. Issued in 2013. WB 82736v2. Downloaded in April 2017 from: http://documents.worldbank.org/curated/en/664441468059651205/pdf/827360ESW0v10P00Box379869B00PUBLIC0.pdf.
84. "Libya's oil production rise overshadowed by Wintershall dispute;" A Raval. The Financial Times May 10, 2017. Downloaded in June 2017 from: https://www.ft.com/content/e7eed8f8-35a7-11e7-bce4-9023f8c0fd2e.
85. Foreign reserves took a plunge from around US$ 110 billion in 2013 to an estimated US$ 43 billion by end-2016. See the Country Report on Libya produced by the Economist Intelligence Unit in March 2017. Downloaded in April 2017. www.eiu.com.

86. "SMEs in Libya's Reconstruction: Preparing for a Post-Conflict Economy;" OECD Report, issued in 2016. http://dx.doi.org/10.1787/9789264264205-3-en. Executive summary accessed in April 2017 at: http://www.keepeek.com/Digital-Asset-Management/oecd/development/smes-in-libya-s-reconstruction/executive-summary_9789264264205-3-en#.WQSfMIiGPIU#page1.
87. Figure based on data from CIA factsheet on Libya. Downloaded in January 2017 from: https://www.cia.gov/library/publications/the-world-factbook/geos/ly.html.
88. Forecasts indicated above are naturally based on a set of assumptions concerning political developments at the national level, as well as international developments, with emphasis on energy prices.
89. Figure constructed on the basis of data published by the Economist Intelligence Unit in its March 2017 country report on Libya. Economist Intelligence Unit in its March 2017 country report. Downloaded in May 2017 from: www.eiu.com.
90. Total youth unemployment as a percentage of total labor force aged between 15 and 24 years, a modeled ILO estimate, reported by the World Bank's Development Indicators Database. See endnote 23.
91. Indeed, the Economist Intelligence Unit's country report issued in March 2017, expects policymaking to remain largely incoherent throughout 2017–19, due to various hurdles, including financing constraints and civil war conditions.
92. World Bank's Country Overview for Mauritania, March 2016. Downloaded in April 2017 from: http://www.worldbank.org/en/country/libya/overview.
93. Mauritania is Africa's second leading exporter of iron ore. It is also an exporter of gold and copper, demand for both of which is expected to witness significant growth.
94. See endnote 23.
95. See endnote 23.
96. In its country report for the first quarter of 2017, the Economist Intelligence Unit expects iron ore prices to recover in 2017–18. However, they are most likely to remain at least around 30% below their 2011–15 averages.
97. The African Development Bank's Strategy Paper for Mauritania for 2016–20 has power supply and agricultural modernization as central themes in its support program. However, the amount of loans to be provided for these purposes may not be sufficient to make a significant contribution toward the country's economic diversification.
98. This despite the fact that many nomads and subsistence farmers were forced into urban centers as a result of droughts in the 1970s and 1980s.
99. The Economist Intelligence Unit, Mauritania's Country Report, 1st Quarter 2017. Downloaded in March 2017, from www.eiu.com.
100. See endnote 92.
101. The indicator used by the World Bank here is the poverty headcount ratio at national poverty lines. According to World Bank's statistics, the poverty rate in Mauritania, estimated as the ratio of the population living at $ 3.1 a day, or less amounted to about 32.5% of the population in 2008, whereas the poverty gap at the national poverty line was estimated at 14.5% in the same year. These figures were estimated based on 2011 United States dollars.
102. Rural poverty headcount ratio at national poverty lines, as percentage of rural population.
103. Rising from levels of about 13% in 2006.
104. The internal political situation has improved, and following two military coups during the present decade, Mauritania has enjoyed relative political stability, with the formation of a national unity government.
105. Mauritania's scores have been consistently low on competitiveness, as measured by the World Economic Forum's criteria. Indeed, at 2.7, its competitiveness score in 2016/2017 was lowest among all 138 countries entered into the ranking scheme. Naturally, it also ranked lowest among all other countries Sub-Saharan Africa as well. See the World Economic Forum's Report for the year 2016–17.

106. Policies aimed at improved responses to the country's challenges would need to emphasize appropriate tax regimes that allow recovery of higher revenues from mineral resources and their reinvestment in activities that eventually generate sustainably higher and more uniformly distributed returns.
107. "The way out of poverty and corruption is paved with good governance." By Sri Mulyani Indrawati. Downloaded in May 2017 from: http://blogs.worldbank.org/voices/way-out-poverty-paved-good-governance.
108. Government debt was estimated to exceed 62% of GDP in 2016, somewhat exceeding the 60% debt-to-GDP criterion imposed by the Gulf Monetary Union. World Bank Memorandum on Economic Outlook for Bahrain's Economy. Accessed in March 2017 at: http://pubdocs.worldbank.org/en/510271475460654402/Bahrain-MEM-Fall-2016-ENG.pdf.
109. World Bank Memorandum on Economic Outlook for Bahrain's Economy. Accessed in March 2017 at: http://pubdocs.worldbank.org/en/510271475460654402/Bahrain-MEM-Fall-2016-ENG.pdf.
110. Bahrain's Country Report by the Economist Intelligence Unit, generated in March 2017. Downloaded in March 2017 from: http://www.eiu.com.
111. Bahrain's oil production which went up in the first quarter of 2016, is expected to decline partly due to the country's commitment to trimming production in line with agreement reached with non-OPEC producers.
112. In particular, new technology-based developments, expected to come online in late 2019, are expected to increase the country's aluminum production capacity by over 50%. Additionally, expansion of oil refinery output projected to go on stream in 2020, will contribute to improved overall prospects for economic growth. See the Economic Intelligence Unit's March 2017 Country Report. Downloaded in March 2017 from: http://www.eiu.com.
113. See endnote 23.
114. Such as World Bank and UNDP databases.
115. "Bahrain's tertiary education reform: A step toward sustainable economic development;" M Karolak. *Revue des Mondes Musulmans et de la Mediterranee*, (131), 163–181. Article downloaded in May 2017 from: https://remmm.revues.org/7665.
116. Some reports indicate that the country plans to increase oil production up to 4 million barrels per day by 2020.
117. Information presented on the website of the Kuwait Foundation for the Advancement of Science, retrieved in February 2017 from: http://www.kfas.org/.
118. Considerably lower rates of growth are forecast by a World Bank memorandum, which covered estimated and forecast growth rates between 2013 and 2018. According to this memorandum, growth rates for Kuwait's main sectors would fall between 2% and 2.4% in 2018. World Bank, Macroeconomics and Fiscal Management Global Practice, and Poverty Global Practice. Downloaded in February 2017 from: http://pubdocs.worldbank.org/en/358541475460667264/Kuwait-MEM-Fall-2016-ENG.pdf.
119. KISR 2004, "Toward a National Policy in the Sciences and Technology for the State of Kuwait;" by A. S. Oteibi and Y. Y. Sultan. Published by KISR in Arabic in September 2004. Document provided in hardcopy by its author.
120. Figure reported by the World Bank within its online World Development Indicators, last updated in February 17, 2016. Data downloaded in March 2016. See endnote 23.
121. Gulf Research Centre's report "The Demographic and Economic Framework of Migration in Kuwait." Downloaded in February 2017 from: http://cadmus.eui.eu/bitstream/handle/1814/32155/GLMM%20ExpNote_01-2013.pdf?sequence=1.
122. Kuwait is often stated as one of the first countries in the Gulf to operate and create a national financial market. Discussions held in 2007, during meetings of the Blue Ribbon Panel requested to review research activity in Kuwait addressed the need to acquire specific STI capabilities in fields such as physics and engineering of financial markets, disciplines that have recently gained recognition as useful tools in this domain.

123. Oil rent as a percentage of GDP went down from an average levels of around 57.8% in the 1970s down to 41.5% during the 1980s and still further down to 30.1 during the 1990s. Oil rent recovered, to reach 35.8% during the first decade of the 20th century and averaged 36.1% of GDP during 2010–13.
124. Lower oil sector investment is expected to slow down hydrocarbon GDP growth to around 1% in the near future from levels of around 4% in 2015.
125. Memo by the World Bank on Oman's Economic Outlook, dated autumn 2016. Downloaded in March 2017 from: http://pubdocs.worldbank.org/en/734141475460791027/Oman-MEM-Fall-2016-ENG.pdf.
126. Based on material posted on the website of the Supreme Council for Planning. Accessed in May 2017 at: http://www.scp.gov.om/en/Page.aspx?I=14.
127. This Vision document was authored in response to a royal decree issued in 1996, in connection with ratification of the fifth 5-Year Development Plan.
128. See endnote 126.
129. See endnote 126.
130. See endnote 126.
131. This is presumably the percentage attained as a result of implementation of the eighth 5 year plan.
132. Oman's government has a reasonable record with regard to leveraging resources by private investors and might well be expected to attain some tangible outcomes within the scope of the present 5-year plan. Information retrieved in May 2017 from 'marketstoday.net' at: http://www.marketstoday.net/includes/download.php?file=rr_18012016114554.pdf&lang=en&s=4464&m=research.
133. Press reports early in 2017 have pointed that the Main Committee of Oman's Vision 2040 held its fourth meeting to discuss the formation of special panels to consider sectoral aspects of the Vision. A National Conference dedicated for reviewing progress achieved in formulating the Vision document is scheduled for the second half of 2018. "Oman Vision 2040 Main Committee discusses formation of sectoral panels;" January 16, 2017. Accessed in May 2017 at: http://timesofoman.com/article/100708/Oman/Government/Oman-Vision-2040-Main-Committee-discusses-formation-of-sectoral-panels.
134. At least during 2016–20.
135. This figure is based on statistics compiled by the World Bank's website dedicated to World Development Indicators. See endnote 23.
136. Figure reported by the World Bank within its online World Development Indicators, last updated in February 17, 2016. Data downloaded in March 2016.
137. "Labor Migration in the GCC Countries: Past, Present and Future;" G Naufal and I Genc. The Middle East Institute (University of Singapore) 3rd June, 2014. Downloaded in May 2017 from: https://mei.nus.edu.sg/index.php/website/publications_TMPL/Volume-9-2-Labor-Migration-in-the-GCC-Countries-Past-Present-and-Future.
138. A 2016 report by the Middle East Institute indicates that the budgets of several institutions were drastically reduced. Thus, the Qatar Foundation's budget suffered a 40% reduction. Additionally, significant cuts in budgets earmarked to supporting academic institutions in Education City. "Qatar Cuts Spending to Cope with Low Oil Prices;" by Giorgio Cafiero, The Middle East Institute, 1st March, 2016. Downloaded in February 2017 from: http://www.mei.edu/content/article/qatar-cuts-spending-cope-low-oil-prices.
139. World Bank's Overview of Qatar last updated October 2016, downloaded in January 2017 from: http://www.worldbank.org/en/country/gcc/overview.
140. At current levels, reserves are expected to last for over half a century hence. Although Qatar's proven natural gas reserves are counted as third largest known reserves worldwide.
141. In particular, Qatar's bid to host the 2022 World Cup has accelerated large-scale infrastructure projects, including underground transport and light rail systems, the construction of a new port, several roads and related sports facilities.

142. According to the Vision Document rapid and uncontrolled expansion might lead to financial vulnerabilities, deterioration public services, low labor productivity, deterioration in project quality, environmental damage, and widening social tensions.
143. See the Global Business Reports' "Qatar Chemicals 2016." Global Business Reports conducts surveys and publishes information and data covering, among others, the hydrocarbon, pharmaceuticals, chemicals, energy industries around the world. See the Global Business Reports' "Qatar Chemicals 2016;" Retrieved in February 2017 from: http://gbreports.com/wp-content/uploads/2016/02/Qatar_Chemicals_2016_IE.pdf.
144. "Country Programming Framework for the State of Qatar;" Jointly Prepared by the Ministry of Environment of the State of Qatar And The Food and Agriculture Organization of the United Nations. Report dated 2012. Downloaded in February 2017 from: http://www.fao.org/3/a-bp590e.pdf.
145. Although press reports dating back a few years ago indicate that Qatar could grow up to 70% of the agricultural products it now imports, the title of a more recent report suggest that soilless farming might provide a solution to food shortage in Qatar. See for example articles in "Soil-free greenhouses to help Qatar grow up to 70% of its veg by 2023;" by L. Walker in Doha News, 25th August, 2015, downloaded from: https://dohanews.co/soil-free-greenhouses-to-help-qatar-grow-up-to-70-of-its-veg-by-2023/ and "Soilless farming suggested as a solution to food shortage in Qatar;" Article published on 17th February, 2017 by Phys.Org, a website dedicated to science and technology issues. Downloaded from: https://phys.org/news/2016-02-soilless-farming-solution-food-shortage.html#jCp.
146. Qatar produced 40.62 million cubic meters of water in January 2015, up 29% from the same month a year earlier. "Qatar races to desalinate, store water to meet nation's growing thirst;" by P. Kovessy, Doha News, March 20, 2015. Article retrieved in February 2017 from: https://dohanews.co/qatar-races-to-desalinate-store-water-to-meet-nations-growing-thirst/.
147. "Qatar races to desalinate, store water to meet nation's growing thirst;" by P. Kovessy, Doha News, March 20, 2015. Article retrieved in February 2017 from: https://dohanews.co/qatar-races-to-desalinate-store-water-to-meet-nations-growing-thirst/.
148. Indicating that Mitsubishi was selected for the task of building the facility, press reports noted that it was the second major contract awarded to the Japanese engineering firm during a very short period. Since, this firm had been awarded a deal worth more than US$ 1 billion to build a power and water production plant designed to produce 2,400 MW and 492,000 $m^3$ daily. "Qatar to build its first reverse-osmosis desalination plant;" News alert published by Utilities Middle East on 9th March, 2015. Downloaded in February 2017 from: http://www.utilities-me.com/article-3407-qatar-to-build-first-reverse-osmosis-desal-plant/.
149. Saudi Arabia's GDP was estimated in 2015 at US$ 637.8 billion, using the official exchange rate of SR3.75/US$.
150. CIA factsheet on Saudi Arabia. Downloaded in January 2017 from: https://www.cia.gov/library/publications/the-world-factbook/geos/sa.html.
151. Saudi Arabia is the world's third biggest military spender. Only 2% of its military spending takes place within its boundaries, with the national defence industrial sector limited to only seven companies and two research centres.
152. The Saudi Arabian Oil Company (earlier name Arabian American Oil Company). An Article by the Middle East Monitor, states that Aramco's crude reserves are estimated at around 260 billion barrels. This is said to constitute 10 times those of Exxon Mobil's, whereas Aramco's daily production of around 10 million barrels is larger than the domestic output of all US oil companies combined. Article by the Middle East Monitor: "International scramble to list Aramco stocks;" A Sloan, 9 September 2016. Downloaded in June 2017 from: https://www.middleeastmonitor.com/20160909-international-scramble-to-list-aramco-stocks/.
153. The document was announced on the April 25, 2016 by Deputy Crown Prince Mohammad bin Salman Al Saud.
154. Saudi Arabia was in fact listed in 15th position worldwide by virtue of its GDP measured by purchasing power parity. This goal may be in reference to GDP measured by a criterion other than purchasing power parity.

155. This goal is to be achieved through attracting regional and international retail investors as well as facilitating ownership by foreigners.
156. This goal will be attained through partnering with private sector to develop and expanding coverage and capacity.
157. Goals related to health care are to optimize past achievements among which the Vision document mentions raising life expectancy from 66 to 74 years and securing hospital 2.2 beds per 1000 of the population.
158. With particular reference to educational goals the Vision document states the need to continue to develop the talents of female students and graduates who constitute over 50% of all university students and graduates, investing in their productive capabilities and enabling them to build their future careers and contribute to the development of Saudi Arabia's society and economy.
159. FDI should, according to the Vision document, go up from its present level at 3.8% to 5.7% of GDP.
160. According to a local investment bank the first three quarters of 2016, total net employment in Saudi Arabia rose by 892,000, compared to 417,000 between 2014 and 2015. However, 95% of these new jobs appear to have gone to non-nationals. At any rate, there are question marks over some of the official employment data. For example, whereas official figures show that the construction sector added 652,000 jobs that went to non-Saudis over the first 9 months of 2016. The fact that extra jobs were created during a period when the sector was going slowing down, seem rather unlikely. Doubts about the reliability of official data linger, particularly over some of the figures provided by the General Authority for Statistics (GAS). Discrepancies are sometimes explained by changes in the way that GAS conducts its surveys. "Rising Unemployment Suggests Saudi Government Reforms Are Failing;" Dominic Dudley, 8th December 2016, Forbes. Viewed in January 2017 at: http://www.forbes.com/sites/dominicdudley/2016/12/08/saudi-unemployment/#3c02f0a736b5.
161. Another source placed this figure at 11.6%, whereas unemployment among women, was quoted as having reached 32.8%. See.
162. In a recent report, the General Authority for Statistics (GAS) indicated that the number of Saudis over 15 years reached 13.8 million in 2016. Of these, only 5.6 million were considered eligible entrants to the country's labour force. Report published by the Saudi Gazette titled "Unemployment, facts and figures;" Barjas Al Barjas, 18th April, 2016. Downloaded in January 2017 from: http://saudigazette.com.sa/opinion/local-viewpoint/unemployment-facts-figures/.
163. The country's National Science, Technology and Innovation Plan discussed in this section addresses acquisition and dissemination of technologies in these as well as several other domains.
164. The UAE economy's openness renders it sensitive to global crises, however. Thus, the country was negatively affected by the global financial crisis of 2008–09. Due to its high exposure to real estate prices, Dubai was hit hardest and had to be bailed out by the UAE Central Bank and Abu Dhabi Government.
165. In its 2013–14 Global Competitiveness Report, the World Economic Forum ranked the country in fourth position worldwide in terms of infrastructure quality.
166. Additionally, fuel subsidies were removed in August 2015.
167. UAE Country Outlook, October 2016. Downloaded in March 2017 from: http://pubdocs.worldbank.org/en/398801475460803726/UAE-MEM-Fall-2016-ENG.pdf.
168. Figures taken from statistical tables available from the World Bank's database which state that they were modeled on ILO estimates. Downloaded in October 2016. See endnote 23.
169. See the CIA Factbook report for the United Arab Emirates. Downloaded in January 2017 from: https://www.cia.gov/library/publications/the-world-factbook/geos/ae.html.
170. However, a figure of 19.5% is quoted for 2003 by See the CIA Factbook report for the United Arab Emirates. Downloaded in January 2017 from: https://www.cia.gov/library/publications/the-world-factbook/geos/ae.html.
171. See endnote 167.
172. "The United Arab Emirates Vision 2021." Downloaded in May 2017 from: http://fgccc.org/wp-content/uploads/2016/08/UAE_Vision_2021.pdf.

173. Abu Dhabi's Vision 2030 document. Posted by Emirate's Economic Council. Downloaded in May 2017 from: https://www.ecouncil.ae/PublicationsEn/economic-vision-2030-full-versionEn.pdf.
174. "UAE pharmaceutical market set to hit $ 5.7 billion by 2020;" The Pharma Letter. 11 October 2016. Downloaded in May 2017 from:.
175. Described by Hvidt, in the case of the Emirate of Dubai for instance, as a defining characteristic of the "developmental state" paradigm. "The Dubai Model: An outline of key components of the development process in Dubai;" Martin Hvidt. Centre for Contemporary Middle East Studies; University of Southern Denmark. Working papers Series No. 12, October 2007. Paper downloaded in June 2017 from: http://static.sdu.dk/mediafiles/Files/Om_SDU/Centre/C_Mellemoest/Videncenter/Working_papers/12WP2007MH1_001.pdf.
176. For example, the Port of Hudaydah, which handled 60% of the country's commercial traffic, was damaged in August 2015 is now reported as operating at 50% of its original capacity.
177. Data for this figure is based on World Bank's World Development Indicators. See endnote 23.
178. Thus, of the UN's US$ 1.6 billion emergency appeal for 2016 only 60% was made available.
179. CIA's Yemen factsheet. Downloaded in January 2017 from: https://www.cia.gov/library/publications/the-world-factbook/geos/ym.html.
180. Country Report on Comoros by the Economist Intelligence for the first quarter of 2017. Downloaded in March 2017 from: www.eiu.com.
181. World Bank's Country Overview for Comoros. Downloaded in May 2017 from: http://www.worldbank.org/en/country/comoros/overview.
182. In Purchasing Power Parity, 2011 exchange rate for the US dollars.
183. A previous household survey dated to 2004, placed the percentage of the population living below the national poverty line at 44.8% and those living below the international poverty line at 13.5%. However, the World Bank's document reporting both results warns that these two household surveys were not comparable, cautioning against interpretations concerning poverty and progress toward its reduction.
184. Report by the International Renewable Energy Agency (IRENA). Report dated 2015. Downloaded in May 2017 from: http://www.irena.org/DocumentDownloads/Publications/IRENA_RRA_Djibout_2015_EN.pdf.
185. This estimate is the population below $ 1.25 per day at purchasing power parity.
186. Estimates from the CIA Factbook on Djibouti. Downloaded in May 2017 from: https://www.cia.gov/library/publications/the-world-factbook/geos/dj.html#cntryFlagModal.
187. Based on a UNICEF Annual Report on Somalia issued in 2013. Based on a UNICEF Annual Report on Somalia issued in 2013. Accessed in May 2017 at: http://reliefweb.int/report/djibouti/unicef-djibouti-humanitarian-situation-report-30-june-2016 and UNICEF's report dated 2014. Accessed in May 2017. At: https://www.unicef.org/appeals/files/HAC_2014_Somalia_-_FINAL.pdf.
188. "The UNESCO Science Report; Toward 2030." Downloaded from: http://unesdoc.unesco.org/images/0023/002354/235406e.pdf.
189. "Humanitarian action for Children; Somalia;" Report by UNICEF, dated 2014. Downloaded in May 2017 from: https://www.unicef.org/appeals/files/HAC_2014_Somalia_-_FINAL.pdf.
190. World Bank's Overview of country performance, dated December 2016. Downloaded in January 2017 from: http://www.worldbank.org/en/country/somalia/overview.
191. Telecommunication firms provide wireless services in most major cities, surprisingly offering the lowest international rates in Africa.
192. Collection of higher revenues is hampered by security considerations, limiting control over all territories, together with poor collection capacity as well as inadequate and absent legal and regulatory arrangements.
193. Remittances are channeled into the country through money transfer services, since the country does not have a formal banking sector. Money transfer services were estimated to have handled around $ 1.6 billion in 2016.

Chapter 3

# Science, Technology, and Innovation Policies and Institutional Landscapes

## Chapter Outline




Science, Technology, Innovation, and Development in the Arab Countries
http://dx.doi.org/10.1016/B978-0-12-812577-9.00003-9

## 3.1 INTRODUCTION

Modern universities and institutions of higher learning in Arab countries such as Egypt and Lebanon, for instance, were responsible for initiating scientific research during the early twentieth century, even before these countries gained their independence. Archaeological and medical research were preferred subjects, the first prompted by outside interest, whereas the second motivated by the need to address a host of local needs.

All such research activities were conducted with little, if any reference to national or even institutional policy regimes as sources of organizing principles and benchmarks for measuring progress or ultimate impact.

Interest shown in science and technology policies in the developing countries at large must be credited to the influence of pioneering projects launched by a small number of international organizations. Among others, the following institutions were instrumental in promoting the cause of science and technology policy as an essential tool for development:

- the Science and Technology Policy Research Unit (SPRU),[1] at Sussex University in the United Kingdom;
- the United Nations Advisory Committee on Science and Technology for Development (UN-ACST);
- the United Nations Industrial Development Organization (UNIDO) and the International Development Research Centre (IDRC).

Reports produced and meetings held on the subject, mainly between the late 1960s and on to much of the 1970s and 1980s helped disseminate these notions, promoting worldwide interest. Developing countries, in Latin America and Asia, some of which had gained their independence two decades or so earlier and were embarking on industrialization with the main purpose of import substitution, were keen recipients of notions advanced by the early pioneers[2] and went on to contribute to the subject based on their own experiences.

Several institutions and experts in the Arab countries must be credited with major roles in promoting interest in science and technology policy as tools for development. Thus, during the 1980s, prompted by widening interest in the subject, the Arab League's Educational, Cultural and Scientific Organisation (ALECSO) organized several conferences specifically designed to promote science and technology policy formulation by the Arab countries and eventually produced a compendium on the subject in association with a group of Arab institutions and experts.[3] Titled, "Strategy for the development of Science and Technology in the Arab Homeland," this compendium was published in 1987. Both as a tribute to this publication and in criticism of institutions responsible for promoting science, technology, and development in the Arab countries, several major recommendations outlined in this publication have yet to be implemented!

The notion of innovation policy as adjunct to science and technology policy owes its popularity to later days during the 1990s. Although very many definitions may be found for innovation in current literature, the principal theme addressed in most development policies concerns innovation as a tool for modifying prevailing patterns in production and services sectors in a manner that promotes economic growth, generates enterprises, and reduces poverty. Innovation as a means for reducing poverty in particular gained momentum since the mid-2000s due to initiatives taken by international aid and philanthropic organizations.

In the Arab countries, interest in innovation as a means for economic growth and social development began to materialize toward the latter part of the 1990s, and several countries throughout the region have in fact formulated national policies and established institutional entities at least partly dedicated to promoting innovation with particular accent on technology-based enterprises.

The following sections will take a brief look at the state of science, technology, and innovation systems, including policies devised by various Arab governments and concerned institutions with a view to promoting scientific research and technology development activities.

## 3.2 THE LEVANT COUNTRIES

Science had a rich past in the Levant's antiquity and its Middle Ages. Terrain that is now delineated by geographical boundaries including Iraq, Syria, Lebanon, Jordan, and Palestine played a major role in the evolution of many scientific disciplines, technologies with numerous innovations in agriculture, industry, jurisprudence, and culture. It may even be argued that success enjoyed by empires that took Mesopotamia and the Nile valley as their seats of power was essentially due to the numerous advances in science, technology, and innovation nurtured across past centuries within the Levant region. Several cities in the region were home to some of the earliest higher learning and scientific research institutions on planet. In short, science, technology, and innovation saw much better times in the past than they have during the past few centuries.

Although attempts were and are still being made to establish institutions of higher learning and research, it may confidently be said that results achieved thus far fall too short of what is needed as bases for healthy socioeconomic development.

### 3.2.1 Iraq's Science, Technology, and Innovation System; Evolution and Recent Developments

Activities related to science and technology were largely confined to Iraq's universities as well as a very small number of specialized technical centers, operating mainly under the ministries of agriculture and industry. With a few exceptions, only limited attempts at innovation were made by universities or research centers beyond their own walls, on the regional or national arena.

Iraq's universities flourished during the 1950s and 1960s. However, starting in the 1970s, internal politics coupled to perceived and actual threats across the country's borders eventually robbed universities of a great deal of their intellectual dynamism at the cost of much reduced educational standards. UN sanctions imposed after Iraq's invasion of Kuwait in 1990, further isolated and impoverished higher education sector throughout the country, whereas during the 2003 invasion of the country, many universities, already running on threadbare resources, were further looted. Violence that followed the invasion cost the universities much more than equipment items educational and research facilities, as hundreds of university professors and administrators were killed and thousands fled the violence that followed the invasion.

Hundreds of millions of dollars were required to rehabilitate campuses, whereas allocated budgets were meager and further diminished through mismanagement and corrupt practices.

Presently, sectarian politics dominate the country's institutions, including those dedicated to higher education and scientific research. Campuses across the country are highly politicized with staff and student organizations competing, often violently, for influence, with focus on religious and sectarian issues, such as enforcing veils for women and separation of the sexes.

#### *3.2.1.1 Science, Technology, and Innovation Governance*

As indicated earlier, science, technology, and innovation (STI) activities in Iraq were mostly confined to research within university departments and a small number of research centers. As such, their governance went through several stages starting with the establishment of the Scientific Research Council (SRC) in 1963. SRC was placed under the care of the Ministry of Higher Education and Scientific Research (MoHESR) when it was established in 1967. Between 1970 and 1976, SRC became an organ of the State Planning Council, reverting to MoHESR in 1976. SRC was able to formulate a national scientific research strategy covering the decades 1980–2000, which was intended for implementation within national 5-year plans, however, SRC was eventually disbanded in 1989.[4]

In 1996, MoHESR established a Department of Research and Development (DRD), consisting of sections responsible for higher studies, scientific affairs, project management, as well as documentation and information management. This department appears to have been quite overloaded with a multiplicity of tasks including[5]

- formulation of policies aimed at promoting science, technology, and innovation within educational and research institutions;
- implementation of special research projects and marketing research output;
- development of curricula for higher studies;
- support and monitoring of scientific societies;
- virtual education, scientific publishing and science libraries;
- follow up on the status of university cadres and implementation of specific programs to improve their professional standing.

DRD achievements are reported as including:

- furnishing a number of laboratories with equipment;
- arranging study tours to well-established overseas research facilities for students enrolled in programs of higher studies as well as their supervisors;
- implementing a program designed to encourage research publications in international journals through a special prize scheme.

DRD is also credited with having launched Iraq's virtual science library and seeking to modernize legislation governing promotion of scientists. However, no indications are available as to whether a comprehensive policy for "promoting science, technology and innovation" was in fact formulated. At any rate, such a policy would have certainly been confined to educational and research institutions run by MoHESR rather than the country as a whole. This is a serious constraint, which would leave out research and training centers run by other ministries as well as efforts aimed at setting up facilities directed at promoting enterprise and employment creation. Additionally, the very fact that such a policy would be the result of activities undertaken at the departmental, at best ministerial, level would reduce its value in comparison to a nationwide initiative, with the multifaceted benefits of such an initiative, in terms of integrating efforts by concerned actors from various backgrounds and with diverse objectives and optimizing outcomes of their individual contributions.

### *3.2.1.2 National Development Policies and STI-Related Plans*

The Iraqi constitution of 2005 addressed a number of issues with direct impact upon higher education and scientific research.[6] Article 34, for example, declares that education is a right for all citizens and that it is to be made available at no financial cost in all its stages, i.e., including higher education. Scientific research was also earmarked for special support. In addition, Article 114, demanded that competencies be shared between federal and regional authorities in the formulation of a nationwide educational policy. It is noteworthy that although this constitution acknowledged a place for private education at all levels, all education in Iraq, including university education continued to be made available by public institutions to all citizens at no cost along the lines adopted under all regimes that ruled Iraq since its independence.

In effect, critics of the constitutions' treatment of higher education and scientific research point out the lack of emphasis on increasing budgetary allocations for scientific research.

Significantly, the Kurdistan Region's Government (KRG) was granted the right to establish its own Ministry of Higher Education and Scientific Research, its right to establishing private institutions of higher education was also recognized by the new constitution, and, in fact several such institutions were eventually licensed and went into operation.

The following section is dedicated to a summary review of aspects concerning STI capacity building in Iraq's current national development plan, with focus on creating e-government capabilities. The essential purpose of this section is to illustrate some of the conceptual flaws and other shortcomings that appear to characterize national policy documents formulated with view to national STI capacity building in the region.

#### 3.2.1.2.1 Iraq's Current National Development Plan

The Iraqi government's National Development Plan covering the 5-year period 2013–17 included the establishment of e-government as one of the principal objectives. The quest of establishing Iraq's e-government is based on the premise that reforming the public sector may only be attained if information and communication technologies are adopted in updating government transactions, allowing simplified procedures as well as improved performance and security as well as stamping out corruption. The plan furthermore indicates that attempts at introducing e-government were the subject of cooperation between the Iraqi e-Government committee and United Nations Development Program (UNDP) since 2009. Objectives listed for the National Plan include:

- convening of an international conference on e-Government held under the logo "Building e-Iraq and e-Governance," in December 2009;
- holding a workshop for trainers in 2010, as a result of which 30 expert trainers were trained;
- launching the e-Iraq portal in July 2011, with information and interactive services concerning the e-government initiative;
- producing a document in December 2011 on exchange and connectivity of information in Iraq;
- implementing a training program for 200 trainers;
- conducting an electronic readiness survey, which indicated that in 2010, over 90% of ministries possessed LAN or WAN networks to which around 63% of computers were connected.

The modest nature of these objectives is patently obvious. Additionally, should the above reflect the actual attainments reached over 4–5 years of work aimed at establishing e-government in Iraq, then the entire exercise must be judged as unambitious, progressing too slowly and possibly, with planning and dedicated resources in need of radical revision. These conclusions are confirmed by the fact that technical obstacles listed by the National Development Plan include:

- the need for larger investments to secure information technology tools;
- inadequate information and communication technology (ICT) infrastructure in the country at large, with limited Internet connectivity, high access costs as well as lack of sustainable and stable computer networks ensuring interoperability among government departments;
- lack of cyber security measures.

The 2013–17 National Development Plan, nevertheless, put forward two major goals for the process of establishing e-government in Iraq, the first of which is "developing a knowledge-based society and closing the digital gap," with means for attaining this goal including:

- formulating legislation in support of e-government, covering e-signatures, e-commerce, cybercrime;
- implementing programs aimed at improving human resource capabilities for dealing with ICT systems;

- increasing investments directed at achieving good governance through widespread adoption of ICTs in public sector institutions;
- providing equal opportunities for the utilization of e-government services in governorates;
- developing an institutional framework for the transition toward e-government;
- spreading awareness of Iraq's interoperability frameworks, standards, and certification procedures.

The second goal set for establishing e-government is formulated as "boosting interaction and strengthening foundations of partnership between development partners in the transformation to a digital economy," with means of achieving this goal essentially focused upon awareness and training campaigns involving training and development centers at universities directed toward decision makers, higher administrations, and leaders in the governorates.

That the development of a knowledge-based society and transformation toward a digital economy are considered as outcomes of an e-government system indicates a major conceptual flaw. Additionally, the 2013–17 National Development Plan does not include numerical indicators to enable progress made toward any of its goals.

In conclusion, although some progress may well have been achieved in terms of awareness raising as well as creation of improved ICT networks and hardware, it would seem that goals set for transformation toward e-government in Iraq as well as the means set aside for attaining them are in need of a radical overhaul.

#### 3.2.1.2.2 Vision and Strategy of the Ministry of Higher Education and Scientific Research in the Kurdistan Region of Iraq

In the field dedicated to the Vision and Strategy on the website of the Ministry of Higher Education and Scientific Research in the Kurdistan Region of Iraq (KRI) has a statement dated 2009, by the Minister conceding that "the current system of higher education, which Kurdistan has inherited from Iraq, is dated and largely dysfunctional." The same statement goes on to indicate that "Whereas student numbers are increasing exponentially, standards within higher education institutions (HEIs) remain below those of the rest of Iraq or neighboring countries."

According to the Minister's statement, the "pyramid-style management structure" allows little transparency and accountability, with a negligible role for staff and students in the management of higher education affairs. The statement goes on to declare that "Internal audit, performance monitoring and quality assurance are virtually non-existing." The Minister's statement, furthermore, considers the existing system as "a major barrier for progress, incapable of healthy evolution … " and declares its intention to reform the system although admitting that the process would be extremely difficult due to enormous challenges.

The reform process that the Ministry intended to embark upon had to address the entire system of higher education, "in a way that no legal or administrative barrier would be considered too sacred to change." With reference to the role of the Ministry itself, the Minister indicated that reforms would address several issues including:

- management structures of Universities, introducing democratic systems, granting staff a greater say in the running of their institution and protecting the students' right to quality education;
- introduction of a system that allows teaching and research quality assessment and performance monitoring, and linking key performance indicators to promotion and pay;
- implementation of auditing modalities to ensure adherence to law and "control of corruption, safe environment, gender equality and protecting the rights of minorities such as non-Kurds";
- reform of funding schemes for student scholarship and research grants, thereby ensuring "compatibility with the Region's future plans and aspirations."

Objectives of the Ministry's strategy are summarized as assisting HEIs in Kurdistan to achieve excellence in teaching and research, to the point of becoming internationally competitive, rendering higher education increasingly market orientated and enhancing their employability. In order to achieve these strategic aims, the Minister's statements promise a number of measures, including:

- reforming management structures of the Ministry and HEIs, introducing greater independence as well as dynamic and democratic management systems that guarantee accountability and respect of human rights;
- adopting measures that ensure Teaching Quality Assurance, based on student and staff evaluations together with modalities for monitoring performance as basis for pay and promotion;
- reviewing the diversity, distribution, and market responsiveness of educational and research specializations within the region's the universities;
- revising research funding and administrative measures as well as ensuring funding schemes for training outstanding academic staff members at international centers of excellence;
- reviewing the Higher Education Law and collaborating with relevant parliamentary bodies on redrafting a modern legislative system.

The following comments relating to the Minister's Vision and Strategy statement authored in 2009 are in order. First, the fact that the statement is still posted in its entirety on the Ministry's website almost 8 years later is significant and probably indicative of limited progress attained in fulfilling its declared ambitions. Indeed, the fact that the Ministry's website is virtually devoid of any information consistent with the calls for accountability and transparency embedded in the Vision and Strategy statement is in itself indicative that such is the case.

Second, although damning statements targeting legal and institutional deficiencies that guided Iraq's higher education and scientific research might be viewed as excessive or motivated by political inclinations, they must nevertheless be taken to carry a good deal of validity. This is particularly because these statements are at the same time self-critical and intended as basis for future action, as well as disapproving of past laws, regulations, and institutional arrangements.

Third, in relation to its call for reform of legislative bases for higher education and scientific research the Ministry seems to have in fact worked with relevant authorities on a draft law to this effect and also went on to adopt a long-term strategy for scientific research in the KRI, covering the years till 2030. The following two paragraphs provide some more information on both the draft reform law and the scientific research strategy.

#### 3.2.1.2.3 Draft Law to Reform Higher Education and Scientific Research in the Kurdistan Region of Iraq

The Ministry of Higher Education and Scientific Research in the Kurdistan Region drafted a new law for approval by the Parliament with quite a few clauses targeting improvements in the region's research capabilities. Among other goals, the draft new law aims at formulating novel strategies and plans to upgrade higher education and scientific research. Openness to the outside world, better funding modalities and focus on human resources as both means and goals of the process are hallmarks of the new draft law. Additionally, the new law demands that modalities for quality assurance in higher education and research are embedded within relevant institutions, whether governmental or otherwise.

The new draft law addresses a number of institutional issues with the intention of improving the manner with which the Ministry supports higher education and scientific research in the Region. Thus, an article in the draft law requires the creation of a "Scientific Research Authority" within the Ministry. The head of this authority is a de facto member of a Council to be established at the Ministry with view to overseeing policy formulation as well as implementation of future plans aimed at revitalizing the Region's activities in higher education and scientific research.

Additionally, the draft law includes provisions for the creation of a department within the Ministry to specialize in funding and evaluation of scientific research projects.

Articles addressing the roles of institutions and individuals concerned with higher education and scientific research may be described as progressive in the sense that they highlight academic freedoms and stipulate independence from political influences of any kind. With regard to funding, the draft law demands that universities dedicate a minimum of 5% of their operating budgets to scientific research.

As to planning higher education and scientific research activities within individual institutions, the draft law requires that an annual "strategy" is formulated by universities and that it be ratified by the senate before implementation. University councils, on the other hand, are charged with the task of ratifying research plans, presumably including funding dedicated to individual research programs/projects.

The draft law appears to place a good deal of the burden for reform on individual universities, charging them with creating appropriate overriding conditions and implementation of modalities to ensure quality through programmed cooperation with well-established universities abroad. In relation to the award of masters degrees, the draft law demands that they are awarded following a course of study of not less than 2 years with at least one third of the credits linked to scientific research activity. A PhD degree, on the other hand, may be awarded only following four, instead of the customary 3 years. The draft law further demands that at least one of the 4 years is spent working at an accredited research institution abroad.

Although it is unclear whether the draft law is headed for ratification, it must be commended with particular regard to a number of intended innovations. However, it should also be stated that it only marks a first step in reforming higher education and scientific research in the KRI. A good deal more is left for ensuing strategies and action plans to attain. For instance, in seeking particular socioeconomic objectives, there is a strong need for effective coordination with other ministries as well as to the business sector in drafting the law and its subsequent implementation modalities. Additionally, no direct mention is made of the need for new institutional forms required for promoting business innovations and technology transfer, such as technology parks, innovation centers, or incubation schemes.

In particular, the issue of arrangements to be adopted in cases where a university, or a research center, partners with business institutions with the aim of undertaking research activity that may result in profit making is not addressed. In this respect, it may be noted that arrangements of this nature are now customary in many developed countries.[7]

### 3.2.1.2.4 Scientific Research Strategy for the Kurdistan Region

While the fate of the above-mentioned draft law are not as yet clear, a strategy for scientific research for KRI, covering the period 2010–30, was formulated in 2010.[8] The objectives of this strategy included:

- tackling problems facing socioeconomic development, as well as to predicting and forestalling such problems through scientific research;
- creation of well-equipped research centers covering a variety of specializations;
- training research cadres possessed of outstanding professional standards;
- improved links, including exchange programs, with prominent research centers around the world.

Based on the above strategy, the Ministry of Higher Education and Scientific Research in the Kurdistan region adopted plans including the following components:

- Establishment of a Project Management Office to promote coordination among the regions research facilities and cooperation with international organizations.
- Enhancing allocation by the regional government as well as improved access to funding from other sources.
- Setting up laboratories equipped to undertake research in areas earmarked of priority concern.

### *3.2.1.3 Iraq's Institutional Landscape*

As indicated above, interest in establishing institutions of higher learning and literary research dates back to the middle of the last century. However, scientific research in Iraq went through many ups and downs throughout the second half of the 20th century. The following paragraphs aim to present a resume of key institutional initiatives undertaken to promote research in the country.

**The Iraqi Academy of Sciences (IAS):** The Academy was established in 1948 to cater for Arabic language studies as well as research in both the social and natural sciences. In 2004,[9] the Academy created departments devoted to research and studies within the pure sciences, encompassing mathematics, computer science, physics, chemistry, biology; and another dedicated to the "applied sciences," including sections with focus on medicine, engineering, and agricultural science.

**Higher Council for Scientific Research:** A landmark legislation[10] issued in 1963, created the Higher Council for Scientific Research, linked to Baghdad University. The Council became operative in 1964. Following several changes in the Council's affiliation it was finally placed under MoHESR auspices in 1980. However, the Council's fortunes began to decline in 1985 and it was finally dissolved in 1989.

**Scientific Research Authority:** In order to compensate for the Council's disappearance, a series of regulatory arrangements were made in the 1990s with the aim of organizing the activities of Iraq's scientific research centers. The Scientific Research Authority was created in the mid-1990s. However, it had little to show by way of substantive output between its founding date and the war in 2003, reportedly due to constraints imposed by the international sanctions on Iraq.

#### 3.2.1.3.1 Ministries Concerned With Promoting STI Activities

At present, scientific research in Iraq carried out mainly by university departments and research centers established under the MoHESR auspices, as well as research centers established by other ministries, e.g., the Ministry of Agriculture. In either case guidelines as well as resources for scientific research are established through policies and regulatory arrangements adopted by concerned ministries.

### 3.2.1.3.2 The Ministry of Higher Education and Scientific Research (MoHESR) in Baghdad

MoHESR is one of the largest ministries in Iraq.[11] Its Directorate of Research and Development, which is overseen by a Deputy Minister, caters for all matters pertaining to scientific research and technology development activities undertaken at Iraq's public universities and their research centers.

The law that established MoHESR[12] invested it with legal means for fulfilling a variety of functions, including formulation of education and research policies, in response to established national objectives. This same law also assigned MoHESR the responsibility of devising plans and specifying requisite human and financial resources and timeframes needed to meet goals of its policies. The Ministry was also required to adopt specific performance indicators with view to monitoring and evaluating outcomes of its activities. Article 4, in the above-mentioned legislation, required the establishment of organizational structures to cater for "Studies, Planning, and Monitoring," "Research," "Caring for External Relations," "Providing Legal and Administrative Support," and "Supervision and Scientific Assessment."

In practice, however, it seems that the objectives of establishing the Ministry, especially as decreed in articles 2 and 3 of the Law, were subjected to somewhat minimalist interpretations with certain high-level responsibilities reduced to mere bureaucratic functions. Such interpretations tend to be explained as aimed at simplifying procedures within difficult conditions met on numerous fronts.

MoHESR is home to at least five commissions and centers including the Iraqi Commission for Computers and Informatics (ICCI), which appears to have performed key roles in capacity building on the information technology front.

An assessment of the Management of the Education System in Iraq[13] conducted in 2010 attempted to characterize issues that hampered the Ministry's operations. Issues were compiled by this assessment with regard to steps that the Ministry will need to take in order to comply more closely with the numerous challenges facing higher education scientific research in Iraq. Frame 3.1 presents brief snippets of this assessment.

**Frame 3.1 Summary Results of an Assessment of Iraq's Higher Education and Research**

Strengths possessed by the national higher education system in Iraq include comprehensiveness of relevant legal framework to the point of inflexibility. Weaknesses, on the other hand, are manifested with particular regard to the performance of the policy and planning and monitoring functions, which suffer poor management and are in need of radical revisions. Structural units concerned with these functions must be restructured with a new scope for their responsibilities.

Quality assurance of higher education institutions and their programs suffers excessive reliance upon external agencies, which often skirt around local conditions and salient national priorities, and produce limited impact.

A Ministry of Science and Technology (MoST) came into being as a result of the dissolution, in August 2003, of the Ministry of Atomic Energy. Around 1500 of the latter ministry's engineers and scientists, reportedly engaged in defense industry projects were transferred to the newly established MoST, which was mandated to "work with public and private industry, governmental agencies, and educational institutions to lead scientific and technological development for the purpose of fostering the restoration and improvement of Iraq's infrastructure and industrial base." The Ministry's mandate covered application of science and technology in a variety of fields including fossil fuel/solar energy, environmental protection, ICTs, agriculture and food, chemical/industrial development. The Ministry is further required to liaise with MoST and the country's universities, in order to ensure optimal use of resources.[14] This Ministry was, however, abolished in 2015, with its functions relegated to the MoHESR.

### 3.2.1.3.3 The KRI Ministry of Higher Education and Scientific Research

This Ministry adopted a short-term strategy with objectives including the establishment of a research center within every university. Indeed, as of 2006, every public university in the Kurdish region was stated as possessing a research center. However, lack of appropriate facilities and crucial laboratory equipment, added to the absence of research plans, hindered their activities, and although public universities sponsored a number of research projects, allocated budgets were invariably insufficient, resulting in failure to meet set objectives.

Research programs supported by the Ministry were intended to raise scientific standards with emphasis on multi- and transdisciplinary projects. International cooperation was also to receive special attention. This is underscored by a recent report posted by the Ministry. The report includes a chapter dedicated to the promotion of scientific research in which the need to foster multidisciplinary collaborative research with the objective of "bringing together researchers from various centers of excellence." The report is also critical of the fact that thousands of the Region's scientists, professors, and research students carry out research, mostly in isolation or in small groups "with little collaboration between them or with the outside world," with the end result that "research funds, capital equipment and institutional infrastructure are fragmented."

This state of affairs is blamed for the paucity of research output and its minimal impact on life in the region. Poor management structures in the region's universities, "inappropriate PhD training pathway, a restrictive and a wholly inadequate Higher Education Law" are also held responsible.

A process of reform with urgent radical changes is considered by the Report as the best means possible for putting an end to the state of affairs, it describes as a crisis situation. With this in mind the initial phase of the reform process would require that research students and their supervisors strengthen their links with the outside world through bilateral visits and joint projects. Sending professors on sabbaticals and students for training abroad are means that would be resorted to in order to expand existing research networks, update knowledge and sharpen skills. The following paragraphs discuss the creation of novel PhD training pathways, an issue that features prominently in the Ministry's plans.

Key to implementing programs aimed at the revitalization of the Higher Education and Scientific Research system in the Kurdistan Region is the Human Capacity Development Program (HCDP) in which the Ministry will send students abroad to world universities and centers of excellence to pursue graduate degrees at the M.A. and PhD levels (Frame 2.3).

#### 3.2.1.3.4 The Kurdistan Institution for Strategic Studies and Scientific Research

The first research institution in Kurdistan, "The Kurdistan Institution for Strategic Studies and Scientific Research (KISSR)" was established in 2008 by the Ministry of Higher Education and Scientific Research in Kurdistan and the Region's Council of Ministers. KISSR was established to undertake high quality research in the sciences and humanities as well as undertake technological development projects in priority area, including environmental protection and monitoring food safety.

By the year 2012, around 110 students are reported as having completed research projects at laboratories on KISSR's four research centers, specializing in "Geology & Organic Petrology," "Microbiology and Health," "Quality Control, Environment, and Agriculture," and "Strategic Studies."

### *3.2.1.4 Status of Selected Research Centers and Units in Iraq*

A report published in 2013 included the results of a survey conducted in 2011, to which 47 research centers and units across the country, but excluding governorates in the Kurdistan region of Iraq, responded to a questionnaire concerning various aspects of their work, resources, and obstacles they had to deal with.

#### 3.2.1.4.1 Financial Resources Dedicated to Research Activity

Fig. 3.1 presents a snapshot of resources dedicated to four broad areas of research activity, namely energy and the environment sciences, medical and life sciences, applied sciences and social sciences and humanities, as well as research on local and regional matters. It is essential to note at the outset that data concerning financial resources available to research centers and units were provided by only 22 institutions. Thus, the distribution presented by this figure may be regarded only as a rough indication of how these four areas appear to share resources dedicated to national research activity. With this limitation in mind, it appears that the medical and life sciences, on the one hand, and the social sciences, on the other, were allotted only 6% of total research resources each, in or around the year 2010. On the other hand, research centers active in the energy and the environment, on the one hand,

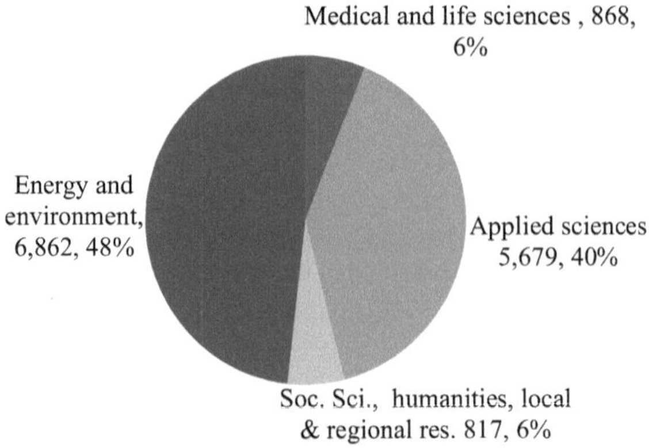


**FIGURE 3.1** Distribution of financial resources across research fields (US$ thousands).

and the applied sciences, on the other, were allocated 48% and 40%, respectively, of all resources reported as spent on all scientific research undertaken by responding centers and units.

With regard to allocations to research in the applied sciences, it is essential to underline the fact that although overall annual allocations seem to be quite meager, considerable disparities are evident in spending appear on various domains. Thus, the Remote Sensing Centre of Mosul University was allocated an annual budget of around US$4.2 million, in or around 2010, although three other research centers and units active in date palm research, polymer research and die design and manufacture were allocated a total of around US$224,000. The same is true within the group of research centers and units active in energy and environmental research, where the Marine Science Centre was allocated a total of over US$6 million, whereas the remaining three research centers and units within this group were allocated a total of US$723,000. Indeed, the budget allocated to a research center specializing in energy and fuel research amounted to a mere US$24,000.

Although the above disparities might be incidental, e.g., due to the fact that particular research centers were established new labs and needed special equipment, the general picture is that inconsistencies seem to exist and that, in general research allocations are quite meager.

### 3.2.1.4.2 Human Resources Active in Research

Fig. 3.2 presents a view of the distribution of Iraq's human capital across fields of research, indicating, perhaps expectedly, that a large proportion, around 32%, of all staff members in the country's research centers and units appears to work within centers and units active in research on the social sciences, humanities. Research facilities dedicated to environmental research, on the other hand, employ around 24% and those active in medicine and

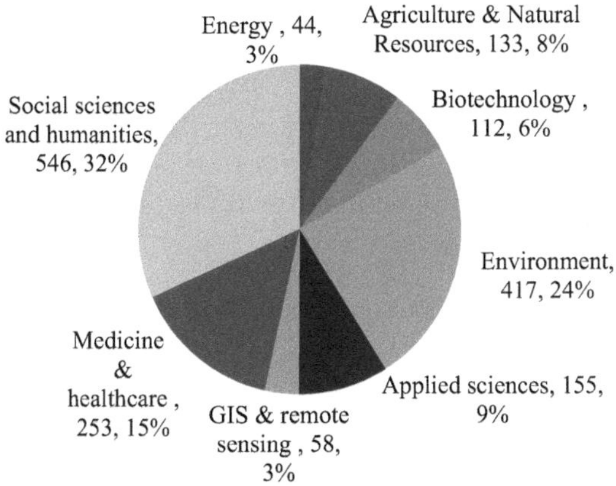


**FIGURE 3.2** Distribution of staff members across fields of studies and research.

**TABLE 3.1 Percentages of Females Among Staff Members With Qualifications at the BSc Level and Above, Among Researchers and With Respect to Totals**

| Research Domains | BSc and Above | Researchers | Total Staff |
|---|---|---|---|
| Medical and healthcare | 67.1 | 52.9 | 60.5 |
| Biotechnology | 47.3 | 44.6 | 48.2 |
| Social sciences and humanities[15] | 47.7 | 39.1 | 50.7 |
| Environmental | 40.8 | 30.9 | 43.4 |
| Applied sciences | 38.5 | 23.4 | 18.1 |
| Geographic information system and remote sensing | 38.0 | 18.2 | 37.9 |
| Agriculture and natural resources | 22.4 | 16.1 | 23.3 |
| Energy | 8.1 | 7.1 | 9.1 |
| Totals | 44.0 | 35.7 | 43.7 |

healthcare, the applied sciences, and agriculture and natural resources employ much smaller percentages, estimated at 15%, 9%, and 8%, respectively.

Table 3.1 presents percentages of female workers within research centers and units in Iraq that took part in the survey. Indicating that, at around 53% of the total number of researchers, females tend to predominate in centers and units dedicated to medical and healthcare.

The proportion of female researchers in research centers and units dedicated to biotechnology, the social sciences and humanities as well as the environment is considerable, estimated at around 45%, 39%, and 31% of the total number of researchers working in these domains altogether. In conclusion, at around 36% female researchers constitute a sizeable proportion of the total volume of human capital engaged in research within Iraqi centers and units, which responded to the survey.

#### 3.2.1.4.3 Obstacles Faced by Research Centers and Units

Table 3.2 presents a summary of responses to survey questions intended to identify the order of importance of various obstacles met by research centers and units in Iraq. As might well be expected, most responding research facilities, assigned a high rank to difficulties met due to low budgetary allocations. The extent of cooperation and coordination came next in importance, whereas initial and further training as well as infrastructural facilities were assigned similar levels of importance as obstacles met by responding research centers and units. Interestingly, management policies were accorded a lowest level of importance as an obstacle to research activity.

### *3.2.1.5 Concluding Remarks*

Iraq lacks a comprehensive science, technology and innovation policy. Promising attempts were reported in 2013[17] of MoHESR taking steps toward formulating a national research strategy and of similar moves made by the corresponding Ministry in the Kurdistan Region with emphasis on researcher training as well as cooperative links with S&T institutions abroad. However, it appears that few concrete outcomes were achieved in either case.[18] Additionally, it would seem that attempts at nurturing innovation in priority production and services sectors with the aim of enhancing the country's competitiveness and growth as well as creating more jobs for its youth have received little or no attention by successive Iraqi governments.

Although merely attempting to draft research strategies is expected to produce short-term benefits, e.g., through exposing participants and their institutions to parameters in need of priority attention and sensitizing decision makers to salient issues; long-lasting benefits may only be attained through consolidated policy regimes, thoroughly researched and debated prior to being adopted and earnestly implemented.

As indicated above, the STI institutional scene in Iraq is also in need of radical overhaul. Thus, with the exception of isolated attempts made in certain fields of agriculture and mining few initiatives appear to have been taken to encourage the higher education and scientific research communities to improve the country's economy. In particular, an exhaustive search in available literature could not identify any serious attempts at establishing science and technology parks, incubation schemes or centers of excellence; all of which now considered essential to bridging STI demand and supply sides.

**TABLE 3.2** Order of Importance Assigned by Groups of Responding Research Institutions to Categories of Obstacles

| Obstacles | All Responding Institutions | All Applied Science Institutions | Selected Applied Sciences | Agricultural and Allied Sciences | Medicine and Healthcare | Energy and Environmental Research | GIS and Remote Sensing | Social and Sciences and Humanities | Average Importance |
|---|---|---|---|---|---|---|---|---|---|
| Budgetary resources | 1 | 1 | 1 | 1 | 1 | 1 | 3 | 1 | 1.25 |
| Cooperation and coordination | 4 | 2 | 2 | 2 | 5 | 2 | 4 | 3 | 3.00 |
| Initial training | 2 | 5 | 5 | 4 | 2 | 5 | 5 | 2 | 3.75 |
| Further training | 3 | 5 | 4 | 5 | 4 | 6 | 5 | 2 | 4.25 |
| Infrastructure | 6 | 4 | 5 | 3 | 3 | 7 | 2 | 5 | 4.38 |
| Access to information | 7 | 3 | 3 | 6 | 7 | 3 | 1 | 6 | 4.50 |
| Critical mass[16] | 5 | 6 | 4 | 4 | 8 | 8 | 6 | 4 | 5.63 |
| Management policies | 8 | 7 | 6 | 5 | 6 | 4 | 7 | 7 | 6.25 |

Order of importance: 1 indicates the highest level of importance assigned to an obstacle, whereas 8 signifies the lowest level.

However, in order to formulate optimal R&D strategies, linking supply and demand sides and harnessing STI efforts to national development policy objectives, there is a strong need for a national STI governing body, suitably situated within top policy and decision-making circles of government. Thus, the establishment of an Iraq science, technology and innovation council (ISTIC), should help set priorities as well as allow collective and rational resource allocation. The lack of an agency equipped with suitable legislative instruments for funding STI activities is also a point in needs of special attention through future initiatives.

It is also essential that legislative and regulatory frameworks that have gone unchanged for decades are finally reformed, infusing greater openness, improved resource levels and distribution modalities and wider scope for innovative inputs targeting socioeconomic development, with employment and enterprise competitiveness as top priorities.

In summary, the backlog of at least four decades of sanctions, all-out war, armed conflict, forced migrations with concurrent mismanagement of efforts at technology acquisition, development and adaptation may only be addressed through a comprehensive STI policy with adequate and sharp-cutting and thorough legislative and regulatory reforms.

### 3.2.2 Jordan's STI System, Policies and Institutions

The following paragraphs are devoted to a review of policy documents and initiatives specifically aimed at STI reforms as well as specific STI national priorities as identified by Jordan's STI community.

Several studies were produced by the Higher Council for Scientific Research (HCST) over the past three decades, with the objective of surveying S&T capabilities. One of the more comprehensive was conducted in the late 1990s.[19] Reporting on the results of this study a report titled "The National Scientific and Technological Requirements and Potential (NSTRP)" outlined a set of policy and strategic measures and drew up lists of priority R&D areas.[20]

The survey designed as basis for the NSTRP study enabled an examination of the requirements of a considerable number of S&T "end-users." Based on the results of this survey,[21] around 3000 institutions were selected for follow-up, including all government ministries, "medium" and "large" industries, selected public sector institutions and public as well as private universities. Conditions faced by both public and private sector institutions were considered and constraints encountered were thoroughly discussed. The study made a number of recommendations aimed at meeting S&T requirements in both the public and private sectors.

Clearly, the benefits of such exercises reside in enhanced communications among partner institutions as well as in the opportunities they provide for rethinking current policies and for the infusion of new ideas into future strategies and action plans.

Policy guidelines and directives based on the NSTRP exercise addressed information, human resources and R&D activity. A number of "sectoral bundles" or capacity-building programs were embedded within these elements, highlighting specific short- and long-term objectives and support toward specific R&D projects related to mineral resources, water, energy, and the environment. Frame 3.2 includes a number of recommendations included in the study that retain their relevance to date. Based on the NSTRP exercise policy guidelines and directives were formulated concerning improved legislation, financial arrangements, coordination, and administration of R&D activity.

### *3.2.2.1 The Science, Technology and Innovation Policy 2013–17*

The latest national STI policy document was produced by HCST.[23] Covering the period 2013–17,[24] it is the first such document to include directives aimed at fostering innovation as well as science and technology. A special committee was entrusted with producing this document on the basis of expert meetings and missions to countries, including Turkey, Finland and South Korea. Areas that received special attention in this latest STI policy document included:

- Institutional frameworks and actors;[25]
- Legislative and regulatory arrangements;

**Frame 3.2 Summary Recommendations of the "National Scientific and Technological Requirements and Potential" Study in Jordan[22]**

A high order of "priority" is accorded to capacity building in (a) management science, (b) engineering, (c) mathematics and computer science, (d) crafts, vocations and trades (e) medical and health sciences. Institutions of higher education and S&T centers were urged to respond to national requirements for basic and applied research in the fields of natural and engineering sciences.

S&T centers and institutions of higher education were requested to improve activities related to the implementation of standards and specifications, quality control and assurance, inspection and testing facilities linked to domestic and international accreditation. Activities aimed at S&T and related managerial know-how, and information and documentation, should also be revitalized.

S&T centers and engineering faculties at Jordan's universities were urged to assist in providing maintenance services and help in formulating plans for the domestic manufacture of spare parts for crucial equipment.

The need for greater assistance by S&T centers and engineering faculties in training the skilled human capital needed in the fields of maintenance is also highlighted.

NSTRP regarded accurate assessment of S&T potential in the national institutions as particularly essential. Based upon this assessment, human capital requirements, hardware needs and financial allocations would be established.

- Human resources;
- Jordan's science, technology and innovation infrastructures and overall environment.

The Policy's vision statement emphasizes a national quest for socioeconomic development and a knowledge economy. Its mission statement, on the other hand, underlines the notion of enabling the country's scientific and technological community to harness science, technology and innovation in the service of the national economy, thereby enhancing standards of living and social welfare of Jordan's citizens. An analysis of strengths, weaknesses, opportunities and challenges facing the country's STI system was undertaken with view to identifying priority actions that will further Jordan's quest for knowledge-based socioeconomic development in the 21st century.

As a result of this exercise, HCST teams of experts identified five strategic objectives that ought to be pursued through implementation of an equal number of national programs during 2013–17. Table 3.3 presents a list of these objectives and programs, the design of which was based on consideration of a variety of issues including steps taken by government in facing a multitude of national economic and social challenges, experience accumulated through formulation and implementation of past science and technology policies. The status of science, technology and innovation in Jordan, compared to that in the region and worldwide was also taken into account. Similarly, availability of political commitment and ability to acquire international assistance were also invoked in arriving at the list of objectives and programmed included in Table 3.3.

Modalities proposed for implementing Jordan's National Policy were designed with emphasis on priorities including the need to guarantee adoption of the Policy's priorities, by the national scientific and technological community as well as those concerned with production and related government activities. Establishing STI knowledge partnerships and networks at the national and international levels was also considered among the above priorities.

Other objectives adopted by this Policy document highlight innovation as basis for entrepreneurship and investment as well as the need to commercialize R&D outcomes and the quest of excellence in education, training and skills acquisition.

A total of 24 specific projects were designed by the Policy's working groups, with view to achieving designated goals. An executive plan, performance evaluation criteria and monitoring measures with lines of responsibility concerning were put forward for each project. The cost of implementing all 24 projects was put at JD10 million, or around US$14 million.

### *3.2.2.2 Jordan's Research and Development Priorities*

More specific orientations for STI activity were set forth in an earlier HCST document titled "Defining Scientific Research Priorities in Jordan for the years

**TABLE 3.3 Objectives and Programs of Jordan's National Science, Technology, and Innovation (STI) Policy Covering the Period 2013–17**

| Objectives | Programs | Budget US$ Thousands |
|---|---|---|
| Consolidated institutional frameworks and channels ensuring collaboration among components of Jordan's STI system as well as streamlined policies and legislative instruments, ensuring their effective implementation, emulating proven models. | Institutional, policies, and legislations | 1993.0 |
| A viable STI infrastructure with advanced equipment, manpower trained in advanced areas of science and technology, with emphasis on partnerships with national firms, international networks, proper use of grants and foreign direct investment. | Manpower and infrastructure | 2521.1 |
| Sustained support for institutes of higher education and research centers, attracting and retaining qualified cadres and national talents, principally through government support. | Government financing for higher education and research centers | 563.4 |
| Enhanced overall productivity, competitiveness and increased support for research and development (R&D) by the private sector, through acquisition of proven advanced technologies and participation in the development of selected technologies in their initial stages of their development. | Productivity, competitiveness, and private sector R&D contributions | 6056.3 |
| Innovation promoted and supported through establishment of innovation clusters. | National innovation | 2574.6 |

Adapted from the National Science, Technology and Innovation Policy 2013–2017. Downloaded in February 2017 from the HCST website. Policy document (Arabic). Downloaded in February 2017 from the HCST website at: http://hcst.gov.jo/sites/default/files/policy-and-stratgy.pdf.

2011–10."[26] As its title indicates, this document aims at identifying R&D issues within two groups with regard to their priority for comprehensive socio-economic development; the first including issues and projects of top priority whereas the second comprises projects of lesser priority. Lists of issues and projects designed in aid of specific sectors and areas of social and economic activities were drawn up on the basis of extensive surveys, utilizing Delphi methodology in certain stages. As indicated in Table 3.4, 148 research topics were identified within 14 domains of social, economic and technology related fields of R&D activity, with over 400 research issues to be addressed within the

**TABLE 3.4** Numbers of Issues and Projects Listed as Top Priorities Within the National Research Priorities Document for the Period 2011–20

| Research Domains | Research Topics | Research Issues |
|---|---|---|
| Antiquities and Tourism | 12 | 28 |
| National Security | 9 | 26 |
| Information and Communication Technology | 10 | 29 |
| Culture, Arts and Media Humanities | 14 | 29 |
| Social Sciences and Languages | 6 | 32 |
| Financial Economic and Administrative Sciences | 10 | 30 |
| Basic Sciences | 8 | 31 |
| Islamic Sciences | 13 | 30 |
| Educational and Teaching Sciences | 10 | 29 |
| Agricultural and Veterinary Sciences | 10 | 34 |
| Political and Legal Sciences | 10 | 30 |
| Medical and Pharmaceutical Sciences | 15 | 28 |
| Water and Environment Sciences | 11 | 30 |
| Engineering Sciences | 10 | 30 |
| Totals | 148 | 416 |

period 2011–20. The following paragraphs present highlights of research topics and issues underscored by this document as top priorities.

Top priority research issues to be pursued within each domain average at around 30 per domain, i.e., area of application or field of technology. Issues assigned top priority cover a wide range research questions within each domain. Thus, national security research projects include, on the one hand, the manufacture of surface-surface missiles with a range of 150–300 km, as well as "[developing] the security services capabilities in investment operations nationally and regionally" and " … [developing] the role of the armed forces and security agencies in national development."

A review of top priority research topics identified within the domains of "Information and Communications Technology," "Agriculture and Veterinary Science" and "Engineering Sciences," is rather informative. Thus notable among top priority topics under "Information and Communications Technology" are research issues related to information security and cybercrime, Arabic language processing as well as research issues designed to promote ICT applications in the management of health, energy, the environment, agriculture and in promoting education and the knowledge society.

Within top priority topics listed under "Agriculture and Veterinary Science," research issues concerned with improving sustainability in various aspects of agriculture and animal husbandry are quite prominent, along with research aimed at infusing the agricultural sector with innovative technologies.

Under "Engineering Sciences," top priority topics are concerned with a host of research issues that are aimed at the development of renewable solar and wind energy systems, desalination and oil shale extraction.

Notable too is concern with uranium mining, presumably from phosphate rock already being extracted for a variety of industrial application as well as export as a raw commodity.

Eight wide research topic areas are considered top priority within the domain labeled "Basic Sciences,"[27] including Materials, Nanotechnology, Environment and health, Energy and the Environment in addition to Mathematics and Statistics. Within the domain labeled "Environment and Health" the HCST Research Priorities Document emphasizes research on pollution including:

- Determination of levels of organic and inorganic pollutants in the cities;
- Transport of pollutants in the air;
- Identifying unconventional sources of pollution;
- Pesticide residues in vegetables.

A highly commendable aspect of the HCST Research Priorities Document concerns its inclusion of a wide range of research topics within social sciences as top priorities. Thus, within the domain labeled "Islamic Sciences," the HCST Research Priorities Document lists a number of research topics clearly intended to encourage research leading to moderate interpretations of Islamic religious edicts. Examples of such topics include: "Foundations of Ijtihad[28] and fatwa[29]," "Methodology of dealing with the Quran," "Methodology of dealing with the Sunnah."[30] A research topic labeled "National Security" under "Islamic Sciences" includes research issues such as "The menace of extremism and fanaticism on national security," "The role of religious education in national security," "Ideological fanaticism: causes and remedies," "Extremism and its impact on the fragmentation of the nation," and "The culture of violence: its sources and means of prevention from Sharia[31] perspective." That such issues are considered top research priorities by Jordan's STI establishment, is a clear indication of its intention to free Islamic teachings and practices from extremist orientations and initiate enquiry and analytical approaches as basis for religious edicts and related practices.

Top priorities within the domain of the "Social Sciences" are similarly forward looking, indicating welcome involvement by the country's research establishment with key socioeconomic problems including unemployment, poverty, corruption, drug abuse and the conditions of women. Thus, top priority research issues in the domain of the "Social Sciences" include, among others:

- Socioeconomic and political consequences of poverty and unemployment;
- Measurement of corruption and favoritism in Jordan;
- Drug abuse in Jordan;
- The impact of poverty and unemployment on marriage, single, and divorce rate;
- Causes of women's harassment;
- Evaluation of anticorruption organizations.

Equally relevant are research issues considered as priorities under the heading "Education and Teaching Sciences," as they include:

- Academic attributes of students, unemployed and employed graduates from universities in Jordan: considerations for relevance;
- Compatibility of higher education graduates with market requirements;
- Continual learning programs designed for coping with the rapid developments on several fronts;
- Compatibility of continual learning programs with local community development needs.

In summary, Jordan appears to award a good deal of attention to promoting R&D activity within specific priority domains. Nevertheless, one of the main shortcomings of Jordan's efforts toward promoting R&D and related activities is the absence published information concerning outcomes of planned activities. Thus, in reporting on outcomes of its 2006–11 S&T plan, the subsequent 2013–17 S&T document merely states that the rate of implementation of the previous plan was estimated at around 50%.[32]

Thus, although it would generally be expected that an institution announcing plans for promoting one of its core activities would attempt to report on progress attained, possibly even provide a critical summary of outcomes, in preparation for similar future ventures, no such reports could be found upon reviewing available sources.

### *3.2.2.3 Jordan's Innovation Strategy 2013–17*

Jordan possesses an active but highly fragmented innovation system with numerous public and private stakeholders, coordination among which is almost absent. Additionally, the country's innovation system suffers inadequate access to relevant information as well as inefficient financial managements.[33]

The principal goal of Jordan's innovation strategy 2013–17 was to transform the country into a sustainable innovation-based economy, through spreading a "culture of innovation, research and development," training specialized human resources and creation of an innovation-friendly business environment.[34]

The Strategy document underscores a number of basic principles regarding innovation and the main requirements for setting up an innovation-based economy. Next, the Strategy's Executive Plan lists a series of activity clusters whose implementation should guarantee achievement of the Strategy's aims, a principal one of which is to establish a National Innovation Centre (NCI). See the following section on "Institutional entities aimed at promoting innovation in Jordan."

Additionally, the Strategy's Executive Plan lists six innovation clusters to tackle a total of 52 projects within the following domains:

- Medical services and pharmaceutical industry;
- Information and Telecommunications technology;
- Clean technologies;
- Architecture and engineering services;
- Education and career guidance services;
- Banking and financial services.

Criteria used in selecting innovation clusters were essentially based on the following attributes:

- Contribution to the GDP;
- Job opportunities created;
- Educational qualifications required;
- Total exports and imports of the cluster in relation to turnover of the clusters;
- Ability to pioneer entrepreneurship;
- Mutual impact/benefits for other clusters.

Fig. 3.3 presents a view of allocations earmarked for the various clusters, indicating overwhelming support for projects within the medical services and pharmaceutical industry cluster, followed by those dedicated to projects

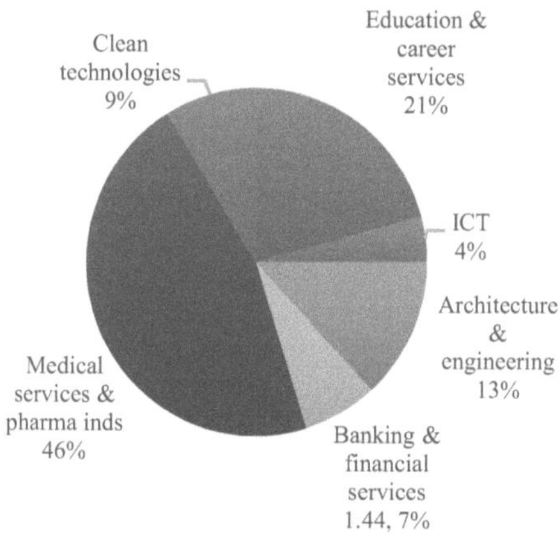


**FIGURE 3.3** Financial allocations to innovation clusters out of a total of US$20.4 million.

within the "education and career guidance services" and "architecture and engineering," clusters. Surprisingly, the "clean technologies" and "ICT" clusters are allowed much lower financial support, at 9% and 4%, respectively, of the total allocations amounting to US$20.4 million.

Innovation strategy's projects are mostly directed toward studies and assessment of various aspects linked to innovation within the above clusters and not a great deal of attention seems to be allocated to financial support, incentives and legislative reforms in aid of improved innovation ecology, e.g., venture capital schemes.

The very fact that HCST is host to NIC and charged with implementation of NIS is indicative of an inherently S&T-based approach to promoting innovation.

### 3.2.2.4 *STI Expenditure by Jordan*

The above-mentioned sum of around US$14 million, allocated for implementation of programs and projects outlined by Jordan's most recent STI Policy might well be expected to fall short of attaining its numerous and cumbersome objectives. An additional sum of around US$20.4 million is pledged by the National Innovation Strategy.

Failure to allocate sufficient funding to STI-related activities is evident upon reviewing previous official planning documents. Thus, a plan authored in response to a Royal decree issued in 2005, with view to launching a series of wide socio-economic reforms in Jordan, proposed a set of numerical targets, regarding R&D funding, scientific publications and patenting activity. As Table 3.5 indicates, R&D spending was supposed to have reached 1.5% of GDP by 2017, whereas patenting

**TABLE 3.5** Actual and Targeted Performance Indicators for Science, Research, and Innovation

| Performance Indicator | In 2003 | Target 2012 | Target 2017 |
|---|---|---|---|
| Expenditure on research and development as a percentage of GDP (domestic funding) | 0.34 | 1.0 | 1.5 |
| Number of internationally published science papers indexed by Thomson Reuters | 485 | 1300 | 2500 |
| Number of patent applications submitted by Jordanians since 2000 | 246 | 1300 | 5000 |
| Number of manuscripts deposited at the National Library since 1994 | 22,550 | 42,000 | 58,000 |

From National Agenda, 2006–2015; "The Jordan we strive for; " page 33. Retrieved in February 2017 from: http://inform.gov.jo/Portals/0/Report%20PDFs/0.%20General/2006-2015%20National%20Agenda.pdf.

activity was to have increased almost threefold in the space of 1 decade and then twofold within the period 2012–17. Based on available information the above-mentioned targets were not eventually realized.

### *3.2.2.5 STI Institutions in Jordan*

Two major Jordanian institutions concerned with science, technology and innovation policy setting as well as implementation of related activities stand out. Namely, the HCST and the Royal Scientific Society (RSS). The following sections present a summary review of their scope of activities.

#### 3.2.2.5.1 Higher Council for Science and Technology

The Higher Council was established in 1987 as a governmental policymaking organization with view to creating a national scientific and technological base in support of Jordan's socioeconomic development priorities. HCST is also mandated with identifying S&T priorities and funding science and technology research and development activities.[35] See Frame 3.3.

Building national scientific and technological capabilities and their continuous development through increasing awareness of applied scientific research are essential elements in the HCST mission.[37] Thus, the Higher Council provides financial support for scientific and technological activities, networking of researchers, research foundations, productivity institutions. Many of its activities are aimed at the acquisition of technological capabilities and their implementation in seeking national development goals.

**Frame 3.3 Research Centers of the Higher Council for Science and Technology**

Several research centers were affiliated to HCST prior to the year 2010. Research undertaken by these centers emphasized human resources development, medical research on specific health issues, research targeting technology acquisition, adaptation and development in areas including nanotechnology, biotechnology and energy, as well as arid zone development.[36] In 2010, however, the National Centre for Research and Development (NCRD) was established, with view to streamlining and harmonizing research that was undertaken within the above domains, with the exception of human resources development and medical research. Hence, NCRD presently covers R&D in objective to develop scientific capabilities in the areas of nanotechnology, biotechnology, energy, Badia research. NCRD is at the same time affiliated to two active centers, namely The National Centre for Diabetes, Endocrine and Inherited Diseases (NCDID) and The National Centre for Human Resources Development (NCHRD) as well as two other centers which are being established, namely The National Center for Innovation (NCI) and the King Abdullah II bin Al Hussein Center for Nano-Technology.

Information retrieved in February 2017, from the NCRD website at: http://hcst.gov.jo/

The Higher Council's vision requires that Jordan achieve international recognition in frontier science and technology, with capability for their implementation in attaining sustainable socio-economic development. HCST's mission includes formulation of the national science and technology policy, its implementation strategies as well as related plans and programs. Additionally, HCST plays a catalytic role in:

- enhancing and accelerating the interaction among the S&T community, production and services sectors, and government;
- transforming science and technology ideas into commercial enterprises through supporting innovation, entrepreneurship, and technology commercialization.

The HCST mission includes setting up the national science and technology policy, strategies, plans and programs, and plays a catalytic role in enhancing and accelerating the interaction among the S&T community, production and services sectors, and government. This is in addition to transforming science and technology ideas into commercial enterprises through supporting innovation, entrepreneurship, and technology commercialization.

#### 3.2.2.5.2 Royal Scientific Society

Jordan's Royal Scientific Society (RSS) was established in 1970.[38] Its vision statement declares its aim to assume a leading position in science and technology, emphasizing quality research, consultations, technical services, and specialized training linked to innovation, with view to the development and prosperity of Jordan as well as providing science and technology linked services to the region. In its mission statement, RSS is also stated as seeking to promote technological change through its research, value-added services, and technology transfer activities it conducts in priority areas. In pursuing various elements of its mission, RSS is stated as capitalizing on forging strategic and productive partnerships in pursuit of building a strong and sustainable knowledge-based economy.

RSS is presently the largest applied research institution and technical service provider in Jordan, with more than 600 staff members working in RSS components listed in Frame 3.4 as well as a host of other technical services departments addressing issues that include marketing, public and international relations, technical services and maintenance, as well as quality assurance.

Major sources of RSS funding include charges for technical services it provides as well as fees for studies and research contracts. RSS also receives some government funding in addition to donations from local institutions, and technical assistance from foreign governments as well as regional and international organizations. Major fields of expertise cultivated by RSS are listed in Frame 3.4.

#### 3.2.2.5.3 The National Centre for Innovation

HCST and the Ministry of Planning came up with the idea of establishing this center while their teams were engaged in elaborating the National Innovation Strategy 2013–17. The notion was that a specialized body was needed to deal with

**Frame 3.4 Components of the Royal Scientific Society and Their Major Areas of Interest**

**RSS research centers**

Information Technology Centre
Industrial Chemistry Centre
Building Research Centre
Jordan National Metrology Institute
Environmental Research Centre
Technology Transfer Centre
Innovation Centre of Jordan
Mechanical Design and Technology Centre

**Areas of interest**

Construction engineering;
Maintenance and repair of medical equipment;
Environmental Studies;
Water and wastewater treatment;
Radiation monitoring;
Air quality management and control;
Information Technology;
Analysis and characterization of materials and food products;
Nondestructive testing;
Quality control and certification.

From Jordan's Government website. Retrieved in February 2017 from: http://www.jordan.gov.jo/wps/portal/

the totality of issues linked to innovation at the national level; because HCST dealt with scientific and technological innovations, rather than commercial and entrepreneurial innovations. A request for the establishment of NCI under the HCST umbrella was submitted to the World Bank in October 2013.[39]

The idea behind establishing NCI was essentially due to the realization that numerous institutional entities concerned with innovation, necessitating coordination of their activities and evaluation of their collective impact. More specifically, NCI is expected to:

- Ensure complementarity, rather than inefficient duplication, with regard to administrative, financial and technical services aimed in support of innovation;
- Coordinate stakeholder activities, including government ministries, donor organizations, scientific and research bodies, universities, research centers, regional and local incubators, start-ups and established entrepreneurs, thereby bringing about improved economies of scale;
- Collect relevant national statistics on the impact of various innovation initiatives, highlighting successes and pointing out areas in need of improvement, thereby enabling authorities to assess progress and devise measures for further enhancing the role played by innovation in the country's socioeconomic development.

In essence, NCI is intended as an advisory and coordination hub for activities related to innovation and private sector development. NCI is also expected to provide legal and regulatory advocacy as well as advisory services to small and medium enterprises (SMEs), whereas collecting feedback to enrich future policy directives and government initiatives aimed at promoting innovation.

#### 3.2.2.5.4 Business Incubators in Jordan[40]

Two incubators were initially established in Jordan during 2005; one at the Royal Scientific Society supporting engineering and industrial projects, and the other at the University of Jordan supporting agro-business projects. One of the key objectives of the early incubator projects was to create a model for further incubation facilities across the country. Several universities in Jordan went on to establish incubation facilities since then, including Philadelphia University.[41] Other universities followed suit with incubators now established at the universities of Yarmouk, Mou'atah, and within the Al Hassan Industrial Estate.

As several incubators were established, several showing tangible outcomes, many more institutions went on to establish their own incubation schemes. As a result, Jordan is now host to a considerable number of incubation and enterprise promotions organizations, governmental, nongovernmental, mostly in partnership with overseas aid and development agencies.

However, as indicated in an early study performed in 2009 by the German Agency for Technical Cooperation (GTZ) on the behalf of the German government,[42] a more solid incubation framework was needed; with stable financing, quality management procedures and permanent evaluation. The GTZ Study's authors indicated that it would be advisable to improve the performance of existing incubators before new ones were created. Focus would need to be placed on services and added-value provided by the incubators as well as improvements to framework conditions, including higher public investments, enterprise-friendly legislation and reduced administrative hurdles. Additionally, the GTZ study recommended that a nationwide network for Jordanian incubators would help facilitate mutual learning and strengthen their position toward public and policy.[43]

#### 3.2.2.5.5 Regional STI Facilities Hosted by Jordan

Jordan has traditionally sought to utilize its international links and location within the Arab region and Middle East in order to perform a central role on many fronts. Over the past few decades, it succeeded in hosting several regional organizations dedicated to economic development. More recently, however, it became home to two regional ventures specifically aimed at promoting science and technology. These are namely the ESCWA Technology Centre (ETC) and the Synchrotron light for Experimental Science and Applications in the Middle East (SESAME) Physics Laboratory.

The ETC was created in 2011 and has since been hosted by Jordan. It is essentially an organ of the United Nations Economic and Social Commission for Western Asia (UN-ESCWA) dedicated to promoting scientific research, technology transfer, adaptation and development as well as innovation with view to furthering socio-economic development in the Commission's member countries, which presently include 18 Arab countries in the Middle East and North Africa.

On the other hand, the SESAME laboratory is dedicated to research on the interaction of intense electromagnetic radiation with matter. The United Nations Education, Science and Culture Organization took steps to establish this facility as far back as 1999, as an independent research laboratory in the cause of promoting peace in the region through scientific collaboration. The SESAME project suffered numerous delays, and upsets, including several construction delays as well as the loss of at least two Iranian researchers, allegedly murdered by Israeli agents in separate attacks. Recent reports indicate that it was inaugurated in May 2017 by King Abdullah II. However, out of the seven light beams it is supposed to have in operation only two have in fact been set up.

### *3.2.2.6 Concluding Remarks*

Documents produced by HCST and other Jordanian institutions concerned with enhancing the country's STI capabilities acknowledge a number of strengths and weaknesses,[44] chief among the former are the following:

- Relative political stability;
- A considerable stock of qualified human capital as well as a youthful population;
- Reasonable legislative and regulatory systems;
- HEIs and research centers with good capabilities in terms of staffing and equipment as well as viable linkages to international sources of STI knowledge.

Clearly, all of the above strengths would have to be further enhanced through a variety of means, including well-elucidated policies and national as well as institutional and local community based initiatives.

On the other hand, the following points have been listed as weaknesses by past policy documents:[45]

- Dearth of initiatives aimed at promoting STI inputs as bases for selected priority production and services sectors;
- Restricted funding for research at the national and institutional levels,[46] coupled to poor financial support for activities designed in support of innovation and entrepreneurship;[47]
- Absence of a dynamic taxing system that encourages higher levels of R&D spending as well as investment in innovative enterprises;
- Lack of networking among institutions of higher education and research centers, often leading to wasteful duplication;
- Expenditure is still directed toward research issues peripheral to national priorities;

- Limited supply of highly specialized scientists coupled to brain drain of promising national talents;
- Limited contributions by universities and research centers toward upgrading production services sectors activities, within both the private and public sectors;
- Lack of experience on behalf of R&D institutions in marketing their research output and technical services;[48]
- Higher education and research institutions still have some way to go before achieving excellence, even within limited priority domains;
- Persistent lack of interest on the part of the private sector in R&D activity;[49]
- Virtual absence of specialized bodies concerned with monitoring and evaluation of STI institutions and their output,[50] with some institutions and their managers highly shielded from all manner of accountability.

In addition to the above weaknesses that are fully recognized by STI policy and strategy documents authored by Jordan's institution, a number of exogenous challenges are also enumerated, including:

- Reduced volume of foreign direct investment (FDI) ventures by international firms, coupled to diversion of investments by GCC countries away from Jordan due to the country's less than adequate business environment as well as political instability within its neighborhood;
- With particular reference to innovation, documents concerned with STI policies in Jordan, underline the small number of technology incubators as well as the lack of specialized policies and strategies designed in support of innovation and entrepreneurship;
- Strong competition posed by recently established satellite branches of prominent overseas universities in neighboring countries, some of which have established their own research facilities.

Recommendations made with view to enhancing strengths and tackling weaknesses as well as challenges have often been made, particularly by international organizations concerned with promoting capacity building in science and technology, such as the United Nations Education, Science and Culture Organization (UNESCO). The most recent report prepared by UNESCO regarding the status of science, technology and innovation in Jordan, dates back to 2013.[51] It included a set of salient recommendation concerning Governance of the National Innovation System Investment in R&D and innovation as well as set of Policy directives aimed at promoting R&D and innovation. Some of the essential recommendations made by this report were directed at the need for:

- increased direct financial contributions, not only in support of public R&D and innovation institutions but also institutions providing technical services to business enterprises;
- making greater use of indirect fiscal measures, including tax credits, based on actual R&D and innovation expenditures;

- utilizing catalytic financial measures, e.g., risk capital, loan guarantees, and equity guarantees;
- achieving an effective balance between competitive and noncompetitive R&D funding, in order to provide steady support for regular R&D activities, mainly within institutions of higher education;
- implementing clear and transparent criteria for both institutional and competitive R&D and innovation funding, essentially by government, but also when such funding is being provided by nongovernmental organizations and aid agencies.[52]

It goes without saying that recommendations such as the above would require considerable financial investment as well as deployment of close management and effective monitoring.

Policy directives to be devised must also achieve a better balance between supply and demand-side measures. Opportunities for carving out balanced directives would be expected particularly in relation to the acquisition and dissemination of e-applications.

Shortcomings met with in Jordan's STI regimes are not always due to inadequate R&D and innovation policies. Rather, they may sometimes be traced to inadequate sectoral policies, in industry and agriculture, to name but two sectors, within which national R&D and innovation do not seem to produce the desired effects in a timely fashion, if at all.

## 3.2.3 Lebanon's STI Policies and Institutions

Lebanon is one of the first few Arab countries to establish a national council. Thus, the first meeting of the Board of Directors of the country's National Council for Scientific Research in 1966 considered the formulation of Lebanon's first science policy. However, ratification of this policy was by competent authorities suffered delays during the early 1970s. The civil war caused even further delays. As a result, Lebanon lacked an officially adopted science policy document although it was probably the first country in the region to embrace the concept. The science policy document drafted by the National Council for Scientific Research (NCSR) addressed the general objectives presented in Frame 3.5. Financial allocations for implementing Lebanon's policy document were set at 1% of GNP, increasing with time up to 3%. However, even the lower percentage has not been achieved to date.

### *3.2.3.1 Lebanon's Science, Technology and Innovation Policy Document*

In 2003, the National Council initiated a fresh policy formulation process, resulting in its "Science, Technology and Innovation Policy (STIP)" document,[53] which was issued in 2006 addressing objectives, stakeholders and problems concerning STI capacity building as a means of meeting socioeconomic challenges

**Frame 3.5 Lebanon's Original Science Policy Efforts**

Principal objectives of Lebanon's first national science policy, outlined in the draft policy document issued by the Council in 1966, were the creation of infrastructure institutions, the development of Lebanon's scientific potential as well as utilizing research results for improving social and economic conditions.

Based on the above objectives this policy sought to pursue the following activities:

*Promoting scientific research* in areas related to industrial and socioeconomic development as well as public health and the environment;

*Providing support and resources* for scientific research including the establishment of laboratories, provision of equipment and training of manpower and provision of research grants and assistantships;

*Rational utilization of natural, human, economic and cultural resources* commencing with relevant surveys and conducting studies and research programs aimed at the discovery of new resources and optimal methods for their utilization;

*Compensating for Lebanon's limited natural resources* with focus on human resource development and promotion of scientific research as a means to improving productivity.

The first S&T work plan based on the above activities included the following components:

*Creating national research centers* such as the Geophysics Centre, the Nuclear Research Centre and the Marine Research Centre were established and staffed by NCSR researchers;

*Establishing national research support facilities* such as The Computer Centre and the Centre for Scientific Documentation and Information;

*Setting up specialist research groups* including a Food Technology Research Group, a Mathematics Research Group, a Solar Energy Research Group and an Atomic and Molecular Physics Research Group;

*Building the research human potential* through a variety of means including the award of scholarships for higher education to high-performers in higher education;

*Providing support for research projects* conducted within public and private universities as well as governmental institutions, on the basis of criteria including socio-economic benefits, mainly for Lebanon, but also with a regional perspective;

*Encouraging multidisciplinary research* involving NCSR researchers as well as other national and international institutions and designed to achieve its established goals.

However, implementation of the NCSR first executive plan met with constraints that included inadequate budgetary resources, unfavorable local conditions for young researchers, limited appreciation of the difficulties and long gestation periods involved in research work.

---

Report produced by the Economic and Social Commission for Western Asia (ESCWA) titled "Science and Technology Policies for the Twenty-First Century," E/ESCW /TECH/1999/4.

that face Lebanon. The document was based on reviews by experts as well as discussions that involved cross-sections of Lebanon's science and technology community.

The STIP document intended for STI to achieve a central position in Lebanon so that "investments in STI will lead to recognizable economic and societal improvements, as well as a stronger environment for research." This was expected to lead to "self-propelling" modalities for further "expansion of both investments in STI and the framework of and the institutions for STI policies."

Initiatives highlighted by the STIP document included the launch of priority research programs, plans for improving university research environment, the creation of centers of excellence and the establishment of a viable information and communications infrastructure. The STIP document underlines the need to make the most of the limited resources available to involve various private and public enterprises and organizations in research activity prioritized with reference to three domains of societal needs, listed together with corresponding actions in Table 3.6.

**TABLE 3.6** Societal Needs and Action Points

| Societal Needs | Specific Actions |
|---|---|
| Basic Sciences, Technology, and Industry | Improve the management of energy, water, and other natural resources through sustainable approaches. |
| | Reduce industrial operating cost (energy, equipment, and maintenance). |
| | Harness information and communication technologies for development. |
| | Improve industrial productivity through enhancing technological content. |
| | Establish new industries targeting national development and job creation. |
| | Increase share in international trade and national export performance. |
| Environment and Agriculture | Stabilize/prevent coastal deterioration through sustainable management. |
| | Utilize economic opportunities posed by modern agricultural products and practices. |
| | Improve nutritional food quality. |
| Medical Sciences and Public Health | Regain Lebanon's position as regional leader in the field of medicine and health care through improvement of (a) higher medical and health sciences education; (b) the quality and cost-effectiveness of medical and health care services. |
| | Create an environment conducive to a flourishing biomedical industry and services sector. |

#### 3.2.3.1.1 The STIP Research Programs

Specific research programs were proposed on the basis of the above considerations in the research domains of activity linked to the "societal needs" listed in Table 3.6, targeting STI capacity building within priority sectors and domains of socioeconomic activity. A list of the domains of activity as well as the research projects is presented in Frame 3.6, in addition to which a program will be designed with the aim of forging links between practitioners in medical and health sciences, on the one hand, and technology, social sciences, and paramedical professions, on the other.[54]

Proposals put forward by the STIP document also included the creation of a knowledge gateway, virtual sectoral centers of excellence, as well as projects aimed at the adoption and adaptation of technical standards. Additionally, resource issues were examined, including funds for commercialization of research outcomes as well as creation and support for enterprise incubation schemes.

The STIP document acknowledged that the STI information and communication infrastructure in Lebanon was in urgent need of upgrading. Concrete proposals were also made to strengthen the links between, on the one hand, the research community, whether in academia or in specialized research centers,

**Frame 3.6 Research Programs Proposed by Lebanon's Science, Technology, and Innovation Policy**

With view to tackling the societal needs listed in Table 2.6, the STIP document includes proposals for implementing the following research programs:

| R&D Domain | Proposed Research Program |
|---|---|
| Basic Sciences, Technology, and Industry | Deployment of information technology within the enterprise sector |
| | Web and Arabized software technologies |
| | Mathematical modeling including financial/economic applications |
| | Renewable energy resources (e.g., chemical, wind, hydroelectric, solar) |
| | Material/basic sciences for innovative applications |
| Environment and Agriculture | Sustainable management of coastal areas |
| | Integrated water management |
| | Technologies for new opportunities in agriculture |
| Medical Sciences and Public Health | Nutritional food quality |
| | Research in molecular and cellular biology with applications in health care and medicine |
| | High quality research in clinical sciences |
| | Forging links between the practitioners of medical and health sciences and technology, social sciences, and paramedical professions |

and the private sector, public sector agencies, NGOs, and civil society institutions, on the other.

Institutional measures and organizational arrangements required to manage the various initiatives proposed by the STIP plan were addressed by the STIP document, whereby NCSR was assigned a major role in coordination of national STI activities. The roles and responsibilities of the various constituent components of the national STI framework were addressed. Some scope is allowed for national and international partnerships on the part of implementing institutions, and legal as well as administrative procedures required to support such partnerships were considered by the STIP document.

Issues pertaining to STIP monitoring and evaluation are also considered, including the adoption of indicators and statistics to evaluate STI efforts as well as coordination modalities between key players in STI policy and in funding STI efforts. The need for a steering committees and advisory boards was highlighted.

### 3.2.3.1.2 Improving Lebanon's Research Environment

With view to improving the research environment the STIP document addresses crucial and long-term issues, such as streamlining procedures and creating a culture of ethical and professional responsibility for improved research performance. Other issues of practical, and perhaps more immediate concerns addressed by the STIP document, include the creation of a Lebanese research network and provision of support for innovative information products.

Linking academia and research centers to private and public sectors as well as to NGOs is also addressed by the STIP document. In particular, partnerships between government ministries, departments and enterprises, on the one hand, and scientists, engineers, universities and research institutes on the other, are considered crucial for achieving meaningful outcomes in economic and social terms.

The STIP document underscores the need for enhancing the availability of well-trained and skilled people and the capacity to undertake endogenous knowledge creation and to facilitate knowledge absorption. Based on these premises STIP proposes to concentrate on tools aimed at creating an appropriate research environment, presented in Frame 3.7.

### 3.2.3.1.3 Funding STIP Implementation

The STIP document ends with a plan of action including formal steps leading to the creation of new institutional entities. An estimated budget of around US$33 billion is proposed for the first phase of its implementation, lasting 6 years.[55] The above budget includes funds required to cover:[56]

- initial preparatory activities within the first of the 6-year period covered by the STIP implementation plan;
- several research programs at the rate of US$250 million per year each, as well as US$100 million each for two incentive programmes;[57]

**Frame 3.7 STIP Proposals for Enhancing the Research Environment**

The STIP proposals for enhancing Lebanon's research environment commence with a central commitment to quality research at Lebanon's universities, entailing promotion of official standards and legislative instruments as well as streamlining procedures to reduce red tape. STIP also proposes to work toward the creation of a culture in which research performance is used as basis for researchers' career development. At the same time the STIP document recognizes the need for improving conditions under which scientists, engineers, universities and institutes work.

Next STIP calls for the establishment of new Centres of Excellence and providing support to existing research centers, with view to attaining quality and sustainability in research and education.

STIP proposals also include specific mention of the need to create incentives and modalities to promote collaboration across disciplines, universities and institutes, in both the public and private sectors.

- the creation of several new research institutions, including:
  - a medical research center which is allocated US$600 million annually, including the cost of implementing one of the research programs in the medical and health sciences;
  - four other research centers at US$200 million per year;
  - sectoral centers of excellence, at US$50 million per year each;
- salaries of personnel involved in the execution of the STIP initiatives.

### *3.2.3.2 Institutions Concerned With STI in Lebanon*

Lebanon's geographic position, its religious diversity, open society, and traditions played an important role in its development as an educational and scientific center in the Middle East. Thus, as indicated above several prominent universities were established in the Country by missionary societies as far back as the nineteenth century, with some setting up research facilities, mainly in medicine and archaeology among other areas of scientific endeavor. Although episodes of political turmoil and civil war hindered the progress of its institutions of learning and research it is still endowed with a relatively well educated population and hosts highly regarded institutions of higher education some with reasonably equipped research facilities enjoying links to various research facilities across the globe.

The following paragraphs will consider some of the principal Lebanese institutions engaged in STI activities with particular regard to promoting socio-economic sustainable development and competitiveness.

#### 3.2.3.2.1 The NCSR and Its Research Centers

The NCSR was established in 1962, as the central public institution in charge of science policy making under the authority of the President of the Council of Ministers and granted administrative and financial autonomy.

**Frame 3.8 Main Functions of the National Council for Scientific Research**

NCSR is entrusted with advisory functions targeting formulation of the general outline of policies directly related to science, as well as related proposals and suggestions to the government and undertaking surveys of ongoing research activities in Lebanon's private and public institutions.

Executive function fulfilled by NCSR are directed at the implementation of the National Science Policy, initiating, promoting and coordinating research activities throughout Lebanon and within the research operating under its auspices as well as research centers within government departments and institutions of higher education.

Through its higher education grants program, NCSR also plays an important part in promoting higher education in a wide variety of STI domains, targeting outstanding graduates of secondary education in Lebanon.

NCSR publishes the "Lebanese Science Journal" and a bimonthly newsletter including recent NCSR activities and notes on topics of interest to Lebanon's STI community.

At its inception NCSR, and along the model of France's "Centre National de Recherche Scientifique," was assigned essential advisory and executive functions summarized in Frame 3.8.

Although initially confined to promoting research in the pure and applied sciences, NCSR mandate was expanded in 2004 to include social, economic and management sciences, law, linguistics, history, and ethics of science as well as the arts.

NCSR enjoys full administrative and financial autonomy. It is managed by a Secretary General appointed by the Council of Ministers, reporting to a Board of Directors. NCSR is charged with disbursing research funding to Lebanon's universities and research centers. In 2013, its budget amounted to around US$6 million, in addition to a grant of around €1 million provided through the European Union's research cooperation programs. Around 80% of NCSR's budget is dedicated to financing the activities of the research centers it was able to establish over the years as well as its own operations that include financing specialized higher education and other activities aimed at human resources development.

With particular regard to STI collaboration with overseas partners, NCSR is the principal national agency charged with implementing cooperation programs including the "CEDRE" Lebanese–French Research Grant Program, the CNRS Lebanese–French Research Grant Program, and the Lebanese–Syrian Research Grant Program. In general, promoting STI cooperation features prominently in many other activities executed by NCSR, including, for example, its management of The Doctoral Scholarship Program.

NCSR enjoys considerable prestige among concerned institutions in Lebanon and the region to the extent that it was awarded a special prize offered by the Islamic Bank for Development in 2013 for contributions made to social and economic development. One of NCSR's recent accomplishments is its adoption and issuance in July 2016 of a "Charter of ethics and guiding principles of scientific research in Lebanon," a much needed document considering the situations, political as well as socioeconomic, under which many scientists and technologists have to work in Lebanon and the region, in general. The fact that this Charter addresses individual as well as institutional responsibilities where scientific research is concerned, is notable. Other articles of this Charter that need to be more widely disseminated concern responsibilities that scientists must hold toward human health and wellbeing with reference to the conduct of research and ensuing publication of research results in both the natural as well as the social sciences.[58]

***3.2.3.2.1.1 The NCSR Research Support Program*** First established in 1962, this program is aimed at providing support for research activities undertaken at Lebanon's universities and research institutions, including the research centers it helped establish as well as a few private sector institutions. Fig. 3.4 presents a view of the distribution of research grants made by the Research Support Program toward various research undertaken by a variety of research Lebanese institutions across five broad disciplines during the period 2007–11. That humanities and the social sciences appear to have received only a small proportion of NCSR research funding might be explained by the fact that it was added to the Centre's mandate only recently, as noted above. The smaller proportion of allocations to research in agricultural sciences stands out, however, and might be only partially explained by the fact that Lebanon has an institute that is dedicated to agricultural research, see the section on the Lebanese Agricultural Research Institute below.

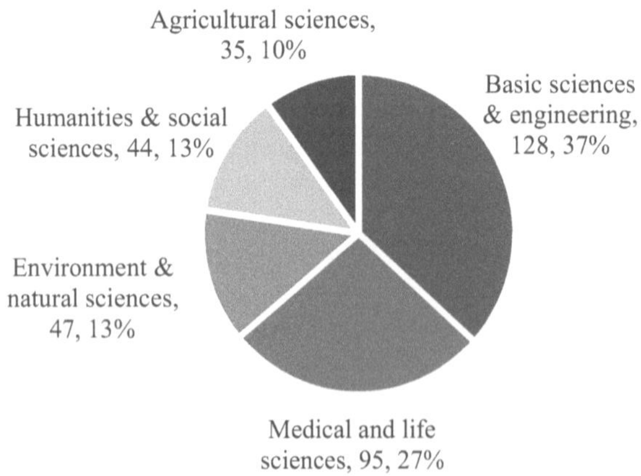


**FIGURE 3.4** Distribution of numbers of projects across five broad research disciplines.

Information available on support by NCSR for research activities over the period 2003–09 indicates that it allocated a total of around US$4.5 million in support of nearly 640 research projects conducted at Lebanon's main universities as well as the NCSR research centers.[59] Fig. 3.5 provides an overall view of the distribution of research funding allocated by NCSR to the main recipients of research funding. Fig. 3.5 shows that, during the above-mentioned period, the American University in Beirut was the largest beneficiary with around 46% of total funding disbursed by NCSR. The Lebanese University came next with around 24% of this total. A similar distribution is observed for the distribution of numbers of projects that received funding during the same period. Thus, as Fig. 3.6 indicates, comparable allocations per project were made across recipient universities. In fact, the values of research funds granted by NCSR per project varied within a rather narrow range, between US$5750 and US$7830, whereas the average grant per project was around US$7000.

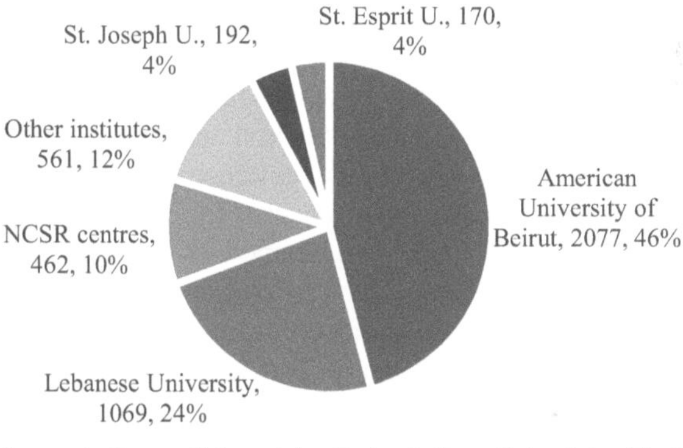


**FIGURE 3.5** Research Grants disbursed by National Council for Scientific Research during 2003–2009 (US$ thousands).

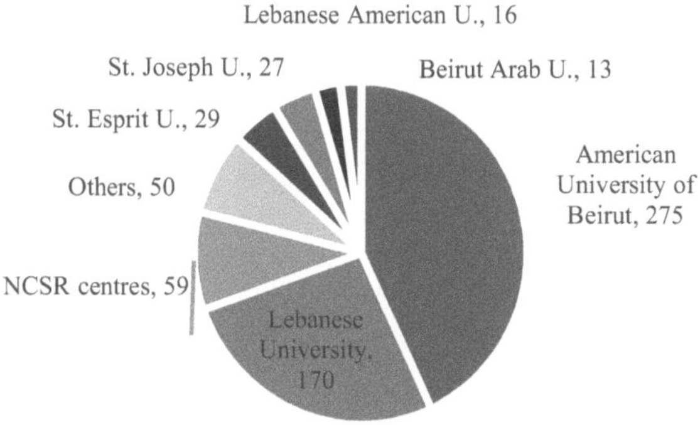


**FIGURE 3.6** Distribution of research project numbers funded by National Council for Scientific Research during 2003–09 across universities and research centers.

In effect, the low values of grants provided by NCSR toward research projects carried out by Lebanon's universities seem more appropriate for rather limited research efforts, such as student's graduation projects rather than any protracted research undertaking aimed at resolving the complex national issues. It may well be the case, on the other hand, that several smaller projects might be directed toward tackling a small number of more complex and long-term research undertakings. However, if such were the case, then issues regarding research planning by recipient institutions, on the one hand, and financial support for research efforts by NCSR, on the other, might be questioned. Because protracted research efforts would be needed to tackle complex issues relating to development priorities, optimal outcomes may not be produced when undertaken under conditions of annual uncertainties regarding continuity of funding.

The argument that the figure of US$7000, referred to above as the average allocation per research project is essentially just that, i.e., an average and that some projects might be awarded much higher sums, would mean that many other projects were awarded much smaller sums than this average.

Fig. 3.7 provides a view of the distribution of funding allocated by NCSR to five main groups of disciplines, namely, (1) medical and life sciences, (2) basic and engineering sciences, including information and communications technology projects, (3) agricultural sciences, (4) environment and natural resources, as well as (5) humanities and social sciences. Fig. 3.7 indicates that the first two groups of disciplines were allocated around 68% of total funding available from NCSR during the period 2003–09. Research devoted to subjects related to environmental and natural resources came third with around 17% of the total, whereas agricultural sciences took fourth position with 10% of the total. This may well be due to the fact that an independent agricultural research institution, LARI, is charged with implementing research agenda that respond to national priorities within the sector.[60]

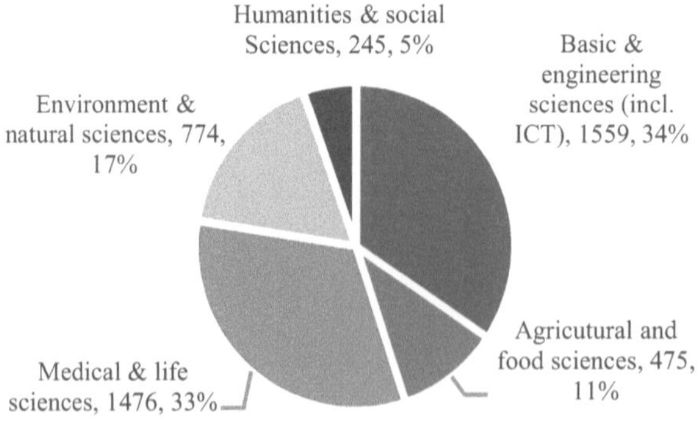


**FIGURE 3.7** Distribution of grants by National Council for Scientific Research totaling US$4.5 million across five broad research areas during 2003–09 (US$ thousands).

Fig. 3.8 presents a more recent view of the distribution of research grants disbursed by NCSR across universities and its own research centers. This figure indicates that, in 2016, the American University of Beirut (AUB) seems to have lost ground to other Lebanese universities as a recipient of most grants, compared to the period 2003–09. Thus, although AUB received 46% of all research funding provided by NCSR during 2003–09, its share went down to 27% during 2016, with NCSR research institutions and other Lebanese universities getting considerably larger shares of NCSR grants.

Fig. 3.9 presents the distribution of grants made by NCSR in 2016 to research projects within five broad disciplines. A total of US$1 million was available for this year and was distributed in a manner that was little

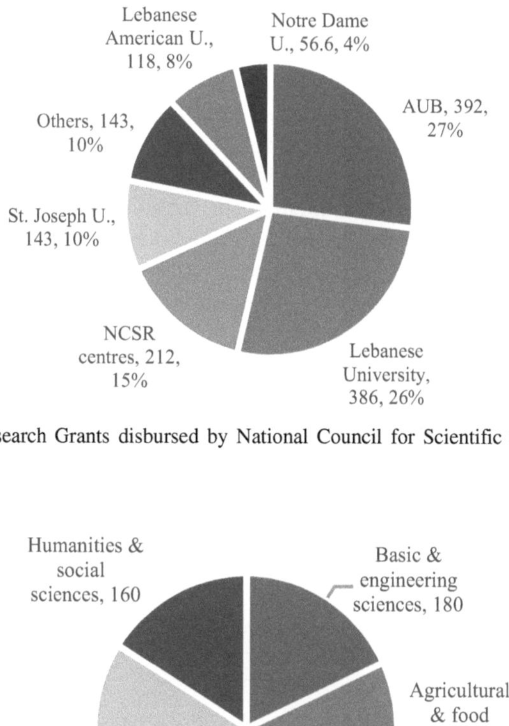


**FIGURE 3.8** Research Grants disbursed by National Council for Scientific Research in 2016 (US$ thousands).

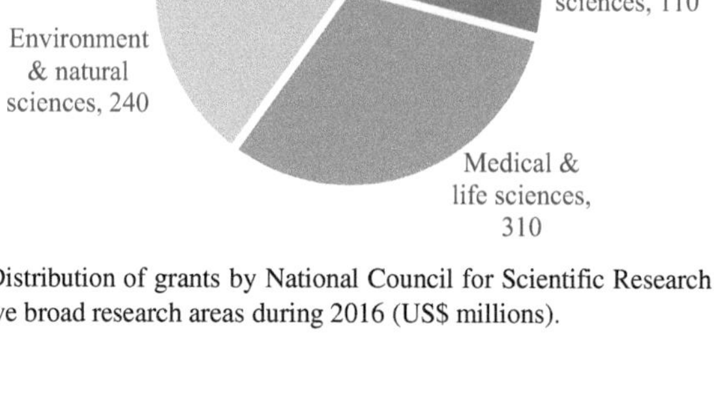


**FIGURE 3.9** Distribution of grants by National Council for Scientific Research to totaling US$ million across five broad research areas during 2016 (US$ millions).

different, with regard to two wide research areas, namely the medical and life sciences, on the one hand and agricultural and food sciences, in the other, from that witnessed during the period 2003–09. However, major gains may be noted with regard to funding research in the humanities and social sciences. Thus, research grants dedicated to the humanities and social sciences went up by more than a threefold increase, whereas allocations to research in the basic and engineering sciences were nearly halved and those targeting the environment and natural sciences went down by around 30%, compared to its share of total research grant allocations made during 2003–09.

With regard to overall growth of research funding provided by NCSR, Fig. 3.10 provides a view of the growth of NCSR research funding over the period 2005–09. The graph indicates a sharp rise in allocations to universities and research centers as of 2007, reaching a total of US$1.8 million in 2009, an approximate threefold increase on its level in 2005.

Fig. 3.11 on the other hand, presents historical trends for research funding across groups of disciplines during the same period. This figure indicates that increased total spending on research funded by NCSR was largely due to the considerable rise in allocations to the largest group of disciplines, namely basic and engineering sciences, which incidentally includes ICT research, as well as a slightly smaller relative increase in allocations to the medical and life sciences. As indicated by this chart, other groups of disciplines witnessed much more modest increases in NCSR research funding.

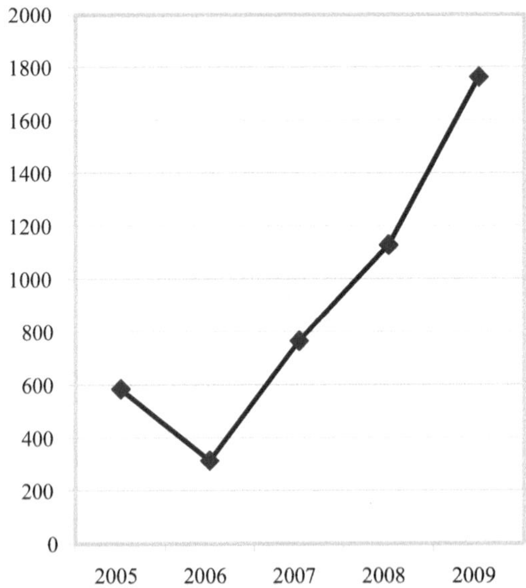


**FIGURE 3.10** Growth of total National Council for Scientific Research allocations (2005–09) (thousands of US$).

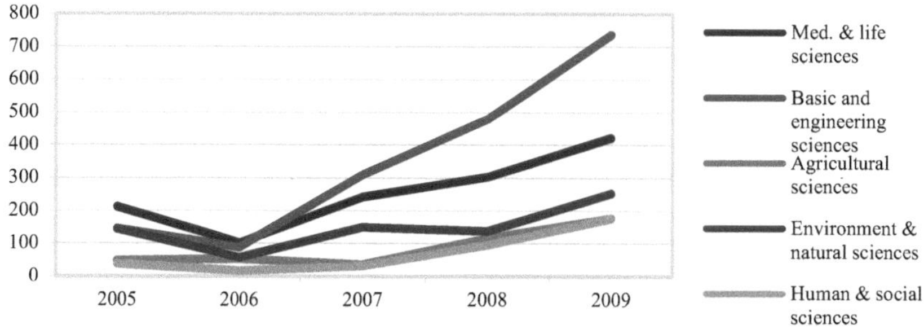


**FIGURE 3.11** Growth of National Council for Scientific Research allocations for research across five broad disciplines during 2005–09 (thousands of US$).

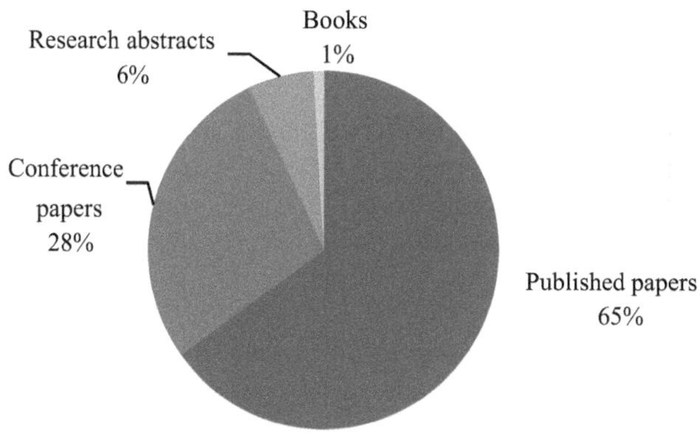


**FIGURE 3.12** Research output produced on the basis of support by NCSR during 2007–13 (% of total).

Fig. 3.12 presents a view of research output produced on the basis of research support by NCSR during the years 2007–13, indicating that, at 65%, published papers constituted the main form of output produced in all five major research areas, followed, at 18%, by papers prepared for presentation at conferences. With research abstracts constituted only 6% of the total and surprisingly small percentage of 1% of all output dedicated to books.

***3.2.3.2.1.2 Human Resources Development Program*** NCSR designed its human resources development (HRDP) program shortly following its establishment in 1962. The program was stopped during the civil war years but was resumed in 1999. During the period 1999–2005 around 600 grants were presented to Lebanese students for graduate studies in Lebanon as well as abroad, covering a variety of principally scientific and engineering specializations.

Since 2002, 25 among the highest scoring students in secondary education examinations are annually awarded grants to study abroad by NCSR. Fig. 3.13 presents a view of the distribution of grants for postgraduate studies awarded

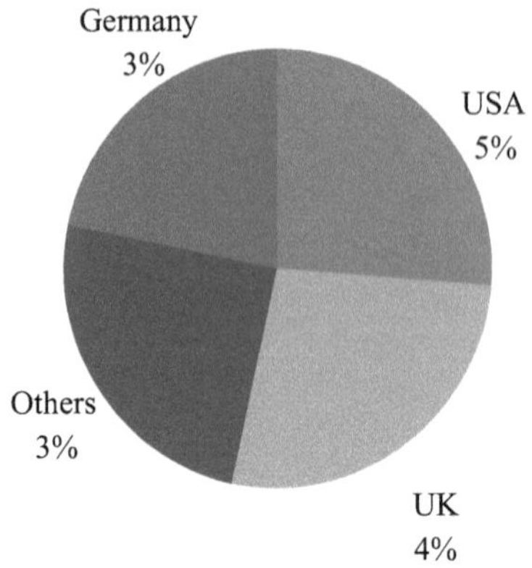


**FIGURE 3.13** Percentage distribution of National Council for Scientific Research grants for higher studies received by Lebanese students across countries (2007–13).

during the above-mentioned period across countries hosting graduate students. France's predominance as a recipient of Lebanon's talented students may readily be observed, as it received 66% of Lebanese postgraduate subsidized by NCSR grants, with grants awarded to postgraduate studies in Lebanon in second position at 19%.

***3.2.3.2.1.3 The NCSR Research Centers*** Since its creation NCSR was able to establish four research centers to undertake research, monitoring as well as provide technical services in geophysics, the marine environment, remote sensing and atomic radiation. These are namely:

- The Marine Research Centre
- National Centre for Remote Sensing (NCRS)
- Lebanese Atomic Energy Commission (LAEC)
- National Centre for Geophysical Research (NCGR)[61]

The NCSR centers are all part of regional and international cooperation schemes, notably with Italy and France, with particular emphasis on environmental aspects of the Mediterranean.

### 3.2.3.2.2 Lebanese Institutions Engaged in Promoting Innovation and Enterprise Creation

Several institutions in Lebanon play an important part in the dissemination of innovation as basis for SME creation, offering direct as well as indirect support for improved economic performance as well as job creation.

***3.2.3.2.2.1 Berytech*** Established in 2002, Berytech grew from a single technology pole, to a conglomerate of three centers through creating: "Berytech Technology and Health" and "Berytech Digital Park," in 2006 and 2013, respectively. Berytech is generally considered a leading actor within Lebanon's technology-based entrepreneurship ecosystem, hosting around 90 start-ups as well as 200 established technology firms.[62]

Berytech currently manages venture capital funds amounting to US$61 million, contributed mostly by Lebanon's major banks for investing in enterprise creation, within the scope of an initiative championed by Lebanon's Central Bank.[63] Records available on Berytech's website indicate technical and financial contributions were made since its creation to over 3000 enterprises, resulting in the creation of 1500 jobs.

Frame 3.9 includes a list of Berytech activities, whereas the following paragraphs present some further information on Berytech's centers and their activities in support of innovation and entrepreneurship.

**Berytech Technology and Health** was launched in 2006, in association with the faculty of medicine at the Saint-Joseph University, to support start-ups and SMEs based on innovations aimed at the health sector, including medical software development, clinical research and consulting services.

Berytech was host to multimedia and ICT clusters at its inception, including start-ups active in software development, film production, animation, website development and multimedia creations, virtual reality and gaming. Activities aimed ICT enterprise promotion evolved into the present "**Berytech Digital Park (BDP)**," which was established through partnership between Berytech, the Ministry of Telecommunications and a Lebanese real estate development firm. BDP plans to host a Incubator/Accelerator Parks designed to facilitate the creation of ICT-based enterprises as well as established ICT firms.

***3.2.3.2.2.2 The Association of Lebanese Industrialists*** The Association of Lebanese Industrialists (ALI) was established in 1943, as an independent entity with a national mandate. Nevertheless, ALI frequently collaborates with the Ministry of Industry on a variety of national initiatives. ALI advocates a policy of balanced industrial development for all regions in Lebanon and acts as a

**Frame 3.9 Activities Undertaken by Berytech With the Aim of Promoting Innovation and Enterprise Creation**

Berytech's facilities are home to enterprises that provide training, support and business counseling, with special emphasis on SME needs. Berytech activities aimed at enterprise promotion and related skill dissemination include incubation awards, summer schools for entrepreneurs, with special courses for engineers, microenterprise acceleration programs, local and international exhibitions, mentoring programs and networking schemes with Lebanese and foreign stakeholders. A notable recent development in this latter field is the creation of the "Bridge to Innovation,"[64] with the aim of encouraging strategic partnerships between Canadian and Lebanese businesses through the transfer of enterprise-related knowledge and technology.

major industrial lobbying group representing Lebanese industrialists in national, regional and international negotiations concerning industrial activity in Lebanon. Additionally, ALI runs special programs in support of quality management and research activities, including: quality management services as well as the Lebanese Industrial Research Achievements (LIRA). The latter program was launched by ALI in collaboration with the Ministry of Industry as well as selected Lebanese universities and research institutes. LIRA organizes an annual exhibition of projects by university faculties and research centers targeting commercialization by investors and industrialists. Exhibits normally include machinery, tooling, software and production line equipment. The program is essentially designed to:

- Encourage university–industry cooperation;
- Match university research activities with industry's needs and problems;
- Attract sponsors for proposed R&D projects;
- Provide a platform for potential new products and services.

***3.2.3.2.2.3 Kafalat*** Kafalat is a Lebanese financial firm whose essential objective is to assist SMEs in accessing funding by commercial banks. Kafalat helps SMEs by providing loan guarantees based on solid business plans/feasibility studies showing the viability of the proposed business activity.

Loan guarantees provided by Kafalat are designed to serve the borrowers, the bankers as well as the Lebanese economy at large. Thus, Kafalat seeks to reduce the risk of the loan for the bank, while, at the same time, guaranteed loans in Lebanese Pounds benefit from special arrangements by the Central Bank of Lebanon, significantly reducing costs for lending banks, allowing them to lend their customers at reduced interest rates.[65]

Seventy five percent of Kafalat assets are owned by the National Institute for the Guarantee of Deposits, whereas the remaining 25% is owned by 50 Lebanese banks. Kafalat operations target SMEs and innovative start-ups active in industry, agriculture, tourism, traditional crafts, and high technology.

In 2006, Kafalat signed a partnership with the European Union and the Ministry of Economy and Trade of Lebanon (MOET) to increase the amount and the size of the loan guarantees beyond the level offered by Kafalat. This partnership was conceived as part of the government's efforts SME strategy and overall promotion of economic development.

### 3.2.3.2.3 Industrial Research Institute[66]

Established in 1950s as a not-for-profit institution with administrative and financial autonomy to undertake industrial research and scientific testing and analysis for the industrial sector in Lebanon, the Industrial Research Institute (IRI) was turned into an organ of the Ministry of Industry in 1997.

Over the years, IRI was able to put together a reliable body of scientists and engineers providing Lebanon's industrial enterprises with expertise covering the country's light industries, particularly within the food and construction industries. IRI staff publish scientific papers, in chemistry, materials science and technology, catalysis, renewable energy, environmental protection and air quality. Some of their publications are coauthored with scientists and technologists abroad, particularly in France.[67]

Additionally, IRI is active in the following areas:

- issuing certification in accordance with rules issued by the International Organisation for Standardization (ISO) to enterprises and other entities
- providing product certification as well as eco-labeling services in accordance with international norms
- certification of experts trained as auditors in a variety of areas including environmental audits;
- conducting routine inspection and testing activities covering a range of industrial products and equipment items.

The Industrial Research Institute is home to the Centre for Innovation and Technology (CIT), which was established as a division within IRI with a mandate aimed at providing innovative inputs to the manufacturing sector in Lebanon.[68] CIT implements the "Innovation Vouchers," initiative, through which industrial enterprise may gain support up to €10,000, allowing it to implement innovative ideas designed to improve its operations.[69]

### *3.2.3.3 Concluding Remarks*

Lebanon was one of the region's forerunners in formulating a science and technology policy. However, despite the passage of many years, and a later attempt at drafting a more up-to-date STI policy, the country is yet to formally adopt a document to guide its efforts toward STI-led economic growth and social development.

Institutions charged with conducting and promoting STI activity in Lebanon were established earlier than many others in the region. Almost all seem to have amassed considerable experience within their fields of competence, which essentially cover traditional, as well as a few novel areas of research and technological development. Yet, individual scientists and technologists as well as managers of the institutions they work in, are unanimous in that resources allocated are barely sufficient for attaining even some of their less ambitious objectives and that at least much better outcomes would be available had more plentiful resources, human as well as financial, been available.

For a variety of reasons, including its limited natural resources, its early hosting of foreign educational institutions as well as the impact of several religious and sectarian constraints, the brain drain traditionally affected Lebanon far more than other countries in the region. Thus the number of people of Lebanese origins around the world, stands around fourfold those presently residing in Lebanon, with numerous examples of those who made their mark in various fields of science and technology in their adoptive countries around the world.

For a variety of reasons, attempts at making use of the enormous wealth that Lebanon's diaspora represent in reviving the national STI scene have generally been disappointing, in terms of their extent, their frequency and their ultimate outcomes.[70] One example, of such attempts is the program "Transfer of Knowledge Through Expatriate Nationals (TOKTEN)." This is an extremely worthy effort on the part of the UNDP, conceived as a means of "tapping on expatriate nationals, who had migrated to other countries and achieved professional success abroad, and mobilizing them to undertake short-term consultancies in their countries of origin."[71] However, the most recent data on results achieved by the this program's Lebanese component indicate that within the period 2006–12, only 30 missions were organized by Lebanese expatriates to 14 institutions in their home country. Apart from the fact that no further information is available concerning activities that took place following 2012, the above figures are simply too small to produce meaningful and long-lasting results for a country's STI system, let alone its socioeconomic development.[72]

Lebanon's STIP document also underscored the need for members of Lebanon's extensive STI expatriate community to take part in high-level steering committees charged with long-term STI planning and monitoring progress achieved, through national STI ventures, practical outcomes have not materialized.

Clearly, making effective use of the country's expatriate scientists and technologists must await well-structured programs with particular accent on cooperation between their adoptive institutions, on the one hand, and counterpart STI personnel and their institutions in Lebanon.

An even more serious hurdle facing STI in Lebanon is inadequate funding of related activities, in both the public as well as the private sectors. Thus, as indicated above, funds made available by NCSR in support of research projects at Lebanese universities and research centers during the 6–7 year period between 2003 and 2009 amounted to no more than US$4.5 million, for a total of almost 640 projects, providing support at an average of US$7000 per research project.

With particular regard to innovation, that over a period of 15 years, funds provided by Berytech, which, for all intents and purposes, is the sole incubation scheme actively supporting innovative SME creation, amounted to only US$70 million, i.e., a little over US$4.6 million per year, and that only 1500 enterprises were created is simply insufficient. The same is true of the IRI Centre for Innovation and Technology is able to offer around €10,000 on competitive basis to businesses with innovative ideas that may result in advancing productivity through implementation of novel technologies.

Most disappointing of all is the sheer lack of openly published monitoring and reporting schemes that are most essential for evaluating progress attained on the basis of official declarations and institutional pronouncements. Indeed, the fact that information posted on several websites seem to be outdated, sometimes by 4–5 years indicates that at least some of Lebanon's STI institutions could improve things considerably based on some simple measures. However, more ambitious outcomes are in need of far more radical reforms. Securing additional financial as well as manpower resources, is certainly a priority goal, if STI projects are to have their desired widespread, in-depth and lasting impact on socioeconomic development. Exploiting Lebanon's regional and international attributes with a view to further and wider cooperation with national as well as overseas partners should provide valuable aid in attaining tangible outcomes. Above all, however, there is a need for Lebanon to adopt novel STI as well as industrial policies together with clear implementation modalities, monitoring, evaluation, and reporting modalities.

## 3.2.4 Palestine's STI System, Its Institutions and Priorities

In so far as a science, technology and innovation system is commonly taken to consist of a set of institutions guided by a commonly adopted set of policy directives; Palestine may not be thought of as possessing an integrated and cohesive STI system. Only a limited number of small institutions are clearly active, notably in higher education, research, technology development and technical services as applied to water, energy, and agriculture, on the one hand, and in ICT systems and enterprise development on the other.

Higher education and research institutions as well as facilities designed to promote business enterprises, with essential focus on ICTs enjoy reasonably strong links with their counterparts as well as aid and donor agencies in the outside world, notably in selected GCC countries and the European Union. Additionally, as the following paragraph indicates, some use is made of Palestinian STI diaspora through a program established under the auspices of the UNDP in occupied Palestine in 1994 with view to promoting the "Transfer of Knowledge through Expatriate Nationals (TOKTEN)."

Palestine's TOKTEN Program is generally credited with several successes in mobilizing skilled expatriate professionals to provide aid in the service of socioeconomic development. The pool of human resources that the Palestinian TOKTEN program might access is enormous, comprising more than 10 million Palestinians residing outside occupied Palestine.[73]

However, the fact that only around 350 Palestinian expatriate experts, were reported as having taking part since the program's inception, volunteering their services in various disciplines in aid of Palestinian development initiatives strongly suggests that the TOKTEN Program is in need of radical revision.

Future UNDP plans for the program will emphasize the transfer of advanced knowledge to Palestinian institutions in a cost-effective manner through practical, demand-driven schemes. To this end, a data bank is being constructed by UNDP, including information about experts among the Palestinian diaspora capable and willing to contribute to the program.

### *3.2.4.1 STI Policy Frameworks*

There are no signs that Palestinian government authorities have in fact formulated an integrated national STI policy. Nor are there indications that such a policy has recently been proposed, by, for example institutions concerned with STI capacity building, including ministries, universities or research centers. Nevertheless, during the past decade the Palestinian Authority Government started to formulate triennial national development plans that touch upon scientific research and technology development.

However, although a document detailing the 2011–13 National Development Plan is available, no trace could be found of the succeeding 2014–16 document, aside from the framework set for its formulation by the Palestinian Ministry of Planning and Administrative Development.[74]

The National Development Plan document for 2011–13 is subtitled "Establishing the State, Building our Future," and includes under the "Social sector national targets" the objective "restoring Palestine as a center of excellence in higher education and academic research." Numerical targets listed under this objective include:

- doubling the number of universities ranked among the top 10 in the Middle East from 1 to 2;
- increasing the number of papers published per staff member at Palestine's universities and research institutes in international journals from 0.7 to 1.5;
- enhancing teacher to staff ratios in Palestinian universities from 1:35 to 1:27.

Allocations earmarked by the National Plan for "promoting research in science and technology" are stated as totaling US$7.9 million over its 3-year span. Additionally, the National Plan sets the establishment of a national fund to support scientific research as one of its priorities. In this context, emphasis is placed upon the need to "enhance and fund scientific research in congruence with the business sector needs." Concurrently, universities were also to be "encouraged to coordinate and network with relevant regional and international organisations."[75] Further reference to research is made with regard to promoting gender equality.

A Palestinian strategy for the agricultural sector covering the years 2014–16 addresses research and technical services and includes an item intended to "improve extension services, plant protection, veterinary services, research and agricultural marketing" as one of its strategic objectives. Under reasons justifying inclusion of this objective, the strategy document clearly underlines "weakness of scientific research and technology transfer activities," however. This document also includes reference to the allocation of US$3 million for setting up a gene bank at the Palestinian National Agricultural Research Center, within the Ministry of Agriculture.[76]

Finally, on the policy front, the above-mentioned framework for the National development Plan 2014–16, set forth by the Ministry of Planning and Administrative Reform, lists the following issues as pillars for the Plan's formulation. In varying degrees, all of these issues entail nurturing scientific and technological capabilities aimed at:

1. Gaining control over and developing the land of the state of Palestine and its resources.
2. Finalizing structuring of the State's institutions and enhancing their functions and capacities to deliver high quality services.
3. Building an effective democratic political system and enhancing societal peace, based on respect for human rights.
4. Developing the national economy and activating the Palestinian private sector.
5. Combatting poverty and unemployment as well as enhancing social justice.
6. Enhancing Palestine's official and unofficial participation in the Arab, regional, and international activities.[77]

### *3.2.4.2 Sectoral Priorities for STI Capacity Building*

Apart from the numerous political issues that engulf the lives of all Palestinians in the West Bank and Gaza, a number of obstacles to socioeconomic development are amenable to interventions based on science and technology inputs. Prime examples of these relate to access by Palestinians to energy and water resources. In both cases, scientific research as well as acquisition and dissemination of new technology inputs at affordable costs would facilitate socioeconomic development.

Although numerous political issues will continue to plague future development in the occupied Palestinian territories, certain aspects of Palestinian's lives are amenable to improved handling of some of their basic needs, particularly in relation to their energy and water requirements.

#### 3.2.4.2.1 Energy

Palestinians rely on Israel for 100% of their fossil fuel and for almost 90% of their electricity requirements. As indicated in Fig. 3.14, their per capita energy consumption is the lowest in the region, its cost is the highest in the entire Middle East.[78]

Solar, wind, and biomass resources are accessible to Palestinians, yet limited efforts have been made to exploit such resources commercially. It has been estimated that using available renewable energy sources would decrease energy reliance on external sources by around 45% as well as improve the Palestinian population's access to energy.[79]

Factors that favor turning to PV systems in the Palestinian territories include:

- High average insolation, with sunshine hours approaching 3000 hours annually;
- Predominance of rural locations with frequent open spaces, which may not be connected to a main grid anytime soon;
- High fuel costs in Palestine render PV systems considerably more attractive than diesel power generation.

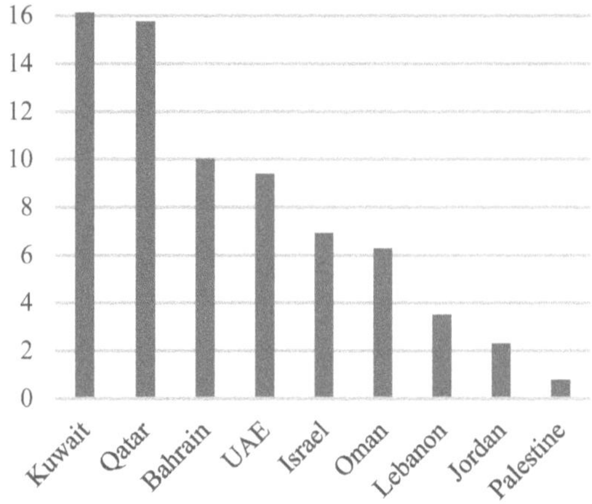


**FIGURE 3.14** Annual power consumption in selected countries of the Middle East (MW/inhabitant).

Several public and private sector institutions in the Palestinian territories are engaged in activities aimed at wider utilization of renewable energy technologies. However, limited information is available on outcomes of their activities. Judging by recent reports it is highly likely that any tangible outcomes fall far short of requirements. At any rate, measures aimed at resorting to renewable sources of energy may be interrupted by Israeli authorities on the pretext that they were constructed without due permission, which in any case is almost impossible to obtain. An example of such an eventuality was recently provided by the confiscation of a solar electricity generating facility donated by the Dutch government to a locality in the neighborhood of Bethlehem, on just such a pretext.[80]

### 3.2.4.2.2 Water

Water is a scarce resource throughout the Middle East and North Africa, with global warming expected to make a bad situation desperate. Added complications in Palestine due to ill-conceived clauses in the Oslo agreement[81] as well as Israel's occupation policies render the water situation extremely difficult in the West Bank, and alarming in the Gaza strip.

The situation in Gaza was better in the late 1990s, when relatively clean drinking water was available on tap in most lodgings. However, over the past decade or so a great deal of water was pumped out of the aquifer beneath Gaza causing water from septic tanks as well as seawater contaminated by poorly treated sewage water to seep in, rendering it unpotable. A World Bank report indicated that only 10% of Gazans have access to safe drinking water.[82] As a result, drinking water has to be bought by all citizens from firms operating filtering and desalination facilities that render water acceptable for drinking and cooking. In certain locations in the West Bank, the weekly cost of filling a domestic water tank amounts to the equivalent of US$75 in localities where the average monthly salary is equivalent to about US$500.

In some areas of the West Bank, daily water consumption per capita is around a quarter of the World Health Organization's recommendations. Demand is particularly acute in the summer and in particular localities such as Hebron, where water is delivered by the town's municipality only once every 3 weeks.

Several ideas for improving the situation have been discussed over the past decade, including seawater desalination. Wastewater and sewage treatment would help save water resources for nondomestic uses as well. For this to be feasible a good deal of investment in infrastructure, technology acquisition, and adaptation would be needed, however.

### *3.2.4.3 Palestine's STI Institutions*

Palestine possesses a limited number of STI institutions. The following paragraphs provide a concise overview of the country's universities and research institutions.

#### 3.2.4.3.1 Universities and Educational Institutions

The Gaza Strip is home to eight universities, one university college and technical colleges, whereas the West Bank sports nine universities, as well as several institutes and colleges, some of which are dedicated to religious studies as well as arts and culture.

Inhabitants on the West Bank had access to educational institutions, in general, and to higher educational institutions, in particular much earlier than their Gazan counterparts did. Thus, relatively well-known colleges of intermediate and secondary education had been established by missionary organizations as far back as the late nineteenth century and well-established higher educational institutions, such as Birzeit and An-Najah universities, have been operating in the West Bank for several decades. The West Bank is currently home to the Quds Open University, which had around 56,000 students enrolled in 2016/2017, within 19 faculties and centers spread over several locations in the West Bank and Gaza. Undergraduate faculties provide instruction in a relatively wide variety of specializations including technology and applied sciences, agriculture, and education.[83]

***3.2.4.3.1.1 Research Institutions*** Twenty-two institutions are listed by Palestine's Central Bureau of Statistics (PCBS) as dedicated to studies and research in Palestine. A closer look at their mission statements reveals that most of these institutions are dedicated to studies and research activities with accent on social, political, community development and even religious affairs. Work undertaken by some of the research institutions dedicated to social and community development targets issues related to water resources and the environment, thereby contributing to improved understanding of resource utilization and consumption patterns and the use of such understanding in future development activities. The following institutions appear to undertake studies and research on topics directly linked to capacity in science, technology and innovation, including related economic aspects.

***3.2.4.3.1.1.1 Palestine Academy for Science and Technology*** Established in 1992, 1 year before the signing of the Oslo Accords in 1993, the Palestine Academy for Science and Technology (PALAST) remained inactive until 1997, when a presidential decree officially confirmed its status and assigned the Academy strategic tasks for advancing scientific research and promoting innovation in Palestine. PALAST was reconfirmed by another presidential decree in 2004.

The Academy is considered the principal science and technology body in Palestine. It is charged with formulating policies, as well as the design and coordination of related programs and projects in support of national development and participating. PALAST opened its first office in Ramallah on the West Bank in 1998 and established another in Gaza in 2000. A third office in East Jerusalem serves as the Academy's headquarters.[84,85]

***3.2.4.3.1.1.2 Centre for Development Studies (CDS); Birzeit University*** CDS is dedicated to development studies, and works with various Palestinian communities, policy makers and local and international institutions and academics to create and disseminate sound knowledge, dynamic ideas, innovative tools and practical solutions that empower individuals and institutions to contribute effectively to the development of our society.

***3.2.4.3.1.1.3 Institute for Environmental and Water Studies; Birzeit University*** IEWS conducts research as well as postgraduate training on environmental and water issues. Its vision statement emphasizes a quest for excellence in education, training and research within the Mediterranean region.

***3.2.4.3.1.1.4 Water and Environment Studies Institute (WESI); An-Najah University*** This institute also undertakes research in the fields of water and the environment and provides services to the local community, including environmental training and equipment at low costs. In the area of water research WESI conducts scientific research on water resources management, wastewater treatment, renewable energy production as well as the impact of agricultural and industrial activity upon water quality. WESI is said to be equipped for conducting tests aimed at assessing groundwater and wastewater quality. On the environment front, WESI is reported as having created a database including environmental and water related parameters enabling improved water and environmental policy formulation.[86]

***3.2.4.3.1.1.5 Health, Development, Information and Policy Institute*** An independent, nonprofit organization, HDIPI provides information services, evaluation, and training in the areas of health, development policy and systems management and carries out policy and planning research aimed at healthcare and related systems development in the West Bank and Gaza Strip. Created in 1989, it has since expanded its operations to include advocacy for better government policies on behalf of women, youth, and the disabled.

***3.2.4.3.1.1.6 The Water and Soil Environmental Research Unit (WSERU); Bethlehem University*** WSERU was established in 1988 by Bethlehem University's Chemistry Department. It has provided advisory and technical services to a large number of municipal authorities in relation to water quality issues. Research activities undertaken by WSERU have helped it set up a comprehensive database on water quality in both the West Bank and Gaza. The Unit runs a graduate program for water and soil analysis. With support from the Ford Foundation, WSERU carried out a comprehensive survey of the water quality of springs, wells, surface, and network water during the early 1990s. Further research with focus on southern parts of the West Bank, has been undertaken with assistance by the Irish Development Aid.

***3.2.4.3.1.1.7 Energy Research Centre*** The Energy Research Centre at An-Najah National University is dedicated to research, development, system design, feasibility studies as well as training in matters concerning conventional and renewable energy resources as well as energy resources management and conservation. It is reported as conducting projects within domains that include wind energy, rural electrification, clean energy and conservation strategies, biogas generation from agricultural wastes.[87]

***3.2.4.3.1.1.8 National Agricultural Research Centre*** The National Agricultural Research Centre (NARC) was established, in 1995, under the umbrella of the Ministry of Agriculture (MoA) as an institute for applied agricultural research in Palestine. Its activities are intended to "improve and increase plant, livestock production and enhance sustainable use and efficiency of natural resources for the benefit of farmers and other producers." NARC runs four principal programs: natural resources, plant production, animal production and capacity building, and technical services. In addition to its research, NARC is equipped to carry out analysis and characterization of water, soil and olive oil, in addition to undertaking extension service and public awareness activities.

***3.2.4.3.1.2 Incubators and Enterprise Promotion*** Incubators and business start-ups are effective tools for innovations within the business sector as well as society, at large. Based on information gathered by WAMDA, a platform of integrated programs in support of entrepreneurship ecosystems throughout the MENA region,[88] the following paragraphs provide very brief outlines on eight incubation and start-up acceleration facilities that have been established since 2010 in the West Bank and Gaza.

Arabreneur is a West Bank business accelerator that specializes in providing support to software developers. Its activities are conducted on the basis of a call for new application concepts that is issued thrice every year. Arabreneur provides training, a start-up fund starting from US$100,000, as well as a range services including regional and international exposure.

The Business and Technology Incubator (BTI) was established in Gaza in 2006 to help its graduates launch their own start-ups and small businesses with both established and novel concepts, the latter under the label "SEED Incubator." BTI project selection is undertaken on the basis of annual business plan competitions; open to aspiring entrepreneurs and university graduates. Around 10 candidates are singled out for support out of each round and provided with start-up funding, office space, logistic and business support, and investor and customer matching services.

The Business Development Centre (BDC) was established by The Business Women's Forum (BWF) in 2006 with the aim of empowering Palestinian women. It runs a preincubation program supporting women entrepreneurs directly in Ramallah, and remotely in Gaza, to develop their start-ups.

Gaza Sky Geeks (GSG) is Gaza's first ICT start-up accelerator and co-working hub. It was established by MercyCorps and Google to support start-ups and connect local entrepreneurs to global networks in 2011. GSG launched two large crowdfunding campaign in support of its operations in 2014 and has enrolled selected entrepreneurs into its incubation program where they receive coaching by local and international mentors and helps match start-ups with investors.

Fast Forward Acceleration Programme (FFAP) was launched in 2013 in the West Bank by The Leaders Organization, a Palestinian NGO. Its focus is on supporting businesses in Internet and mobile technologies. FFAP receives support from the IBTIKAR fund, GIZ, and e-Zone. Its premises constitute a mini technology park and a forum for community discussions on entrepreneurship. Entrepreneurs with sound business ideas are provided with an initial investment of around US$20,000, business coaching, office space and support services, and the opportunity of linking to likely investors.

The Mobaderoon Program, which was established as one of the first incubation schemes in the Gaza strip in 2010, is rather unique in that it supports entrepreneurs in a wide range of sectors ranging from ICT to hardware development and agricultural products. Mobaderoon is run by the Islamic University of Gaza with support from a number of development agencies, and organizations including the Arab Fund for Economic and Social Development and the Islamic Development Bank. Accepted candidates are provided with start-up funding, office space, logistics, mentorship and market access as well as assistance in registering any intellectual property rights and business brands that their ideas or products might carry.

The Palestinian ICT Incubator (PICTI) was launched in both Ramallah and Gaza as an information and communication technologies incubator. Following establishment of its Business and Innovation Centre in 2012, however, it moved over toward acceleration and product commercialization. PICTI has links to the Palestinian IT Association (PITA) and has implemented business competitions with support from international organization and firms such as Intel and Oxfam. Prospective entrepreneurs are guided through a series of training and assessment exercises before only a few are selected, whence they receive financial and legal support in addition to further training by a PICTI network of experts. Some PICTI candidates are chosen for induction into an acceleration program and are provided with support as they seek capital investment.

The University College for Applied Science in Gaza launched the UCAS Technology Incubator (UCASTI) with support by several organizations, including the Kuwait Fund for Arab Economic Development, the Islamic Development

Bank and Oxfam. UCASTI accepts entrepreneurs and university students with potentially successful business ideas, offers them training, seed funding and logistical support.

#### 3.2.4.4 Concluding Remarks

Organized scientific and technological development activity under conditions of occupation, political instability, economic pressures, and social tensions is, to say the least, a daunting undertaking. Yet this is exactly what Palestinians need to do in order to resolve a host of problems, including water scarcity, food shortages, and uncertain energy supplies.

Achievements over the past two decades emphasized institution building and rehabilitation, with focus on higher education. Research and technology development facilities have also been established, often with donor support, with particular focus on the water and energy sectors, on the one hand, and with regard to enterprise creation, with emphasis on ICT applications, on the other.

Yet, scientific, technological development and innovation activities in Palestine are nowhere near the levels that are needed in order to secure reasonable living conditions in certain locations, let alone meaningful overall development. A great deal more could, and should, be done.

Extending the roles played by international volunteers and NGOs in humanitarian relief during various stages of the Palestinian crisis toward providing expertise on priority issues is of immense importance, particularly with regard to much needed water and renewable energy resources. In doing so, care must be taken to engage local institutions and experts in order to allow meaningful adaptation of novel inputs to local conditions on the one hand, and to create enduring capabilities for future evaluation and development of future innovative inputs, on the other.

### 3.2.5 The State of STI in Syria[89]

Significant steps were taken as far back as the early 1970s in Syria toward establishing endogenous scientific research and technological development capabilities. Advances were undoubtedly made on the basis of such efforts, some eventually translating into tangible gains for national development, particularly in domains such as agricultural production. Thus, there is little doubt that the advances made by Syria toward near self-sufficiency in agricultural production were indebted to government policies and the creation of agricultural research stations and extension services, on the one hand, and to cooperation with regional and international agricultural and rural development research activities, on the other. National R&D efforts as well as steps taken toward disseminating novel agricultural practices, received a good deal of support from initiatives such as those promoted by the International Centre for Agricultural Research in Arid Areas (ICARDA), established in 1977, and the Arab Centre for the Study of Dry and Arid Zones (ACSAD), established in 1968.

Political developments during the eighties and nineties saw greater focus on R&D efforts in domains of little relevance to sustainable development and job creation. Sanctions imposed by the United States on dealing with the Syrian government aggravated this tendency, with the result that even more resources were diverted away from areas crucial for economic growth and social development.[90] In theory, sanctions were confined to equipment and software items with more than 10% of their value originating from the United States. However, measures taken by providers of equipment and software firms operating out of Europe, as well as a number of other countries, often went much farther for fear of retributions by the United States. At any rate, it is unlikely that these sanctions in fact stemmed the inflow of needed equipment and software. Thus, it is widely believed that sanctions against Syria were only partly effective, as its government managed to secure most of its essential requirements through third parties, both regional and international. However, sanctions, quite expectedly, created fertile grounds for all manner of corrupt practices ultimately harming ordinary citizens most.

At any rate, national institutions involved in STI activities suffered additional pressures, due to outdated practices that many other institutions in the country were inundated with, thereby curtailing their independence and ability to render positive contributions to national development.

At the very outset, it is essential to state, that activities undertaken by STI institutions in Syria throughout the past two decades fell far below the country's potential, both in terms of quality and quantity. Indeed, a number of STI indicators point out that Syria had been lagging behind neighboring countries on a number of fronts for the past three to four decades.[91]

Reviews of scientific research and development activity undertaken in Syria prior to the onset of its current tragedy, compared to its past performance, as well as that of other countries in the region, indicate meager progress. For example, the number of scientific papers published in international journals seems to have increased from 828 in 2007 to 897 papers in 2008, whereas the number of researchers went up from 5158 in 2008 to 5218 researchers in 2009. With regard to innovation and innovative activities undertaken for the benefit of enterprise and job creation, several reports issued by international bodies, notably, the World Economic Forum, also indicate that Syria occupied a less than satisfactory position even in comparison to other developing countries, for a number of years before the onset of political turmoil and the current armed conflict.

### *3.2.5.1 Syria's STI Policies and Strategies*

As in many countries around the world, issues related to STI policies and their implementation, are of concern to many ministries and public, as well as private, sector institutions. However, in the case of Syria the fact that limited cooperation and coordination takes place among concerned bodies, with regard to STI activities is a shortcoming that engenders considerable waste of efforts and resources. Additionally, limited data collection capabilities, and inadequate indicators, conspire to further obscure evaluation of progress made on various tracks.

At any rate, no less than three major STI policy/strategy documents were authored during the past two decades, all of which acknowledging the less-than-satisfactory state of affairs in relation to many aspects of scientific research, technology dissemination and innovative enterprise creation. Novel institutions were established and fresh initiatives launched to promote R&D and innovative enterprise creation. However, whereas some of these institutions and initiatives may be credited with positive contributions, little could be achieved to mend things on the broader, national, scale. Indeed, several such developments may be regarded as classic examples of doing too little too late. Since, although strategies were drafted, institutions established, and initiatives launched, little was done in terms of monitoring performance, in the first place. Second, influential networks of power centers with business connections and agenda aimed at importing technology inputs and intent upon sidestepping local STI institutions gained the upper hand to such an extent that meaningful national initiatives became progressively fewer and gradually enjoyed less support as time went on.

At least three documents are known to have been drafted with national directives for future STI activities in Syria:

- Syria's Vision 2025;
- The National Science, Technology, and Innovation Policy;
- The ICT Strategy for Socioeconomic Development.

However, with the exception of the ICT Strategy document,[92] no indications are available as to whether any of the past documents were in fact officially adopted as basis for future action. It may also be noted that contributions by private sector institutions toward these documents were at best minimal. This said, however, it is also essential to state that limited consultations were indeed held at the behest of international organizations concerned with promoting innovation policies, as well as setting up enterprise incubation and creation schemes. The following paragraphs provide a summary of the above document's content and salient recommendations.

#### 3.2.5.1.1 Science, Technology and Innovation in Syria's "Vision 2025"

Syria's 10th 5-year plan outlined a vision for science and technology covering the years till 2025. This vision was essentially based on a study[93] carried out in 2007. The study's findings with regard to the state of the national STI system at the time included both negative and positive conclusions. On the positive side, Syrian S&T institutions were found to be most successful when their activities were confined within specific domains, with well-defined objectives and effective coordination modalities. Thus the fact that Syria enjoys a good measure of food security is credited to its success in R&D and technology dissemination services dedicated to agricultural development. On the negative side, however, the study's conclusions underlined the following issues.

- The absence of an accredited national S&T policy and a strategy for its implementation, including research, human resource development, access to and acquisition of technology.
- That fact that S&T institutions in Syria failed to implement evaluation schemes, therefore, depriving management of a mechanism for performance improvement.
- Private sector's failure to create demand for R&D activities, depriving national S&T institutions of motivation as well as additional financial support.
- Lack of sufficient resources dedicated to long-term programs targeting scientific research and technology development.
- Inadequate legislative and regulatory arrangements that govern S&T institutions' activities, often limiting their ability to making use of even the meager resource they possess.
- Limited investment in human resources, with subsequent failures in supporting continuous education schemes.
- Underdeveloped research cooperation and overreliance on factors such as personal initiatives, rather than an institutionalized efforts.
- Negative effects of the "brain drain" due to limited incomes that may be attained in the domains of education, scientific research and technological development.

The study also enumerated the following obstacles that faced its future development:

- continuing deterioration in regional security;
- the boycott imposed by the United States on the export of technology goods to the country;
- resistance to change on the part of various ministries and establishments leading to limited cooperation with regard to attainment of set goals;
- delays in promulgating legislative arrangements related to the objectives set by the 10th 5-year plan.

According to this study, the essential objectives to be targeted by the national STI system could be summarized as:

- Harnessing science and technology in the service of socioeconomic development;
- Enabling the national economy to evolve so as to face global competition, enhance utilization of its comparative advantages, improve the quality of its products and achieve sustainable development;
- Linking efforts aimed at human resource development to essential requirements of production and services sectors;
- Inducing radical changes within social and economic frameworks with a view to enabling the country's transformation into a knowledge economy based on the utilization of innovative science and technology inputs, with particular accent on ICTs in education as well as the production and services sectors.

The study introduced a preferred scenario for the future, emphasizing the need to take up novel technology inputs, essentially ICTs and selected biotechnologies with the aim of improving productivity as well as enhancing product quality and sustainable resource utilization. This scenario involves hybridization of selected key areas of economic activity, in agriculture, industry, and service sectors, with high-technology inputs, including information and communication technologies and biotechnologies with the aim of improving productivity, enhancing output quality and sustainable resource utilization. Domains singled out for special attention as recipients of STI inputs included: agriculture, energy generation, public health, water resources, industry, and the environment.

Frame 3.10 presents numerical goals set by the tenth 5-year plan for advancing national STI capabilities, including total national spending on R&D as a percentage of GDP, measures to ensure enterprise contributions to R&D efforts and targets for STI manpower training and productivity.

#### 3.2.5.1.2 Syria's National STI Policy

Work on the most recent National Science, Technology, and Innovation Policy (NSTIP) document was initiated in 2008 within a project managed by the Higher Commission for Scientific Research and a draft was completed by 2012. During its implementation the project's guidelines were amended limiting attention to selected "strategic sectors," with eventual implementation designed to take place through well-defined work programs to be carried out within three time frames: short (2 years), medium (5 years) and long (10 years). It bears emphasis that although many of the issues raised in the NSTIP document will have become outdated once the conflict situation in the country is over, it is highly likely that several issues it raises would still be highly relevant once the dust has settled down in Syria.

**Frame 3.10 Sectoral Priorities and Numerical Goals Set by Syria's Tenth 5-year Plan**

Numerical targets set by the plan included an increase in resources dedicated to R&D to an equivalent of 1% of GDP, by the end of 2010 and to 2% by the year 2020.

Additionally, public and joint public–private sector enterprises are required to dedicate a minimum of 5% out of their annual budgets to R&D activity.

With regard to human resource development, the 10th 5-year plan requires that the number of masters and doctorate degree holders amount to a minimum of 10% relative to first degree graduates in branches of the applied sciences. Furthermore, the ratio of students opting for scientific and literary branches in secondary education and at universities is set at a minimum of 60:40.

In relation to S&T productivity, the plan stipulates that researchers holding doctorate degrees must publish at least one paper or book annually.

According to this NSTIP document, challenges that retarded the country's economic performance over the first decade of the century, included declining oil and gas revenues, the crisis suffered by the agricultural sectors, rising unemployment and emerging threats to the competitiveness of Syria's exports due to regional actors taking up more advanced technologies in their production and service operations.

The NSTIP document must be credited with pointing out that measures aimed at resolving such issues hardly progressed beyond official statements and were rarely translated into effective initiatives required by the challenges faced by the country. Thus, limited or no action was taken on several government declarations in the past, pointing out the need for:

- Planning aimed at reform and "vertical expansion" in the agricultural and tourist sectors, with view to enhancing their revenues;
- Diversification of economic activities, through an improved climate for entrepreneurship and foreign direct investment;
- Improving the performance of the country's public sector;
- Support for the SME sector with view to creating a wider range of employment opportunities.

The NSTIP document also described a wide range of measures that are needed in order to reform the national STI system, enabling it to meet the country's numerous challenges and future needs.

In particular, issues listed by the NSTIP document as hindering progress in "transforming the country into a knowledge economy" included prevailing inflexibility of the country's political system and its organic links to centers of decision-making in which corrupt practices; administrative, financial and professional have taken hold. However, although moves toward transforming the country into a knowledge economy are considered in various parts of the NSTIP document, no concrete indications are made as to how specific objectives relating to such moves might be fulfilled. Frame 3.11 presents the policy's vision, priorities, directives, and selected numerical goals.

**Frame 3.11 Syria's Policy for Scientific Research, Technology Development and Innovation (NSTIP)[94]**

**The NSTIP vision**

The policy's vision is summarized as "constantly seeking to establish STI institutions that adopt excellence and a pioneering attitude as essential modalities for their operations."

**Priorities**

Areas singled out for attention were listed as:

- Creating STI human capital;
- Promoting and fostering innovation;

**Frame 3.11 Syria's Policy for Scientific Research, Technology Development and Innovation (NSTIP)[94]—cont'd**

- Enhancing national STI capabilities;
- Improving scientific and technological infrastructures including information networks;
- Capacity building in a range of advanced technologies as bases for strategic partnerships.

With the above in mind, the following 15 sectors and action areas were earmarked as priorities: water resources, energy, agriculture, industry, ICTs, the environment, healthcare, capacity building in selected advanced technologies, transport, tourism, construction, social and cultural development, financial resources, and local community development. The first eight sectors/areas were accorded top priority.

**NSTIP Directives**

The NSTIP document goes on to enumerate the following measures, whose implementation would enable the country to face challenges as well as bring about a more conducive environment for revamped STI capabilities:

- Setting up a legislative environment with effective incentives as well as checks and balances;
- Restructuring production and services sectors to enable dynamic responses to emerging needs and opportunities;
- Enhancing capabilities of the national labor force and improving working conditions;
- Building channels for the exchange of information among all concerned parties and consolidating information into viable knowledge for effective decision making;
- Promoting cooperation among parties engaged in socioeconomic development programs;
- Developing abilities for creativity and innovation;
- Effective sharing of specialists in various scientific disciplines and branches of technology between various components of the national STI system;
- Adopting dynamic approaches whereby certain activities are implemented in parallel, rather than in series;
- Creating public awareness in support of technological change, with emphasis on renewable energy resources and ICTs.

**Numerical goals[95]**

- Dedicating 1% of the country's GDP to scientific research by the year 2010 and 2% by the year 2020;
- Earmarking 5% of the budgets of institutions in the public, private and joint sectors to R&D activities, and dedicating part of these institutions' proceeds to finance R&D with national research centers;
- Increasing the percentage of PhD and masters degree holders relative to the entire population of undergraduates in applied science and engineering disciplines up to 10%;
- Ensuring that the proportion of students opting for scientific studies within the secondary and tertiary stages relative to those taking up literary disciplines falls between 40% and 60%.

In its treatment of innovation, the NSTIP document is more focused on initiatives related to endogenous R&D activity, with little emphasis on improving the performance of SMEs based on well-adapted off-the-shelf technologies. In common with many other documents authored by other Arab countries, little attention is paid to special needs of marginalized and impoverished local communities.

The NSTI document highlighted the need for intensive cooperation and coordination among concerned ministries and public sector institutions, on the one hand, and with private sector enterprises, on the other, with the aim of enhancing the STI content of their operations. Ensuring optimal outcomes is deemed to require the establishment of a Higher Committee to oversee implementation of NSTIP and its subsequent revisions, a management committee to coordinate implementation activities, in accordance with set priorities, resource allocations and outcome criteria. Additionally, NSTIP also proposes that an evaluation team is established to ensure adherence to priorities and budgetary allocations. Committees of experts are also to be set up to oversee NSTIP implementation within specific sectors and cross-cutting domains, such as water resources management among other environmental issues.

Monitoring measures recommended by the policy document emphasize the need for annual reports to detail progress made on various tracks as well as amendments needed in order to accommodate salient factors and emerging situations.

#### 3.2.5.1.3 Syria's ICT Strategy for Socioeconomic Development

The ICT strategy's vision is essentially aimed at the development of the ICT sector through plans to be implemented in collaboration between public and private sectors as well as civil society institutions under the supervision of the State. Principal objectives of this strategy included turning Syria into an information society during the first quarter of the present century. Other objectives emphasized the need to utilize ICTs in order to attain the following goals:

- Integrating the country into the global knowledge-based economy, and improving the performance of the services sectors;
- Enhancing the effectiveness of administrative procedures and services in the public sector;
- Improving competitiveness and profitability of production and services sectors, thereby enhancing the performance of the national economy;
- Advancing national performance in education vocational training and continuous learning;
- Providing legislative frameworks for the development and utilization of ICTs;
- Supporting social interaction and subsequent participation in public life by all citizens;
- Ensuring simplification and transparency in government transactions;
- Creation of a national capabilities for the development of software digital content;
- Encouraging production and dissemination of knowledge with emphasis on cultural priorities.

It is of interest that Syria's ICT Strategy lays down a set of conditions as background for its implementation. In fact, an opening statement of the section titled "background for Action program" in the strategy's document makes the assumption that work programs and projects aimed at the development of the ICT sector would run in parallel with:

- A comprehensive program of economic and financial reforms based on thorough analysis of past successes and failures;
- A strategy for sustainable economic development, allowing contributions by all socioeconomic actors;
- A far reaching administrative reform program enabling more effective contributions by government ministries and other public institutions toward economic development with serves rendered to citizens, as well as production and service sectors, coupled to a more open approach toward the world at large, at the core of their operations;
- A comprehensive program of political and social reform, allowing greater and more effective contributions by all citizens in decision making, as well as building confidence between citizens and government institutions.

Conditions set forth for the ICT Strategy's implementation include the need for direct contributions by a variety of public sector institutions, including banks – all of which were state owned and run at the time of the Strategy's formulation – as well as the chambers of trade and commerce.

The overall framework put forward for implementing Syria's ICT strategy for sustainable development included two "backbone programs," to be launched by MCT and executed in collaboration with concerned parties with the objective of restructuring and rebuilding the national ICT sector as well as five national initiatives essentially concerned with attaining specific objectives within the following domains. Notably the objectives of programs and initiatives designed for implementing this strategy included:

- Restructuring the telecommunications sector;
- Wider dissemination of access to ICTs and the internet, within rural as well as urban locations throughout the country;
- A comprehensive program for computerizing public sector institutions;
- Setting up ICT-related incubation schemes and centers of excellence;
- Promotion of Arabic content on the Internet through cooperation with the concerned parties.

Syria's ICT strategy document also included an initiative dedicated to transforming the country into a knowledge society, with components aimed at adopting the declaration of principles of the Geneva phase of the World Summit on the Information Society and the recommendations of the UNDP Human Development Report 2003.

It is essential to emphasize that although the ICT Strategy document appears to be the most comprehensive attempt authored thus far with the intention of promoting specific STI capabilities in the country, it addressed a range of STI application areas that fell beyond issues merely concerned with acquisition and dissemination of ICTs and capacity building in these technologies' utilization. Thus, in addition to promoting ICT services and related enterprises, some of the initiatives advocated by the strategy aimed at overall capacity building in science and technology, education, research as well as the creation of technology-based enterprises. Another positive attribute of this Strategy is that it underscored basic conditions for future socioeconomic development, such as transparency, shared decision making, without which its objectives could not be achieved.

### *3.2.5.2 STI Institutions in Syria*

Syria's government established during the second half of the last century a variety of institutions with mandates including creation, acquisition, dissemination and implementation of scientific and technological knowledge. Thus, a Supreme Sciences Council was established as early as the 1950s and the Scientific Studies and Research Centre and the Higher Institute for Applied Science and Technology came into being in the 1970s. The first decade of this century witnessed the establishment of a "Ministry of State for Technology Transfer and Development" as well as a "Ministry of Communications and Technology" and a "Higher Commission for Scientific Research" was set up under the auspices of the Council of Ministers. A number of both government institutions and NGOs were also created during the first decade of the century with particular emphasis on providing support for innovation and related enterprise creation. Despite several successful developments produced or catalyzed by many of the above institutions their impact remained limited, largely due to overbearing central planning. Yet the impact of many of these institutions remained limited due to a lack of a comprehensive STI policy and the fact that implementing a variety of STI initiatives was undertaken by institutions, including ministries, with limited cooperation and coordination both among themselves and with public and private enterprise institutions.

Concurrent with steps taken toward economic liberalization, several entities were created with view to promoting innovation as well as enterprise and employment creation, particularly during first decade of the present century. Joint initiatives were launched through cooperative ventures with the European Union and other international bodies. Nevertheless, bureaucratic practices inherited from decades past coupled to lack of appropriate measures for monitoring and accountability have tended to diminish the impact of even well designed and reasonably funded initiatives. Thus, despite attempts to create fresh STI institutions, as well as author and debate new STI strategies, including some of the more recent national development plans, much remains to be done, by a variety of actors.

#### 3.2.5.2.1 The Higher Commission for Scientific Research

The HCSR was established in 2005. Its charter emphasizes its role as the top R&D governing body and its direct link to the Council of Ministers. HCSR is essentially made up of three major components: the Supreme Council for Scientific Research (SCSR), which is headed by the Prime Minister, with its membership including several ministers, presidents of public universities, heads of national research as well as heads of chambers of commerce, agriculture and industry, a board of directors and a general secretariat.

HCSR occupies a central position within the national scientific research arena, requiring it to coordinate scientific research activities throughout the country with its mandate including elaboration of comprehensive national policy directives for scientific research and technological development as well as strategies at sustainable development objectives. HCSR is also required to coordinate research plans undertaken by national scientific research bodies, propose the establishment of novel research institutions, as well as draft plans for research spending.

Frame 3.12 presents a list of projects that HCSR is reported as having undertaken, whereas pie charts in Figs. 3.15 and 3.16 present, respectively, the distribution of planning activities across objectives and tools designed for their implementation. Fig. 3.15 indicates that activities aimed at (a) implementing the 10th 5-year plan, (b) resolving immediate problems, and (c) technology transfer receive equal weights, of around 20% each. This distribution raises questions such as why technology transfer is addressed as separate from achieving economic objectives or resolving immediate problems, some of which at least must have economic roots or implications. Furthermore, it might have been assumed that the objectives of the 10th 5-year would in fact include achieving economic objectives as well as technology transfer activities, whereas they are dealt with within separate categories. Additionally, at 14% it seems that a rather large number of issues are grouped within the category "other objectives."

**Frame 3.12 Higher Commission for Scientific Research Activities**

A report issued toward the end of 2011 included a set of projects that Higher Commission for Scientific Research (HCSR) intended to implement in the immediate future, including:

- development of a national information system with details of scientific research undertaken in the country;
- coordination of activities undertaken by scientific research institutions in Syria;
- design of regulations to encourage and evaluate researchers;
- formulation of novel cooperation modalities between research institutions and beneficiary production and services sectors;
- setting up national networks within HCSR for research in renewable energy sources, agricultural production and the environment;
- establishment of consultative committees within HCSR to propose, monitor and evaluate research activities in the fields of water resources management, industry and communications;
- creation of a national nanotechnology committee.

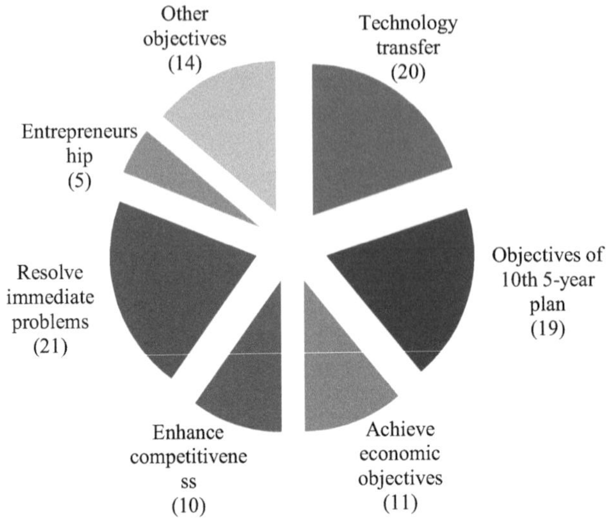


**FIGURE 3.15** Economic and social objectives of Syria's research plans (percentage weight).

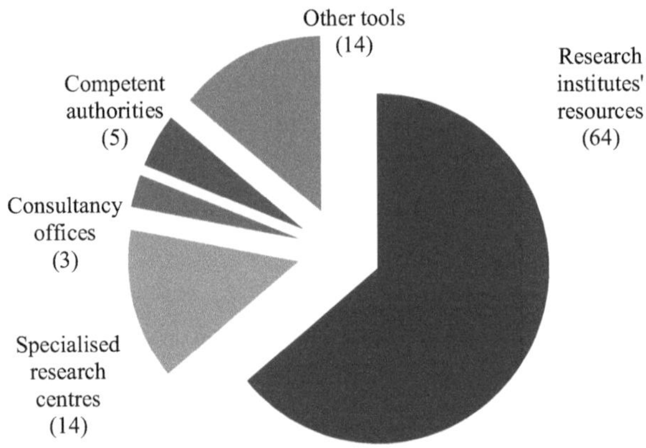


**FIGURE 3.16** Tools for implementing research plans (%).

Fig. 3.16 indicates that the majority of activities required for the national research plan's implementation are handled by research institutions using their own resources. No scope is apparent for special initiatives designed to address specific issues, e.g., regarding transfer of critical STI inputs and their dissemination within production or services sectors.

Main challenges facing HCSR are stated as including:

- lack of an enabling environment for scientific research at the national level;
- the need to reform regulatory arrangements in order to attract and retain talented researchers and maintain improved links with end users.

Additionally, it is likely that HCSR does not appear to have been allocated sufficient manpower and financial resources in order to enable implementation of its tasks as well as might have been desired.

#### 3.2.5.2.2 Other Institutions Concerned With Scientific Research in Syria

Although not directly involved in STI policy formulation, several other institutions undertake a variety of activities that impact research and development as well as technology acquisition and dissemination. Some ministries, for example, are known to possess their own research and development facilities. Prime examples include the Ministry of Industry with its Industrial Testing and Research Centre and the Ministry of Agriculture and Agricultural Reform's General Authority for Scientific Agricultural Research (GASAR). Additionally, chambers of commerce, agriculture and industry often play supporting roles in STI promotion, each within its own domain. Emphasis is generally placed on training of human resources and acquisition of specific technologies, essentially embedded in machinery and processing technologies collaboration with their regional and international providers.

As indicated above, Syria is also the initial home to two international agricultural research centers and often address issues pertaining to agro-food processing as well as local social and economic development, namely, the ICARDA and ACSAD. Prior to the outbreak of hostilities in Syria, the first of these two institutions operated research facilities and field stations, principally in Northern Syria, close to Aleppo. ICARDA enjoys an internationally recognized track record in agricultural research and has pioneered many innovative technologies for the region. Its cooperation with Syrian agricultural research centers is generally quite effective and it engages in frequent capacity building exercises for national as well as regional and international groups of agricultural engineers and decision makers.

#### 3.2.5.2.3 Institutions Dedicated to Innovation and Enterprise Promotion

For nearly two decades prior to the ongoing armed conflict, Syria witnessed a steady growth in the number and variety of institutions aimed at promoting innovation and enterprise creation. In particular, the Syrian Enterprise and Business Centre (SEBC) and the Syrian Trust for Development (STD), which were established in 2006 and 2007, respectively, may be credited with playing a significant role in promoting innovative enterprise.

Both SEBC and STD are funded by the European Union to implement programs aimed at enhancing the competitiveness of Syrian enterprises and their adherence to quality standards, which often result in the adoption of innovative practices. More specifically, SEBC activities are aimed at:

- enhancing business development services;
- facilitating SMEs access to finance and export market information;
- strengthening SME support institution.

Table 3.7 provides a summary view of SEBC projects[96] and programs as well as their individual objectives, durations and budgets. The low level of resources dedicated by SEBC, to two of its principal activities, the SME Support, and the Quality support programs, see Table 3.7, is essentially representative of the general situation with regard to supporting enterprises and innovation throughout.

STD was established in 2007 as a nongovernmental, nonprofit organization with the aim of assisting Syrians intending from all walks of life to play a role in the country's future. STD often partners with local community authorities, United Nations organizations, other NGOs and charities as well as government and the private sector institutions in implementing programs in three focus areas: education and vocational training, rural development, as well as promoting national and local cultural and heritage-related activities.[97]

A detailed evaluation of the performance of efforts by the above programs falls outside the scope of the present work. Suffice it to say, that they would essentially need to be multiplied several times over before they might begin to make truly tangible impacts on innovation as well as enterprise and job creation in Syria.

**TABLE 3.7 Programs and Projects Run by the Syrian Enterprise and Business Center**

| Program | Objectives | Duration | Funds (€ mill) |
|---|---|---|---|
| Small and Medium Enterprise (SME) Support Program | Provide a full range of services to the private sector and the government. | 2006/2010 | 15 |
| Syrian Quality Program | Strengthening quality management capabilities and infrastructure in Syria. | 4 years | 12 |
| Business Environment Simplification Program | Contributing to the Syrian economic reform program. | – | – |
| Trade Enhancement Program | Facilitating Syrian trade flows to enhance globalization benefits. | – | – |
| Enterprise Europe Network—Syria | Offers a "one-stop shop" to meet all the information needs of SMEs and entrepreneurs in Europe and partner countries. | – | – |

From SEBC website. Accessed in June 2016 at: http://www.sebcsyria.com/web2008/index.php.

#### 3.2.5.2.4 STI Support Institutions in Syria

Several institutions were established, mostly during the 1960s and 1970s, with view to providing services concerning intellectual property rights and quality standards. Thus, Syria established its national Authority for Standards and Calibration (SASMO), which offers both paid and unpaid services to various enterprises throughout the country.

With regard to intellectual property rights, Syria promulgated its first article of legislation regarding IPRs back in 1946. According to this law, and its later amendments, patents are valid for a renewable 15-year period provided that an invention is commercially utilized within a set period of time. Syria became signatory to the Paris Convention and a member of the World Intellectual Property Organization (WIPO) in May 2004. The Syrian Association for Intellectual Property (SAIP) was established as an NGO in 2005 using a grant by the UNDP. SAIP gained the status of an observer at WIPO. A law approved in February 2001 included computer programs, books, films, and other creative works as protectable under Syrian jurisdiction. Nevertheless, it is widely known that Syria's IPR system provides insufficient safeguards for protecting and enforcing IPRs. In principle, the Ministry of Economy and Trade processes patent and trademark registration, whereas the Ministry of Culture is in charge of copyrights. Nevertheless, music, software, videos and games on sale in the country are more likely to be pirated copies than originals. An estimated 80% of films sold in the home video market are pirated.[98] Indeed, legal vacuum has long persisted in relation to copyright.[99]

Trademarks are registered for a period of 10 years and the first applicant is regarded as the proprietor of the trademark.[100] Legislation aimed at regulating registered trademarks, industrial models and logos was passed in 2007. Industrial drawings and designs as well as geographical indications are protected by this legislation.

### *3.2.5.3 Scientific Research and Its Outcomes 2008–09*

The following paragraphs summarize the state of STI in Syria during the years 2008–09 as reported by the head of Syria's Commission for Scientific Research (HCSR),[101] indicating that STI activity had been suffering a number of problems even before the breakout of hostilities in 2011. Subsequent sections will provide a concise summary of institutional arrangements and policies that shaped and might continue to influence STI in Syria.

#### 3.2.5.3.1 PhD Researchers and Projects

As indicated in section 3.2.5, the total number of researchers holding PhD degrees in Syria and working within the country's universities or at research centers and government institutions was estimated at 5158 in 2008 and 5218 in

2009. PhD degree holders among researchers in Syria are deemed by HCSR as inferior in number to what might be expected on the basis of national aspirations, actual requirements and global indicators.

Furthermore, it is essential to note that the majority of researchers holding PhD degrees, worked within the country's universities. With regard to PhD research carried out by universities, the HCSR statement indicated that the majority of such research activity was undertaken at Damascus University, although other universities throughout the country were in charge of implementing less than 50% of university research projects. Noteworthy is the considerable increase, of around 38%, in the number of projects undertaken in 2009 compared to those in 2008, according to information provided by HCSR. See Fig. 3.17.

The chart in Fig. 3.18 provides a view of the distribution of research projects undertaken at Syria's public universities, across disciplines during 2008 and 2009. Research projects carried out in the engineering and agricultural sciences were predominant in both years. However, projects in medicine and pharmacology, on the one hand, and basic sciences, on the other, appear to have at least doubled from 2008 to 2009.

It is also worth underlining that in 2008 twice as many research projects were undertaken on religious and legal affairs than on information and communications technology, and that this ratio increased to 2.7 in 2009.

### 3.2.5.3.2 Research Publications

Publications made by researchers at public universities as well as other institutions under the supervision of the Ministry of Higher Education, in 2008 and 2009, in both national and international journals, amounted to 1890 and 2916, respectively, an impressive increase of over 150% in just 1 year. The number of publications by research institutions other than those operating under the Ministry amounted to 1530 and 1506, in 2008 and 2009,

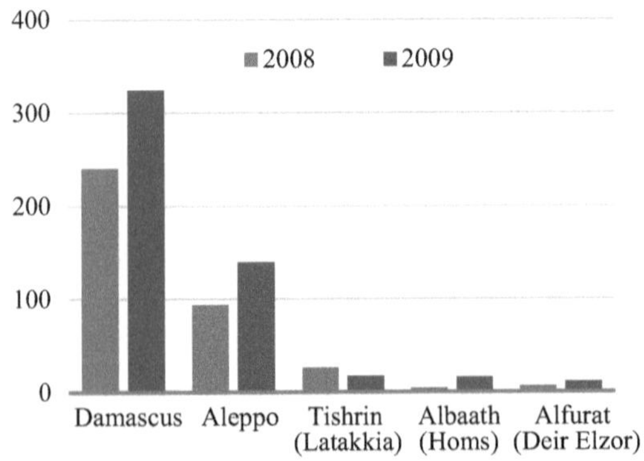


**FIGURE 3.17** Number of PhD projects at public universities in Syria in 2008 and 2009.

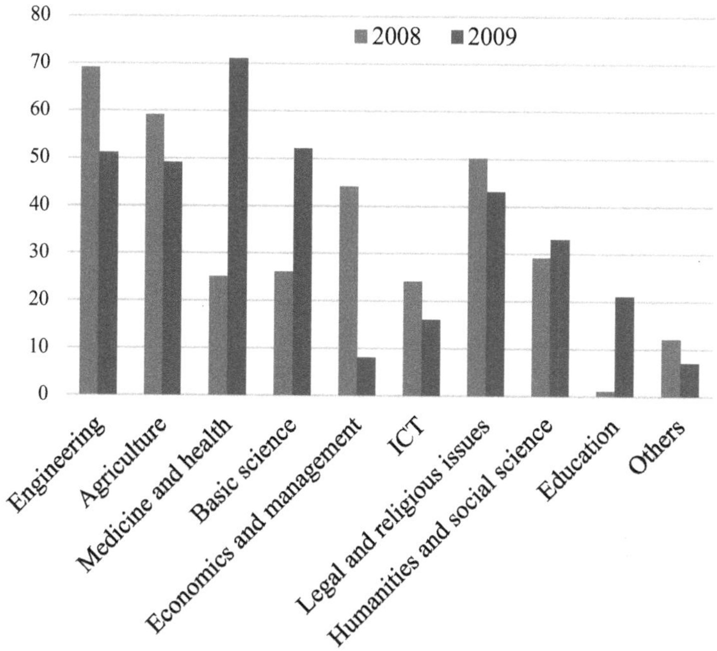


**FIGURE 3.18** Distribution of research projects undertaken by public universities across disciplines.

respectively. A high proportion of these publications is dedicated to agriculture and related sciences, nearly 80% and 84%, in 2008 and 2009, respectively. This preponderance of scientific research publications in agriculture might be explained by the fact that Syria is host to two agricultural research organizations, namely, the above-mentioned ICARDA and ACSAD. Other factors might include genuine interest on the part of concerned university faculties in resolving some of the problems faced by Syria's agricultural sector, on whose performance the well-being of a considerable proportion of the population depended. However, this does not tally with the obvious decline in value added to the country's GDP by the agricultural sector referred to in the section on Syria's economy.

#### 3.2.5.3.3 Registered Patents

Only 65 patents were registered in 2008 by various research institutions in Syria, most, presumably, with national patenting authorities. An even smaller number, 49 patents, is reported for 2009. The above-mentioned HCSR presentation calls for looking into the small number of patents registered, quoting, lack of an innovation policy, and "lack of conviction" as likely reasons for the small number of patents registered by Syrian research institutions. At any rate, it is noteworthy that nearly half the patents registered in both 2008 and

2009 fell within engineering sciences, while around 20% were registered in medicine and pharmacology, on the one hand, and basic science disciplines, on the other. By contrast, the proportion of patents registered in agricultural disciplines was particularly small, around 1.5% in 2008 and 6% in 2009. Considering the large proportion of research projects undertaken in agricultural disciplines, compared to those carried out in other areas, as well as the fact that Syria, as indicated above, is home to two international agricultural research organizations, this indicates that the large majority of such projects were not geared to patentable outcomes.

#### 3.2.5.3.4 Research Priorities

With regard to research prioritization, the HCSR presentation outlined a process whereby the importance of scientific research areas is established on the basis of levels of importance allotted to economic, scientific, strategic and social attributes of the discipline or research area, set at 36, 26, 25, and 13, respectively. A survey should then be used to determine the weights of various factors that determine benefits resulting from a given research project.

As a result of manipulating weights given to these factors and the relative importance of the above attributes, a ranking of research areas is obtained. Table 3.8 presents the results of the exercise undertaken by HCSR, indicating that research topics linked to agriculture, energy, industry, health, and water, occupy the five top positions, whereas issues related to local and regional development, finance, tourism, population, and the environment occupy five positions allocated to the lowest five ranks, in that order. According to this scheme, the environment comes as the very last area in need of scientific research, among the 15 domains of scientific research ranked by HCSR. The notion of looking at certain research domains, such as the environment, water resources, etc., as cross-cutting, i.e., impacting several areas of socioeconomic activity and hence singling them out for special treatment does not appear to have been considered by HCSR.

Furthermore, the HCSR presentation indicates that goals set for each one of the sectors would be proposed within future plans and programs and that these goals would need be compatible with the 11th 5-year national plan, with set activities designed for implementation within three stages, in order to meet short, medium and long term overall development objectives.

#### 3.2.5.3.5 Research Expenditure

Resources allocated to scientific research in Syria were estimated, by HCSR, at US$83 million in 2009. This is a threefold increase relative to the amount actually spent on scientific research the previous year.[102] However, by far the highest proportion of resources dedicated to scientific research

**TABLE 3.8** Higher Commission for Scientific Research Ranking of Research Domains With Respect to Economic, Scientific, Strategic, and Social Importance

| | |
|---|---|
| Agriculture | 4.62 |
| Energy | 4.28 |
| Industry | 4.05 |
| Health | 4.03 |
| The water resources and irrigation | 3.96 |
| Telecommunications | 3.96 |
| Building of enabling capacity | 3.91 |
| Building and construction | 3.83 |
| Transport | 3.70 |
| Social and cultural development | 3.66 |
| Local and regional development | 3.63 |
| Finance | 3.62 |
| Tourism | 3.58 |
| Population | 3.50 |
| Environment | 3.25 |

was allocated to, and spent by, research institutions outside the realm of higher education, with their share amounting to around 7.7% and 9.5% in 2008 and 2009, respectively. See Fig. 3.19.

#### 3.2.5.3.6 Research Cooperation

HCSR's evaluation of cooperation among institutions, concerned with conducting scientific research within economic and social sectors, is summarized in a pie chart indicating that, with regard to the total number of cases evaluated, effective cooperation took place in 19% of these cases, whereas insufficient cooperation took place in 74%, and no cooperation whatsoever took place in 7% of the cases.

### *3.2.5.4 International Research Cooperation*

A report,[103] prepared for the European Union in 2005, listed a number of shortcomings suffered by Syria's research system[104] but considered that international

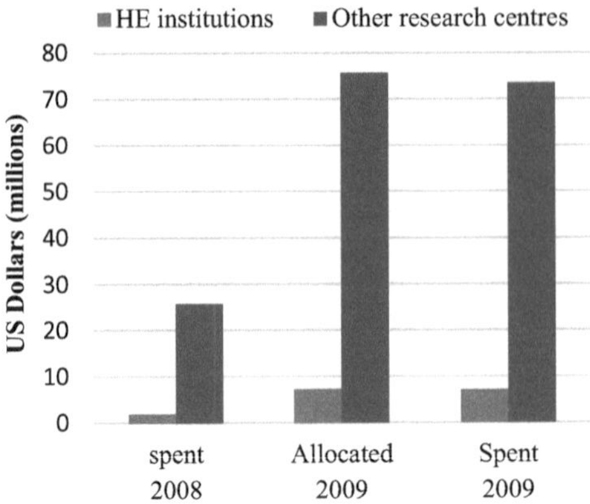


**FIGURE 3.19** Funding allocated and spent on scientific research in Syria 2008–09.

cooperation has been an asset of the present research system in Syria. In particular, capacity building efforts funded by international organizations, mainly the UN and EU, and bilateral aid, with Germany, Japan, as well as other countries, were considered extremely useful as bases for modernizing the research infrastructure.

Research cooperation with other countries in the region is also addressed by this report, citing the example of the "Scientific Research Cooperation Programme between Lebanon and the Syrian Arab Republic." This program was launched in 2001 with the aim of promoting scientific cooperation between Lebanese researchers and their Syrian counterparts. Reports have also been made of recent attempts at forging closer cooperation between Syria's research institutions and well established research establishments in Turkey and Iran.

In order to reap benefits through cooperation, this report calls for facilitating visits to partner and counterpart organizations in the EU and other countries in the region and to participate in conferences and seminars. Other interventions suggested by the report include adoption of systematic policies and changes to regulatory frameworks. Catalytic interventions would certainly be needed in the short term, however.

Concluding remarks by HCSR include the observation that the "STI system in Syria is not functioning well." With regard to STI indicators HCSR points out problems in data collection due to lack of information, limited motivation and inadequate understanding of the role that indicators play in the design and implementation of research policies.

### *3.2.5.5 Innovation in Syria*

Three reports were produced during the past 15 years on innovation in Syria. Two were authored by European experts between 2003 and 2009, and the latest was finalised in 2011 by a World Bank team. The following is a summary of the main findings and recommendations of these reports.

The report sponsored by the European Union's Medibtikar program in 2003,[105] examined the state of Syria's STI governance with emphasis on innovation as defined through EU and OECD approaches.

According to this report, a traditional view of innovation was held by the business sector; whereby innovation inputs were viewed mostly as embedded into machinery and processes, with a limited role for indigenous entrepreneurial initiatives.

In reference to innovation stakeholders, the report indicated that, for the most part, principal stakeholders of the national "innovation system" in Syria engaged in sporadic dialogue. Efforts by national research institutions were similarly fragmented with individual researchers' and institutional loyalties outweighing national interests.

In particular, the Medibticar report considered that private sector entrepreneurs active in production and services sectors should receive far greater support, from national federation and chambers of commerce, industry and agriculture in line with viable innovation policies. The fact that these federations and chambers took passive, at best ad hoc, postures meant that little or no progress could be achieved on the innovation front.

The report finally highlighted the need for formulating a national innovation strategy that would help establish a common vision, reference points and milestones with positive reflections on STI governance in the country. Adoption of "innovation policy cycles" was needed with the aim of revamping STI governance in Syria. The report underlined the need for inculcating a set of attributes found to be essentially lacking or in need of much greater attention, such as clear responsibilities, improved coordination, an active system of accountability and an evaluation-based culture, effective stakeholder involvement, and an international outlook in support of further improvement.

A study of Syria's national innovation system was carried out in December 2009 by the German international aid agency, GTZ.[106] The study was aimed at identifying future directions for enhancing the competitiveness of local SME as well as the capabilities of selected NIS actors. In its concluding section the GTZ study presented a set of recommendations for implementation in the near future.

The GTZ study was based on an expert opinion survey,[107] utilizing determinants grouped within four levels: a macro level in reference to innovation policy, intermediate levels relating to institutional and programmatic support and a microlevel that addressed capacity building. Thus, national and regional innovation policies and master plans are among determinants included under the macro level. Determinants under the institutional level included technology transfer centers, technology parks and clusters as well as business promotion agencies, although determinants at the programmatic level included funding programs for applied R&D, the design and adoption of innovation-friendly regulations as well as promotion of an entrepreneurial culture and entrepreneurship training. Determinants at the microlevel included action at the level of university curricula and innovation related R&D, incubation schemes.

One of the more significant conclusions of the GTZ study at the policy level was that innovation in Syria was in early stages of development, hence, a national innovation policy was needed. Elements for such a policy were available, albeit as scattered proclamations, policy directives, laws and regulations as well as enunciations made in 5-year plans.

STI foresight exercises, which would constitute means for the prioritizing and planning activities aimed at promoting innovation, are yet to be undertaken. Nevertheless, many experts from academia and industry appeared to be involved in discussions on future technological developments and challenges.

With regard to the institutional support for innovation, the survey revealed that business promotion agencies set up by the Syrian government, aided in part by international donors, were highly regarded as tools that support firms in relation to the adoption of innovative inputs. Innovation service providers were found to operate in a range of areas. However, institutions such as technology parks, technology transfer centers and clusters, are practically nonexistent, or in early stages of development.

In relation to programmatic support for innovation, determinants related to education and entrepreneurial support were rated fairly highly by experts taking part in the survey, which further indicated that a number of programs, initiatives as well as institutions that support entrepreneurs have been launched in a number of areas.[108]

Concerning public funds for R&D, barriers against access to such funds were considered to be too high in practice and demand by far exceeded available funds. Significantly, Private investments in R&D are still a very rare exception rather than the rule.

At the level of innovation capacity in Syria, a great deal of emphasis is placed on the role that should be played by the country's universities. However, traditional curricula, limited research activity, and poor links to industry tend to limit their universities on the national innovation scene. And, although promising approaches could be identified, the survey concluded that it could be sometime before their impact is felt.

Despite the fact the development of an innovation culture in Syria was being championed by groups of entrepreneurs based in major cities, SMEs were not yet in a position to act as drivers of national innovation. This is essentially due to the fact that only very few innovative SMEs have been created.

A set of recommendations was put forward by the GTZ study, emphasizing the need for:

- drafting and implementing a national innovation policy with derivative strategies involving all relevant actors;
- nurturing an innovation friendly culture while at the same time improving synergy between the research community and production and services sectors;
- improving framework conditions for entrepreneurs;
- increasing the capacities of institutional innovation supporters;
- establishing a series of innovation support structures, including innovation networks, technology transfer centers and clusters.

Experts interviewed within the scope of the GTZ survey considered technology transfer as one of the most critical challenges for the Syrian economy, with infrastructural arrangements and budgets dedicated to technology transfer and dissemination in Syria woefully wanting, even in comparison to some neighboring countries. In particular, budget allocations for technology transfer were insufficient for meeting actual demand.

Ultimately, potential improvements in technology transfer were bound to be directly linked to government's readiness to provide public funding for programs that facilitate technology transfer both from outside sources as well as between industry and academia. Demand oriented programs, coupled to effective management, held the promise of considerable benefits. However, in order to reach fruition efforts along these lines would require considerable allocations as well as astute and transparent coordination, monitoring and reporting modalities.

No science or technology parks were in evidence in Syria.[109] With many essential ingredients for successful technology parks, including innovative companies, centers of research excellence, a network of auxiliary and support services, legal, financial as well as technical support schemes more or less absent. Hence, the GTZ document considered that technology parks may be a more likely option in the medium, rather than immediate or short term. Alternative institutional forms such as incubators and technology transfer centers, attached to universities and other higher education institutions were deemed as standing a better chance of promoting innovation in the short term.

In concluding this section, it is essential to emphasize the role that might be played by micro, small and medium enterprises in promoting and disseminating innovative inputs within the country's economy. Thus, the number of private SMEs in Syria is estimated as exceeding 300,000. Syria's SMEs are stated as responsible more than 60% of the country's GDP.[110] Difficulties met by these SMEs include accessing skilled manpower and the ability to produce innovative output with wider market outreach towards both regional and international markets.

Indeed, innovation management is in fact underdeveloped throughout the country. However, innovation is not the only challenge currently faced by the country's SMEs. Other challenges include limited access to novel technologies and to capital for renovation and upgrading, Responses to this situation must come in several and varied forms with emphasis on radical regulatory and legislative reforms as well as arrangements facilitating access to venture capital. Strategic approaches with focus on enhancing SME capabilities might include the establishment of a small and medium enterprise agency to strengthen links between SME and other components of national innovation system. Objectives and tasks set for such an agency should guarantee delivery of a range of innovation-related inputs and services dedicated to SMEs, with well-defined interventions, specific goals and monitoring modalities.

#### 3.2.5.5.1 The Rapid Innovation Action Learning Initiative in Syria

In 2010 a World Bank Team of experts undertook a mission to Syria within a global initiative[111] referred to as the Rapid Innovation Action Learning Initiative or RIAL. Based on a background study as well as extensive meetings with concerned officials, the team took a close look at Syria's innovation system and its enterprise promotion schemes.

Conclusions of the RIAL mission's report indicated that, insofar as innovation was concerned, Syria was underperforming and that although this might have partly been due to external factors, including persistent regional political tensions, more inherent causes might be blamed. However, rather than dwell on causes of problems facing innovation in Syria, the RIAL Team's attention was devoted to producing a set of recommendations with view to gradually enhancing the performance of the country's innovation system.

The RIAL Team was thus able to identify several causes that should enable Syria to attain higher grounds, including its human capital, both at home and abroad, available natural resources as well as the fact that the government's intention to transform the country into a knowledge economy by the year 2025. Because attaining such an objective would necessitate wide-ranging innovations the report emphasized the need to adopt gradual and measured steps leading to promoting innovation and entrepreneurship within reform and liberalization measures that were already being taken by Syria's government.

In essence, steps had to be taken at several levels and on a variety of fronts, in order to remodel economic regulations, the legislative environment and the educational system, following intensive debate, aimed at fine-tuning the recommendations produced by the mission's report.

It is notable that the final statement in the report's conclusions underlined the need to ensure that the proposed national innovation policy and the design of its implementation strategies and action plans would deliver benefits as widely as possible, because, otherwise gains attained would be quite limited and ephemeral.

### 3.2.6 Concluding Remarks

Violence in Syria has claimed hundreds of thousands of lives. By early 2015 political upheaval and armed struggle had destroyed more than one third of the country's capital assets and cut back its GDP to where it was more than a quarter-century ago, turning millions of the country's inhabitants into refugees.[112] The human and economic cost of the conflict is on the rise with no end yet in sight.

Analyses of the immediate causes of the conflict have tended to blame a wide range of factors. However, it seems certain that the very roots of the still unfolding Syrian disaster are linked to repeated failures on several fronts. In particular, decades of short-sighted social and economic development policies and inadequate responses to new global and regional realities may confidently be quoted as essential popular resentment.[113]

Essential bases for progress, embodied in good governance, quality education and transparent efforts at socioeconomic development had simply gone lacking for too long, precipitating the humanitarian disaster that befell the country and its people with echoes reverberating within the region, even around the entire globe. Needless to say, all the above shortcomings and failures are linked both directly and indirectly to inadequate assimilation, dissemination, and implementation of knowledge, a good deal of which related to science, technology and innovation. This is attested by numerous reviews particularly targeting the state of scientific research and technology development in the country.

Inability to adapt and utilize existing, as well as generate new, STI knowledge in order to tackle many of the problems that the country faced, the manner and rate with which STI inputs were incorporated in various areas of socioeconomic activity, are all implicated. The scale of social and political reforms that are certainly needed on many other fronts and at various levels of national, provincial and institutional governance may not be underestimated. Thus, to recall the concluding remarks in the above mentioned HCSR presentation the "STI system in Syria is not functioning well." This same presentation, went on to detail a series of problems including difficulties encountered in compiling and evaluating STI performance, insufficient understanding of the role that indicators play in the design and implementation of research policies as well as inadequate measures for motivating researchers, in general.

Nothing in other literature sources would seem to contradict the above observations. In particular, the effectiveness of steps taken by HCSR to orient scientific research in response to the country's most needy sectors or domains of research activity appear to have gone largely unheeded, at best producing little or no impact in actual reality. A prominent example of this state of affairs is provided by scientific research in agriculture. Thus, as indicated above, agriculture was ranked as a top priority according to the

HCSR exercise, and the number of scientific publications, both within the country and in regional and international journals, by far exceeds publications in all other areas of scientific research. However, this immense interest does not seem to have been able to reverse or mitigate the constant decline of value added due to the agricultural sector, measured as a percentage of GDP. See Figure 2.9 in Chapter 2. And, although it must be granted that value added due to agriculture, may have been depressed by a variety of factors, it would be correct to assume that scientific research conducted in aid of this sector was unable to stop the decline in its contributions to the national economy.

Steps taken to promote research, technology transfer and innovation with view to sustainable development were severely curtailed by 2012, with the exception of activities designed to serve priorities dictated by the on-going conflict. However, despite past failures and current diversions, an examination of past attempts at harnessing STI inputs for sustainable development should be of immense value.

As to the future, it is certain that nothing short of a complete overhaul of the country's political, social and economic institutions would succeed in putting Syria back on the map with promise of a better future for its people, its neighborhood and the world at large. In achieving such ends, science, technology and innovation are destined to play an immensely crucial role.

Making proper use of some of the positive developments that seem to have taken place in the past, requires greater focus on measures aimed at:

- Ensuring that more comprehensive policy reforms are undertaken targeting all aspects of STI activities including education in all its stages. Active participation by all stakeholder organizations must be guaranteed in drafting new policy regimes, so that they may take ownership of, and commensurate responsibility for, subsequent implementation, monitoring, and evaluation.
- Modernizing research organizations, their operational modalities, linkages as well as their evaluation methodologies. Research institutions will need to take a more proactive role in overall development with clear agenda objectives. At the same time, new laboratories and pilot facilities will be required to cater for innovation in crucial sectors with modern equipment and well-trained personnel.
- Extending infrastructures crucial for carrying out as much STI activities as possible at the local community level as well as redirecting sectoral development initiatives in a manner that would prioritize higher levels of competitiveness and job creation.
- Providing STI activities with much higher levels of material resources and access to infrastructural arrangements, with particular emphasis on STI activities in priority sectors and local community components.

- Introduce project-based funding in parallel with institutional funding; Design and implement flagship schemes to promote areas of strength with particular emphasis on European and regional clients/partners; Create and sustain new clusters.
- Promoting more intensive and much wider STI cooperation and collaboration within the national, regional and international arena with particular emphasis on manpower training as well as resource sharing. At the national level, full attention must be allocated to enhancing university–industry cooperation. This would have to proceed through examining:
  - current as well as feasible future capabilities of public and private sector universities to cater for the STI needs of production and services sectors;
  - needs and requirements for establishing new educational and research specialties and related facilities with view to improved response to market needs.[114]
- Implementing measures to ensure adherence to established norms concerning quality issues, related to environmental and health impacts. National initiatives would be needed for joint efforts by research organizations, universities, concerned public authorities as well as civil society organizations in full cognizance of regulatory environment, in order to identify barriers to change as well as measures needed to conform, derive objectives, design modalities and milestones for modern STI responses to environmental and health issues.
- Making full use of information and communications technologies, with particular emphasis on areas such as education, research and entrepreneurship. Particular attention would have to be allocated to improving the scope and quality of vocational training and life-long learning through ICT mediated means as well as traditional methods. Emphasis must also be placed upon attaining broad strategic frameworks for coordination among all concerned actors and stakeholders with accent on the highest quality in training activities.

An area where direct government action is needed, aided as far as possible by concerned regional and international organizations, is in securing more plentiful resources for STI activities within a well-defined system of priorities and with modalities in place that facilitate the transfer of results into commercial products and processes and extending full and long-term benefits at the level rather than rapid returns for businesses, national regional and international.

Other areas in need of urgent action include building sustainable bridges between academic institutions and priority sectors of the economy. Although novel institutional forms designed in aid of innovation and technology transfer have already started to appear on the STI landscape, there is still need for many more incubators to cater for a much larger number of entrepreneurs. Industrial clusters and innovation support institutions are yet to make an appearance and must be placed as close as possible to the top in the list of future priorities. However, many of the above developments would produce little impact without the benefit of innovation-friendly legislation and regulatory arrangements.

Despite the turmoil, bloodshed and destruction it has witnessed, Syria is still in possession of considerable resources, chiefly due to its creative and youthful population, both at home and abroad. There is every reason to believe that, equipped with a viable national STI system Syria would be able to achieve resounding successes in handling the multitude of challenges facing its future development.

## 3.3 ARAB COUNTRIES OF THE NILE VALLEY

Egypt was one of the first countries in the Arab region to build industrial and technological capabilities back in the nineteenth century. Both Egypt and Sudan were leaders in establishing the region's first institutions of higher education, very early in the twentieth century.[115]

Although both countries can boast isolated successes in specific areas of scientific research and in technology development, it may confidently be stated that STI institutions in neither country did go on to fulfill the promise they held, particularly with regard to renovating their respective economies, creating jobs and ameliorating poverty.

The following sections address science, technology and innovation systems in both countries with emphasis on policy frameworks and institutional settings.

### 3.3.1 Egypt's STI System

Overregulated and heavily centralized governance procedures coupled to constantly shifting sectoral strategies, are known to be obstacles to capacity building in science, technology and innovation in a variety of developing countries. Egypt has been no exception. Rather, despite isolated successes, many indications would point to failures by the country's research system in effectively tackling Egypt's socioeconomic priorities. The following sections are dedicated to a review of the country's STI governance, its recent policy initiatives, major institutions and funding modalities.

#### *3.3.1.1 Egypt's STI Governing Institutions*

Following several restructuring exercises undertaken in 1998 and more recently in 2006, Egypt's STI system consist of the following institutional entities:

- The Ministry of Scientific Research (MOSR);[116,117]
- The Higher Council of Science and Technology (HCST-E)
- The National Council for Education and Scientific Research (NCESR);[118]
- The Academy of Scientific Research and Technology (ASRT);
- The Science and Technology Development Fund (STDF);
- Independent research institutes as well as research centers within universities;
- Several technology and innovation initiatives.

Naturally, several other institutional entities contribute to the STI system's operations. These include the Ministry of Higher Education as well as several other ministries and supervisory bodies referred to as "Supreme" Councils that undertake coordination activities among the various national research centers and universities,[119] including the Supreme Council for Universities.

The Ministry of Scientific Research (MOSR) oversees a number of constituent research facilities, including the Academy of Scientific Research, the National Research Centre, the Remote Sensing Centre, the Petroleum Research Centre as well as several longstanding centers concerned with medical research, such as the Eye Diseases Centre and the Theodor Bilharz Research Centre.[120] MOSR funds and supports Scientific Research and Innovation through several programs targeting:

- Innovation funding;
- Small-business innovation research;
- New technology-based start-ups.

MOSR is also reported as having established and sponsored specialized networks and clusters in order to promote new technologies linked to particular industrial applications.

The Higher Council of Science and Technology was created in 2007, as a result of a restructuring of the country's STI system. Egypt's HCST is chaired by the Prime Minister and charged with strategic planning "for the utilization of scientific research for economic and societal development."

The Academy of Scientific Research and Technology (ASRT) was also restructured to perform an advisory role with regard to STI policies as well as assessment and evaluation. Additionally, tasks that ASRT undertook, in relation to funding STI activities, were transferred to a newly established Science and Technology Development Fund (STDF), which was charged with:[121]

- Funding scientific research and technological development in accordance with the priorities defined by Egypt's HCST;
- Supporting the full cycle of scientific research and upgrading knowledge and technology-based products;
- Promoting the dissemination of data and information on science and technology;
- Performing a regular and thorough evaluation of science and technology indicators including research papers, patents and their impact on the economy for submission to HCST in Egypt.

Further details concerning ASRT are included in Section 3.3.1.3 below, on Egypt's STI institutions. At any rate, the above arrangements are credited by official documents with stimulating Egypt's S&T activities through funding research undertaken by Egyptian scientists as well as forging partnerships with their counterparts in advanced countries in order to "keep track with quickly advancing technology."[122]

### 3.3.1.2 *Science and Technology Policies*

Several attempts have been made at revitalizing national science and technology during the past 2 decades. The latest was made in 2006, on the basis of an overall evaluation of the national S&T system, following which Egypt launched the "Decade for Science and Technology 2007–16" with the aim of strengthening national S&T capabilities through, among other means, fostering cooperation with developed countries.[123] A scheme titled "Developing a Scientific Research Plan 2007–16" was drafted emphasizing S&T governance, increased funding and improved training for human resources. Principal objectives of this plan included developing a complete value chain from research to commercialization and disseminating an S&T culture throughout society. This Plan adopted a sector and technology-oriented approach. However, in February 2012, a new strategy was announced, with commercialization of research outcomes as its primary objective.

An even more recent document targeting revitalization of Egypt's scientific research and innovation scene, was included within Egypt's Vision 2030. Thus the third "pillar" of this vision is solely concerned with "Knowledge, Innovation, and Scientific Research." The following paragraphs will consider this document with particular reference to its principal objectives, indicators to be used in monitoring progress of its implementation and its main programs.

#### 3.3.1.2.1 Knowledge, Innovation and Scientific Research in Egypt's Vision 2030

Egypt's Vision 2030 document was published in May 2016 by Egypt's Ministry of Planning, Monitoring and Administrative Reform. The vision document is intended as a comprehensive strategy for sustainable development based on 10 pillars including one specifically dedicated to Knowledge, Innovation and Scientific Research. Clearly, all the other nine pillars[124] should theoretically receive support from improved access to, and better utilization of, knowledge and scientific research capabilities, which constitute the focus of the Third Pillar. This is in fact acknowledged in the introduction to the section on that pillar, with the following essential components:

- A general environment, with the State playing a primary role in creating conditions that nurture innovation, and empower the economic system through improved policies, legislation, and principles of governance.
- An infrastructure that is supportive of creative production and services activities, with standards and regulatory arrangements that are supportive of innovation and scientific research.

- Educational and scientific research activities encompassing research organizations and institutions, technical education and vocational training institutions as well as HEIs.
- Industries including large-, medium-, and small-scale firms taking part in activities that promote innovation and actively utilize R&D outcomes.
- Socioeconomic development creating demand for innovation, knowledge, and scientific research to fulfills needs.
- Intermediary institutions and arrangements with particular focus on linking industrial innovation, research and development on the one hand, with academic and research institutions, on the other.

The Third Pillar is to enable conversion of knowledge, innovation, and scientific research into outcomes that satisfy "the developmental needs of the State during the next 15 years." Thus, through "a comprehensive system ensuring the developmental value of knowledge and innovation" challenges would be overcome and national objectives attained. Three essential objectives of the Third Pillar are enumerated as:

- Creating stimulating environment for the acquisition, adaptation and production of knowledge, through improved legislative and infrastructural arrangements and a conducive investment climate;
- Linking applications of knowledge, scientific research and innovation outputs to national priorities, through devising means whereby sectorial priorities may be met and challenges overcome;
- Developing of an integrated national innovation system.

According to the Vision document, challenges that had to be overcome in the process included poor funding and the need to raise awareness of the importance of knowledge, innovation and scientific research in order to attain higher rates of sustainable development and encourage more young people to take up careers in scientific research and innovation as bases for entrepreneurial initiatives.

Table 3.9 lists 24 performance indicators that the Third Pillar's strategy aims to use in seeking the above goals, together with estimates of their current values as indicated by reports issued by the World Intellectual Property Organization (WIPO) and the World Economic Forum (WEF)[125] as well as the values to be attained in the years 2020 and 2030. A number of issues may be raised with regard to these indicators. That they are almost totally based on ranking systems designed and implemented by international organizations, rather than on tangible attainments linked to national priorities, is noteworthy. And, although it might be argued that most of the above indicators could eventually be linked to national priorities, their adoption would undoubtedly complicate issues since international comparisons involve numerous exogenous factors over which the national STI system has no control whatsoever.

**TABLE 3.9 Indicators to Be Used in Monitoring Progress of Implementing the Third Pillar of Egypt's Vision 2030**

| Indicators[a] | Current Value | Value by 2020 | Value by 2030 |
|---|---|---|---|
| **1.** Global Innovation Index (GII) | 99 | 85 | 60 |
| **2.** Innovation efficiency ratio | 0.8 | 0.85 | 1 |
| **3.** Companies' innovation capacity index | 132 | 100 | 60 |
| **4.** Knowledge impact GII subindex | 89 | 80 | 60 |
| **5.** Knowledge transfer GII subindex | 69 | 60 | 30 |
| **6.** Innovative products and services GII subindex | 98 | 85 | 60 |
| **7.** Information and communication technology GII Subindex | 73 | 50 | 30 |
| **8.** Public infrastructure GII subindex | 121 | 100 | 60 |
| **9.** Environmental sustainability GII subindex | 65 | 50 | 30 |
| **10.** Knowledge contribution GII subindex | 71 | 50 | 30 |
| **11.** Intangible assets GII subindex | 89 | 80 | 60 |
| **12.** Digital creativity GII subindex | 74 | 60 | 30 |
| **13.** Innovation linkages GII subindex | 70 | 60 | 30 |
| **14.** Quality of scientific research institutions | 135 | 100 | 60 |
| **15.** Input Credit facilities GII subindex | 123 | 100 | 60 |
| **16.** Investment GII subindex | 138 | 100 | 60 |
| **17.** Trade and competition GII subindex | 124 | 100 | 60 |
| **18.** Legislative environment GII sub-index | 131 | 100 | 60 |
| **19.** Business environment GII subindex | 105 | 90 | 60 |
| **20.** Education GII subindex | 53 | 50 | 30 |
| **21.** Higher education GII subindex | 102 | 90 | 60 |
| **22.** Research and development GII subindex | 50 | 40 | 30 |
| **23.** Knowledge absorption GII subindex | 119 | 100 | 60 |
| **24.** Knowledge workers GII subindex | 69 | 50 | 30 |

[a]*With the exception of indicator 2 all indicators in this list refer to rankings allocated to Egypt in reports issued by the World Economic Forum's Global Competitiveness Report 2013, and the Global Innovation Index 2014.*

From "Sustainable Development Strategy; Egypt Vison 2030. Third Pillar: Knowledge, Innovation and Scientific Research;" Downloaded in March 2017 from: http://sdsegypt2030.com/wp-content/uploads/2016/10/4.-Knowledge-Innovation-Scientific-Research-Pillar2.pdf.

Another issue that might be raised with regard to indicators proposed for charting progress toward implementing the Third Pillar strategy is their bias toward innovation and business development. Thus, even though some of the indicators refer to educational, environmental, and scientific research developments, they do so in reference to outcomes linked to innovation and business development rather than specific social welfare and overall socioeconomic development issues.

This is understandable, considering that these indicators belong to reports that are essentially concerned with innovation as basis for business competitiveness, namely the Global Competitiveness Report and the Global Innovation Index Report. A more suitable choice of indicators might have been a mix including some of those now to be used, essentially based on innovative business development, on the one hand, and the customary indicators that are normally used in charting progress of implementing scientific research and technology development strategies,[126] on the other.

Another advantage of simpler and more straight forward indicators for measuring progress toward scientific research and innovation is due to the fact that compound indicators can carry distortions that render their application within certain national environments unreliable.

In addition to the above mostly compound indicators that the Third Pillar strategy is supposed to use, a set of 13 new indicators are suggested as supplementary means for gauging progress of the Third Pillar's strategy. Table 3.10 presents a list of these indicators. At least some of these indicators appear problematic with particular regard to the attributes they are intended to measure and methods that might be used for qualitative and quantitative estimation of both

**TABLE 3.10 Suggested Additional Indicators for Use in Monitoring Progress of Implementing the Third Pillar of Egypt's Vision 2030**

1. Ratio of knowledge economy contribution to GDP
2. Social return on investment in innovation
3. Ratio of local content to gross national manufacturing
4. Sectorial ratio for local component to gross national manufacturing
5. Number of small and medium companies contracting with government
6. Number of international awards in the innovation field
7. Number of newly founded companies in each sector
8. Ratio of companies practicing innovation and research and development (R&D) for each sector
9. Number of patents per sector
10. Ratio of sectorial budget expenditure on innovation, research, and development
11. Ratio of companies' contributions to total sectorial expenditure on innovation and R&D
12. Ratio of international funding for innovation, research and development to total funding
13. Ratio of foreign funding to small and medium companies

their initial as well as interim and final values. Two of these indicators may serve to illustrate the above.

- *Ratio of knowledge economy contribution to GDP*: which is intended to measure "how far the economy of research and development is contributing to GDP, representing the importance of the knowledge economy in Egypt's priorities;" which will be evaluated on the basis of counting all "expenditures on innovation and R&D" from all possible sources.[127]
- *Social return on investment in innovation*: which is designed to measure "the impact of expenditure on innovation on Egyptian society [turning] it into financial returns in order to ensure the development of an environment that encourages innovation." In order to evaluate this indicator, the Third Pillar's strategy document proposes the use of a "combination of qualitative and quantitative indicators, measuring investments returns."

Arriving at accurate assessments of progress based on the above two indicators taken as mere examples would prove immensely difficult given the state of affairs regarding Egypt's economy and the even the more complex state of Egypt's society, particularly considering the dearth of research on social issues precipitated by political as well as economic considerations.

Challenges expected to face implementation of the Third Pillar's strategies are classified into three main groups, with regard to their impact and relative ease with which they might be overcome. Within the first group, the Third Pillar's strategy document lists inadequate legislative tools and regulatory arrangements, on the one hand, and poor "coordination between social needs and innovation," on the other. Both of these challenges are considered relatively easy to tackle, with overcoming them assured of high impact.

Challenges within the second group, which are also considered to be relatively easy to overcome, include:

- Inadequate sectoral planning;
- Poor linkages connecting knowledge to innovation;
- Insufficient incentives for innovation;
- Medium and small companies' inability to produce and market innovation.

However, addressing these challenges was considered to result in medium impact, in terms of outcomes achieved.

Within the third group two challenges, characterized by their high impact, resilient nature and the time it would take to overcome, are identified as a poor "innovation culture of society"[128] and inadequate "awareness of the importance of intellectual property and its protection."

Thus, these challenges seem to be viewed as inherent characteristics of the national science, technology and innovation system, on the one hand, and intricately linked to components of the national socioeconomic system, on the other. Although in fact, they are essentially mere products of inadequate legislative tools and regulatory arrangements, ill-equipped and inadequately financed educational and vocational training institutions as well as a mismanaged economy suffering the impact of corrupt practices at various levels.

Additionally, it would appear that emphasis on intellectual property rights (IPR) protection might have been taken up in order to please possible lenders and donors, whereas the various problems that presently plague Egypt's sustainable development reside almost entirely elsewhere. In-depth analysis might well be needed in order to accurately estimate the extent of damage caused by lax IPR regulations to sustainable development initiatives. It is highly likely that such an analysis would reveal that many other issues should be awarded precedence.

Preoccupation with IPR regulations is further manifested in priority programs designed in implementation of the Third Pillar's strategy. These programs are divided into two categories, with the first including two programs aimed at reforming legislation and governance of activities related to knowledge and innovation. The second group has constituent programs intended to support achievement of the strategy's objectives.

Table 3.11 summarizes the essential objectives of programs in both categories indicating that at least three of these programs were to be initiated in 2016 and that were two programs were expected to be concluded within the next few years. Thus, the program intended to review and "implement legal reforms related to knowledge and innovation," including legislative frameworks concerning higher education, scientific research, investment, tax incentives systems, governmental expenditures, and intellectual property protection was scheduled for completion by 2017, whereas the program targeting promotion of private-public partnerships in support of innovation would be finalized by 2020. Considering that work on both programs may have been initiated in 2016, it would seem that at least the 2017 deadline set for finalizing the former program might be too optimistic, in view of the substantive scope of this program's activities and their spread across a variety of sectors and application areas.

Judging by available information regarding programs within the second category, it would seem that far better definition of their particular goals is needed if their implementation is to be appropriately monitored and positive outcomes ensured.

At any rate, it does not appear that any of the programs proposed as means for implementing the Third Pillars strategy is aimed at tackling head-on some of the principal science and technology related development priorities, such as the environment, water and food security. Egypt is known to suffer serious water scarcity—that is forecast to become even more acute as Ethiopia finalizes its dam—and is threatened by desertification and permanent soil damage. Additionally, the development of new and renewable energy technologies and moves away from dependence upon hydrocarbon resources in power generation have repeatedly been identified as a national policy priorities.[129]

**TABLE 3.11 Essential Programs Designed With View to implementing Egypt's Knowledge, Innovation and Scientific Research Strategy**

| | | |
|---|---|---|
| **Legislation and Governance** | | |
| Legal reforms related to knowledge and innovation | Review legislation related to knowledge and innovation in sectors, including higher education, scientific research, investment, tax incentives systems, government expenditures, and IPR schemes. | Review and amendment of legislation "is scheduled to be performed by the year 2017." |
| Developing and restructuring the knowledge and innovation system | Enhancing the efficiency of knowledge and innovation management systems through rational governance ensuring coordination and integration of strategic priorities for sustainable development. | To start by 2016 and to be fully implemented by 2030. |
| **Programs to Support Achievement of the Vision and Strategic Objectives** | | |
| Promoting an innovation and knowledge culture | Encouraging creative thinking and the development of an innovation culture among youth and community through achieving integration of efforts through motivation and awareness initiatives. | To start by 2016 and to continue until 2025. |
| Stimulating innovation activities by SME | Creating a conducive environment for innovative SMEs in priority areas, maximization investment returns and competitive advantage. | To start by 2020 and to be fully implemented by 2030. |
| Public–private partnerships to support and stimulate innovation | Supporting partnerships between the public and private sectors to maximize the volume of investment in innovation and the creation of conditions to ensure ample returns and integrating knowledge output with market needs. | To start by 2016 and to be fully implemented by 2020. |

From "Sustainable Development Strategy; Egypt Vison 2030. Third Pillar: Knowledge, Innovation and Scientific Research;" Downloaded in March 2017 from: http://sdsegypt2030.com/wp-content/uploads/2016/10/4.-Knowledge-Innovation-Scientific-Research-Pillar2.pdf.

### *3.3.1.3 Principal STI Institutions*

Egypt was one of the first few countries in the region to establish institutions specializing in the promotion of scientific and technological capabilities and the formulation of related national policies, legislative and regulatory arrangements. Most of these institutions suffered reversals over time, in terms of their scope of activities and budgetary allocations even prior to the unrest in 2011 and thereafter.

At any rate, Egypt now hosts the largest number of research centers in the region, totaling 219 research centers under the auspices of ministries, and

114 centers at universities as well as 14 specialized government research centers.[130] A sizeable community of researchers, estimated at 98,000 theoretical and applied scientists, is responsible for considerable research output, mainly publications in local, regional as well as international journals. The following is an outline of some of the more recently created institutional STI entities. Further information on overall research output that materializes in publications within international journals is presented in chapter (IV).

### 3.3.1.3.1 Academy of Scientific Research and Technology and Its Initiatives

The ASRT which was established in September 1971, effectively shaped Egypt's science and technology scene throughout the past four decades. At its inception, ASRT was assigned responsibilities directed at promoting science and technology and their role in national development. ASRT was initially linked to the Council of Ministers, functioning under the Prime Minister's patronage. Following several interim changes of affiliation, particularly in 1998, it became affiliated to the Ministry of State for Scientific Research.

The academy is managed by a council with appointed members as well as representatives of major institutions concerned with science, technology and innovation.[131] However, at its inception, the main intention was for the Academy's members to be elected to its membership in order for it to act as the principal scientific advisory body, and an independent arbiter on scientific affairs, in the country.

Frame 3.13 presents a list of the ASRT principal functions. Programs that the Academy implements with the particular intention of promoting innovation include the National Program for Technological Specialized Incubators "INTILAC," which is aimed at incubating innovative technological projects for periods of up to 1 year. Once selected, INTILAC tenants with promising projects are presented with a seed fund of up to LE150,000, in addition to training as well as access to technical services and advisory support within a reportedly conducive ecosystem that allows effective networking with prospective customers and investors. Around 12 incubation facilities were billed as active in various geographic locations in specializations that included hardware, multimedia, education, agribusiness, textiles, leather, jewelry and furniture, among others.

Jesor is another ASRT initiative aimed at establishing solid and sustainable cooperation between Egyptians scientists, outside and inside Egypt. This program is aimed at mitigating the impact of the brain drain through engaging expatriate Egyptian science and technology professionals in collaborative research and innovation projects. Transfer of latest know-how and cutting edge technologies is at the focus of the Jesor initiative, with up to LE1 million for successful candidates.

**Frame 3.13 Main Functions of Egypt's Academy for Scientific Research and Technology**

The Academy performs functions that may be summarized as

- Encouraging research in the basic and applied sciences and supporting scientific research centers working in state-of-the-art research.
- Identifying scientific research excellence and promoting its applications through a system of grants and awards.
- Planning scientific research and technological development programs of interdisciplinary character within national priority domains, providing financial and material resources.
- Raising national awareness of the importance of scientific research and technology toward sustainable development and promoting a culture of science among national citizens, through support for science exhibitions.
- Promoting cooperation between Egyptian scientists and experts from universities, research institutions, the private sector, NGOs, and prominent Egyptian scientists in diaspora.[132]
- Providing support to ministries and research institutions toward creation of an integrated system of scientific research in response to the country's development needs and the quest for a knowledge-based economy.
- Encouraging female and youth participation and taking leadership positions in science and technology.
- Planning and implementation of a national system aimed at monitoring scientific and technological information with emphasis on monitoring national scientific capabilities and achievements in various fields of science and technology as well as ways and means for to disseminating such information and facilitating its use.

ASRT also undertakes coordination and cooperation functions both within the country through helping organize and support national scientific and professional associations and unions, as well as organizing cooperation agreements with counterpart organizations across the world.

ASRT annually awards State Prizes totaling LE750,000 as well as its own special prizes to scholars in fields including the environment, agricultural sciences, engineering and development and medical research. In addition ASRT produces publications aimed at popularizing science and technology and a quarterly magazine that reports on everyday news of the institutes of research.

### 3.3.1.3.2 Zewail City of Science and Technology

Zewail City is billed as a nonprofit, independent institution of learning, research and innovation. It was officially inaugurated on November 1, 2011. However, the City's charter, outlining its mission, legal status, its constituent parts, as well as relevant financial and administrative structures was issued a year later in December 2012.[133]

The Zewail City currently has seven research institutes, with plans underway for establishing five more. Established research institutes cover a range of novel science and technology activities, including stem cell research and regenerative medicine, nanotechnology, photonics, smart materials as well as energy technologies. Zewail City is also home to institutional entities intended to promote research in basic science and fundamental physics as well as institutions dedicated to promoting virtual education and research in learning technologies. Additionally, Zewail City sports a center for economics and development as well as an institute for economics and global affairs. However, out of the five articles listed by Nature Index as attaining top Altmetric scores[134] published by researchers at Zewail City during 2015, four were dedicated to high energy physics and the Higgs boson, in particular.

### 3.3.1.3.3 Egypt's STI Observatory

The Egyptian Science, Technology and Innovation Observatory (ESTIO) was established in February 2014 by the Academy of Scientific Research and Technology.[135] The observatory targets the following objectives:

- providing decision-makers with up-to-date information on global scientific and technological trends;
- monitoring the evolution of national STI capabilities using established and well-documented methodologies;
- formulating country-specific STI indicators in measuring progress toward building national scientific and technological capabilities;
- conducting comparative analytical exercises with view to establishing progress achieved by Egypt relative to other countries around the world;
- promoting effective data collection and sharing capabilities by Egypt's STI institutions.

Little evidence is available of the observatory's activities however, with its dedicated website providing basic information concerning its vision and mission and devoid of publications of statistical data. Yet, it is doubtless that once properly activated ESTIO should perform valuable services in overhauling the country's STI capabilities.

### 3.3.1.3.4 Technology and Innovation Centers

Several centers have been set up over the past three decades by Egyptian authorities with the aim of promoting technology-based innovative ventures with some emphasis on enterprise and job creation. The following paragraphs provide a summary of selected ventures.

**The City for Scientific Research and Technology Applications (SRTA City),**[136] was established by presidential decree in the early 1990s as Egypt's first Science Park in an area dedicated to industrial development close to Alexandria. When inaugurated in August 2000, SRTA City already had facilities dedicated to genetic engineering and biotechnology, informatics, new materials as well as arid land agriculture research. SRTA presently comprises 12 research centers. Projects undertaken during later years included:

- a 1 MW solar power and desalination plant;[137]
- applications of information technology tools to recreate Egypt's cultural heritage online.

**The Smart Villages Development and Management Company (SVDMC)** was founded in November 2001 as a Public–Private Partnership (PPP) investment initiative with a mandate to establish and manage a chain of technology clusters and business parks. One of the SVDMC first creations is a communication and IT business park located in the western suburbs of Cairo. This is essentially an agglomeration of technology businesses, with a cluster of nearly 130 local and multinational firms as well as offices of the Egyptian Ministry of Communications and Information Technology. Other SVDMC business parks were planned at Damietta and Alexandria suburb in 2010. The Smart Village business parks currently employ more than 40,000 people, and that number was expected to grow to 100,000 by 2015.[138]

Projects aimed at reviving regions in the neighborhood of the Suez Canal announced in 2014, have included plans to construct three ICT parks in strategic locations. These include a park dedicated to information and communication technologies to be built between Port Said and Ismailia, making use of the New Suez Canal University in Ismailia, as well as the **Ismailia Technology Valley (ITV)**, which was launched nearly two decades ago, abandoned and is now being developed. Technologies and application areas destined to receive particular attention at ITV include renewable energy, medical and health care industries, computer systems and software development, microelectronics, and biotechnology.

Two smaller ICT parks are also planned for Port Said, on the North Coast, and Ain Soukhna, on the Red Sea. The latter is expected to include an ICT manufacturing park.

Through the above and other similar projects, Egypt's government aims to attract international firms willing to outsource and offshore ICT services and cloud computing to the new technology parks. The number of IT companies operating in the country has expanded significantly in the last 5 years, from 2600 to 4600, according to SCZone.[139]

It may be noted that the latest information regarding other Smart Village projects in locations such as Alexandria and Damietta indicate that they had not been established by 2013.[140] Additionally, technology-based innovation ventures established by the Smart Villages Company, appear to be more inclined toward hosting technology-related enterprises both from Egypt and abroad, with real estate and business promotion as priorities.

### 3.3.1.4 Research Funding in Egypt

For nearly 30 years, Egypt the Arab world's most populous country spent a maximum of around 0.2% of its GDP on research and development activity,[141] far less than the world's average, estimated at 1.2% and less than what is spent by countries in sub-Saharan Africa, which averages at about 0.3%.[142] Research funding allocated to institutions based outside the country's major cities is considered a rare occurrence.[143] Contributions to the country's research efforts are mainly due to government allocations. The private sector's contribution to R&D is negligible however, with few reliable data in any case.

Funding of research activity in Egypt is provided by three institutions, namely the Ministry of Finance, through officially approved budgets designed to cater for running and limited investment expenses,[144] and the Science and Technology Development Fund (STDF) as well as a separate program that supports research and development aimed at industrial development (RDI).

STDF is reported as having made grants totaling the equivalent of around US$90 million since its inception in 2008 and the end of 2010, and to plan distribution of around US$33 million in 2012.[145] Funds provided by STDF are dedicated to research as well as incubation projects on competitive bases. STDF receives considerable funds in addition to its government budget through grants provided by Germany, France, the United States and Japan as well as the European Union. Table A.3.1 presents a summary view of sums and durations linked to the various grant programs managed by the Fund.

#### 3.3.1.4.1 The RDI Program

RDI constitutes a separate funding source for research and innovation. It was designed to encourage applied research that is directly useful in promoting industrial development. RDI was allocated a grant of €31 million from the European Union between 2007 and 2015, in order to support the development of ideas from the drawing board to commercial products.[146] A network of 30 focal points is supposed to have been established in order to guide researchers intending to apply for grants from the RDI or the European Union's Framework Programme (FP7). Measures designed for monitoring and evaluating RDI projects are planned for implementation in two phases. With the first aimed at reviewing evaluation methodologies already in use in Egyptian ministries and R&D centers, whereas in the second phase selected Egyptian organizations would be charged with implementing and then disseminating best practices based on a thorough worldwide survey of measures designed for research evaluation.

### 3.3.1.5 Concluding Remarks

A review of science, technology, and innovation in Egypt indicates that STI institutions and their initiatives suffered several setbacks during the past

2 decades, along with other institutions and initiatives throughout the country. Some setbacks could be a traced to the turmoil that engulfed the country and the entire region following the events of January 2011. However, there is little doubt that many had their origins within previous policies and ill-conceived measures taken for their implementation, further complicated by corruption and cronyism that prevail in almost all walks of Egypt's social and economic life.

Nowhere are these setbacks apparent than in Egypt's output of patents, and this although enhancing commercialization of research outcomes through promotion of university–industry collaboration had repeatedly been declared as a principal preoccupation for policy and decision makers in reforms at the outset of the present century and then, even more forcefully in 2007. As indicated in Chapter IV, patenting performance by Egypt fell short of that exhibited by other Arab countries, some of which recent arrivals on the regional STI scene.

A special program has been designed by the STDF grants scheme to cater for research proposals put forward by firms in all domains, and infrastructural arrangements have been established to support public-private partnerships. However, lack of statistics on how many grants were in fact made because STDF was established and concerning total volumes of its annual, actual and projected grants as well as their distribution across fields of application, render it impossible to judge the program's effectiveness.

On the other hand, a review of the Table A.3.1 indicates that the volume of available allocations generally falls far short of achieving anything resembling radical overhaul of the country's STI capabilities. One must obviously add to this shortcomings recently revealed with regard to taxing research centers established by universities and foreign aid agencies (see Frame 3.14).

**Frame 3.14 Research Funding in Egypt**

A recent press article[147] by an Egyptian professor addressed deep contradictions between the official position taken by government with regard to scientific research in Egypt and actual realities. Research funding, for instance, was still far short of the 1% of GDP promised decades ago.[148] However, the article, considers that it is one thing for the government, hard pressed to meet other more urgent necessities, not to keep its promises regarding allocating funds for research, but quite another for it to impose heavy taxes on research centers, established by universities and foreign aid agencies. Figures quoted by this article indicate that a variety of centers, which do not receive any government funding in support of the research they conduct on a broad range of often crucial issues for the country's development, are required to pay 30% of all the income they receive as taxes. Additionally, these centers are required by law to turn in to the national treasury 50% of funds in their possession at the end of each fiscal year. This, the article notes has caused many such centers simply to stop operating, depriving the country of valuable research output.[149]

Yet the design of grant initiatives reveals a number of positive orientations, evidenced, for example by programs designed to help reintegrate expatriate scientists wishing to return home into national STI institutions as well as programs aimed at promoting collaboration with developed country scientists and STI institutions.

What is still missing in all reported endeavors designed to enhance STI capabilities, both in terms of overall intensity as well as focus on chronic socioeconomic problems is effective and transparent monitoring and evaluation of procedures and outcomes. Instituting measures that guarantee such monitoring and evaluation measures is never an easy task. However, little progress may be attained in their absence. An indication of the lack of the required level of transparent monitoring and evaluation is provided by the fact that none of the websites of all STI institutions consulted in preparing this material carried reports outlining results achieved in relation to previously established goals, analyzing shortcomings and means devised to forestall future challenges or attain more effective successes.

In short, overregulated and heavily centralized governance procedures, as well as the lack of clearly articulated sectoral strategies, are major obstacles to the creation of a mature public research system that may eventually contribute to tackling Egypt's socioeconomic priorities. Additionally, there are indications that Egypt's researcher community is aging, and in need of new blood, with larger proportions of recent graduates lured by fields of study, such as investment finance and management. On the other hand, older university staff members are often overloaded with teaching assignments and heavy administrative duties, leaving them little time for research activities.[150]

The past two decades, nevertheless, did witness the creation of several new research and innovation centers, providing fresh hope for a better future for science, technology and innovation in Egypt. Yet prevailing norms and deeply rooted practices might delay, possibly even inhibit, positive outcomes in the near future. This is essentially due to major constraints, including meager funding for research goals closely linked to endemic socioeconomic problems combined with a strong tendency toward immediate commercial gains, essentially for multinationals and associated business enterprises.

A recent example is the Zewail City of Science and Technology (ZCST), which encompasses a university, several research centers and a technology park. However, activities undertaken at ZCST as well as other institutional entities, such as the country's innovation parks may not always be directly relevant to commercialization of research outcomes, nor the resolution of longstanding socioeconomic problems. Indeed, a good deal of the research projects being tackled appear more strongly geared to frontline research issues with applicability far in the distant future and linked to career development paths of their staff members.

## 3.3.2 STI in Sudan

The science, technology and innovation system in Sudan suffers many of the shortcomings that afflict its other Arab and African neighbors as well as share some of the strengths that they also possess, essentially in terms of its human capital. The following sections provide some insight into recent policy initiatives, R&D activities as well as institutions dedicated to STI capacity building with primary objectives concerning agricultural development as well as acquisition and dissemination of novel technologies, with ICTs occupying a central position.

### *3.3.2.1 Recent STI Policy Initiatives*

A report, published in November 2016, by the United Nations Economic and Social Commission for Western Asia (UN-ESCWA), indicates that Sudan's Council of Ministers approved in October 2016, a new STI policy, which at the time of the report's publication was in the "process of being translated into strategies and plans of action with clear priorities."[151] The Ministry of Higher Education and Scientific Research was mandated to lead the implementation of this policy with the overriding role of promoting future sustainable development, and with emphasis on:

- Alleviating poverty, through improved productivity and reduced production costs based on R&D efforts targeting specific sectors.
- Assigning priority status to science, technology and innovation in national strategies and action plans, with precedence to priorities identified among sustainable development goals (SDGs) and corresponding resource allocations while highlighting problems caused by climate change and measures aimed at attenuating its impact.
- Supporting scientific research oriented toward optimal utilization of available natural resources.
- Encouraging ICT service providers to expand coverage into rural and remote areas and to reduce the cost of access and future development of ICT-based services.
- Strengthening links between society's components and building viable partnerships between research, industry and services organizations.
- Encouraging publications in prestigious international scientific journals and enhancing the quality of local scientific journals.
- Developing capabilities in foreign language training in order to facilitate absorption of best STI practices.
- Promoting intellectual property rights protection for researchers and research institutions.
- Collaborating and sharing experiences and expertise with countries in the region and worldwide.
- Building bridges to enable national experts in the diaspora in fields of crucial importance to Sudan's STI policy.

- Strengthening cooperation with regional and international organizations concerned with promoting STI capacity.
- Supporting and strengthening scientific and research institutions.
- Developing science, technology and innovation curricula in public education, particularly at the undergraduate and graduate levels.

The above-mentioned report also indicates that a previous review of Sudan's STI policy directives had been undertaken with emphasis on a number of issues, indicated in Frame 3.15.

Not unlike policies in other Arab countries, and some developing countries around the world, policies for higher education and scientific research in Sudan are virtual entities, rather than sets of well-researched directives, based on detailed analysis of salient socioeconomic realities and future aspirations. Thus, a document authored in 2016, on Sudan's STI policy, indicated that policies devised by the Ministry of Higher Education (MOHE) and those of its STI arms were updated during conferences "with the help of other stakeholders from different ministries" setting "national SDGs in terms of knowledge generation, capacity building and suaveness in applying innovative technologies." No mention is made of preceding steps including analytic examination of successes and failures, in reference to measurable benchmarks, changes at the national, regional and international scenes, to be reflected in fresh directives and implementation modalities.

Yet to be fair, this same document also indicates that transforming Sudan into an innovation-based economy would not be possible without STI policies being adopted within legislation "that governs all R&D sectors across the different line ministries." This same document also calls for "an autonomous funding system [that] fosters scientific research and innovation at the national level," while taking local requirements and priorities into account.

**Frame 3.15 Points Emphasized in Reviewing STI Policy Directives in 2009**

The report referred to in the adjoining text also indicates that past STI policies had been subjected to a review, in 2009 with assistance by the United Nations Educational, Scientific and Cultural Organization (UNESCO). Major issues considered in this review concerned the need to:

- enhance the role of knowledge in the economy;
- strengthen partnerships between research, development, and innovation at universities and R&D centers on one hand, and between industries and services organizations, on the other;
- stress the impact of innovation in the sustainable development;
- promote national demand on R&D and innovation;[7]
- increase public STI awareness;
- transfer of knowledge into wealth through R&D and innovation;
- develop novel STI curricula with the objective of transforming graduates from job seekers to opportunity creators.

### *3.3.2.2 Sudan's STI Institutions*

Higher education and scientific research activities in Sudan date back to the first decade of the twentieth century. The first higher educational institution, the Gordon Memorial College (GMC) was established in 1902 and the Agricultural Research Unit (ARU) was created in 1904. Several higher education and scientific research institutions were created in later decades with an enhanced tempo of institution building since the 1990s. A Ministry of Science, Technology and Innovation (MOSTI), was established in 2001 and a Ministry of science and telecommunication was later created in 2011. However, MOSTI was dissolved only 4 years later in 2015. And in line with the mandate granted to the MHESR to lead the implementation of the country's STI policy directives, national STI institutions, previously operating within MOSTI, were placed within the MHESR administrative structure. The following paragraphs provide summary information on activities undertaken by MHESR and other institutions active on Sudan's STI landscape.

#### 3.3.2.2.1 The National Council for Higher Education and Scientific Research

One of the more recent developments on Sudan's institutional STI scene has been the creation of this council, whose membership includes around 90 officials and experts; including the Minister of Higher Education and Scientific Research, heads of public as well as private universities.[152] However, it does not appear that the ministries concerned with scientific research, such as the ministry of industry, for example, are represented on this Council.

The main functions of the Council revolve around:[153]

- Formulate "overall policy" for higher education and scientific research as well as associated plans, in accordance with national development policies, and monitoring this policy's implementation.
- Coordination between the Ministry and the country's universities, while maintaining a measure of independence for the universities, including setting their curricula and managing their own affairs, while ensuring "implementation of higher education policies, in accordance with decisions taken by the State, represented by its Head and its institutions."
- Monitoring the performance of higher education and scientific research institutions.

The same statement emphasizes the role that the Council has in allowing universities to continue their missions and grow, while coming up with rational responses to "criticism leveled at the revolution in higher education and scientific research."

The Council established 10 consultative committees specializing in studies within fields, including agriculture, veterinary, environmental and natural resources, engineering, education, medical sciences, social and economic affairs, basic sciences, legal and polity affairs, computers and informatics, the humanities and scientific research.

The above committees are entrusted with providing specialist advice and assistance to the Ministry on matters relating to carrying out studies and implementing research activities in the above domains and may recommend the establishment of new colleges within public and private universities.

No reliable figures are available with regard to expenditure by MHESR on its various research activities undertaken at the country's public universities. However, figures quoted by the above-mentioned UN-ESCWA publication appear to indicate an annual average of around US$17.2 million over the years 2009–16.[154] A meager sum by all accounts, considering considerable number of R&D units, students in higher education as well as challenges faced by the country at large.

### 3.3.2.2.2 Sudan's Government Research Centers

The following paragraphs provide summaries concerning institutions established by Sudan's government with view to building scientific and technological capacity. Attention is drawn to the fact that two of these institutions, namely the Sudan Academy of Sciences and the Nile Centre for Technology Research were established in 2004 and 2007, respectively.

**The Sudan Academy of Sciences (SAS)** was established with view to assisting government in implementing STI policies and related initiatives as well as contribute to transforming Sudan into a knowledge-based society, providing related advisory services to Sudan's government on STI issues. SAS has the status of a public university for postgraduate studies. Activities it undertakes target capacity building of STI staff in Sudan. The Academy covers a diversity of fields, organized within 10 "Councils" that take care of postgraduate research and coursework within domains including: animal resources, engineering sciences and industrial technologies, biological sciences, advanced technologies and environmental sciences, nuclear science and engineering, economic and social studies, the humanities, community services, agriculture, diplomacy and political studies, as well as energy and earth sciences.[155] Around 39% of the SAS graduates between 2008 and 2014 specialised in economic and social studies, whereas almost 21% opted for qualifications in biological sciences, advanced technologies and environmental sciences and around 11% graduated in specializations relating to agriculture and animal resources. Incidentally, at 0.12% of the total, the percentage of those graduating in specializations linked to ICTs appears particularly low.[156] This may be explained by the fact that another institution, namely the Nile Centre for Technology Research, caters for R&D in these specializations.

**The Nile Centre for Technology Research (NCTR)** was established to undertake research and development in ICT, high-tech, microelectronic, communication navigation and geographic information systems. NCTR runs 3R&D facilities at three of the country's principal universities, namely the University of Khartoum, the Sudan University for Science and Technology and Karary University, with the aim of promoting applied research in the above domains. Practical successes achieved by NCTR appear to have been evident in the area of encryption algorithms as well as ICT applications in financial transactions. NCTR appears to have established partnerships and cooperation arrangements with a host of multinationals.[157]

**National Centre for Research (NCR)** is a multidisciplinary applied research organization whose activities have included work in a diverse range of domains, including materials technology, electronics, remote sensing and biotechnology. Recent NCR output is reported as resulting in the development of systems for pulp and paper production based on local cellulosic raw material, improved building materials, breeding of drought resistant crops for dry-land agriculture as well as measures aimed at increased awareness and control of endemic diseases e.g., malaria, bilharziasis, and leishmaniasis.[158]

**National Research Institute for Food Science (NRIFS)** undertakes food science and technology research and development activities in collaboration and with funding support from industrial as well as and government departments. NIFSR projects are essentially aimed at improving food production processes as well as food safety and quality.

**National Research Institute for Energy (NRIE)** has been implementing renewable energy development programs one of whose aims is to disseminate the use of solar panels in rural area. The institute has established a solar cell production plant and runs R&D, demonstration, testing, calibration, and certification in the field of solar technologies facilities.

**Sudan Atomic Energy Commission (SAEC)** was founded in 1996 on the basis of two previous entities concerned with atomic energy and nuclear technologies. The Commission coordinates activities concerning the civilian uses of atomic energy as well as protection from radiation hazards. SAEC has three components specializing in: Radiation Safety, Nuclear Application for Biological Science and Nuclear Chemistry and Physics.

#### 3.3.2.2.3 Research Centers Established by Sudan's Ministries

Ministries in Sudan established a number of research facilities with view to enhancing their operations. Principal among such ministries are the Ministries of Agricultural Resources and the Ministry of Livestock, Fisheries and Rangelands. However, other ministries, including the Ministry of Industry also established its own research facilities. The following is a summary regarding these centers.

**The Agricultural Research Corporation (ARC)** was set up by Sudan's Ministry of Agricultural Resources (MAR) research corporation in order "to become a center of excellence in agricultural research by developing skilled manpower and conducting quality research."[159] The ARC main objectives include:

- promoting self-reliance with regard to food security;
- improving agricultural productivity and the livelihood of producers of agricultural products;
- encouraging production of surplus crops for export purposes;
- conserving natural resources through efficiently exploiting production inputs;
- acquiring and disseminating new and proven technologies with view to enhancing increasing farmers' income.

ARC possesses 21 field stations distributed over all Sudan with units covering specializations including tissue culture and biotechnology applications as well as issues linked to specific crops such as a gum Arabic. ARC facilities are reported as including research centers charged with implementing a considerable number of national programs and research projects.[160] ARC employs around 3000 staff, the large majority of whom are nontechnical.

**Livestock Resources Research Corporation (LRRC)** was set up by the Ministry of Livestock, Fisheries and Rangelands in 1996.[162] LRRC is said to be an active research institution with its output contributing over 40% of all agricultural research output and around 30% of all research output produced by Sudan. An interesting angle in the research undertaken by LRRC is expressed in its vision and mission statements, which indicate fighting poverty among its principal aims along with ensuring food security for the country at large. LRRC employs around 1270 workers, of whom 330 are researchers and 400 are technical staff.

Frame 3.16 presents a list of research centers established within the above two research Corporations, ARC and LRRC.

**Industrial Research and Consultancy Centre (IRCC)** was established by the Ministry of Industry (MOI) in order to support higher growth and productivity within the industrial sector. IRCC is a nonprofit scientific governmental organization[163] whose principal activities include building research and technology development capabilities in certain industries as well as training aimed technology transfer. Applied research conducted at IRCC is reported as targeting product development, technology transfer and production line design as well as industrial process modeling. Promoting SMEs with focus on priority areas of clear benefits for economic diversification is also listed among its areas of activity. IRCC publishes an industrial research journal.[164]

### *3.3.2.3 R&D Activities in Sudan*

In common with many other countries R&D activities in Sudan are considered as the backbone of national science, technology and innovation endeavors. A paper aimed at evaluating Sudan's S&T indicators in 2010,[165] concluded that most R&D activities in Sudan occurred within the public universities, with the private sector and industrial concerns making only minor contributions. Based

**Frame 3.16 Research Centres Established by the Ministries of Agricultural Resources (MAR) and Livestock, Fisheries and Rangelands (MLFR)[161]**

**MAR Research Centres**

Agricultural Plant Genetic Resources (PGR) Conservation Research Centre
Crop Protection Research Centre
Forestry Research Centre
Land and Water Research Centre
Cereal Research Centre
Oil Crop Research Centre
Horticultural Crop Research Centre
Socioeconomic and Policy Analysis Research Centre
Dry land Research Centre
Water Harvesting Research Institute

**MLFR Research Centres**

Veterinary Research Institute
Animal Production Research Centre
Wildlife Research Centre
Fish Research Centre
Veterinary laboratories States
Animal Production Plants States
Camel Research Centre

on a review of literature published by international organizations as well as a limited number of interviews with business people and industrialists, this paper concluded that Sudan lagged "behind the world's developed and leading developing countries in terms of the same input and output indicators."[166] In this paper's estimation, this was essentially due to a combination of meager resources dedicated to ensuring viable science and technology inputs as well as inadequacies inherent in the country's economic system.

More specifically, problems hindering R&D were found to include:

- Inadequate financing by the public sector and almost total lack of funds from the private sector.
- Limited management and organizational abilities, coupled to poor coordination and inadequate networking/cooperation between universities and HEIs on the one hand and the productive sectors, on the other.
- Poor R&D infrastructures in addition to a limited numbers of researchers and trained technicians in most R&D fields.

Suggestions gathered in relation to improving the state of R&D were centered on rectifying the above shortcomings with emphasis on improved funding and human capital engaged in R&D activity, in terms of numbers and quality of training. Thus, Sudan needed to make greater investments in both financial and human resources as well as follow paths taken by several advanced and developing countries in encouraging demand-side approaches to R&D endeavors. However, this paper also emphasizes that in order to be successful, such investment moves would have to be guided by well-defined policies designed to vitalize the country's entire economy, promoting national production and services activities. In particular, the paper considers that partnerships between the public and private sectors should help bridge deficiencies, on both the funding as well as the human capital fronts.

Implementing effective S&T cooperation and pooling resources wherever possible with other Arab and African countries were also likely to enhance S&T capacity building in Sudan, ultimately contributing to "long-term harmonious development" of the country.

#### *3.3.2.4 Innovation and Enterprise Incubation Schemes*

One of the more prominent attempts aimed at promoting business innovation in Sudan appears to be the establishment by its government of the Africa City for Technology (ACT). One of ACT's strategic aims is diversification of the national economy through technology-based spin-offs. ACT strives to link R&D institutes, universities and high-tech industries in order to enhance STI and knowledge-based economic performance rather than resource-based economic activity. The ACT intention of assuming a regional role in promoting STI and knowledge business applications is evident. The scope of its actual and prospective projects covers a wide variety of domains; including space technology, mining, multimedia and smart classrooms. ACT also appears intent upon setting up enterprise incubation schemes.[167]

Another incubation schemes, which appears to be dedicated to microenterprise creation include a 3-year project launched by the Sudan University for Science and Technology (SUST), with assistance from the Islamic Development Bank and the Central Bank of Sudan. The SUST project's objectives includes setting up a number of incubators with the intention of providing 4000 graduates with technical as well as management skills with focus on microbusiness activities, including computer maintenance, dressmaking, manufacture of various items such as furniture, leather items, processing milk and other animal product.

#### *3.3.2.5 Concluding Remarks*

Many of the drawbacks that are prevalent in other Arab and developing countries with regard to STI capacity building activities are evident in Sudan. Thus, although the national STI policy referred to above does include a list of domains earmarked for special attention, no evidence of a self-contained policy document with explicit objectives to be attained within these domains. Less obvious is the picture with regard to resources to be earmarked for implementing priority directives of this policy.

Apart from a certain degree of activity in institution building over the past decade, no doubt complicated at times by changes of allegiance and shifting of responsibilities for various STI related tasks within and between ministries, there is little to indicate that modalities being developed are equal to evident socioeconomic challenges facing Sudan. This is particularly the case with regard to the performance of the country's agricultural and industrial sectors.

Nevertheless, one of the main strengths that Sudan possesses, despite so many years of internal strife, sanctions and inadequate policies, is its human capital. Thus, even at the hardest of times, a good deal of attention was always accorded to training and capacity building. A sizeable proportion of staff at Sudan's university and research centers are trained overseas in high ranking universities, previously in the United Kingdom and the United States, but more recently in countries such as Malaysia, China and South Africa.

Sudan has the potential for enormous progress positively reflecting on its own people as well as its neighbors. Thus, despite hard times the country went through over the past decades, Sudan could be a regional leader in agriculture and related technologies and industrial applications. Given sound policies in addition to its geographical position, its emerging sensitivity to environmental sustainability, Sudan's future is ascertained as the "food basket" for its Arab and African neighbors. Even in higher education, the fact that it still maintains a number of university institutions with a tradition of excellence must be considered as bases to build upon.

## 3.4 THE ARAB COUNTRIES IN NORTH AFRICA; ALGERIA, LIBYA, MAURITANIA, MOROCCO, AND TUNISIA

Traditional institutions of higher learning, mainly in Islamic sciences and traditional medicine, were hosted by all the Arab countries in North Africa in the past. Fez in Morocco, Al Qairawan in Tunisia, Timbuktu in Mauritania and Cosantina in Algeria and Tripoli in Libya, were home to centers of learning and libraries whose history goes back many hundreds of years.

More recently, influenced no doubt by links established during colonial presence of the French in North Africa, Algeria, Morocco, Tunisia as well as Mauritania, have tended to maintain close links to France in particular and Europe, in general. These links remain prominent today with dense two-way traffic by students and research scientists. More recently, however, closer links have been forged by Tunisian and Moroccan Science, technology and innovation institutions with the counterparts in other countries including the United States and Canada.

The following sections are devoted to a review of STI policy regimes, institutions and related activities in all five countries.

### 3.4.1 STI in Algeria; Policies, Institutional Settings and Future Prospects

Algeria made significant strides toward building its science and technology system, particularly in the decades immediately following its independence, with nearly six major universities established between the years 1965 and

1975, namely, Oran Es-senia (1965), Cosantina University (1969), Houari Boumedienne University of Sciences and Technology (1974), Tlemcen (1974), Oran University of Sciences and Technology (1975), and Annaba University (1975).

Yet the country suffers severe lack of human capital in various fields of science and technology. A 2012 report indicated that Algeria has 480 active researchers per million citizens, compared to the global average of 1080.[168] For this reason, a strategy was being formulated with view to tackling the causes of the brain drain, such as administrative bureaucracy and low researcher incomes. Additionally, Algeria is reported as attempting to open its research facilities to researchers from abroad, principally from other developing countries, with the aim of filling existing gaps in human capital.[169]

### *3.4.1.1 STI Policy in Algeria*

Algeria does not appear to have worked out an integrated national STI policy, in order to pave the way toward implementing scientific research and technology development strategies and the multitude of interrelated activities needed to attain concrete outcomes. However, a 2012 report, titled "Algeria to open research sector to foreign scientists";[170] published by a network dedicated to promoting science and technology for development, indicated that Algeria did in fact formulate two science strategies covering the years 2008–12 and 2013–17. With increased spending and university reforms high on their list of goals. However, a search for further details concerning the objectives of these strategies and means for their attainment proved unsuccessful. Available literature on the other hand does point to efforts aimed at the formulation of environmental and agricultural strategies, which would be bound to include goals involving enhanced R&D capabilities as well as possibilities for related innovative activities.

In effect, it is more appropriate to speak of a virtual or de facto policy, i.e., the sum total of official decrees, governmental regulations and legislative measures as well as ministerial and corporate strategies and related plans. This is clearly a very poor substitute for an all-encompassing and a coherent policy with directives and goals ensuring optimal performance at the national as well as sectoral levels.

Naturally, the absence of a global and strategic vision of technological development would ultimately contribute to behavioral and institutional rigidities, on the other hand, as well as wasteful laxity, on the other. With both trends having negative impacts upon STI activities in Algeria.

Legislation passed in 1998,[171] was intended as a turning point, in terms of support for scientific research, technological development, and innovation. This legislation included scope for the creation of specialized institutions, enhanced funding, and direct incentives. Nevertheless, it failed to boost innovation, particularly with regard to the industrial sector. Reform of the 1998 law, was geared toward more emphasis on innovation than research. It nevertheless, pledged increased funding, of up to 1% of GDP, by 2011.[172] Similar goals were expressed in an article published in 2012,[173] however, no information as to whether this had indeed been enacted.

Reasons given for inadequate outcomes based on original legislation as well as later reforms include limited involvement by researchers, weak demand for innovation by domestic firms. With particular regard to this latter point, the fact that incentive systems were implemented in the absence of public procurement policies targeting domestic firms, thus, creating limited demand for homegrown innovative products and services. Additionally, a fund established with the intention of promoting national research[174] did not quite cater for the specific needs of industry, neither did it provide the substantial support for support industrial R&D.[175]

Thus, Algeria's high technology exports generally amount to no more than US$32 million, around or around 0.2% of the country's manufactures exports and a much smaller percentage of the country's annual GDP.

### 3.4.1.2 Moves to Transform Algeria into a Knowledge-Based Economy

Efforts, mainly studies and expert meetings aimed at transforming Algeria into a knowledge-based economy (KBE), were undertaken by a variety of institutions,[176] with support by the World bank Institute. The National Economic and Social Council (CNES)[177] took up the issue and conducted meetings during which possible strategies were discussed, highlighting major benefits, such as reduced dependence upon hydrocarbon wealth, and creation of infrastructures allowing fresh value-added activities as well as employment and enterprise opportunities. Key players from academia, industry, central administration and the country's Diaspora became involved in a series of activities aimed at debating strategies and the need to overcome major handicaps including quality of knowledge related activities, at the country's universities and research establishments, incentives to be provided through legislative instruments as well as issues pertaining to governance. Indications are that the essential need to address all of the above issues within a coherent policy framework has not been adequately recognized as yet. And no steps toward formulating such a policy appear to have been taken.

### 3.4.1.3 STI Institutions

Algeria sports a relatively large number of higher educational institutions, including 50 universities, 20 university centers, 10 research centers as well as around 40 higher educational institutions established in analogy to the French "écoles nationales," "écoles supérieures," and "écoles normales supérieures."[178] Academic staff working at these institutions undertake scientific research and consultancy activities, the results of at least some of which are published in both national and international journals. Additionally, several government research centers undertake research and provide technical services within a variety of domains.

Research centers established by public sector institutions in Algeria cover a relatively wide range of specializations, including agriculture, fishing and aquaculture, advanced technologies, astronomy, astrophysics, and geophysics, chemistry, hydrocarbon resources, nuclear engineering and radiological studies, pharmaceuticals, scientific information, seismology and earthquake studies, space technologies. With the exception of research centers concerned with hydrocarbon resources, which are run by the national hydrocarbon corporation,[179] most other research centers operate under the aegis of the Ministry of Higher Education and Scientific Research, and a small number of other ministries, including for example the Ministry of Agriculture, Rural Development and Fishing.

As shown in Table (A.3.2) a number of large Algerian public sector companies active in a variety of fields, including hydrocarbon resources, iron and steel, electronics, chemistry, food and agriculture have established research and development centers, others have simply set up research units. The largest and most established of the research centers are those established by SONATRACH, which was founded in 1960, and was transformed into a Central Hydrocarbons Laboratory in 1988.

Operations of the above semipublic research centers and units are naturally closely linked to those of the parent company. However, there have been instances where research centers did evolve and emerge as research entities in their own right. A prime example of this trend is the research center established by pharmaceuticals corporation SAIDAL. There are also cases in which research centers established by companies have faltered, an example of such situations is the Department of Applied Research of the steel corporation SIDER.[180]

### The General Directorate for Scientific Research and Technological Development (DGSRDT)[181]

This Directorate implements, the national policy for scientific research and technological development as defined by an Act issued in 1998. In this capacity, the Directorate is responsible for programming, evaluation, institutional organization, human resource development, university research, technological development and engineering, research in the social and human sciences, scientific and technical information, scientific cooperation, the exploitation of research results and infrastructure development.

DGSRDT is made up of four directorates responsible for:

- Programming of research, evaluation and foresight;
- Administration and financing scientific research and technology development;
- Technology development and innovation;
- Development, scientific and technical services.

In addition to the above institutional entities several monitoring, control and evaluation bodies were also established during the late 1990s. These include the National Council for Scientific Research and Technological Development (CNRST),[182] the National Committee for the Evaluation of University Research Projects (CNEPRU)[183] and the National Agency for the Valuation of Research and Technological Development (ANVREDET).[184] This agency was established in 1998, under the supervision of the Ministry of Higher Education and Scientific Research. Its mission centers on selecting and supporting innovative projects developed within higher education institutions (HEIs). The Agency is credited with creating a National Scientific Research and Technological Development Fund as well as the organization of expert meetings in support of its operations. However, in the opinions of several observers none of the above institutions may be credited with major successes, least of all with regard to commercialization of research outcomes.

Centres have also been established with view to conducting studies and provide technical services in support of the industrial sector. Examples listed in Table (A.3.2) include the Centre for Studies and Technological Services for the Materials and Construction Industry (CETIM)[185] and National Centre for Technologies and Consulting (CNTC).[186] These centers would surely be expected to possess potential for supporting research, technology development and innovation within industry. However, they are stated as having had limited government support, which has forced them to confine their activities mainly to testing and technical services.

### 3.4.1.4 Algeria's National Innovation System

With regard to institutional entities established specifically with promoting innovation in mind, Algeria set up the Cyberpark of Sidi-Abdallah (CSA) with the aim of training as well as the development of ICT applications for a variety of purposes. CSA facilities were designed with view to attracting multinationals active in ACT domains as well as commercialization of outcomes of research from universities and research centers. However, available information does not indicate that any of the above institutions may be credited with major successes, least of all with regard to commercialization of research outcomes.

In effect, innovation in Algeria is encumbered by outdated legislation, with most institutional entities suffering weaknesses that include inadequate funding and managerial practices, with focus maintained essentially upon commercializing academic research. Indeed, some observers believe that scientific research and technological development institutions are in need of extensive restructuring while others affirm that it is immature to speak of an Algerian innovation system, with firms supposed to be pivotal players in such a system "largely not using knowledge and generating little innovation." The environment surrounding STI activity is judged as lacking incentives, with tendencies toward compartmentalization and confinement.[187]

#### *3.4.1.5 Concluding Remarks*

For all the initial dynamism exhibited in postindependence days and all the resources it has at its disposal by virtue of its oil wealth, Algeria's performance on the STI stage is a phenomenal disappointment. The main reasons for this are essentially due to inadequate governance, and absence of viable guidelines in the form of well researched and widely debated policies. Such shortcomings are by no means confined to the STI domain. Thus, as indicated in sections concerning Algeria's economic performance and its performance with regard to a variety of indicators, great efforts are required with view to a radical overhaul of institutional, legal and regulatory arrangements.

### 3.4.2 STI Activities in Libya

Many indications point to the conclusion that Libya's STI system was largely dysfunctional even before the onset of the country's current troubles began in 2011. With few, and often outdated information on basic and higher education, little data was available from usual sources concerning most aspects of the country's STI system. On the other hand, it must be acknowledged that efforts were made by successive Libyan governments over the past decades to promote human capital through overseas study tours, which are mainly undertaken at Arab and European universities.

#### *3.4.2.1 Libya's STI Policy*

With specific regard to Libya's STI policy regime, there are no indications of a policy guiding Libya's science, technology and innovation efforts at any time prior to or following the events of 2011. Indeed, it is almost certain that all STI-related activities were essentially guided by Presidential decrees and other official announcements in reaction to various, innate as well as external demands. An attempt is known to have been made in 2013 to draft a science and technology strategy in collaboration with the UAE based Arab Science and Technology Foundation. However, there are no indications that an actual strategy document ever materialized.

#### *3.4.2.2 Libya's STI Institutions*

Frame 3.17 presents an outline of the development of the country's higher education, highlighting its youthfulness, with faculties of science, engineering and agriculture established less than 50 years ago. Libya's Ministry of Higher Education and Scientific Research (MHESR) governs all higher education and public sector research activity.

At least on paper, research is considered an essential part of academic staffs' responsibilities and universities are expected to contribute to resolving social and economic issues as well as technological development through

**Frame 3.17 An Outline of the Development of Higher Education in Libya**

Libya's first university was established in 1955 shortly following independence. The Faculty of Arts, which was located in Benghazi, became the nucleus of higher education in the country. Between 1962 and 1967, faculties of economics, commerce, law, sciences, agriculture, engineering and education were established in Benghazi and Tripoli, with faculties of medicine, Arabic language and Islamic studies, petroleum engineering and mineralogy following suit in the 1970's.

The Libyan University was split into two separate entities: the University of Tripoli, and the University of Benghazi and more universities were later established in other parts of the country.

By 2012, Libya's institutions of higher education consisted of 12 public and 5 private universities with a total of 160 faculties, in addition to 16 technical faculties and 81 technical and vocational centers.

scientific research. The "Libyan Authority for Scientific Research and Technology (LASRT)," is charged with supervising scientific research and technological activities undertaken by the Ministry and institutions under its supervision. Many university staff members participate in research activities carried out by these centers. On the other hand, cooperation between higher education institutions and the business and industrial sectors is stated as extremely limited. Additionally, criticism directed at Libya's higher education system centers on its inability to meet the needs of these sectors and respond to changes in the country's economy. Industry is not represented on management or advisory boards of higher education. Nevertheless, some university faculties are reported as maintaining links with industry and service sectors through students' projects and internships.

Libyan universities maintain cooperation links with Arab, African and European institutions of higher education. Additionally, MHESR regularly provides grants allowing top performers within the student body to continue their higher education abroad, mainly with view to obtaining Masters and PhD degrees. In 2012, more than 10,000 students were reported as studying abroad in more than 30 countries, in fields including medicine, engineering, basic sciences, social sciences, and economics.[188]

No information was available on the MHESR website concerning current scientific research activities whereas the link to LARST was active it only led to information indicating that this national authority comprised six centers concerned with biotechnology, medical research, tumor research, monitoring climatic change, economic research, translation and study of languages, urban and regional planning. However, none of the links associated with these centers was in fact active.

### 3.4.2.3 Research Activities in Libya

Research output by Libyan academics is low, even by developing country standards as pointed out by a survey of published medical literature conducted around 2008–09, which indicated that the average annual production rate for research publications at the Al-Fateh Medical University, for example was 1.4 articles per 100 academic staff members.[189] Indeed, historical records of papers published in scientific journals, expenditure on scientific research and numbers of researchers per million inhabitants, are particularly scarce for Libya, although they are quite accessible for many of its African and Middle Eastern neighbours.[190]

Four major reasons are normally put forward in explanation of this situation:

- Limited funding is allocated to research activity at the country's universities;
- Lack of scientific infrastructure and technical cadres with inadequate expertise;
- University lecturers are overloaded with teaching duties leaving them little time to undertake and supervise research activities;
- A relatively high proportion of scientists and engineers are lost to the brain drain.

Concerning the first point, sources such as the Human Development Reports of the UNDP, the UNESCO as well as the World Bank's statistical resources hardly ever include estimates of Libya's allocations to scientific research. On the other hand, it seems that resource allocations by Libya's government to the entire education sector were below average values dedicated by other countries with comparable human development rankings, years before the onset of the country's current troubles. Thus, although Saudi Arabia and Kuwait, both of which within the same group of high Human Development Index values, dedicated 5.7% and 3.8% of their GDP to education, respectively, Libya managed to dedicate only 2.7% of its GDP, i.e., less than half of what Saudi Arabia did and considerably less than Kuwait did.[191] Incidentally, Libya's expenditure on education was a mere 60% of the average for all 82 countries with high Human Development Index values, which is estimated at 4.5% during 2000–07.

Low levels of expenditure on the education sector might additionally explain the second and third points singled out as reasons for Libya's low scientific output, concerning mediocre infrastructure, inadequately trained technical staff, on the one hand, and high teaching loads, on the other. However, other factors must also be examined, including limited availability of associate or junior lecturers and research assistants.

With regard to the fourth point, i.e., concerning the country's brain drain, the Libyan government was known to invest rather heavily in training postgraduates in Europe, Canada, and the United States, and this strategy is likely to have enhanced the risk of losing trained scientists to academic and business institutions abroad.

Information concerning initiatives taken to link up to Libya's Diaspora is rather scarce and for the most part unreliable. And although information is lacking on the outcome of relatively recent initiatives along these lines, it is likely that such initiatives might have met limited success due to the repressive nature of Libya's political system and regulatory arrangements in force prior to 2011. At any rate, more innovative solutions will be needed to get the most out of such initiatives, throughout the region rather than just in Libya.

#### *3.4.2.4 Libya's Innovation and Enterprise Promotion Facilities*

Innovation and enterprise promotion policies in Libya are essentially embryonic. The country's 2006 Economic Development Strategy considered SME promotion as means to "catalyze innovation, risk-taking and business formation in Libya and to create the next generation of Libyan entrepreneurs."[192] A national program under the title "Establishing a Libyan Incubator for Technology and Innovation (ELITE)" was launched in 2006 with the aim of setting up incubators in Tripoli, Benghazi and Sabha to cater for innovative enterprises in engineering, biotechnology, renewable energy technologies as well as information and communication technologies.

A steering committee was also established to oversee and coordinate activities, surveys and analytical studies of entrepreneurship in Libya were also supposed to have been undertaken and their results presented and debated at special conferences held at four Libyan universities. However, no information is available regarding whether any of the proposed incubators did in fact become operational.[193]

Several institutional entities have been established in order to promote SMEs and entrepreneurship, in general, including a "National Economic Development Board (NEDB)," which was set up in 2007, within the Prime Minister's Office, and a National SME Program was launched in the same year. In 2012, this program was renamed as "Libya Enterprise," and placed within the Ministry of the Economy, with a mandate including promotion of entrepreneurship, creation of a conducive environment for SMEs as well supporting business start-ups and establishing incubators.

Criticism of this program has focused on the fact that it did not sufficiently take into account challenges that young entrepreneurs and ethnic minorities faced and that it failed to adequately consider regional disparities. However, some progress was made with a few incubators and business centers established in a few locations.[194]

More recently, Libya's 2030 Vision Committee, which held its first meeting in May 2013,[195] focused on SME promotion as an engine for driving the economy forward and an SME Policy Unit was created in 2014 within the Ministry of the Economy and charged with conducting activities that included policy evaluation, promoting and monitoring international relations. A total of 13 enterprise centers were launched, 10 along the Mediterranean coast and three in interior and southern locations. These centers include one center dedicated to women and another for persons with special needs. Reports indicate that half of these centers were still under construction in 2014, whereas activities of those that reached operational status were curtailed due to the ongoing conflict.[196]

With regard to funding dedicated to SME promotion, the Libyan government set up five regional funds in 2013, totaling around US$150 million each. No indications were available as to the period of time for which these sums were allocated. Unless such sums are intended for annual expenditure, they must be considered derisory in view of actual needs, particularly considering the extent of damage sustained by existing enterprise infrastructures.

Development of a dynamic SME sector with focus on innovation as basis for its operations should become a vital ingredient in Libya's future strategies, targeting economic diversification and job creation. Care must be taken at the same time to establish oversight structures through which public and civil society institutions might collaborate with the business sector in order to guarantee optimal benefits for all segments of society. This is particularly important in view of past experience with transitions made by countries following a long history of state dominance, widespread corruption and mismanagement of state enterprises. The purpose of oversight and collaborative arrangements may not be taken as restricting enterprise activities. Rather, it is to empower SMEs allowing them to partake in reconstruction efforts while ensuring equitable outcomes and guarding against excesses that are known to characterize similar situations elsewhere.

### *3.4.2.5 Concluding Remarks*

Libya's efforts toward building sustainable STI capacity prior to 2011 faced many shortcomings. Policies, designed to guide STI and overall development efforts in the past, were far too often substituted by presidential decrees and ad-hoc announcements. Well-researched and thoroughly debated policy designs will be needed instead, to shepherd the country's development during postconflict reconstruction with focus on:

- promoting overall sustainable socioeconomic development;
- enhancing productivity and growth in various sectors of the economy;
- harmonizing output from the education system with job market needs; and
- creating new employment opportunities for the country's youthful population.

Along with many of the country's other institutions, universities and scientific research centers in Libya have suffered severe setbacks since 2011, adding to mismanagement, favoritism and corruption they had to endure over many decades. Damage recently sustained by the country's STI capabilities with regard to the flight of human capital has been far more detrimental than destroyed physical assets and infrastructure. Additionally, few and embryonic as they were, programs aimed at promoting innovation, coordination among

Libyan STI institutions and cooperation with their counterparts abroad, are known to have witnessed many a reversal.

However, although its STI institutions may not be set as an example for other countries, and although Libya is destined to face further difficulties for several years, if not decades, ahead, the country and at least some of its institutions did make some progress on certain fronts in the past. Several institutions and regulatory arrangements may fruitfully be reformed and reformulated.

One of the causes of optimism regarding the prospects for higher education, scientific research and technological development in Libya is the large number of postgraduate students in all disciplines who have recently trained or are now finalizing their training, at universities abroad, particularly in Europe and North America. Thus according to UNESCO statistics, published in 2016,[197]over 5700 Libyan students were pursuing their studies abroad, with nearly 70% enrolled in postgraduate studies within universities in Europe and North America. This would naturally require well designed and executed initiatives on the part of Libya's STI institutions for making optimal use of returning scientists and engineers as well as forging strong links with those who may opt to remain abroad.

### 3.4.3 Mauritania's STI System; Policies, Institutions and Research Activity

Mauritania's education sector made significant advances since its independence in 1960. This is particularly evident in improved basic education as well as a range of infrastructural arrangements. Additionally, the country witnessed significant expansion in its higher educational facilities. With particular regard to STI capabilities, several instances may be quoted of innovations introduced by private enterprise, in collaboration, for the most part with international firms and, perhaps less often, with overseas research institutions providing management skills in the transfer and implementation of particular technology inputs. However, a great deal more would be needed on a very wide front if the country is to achieve meaningful socioeconomic development.

#### *3.4.3.1 Mauritania's STI System*

A review of Mauritania's STI system, conducted in 2010 by the United Nations Conference on Trade and Development (UNCTAD), concluded that[198]

- The country's science and technology capabilities were currently inadequate to address the challenges it faced.
- Mauritania never formulated a national STI policy, possibly due to the fact that the country lacked a single public body, a ministry, department or advisory body/council mandated to oversee STI affairs.

- The abilities of enterprises as well as the public sector to effectively harness scientific and technological capabilities toward innovative products and services were also deficient, essentially due to serious gaps within the national innovation system (NIS).
- Extensive and rapid human capital building was essential in order to build requisite knowledge for development and related skills, with particular regard to scientific and engineering disciplines relevant to the country's priorities.
- More rapid and widespread technology diffusion and improved absorptive capacity were required to stimulate technological upgrading and improve innovative performance.
- Higher investments in infrastructures related to STI performance were needed to overcome severe bottlenecks to production and services sectors, including transport and trade.
- In many instances, such bottlenecks were not due to lack of scientists or business innovators—important as such factors are indeed—rather, and due to the fact that one is dealing with a least developed country situation, it was merely dearth of engineering, managerial, and organizational capabilities.

With specific regard to Mauritania's current development strategies, the above-mentioned review considered that their weaknesses were manifested in government's neglect of the need to acquire effective scientific research and technology development/adaptation capabilities and a lack of focus on building an effective national innovation system.

The Review recommended that prompt action be taken to remedy this through developing a realistic STI policy, taking into account national priorities and prevailing constraints. This policy would need to be integrated into the overall national development policy. The Review further warned that if such steps were not taken then development efforts would most likely fail to "deliver a dynamic economy" with reasonable growth patterns with benefits for wider social development.

The UNCTAD Review finally concluded by stressing the need for "policymakers representing all key Government ministries and other key stakeholders" to initiate discussions toward formulating a national policy aimed at improving Mauritania's STI capabilities at the earliest opportunity,[199] integrating this strategy with the broader national development strategy. At the same time, steps will have to be made toward building a stronger innovation system, capable of responding to national priorities.

### *3.4.3.2 R&D Institutions and Their Activities*

Mauritania has established 18 public research institutes (PRIs), half of which operate under the Ministry of National Education and specialize in various areas of science, technology, administration and agriculture, as well as linguistic and religious studies. These centers were reported in 2007 as employing a total of around 370 scientists.

The remaining research centers are more specialized, operating under a variety of government ministries,[200] and engaged in addressing national priority issues along sectoral lines, e.g., on fisheries, agriculture, livestock, water, environment, health, geology and archaeology. These latter research centers, are reported as having received significant aid, totaling around US$5.7 million, from over 12 different donors and international development agencies in 2007. They were also reported as employing a total of over 130 scientists in that year.

### 3.4.3.2.1 Agricultural and Fisheries Research

Three centers appear to possess the bulk of available resources, in terms of both personnel and funding. These are namely

- The Mauritanian Institute for Oceanographic Research and Fisheries (IMROP)[201]
- The National Centre of Agricultural Research and Development (CNRADA)[202]
- The National Institute of Livestock and Veterinary Research (CNERV)[203]

However, although no official data on total R&D expenditure are available, all public research institutions are known to suffer insufficient and uncertain financing.[204] Frame 3.18 presents some further information on the issue of

**Frame 3.18 Funding and Human Capital Shortages in Mauritania's Agricultural and Fisheries Research**

Crop and livestock research play a relatively small part on Mauritania's STI scene, essentially due to the country's arid climate. On the other hand, greater attention is awarded to fisheries research. Thus, in 2014, IMROP, accounted for the bulk of the country's agricultural research spending. However, funding allocated to IMROP, fluctuated considerably during the past decade, particularly due to termination and renegotiation of various treaties on fisheries with the European Union.

Research output by CNRADA and CNERV and, presumably their impact on Mauritania's crop and livestock sectors has also been seriously affected by funding shortages. However, a World Bank project is being prepared to address some of these centers human resource challenges.

A World Bank-funded Agricultural Services Program implemented during (1995–2004) included plans for merging CNRADA and CNERV into a single national agricultural research institute (NARI) and for drawing up a national agricultural research plan. Neither plan materialized, however. And now, 12 years later, the country is still in urgent need of an effective national S&T policy strategy, improved coordination of agricultural research, and further financial and human capital support for its crop and livestock research institutes.

In summary, only strong political will, coupled to adequate funding, would succeed in revitalizing the country's agricultural research.

From Agricultural Science and Technology Indicators of the International (IFPRI). Website accessed in May 2017 at: https://www.asti.cgiar.org./mauritania

human capital and funding shortages suffered by Mauritania's agricultural and fisheries research.

Not unlike many other research centers in the Arab region and in Africa, Mauritania's research centers generally collaborate far more with research institutions abroad than their national counterparts. Only two research centers are reported as having closely collaborated with, private sector associations. Thus, CNRADA is known to have collaborated with the Mauritanian Association of Agriculturists and Stockbreeders, whereas CNERV has conducted research upon the request of the Association of Milk Producers, the National Group of Pastoral Associations as well as a number of private enterprises.[205] Additionally, these two research centers appear to collaborate with other African research centers in Senegal, Mali and Morocco, among other countries.

Concerning overall national expenditure on R&D, available information indicates that Mauritania's combined public sector R&D budget in 2004 was equivalent to 0.4% of GDP, which represents around 1.1% of total national expenditures.[206] It is highly likely that current expenditures on R&D do not substantially differ from these much earlier figures.

Little information is available on outcomes of research activity. However, with publications output reported by the World Bank's database at a total of two papers in 1999[207] and 16 papers in 2013,[208] research output may not be regarded as in anyway significant.

#### 3.4.3.2.2 Research at the University of Nouakchott

The University of Nouakchott is the Mauritania's main university institution. It is reputed as having played a central role in human resources development. Along with the above-mentioned research institutions, it also performs an important role on the national research scene. The university also takes part in a range of international collaborative research programs. However, its contributions to the development of a strong national innovation system have been quite limited. Thus, it appears to have established limited or no linkages to private and public sector enterprises. Neither has it acted as a source of significant innovations itself. In other words, The University of Nouakchott is essentially a traditional educational institution whose principal mission is to provide educational services to its students.

### *3.4.3.3 Concluding Remarks*

In line with conclusions reached by several past evaluations of Mauritania's national STI system, available STI capabilities are essentially inadequate for tackling most challenges that face the country at present and even more so in the near and medium futures. Additionally, abilities exhibited by enterprises and public sector institutions, to effectively acquire and adapt technological inputs brought in by multinational firms in the mining and oil industries, are deficient.

Serious gaps within the national innovation system (NIS) remain and extensive human capital building is urgently needed in order to build, adapt and disseminate requisite knowledge and related skills for development on a wide front, effectively including all production and services sectors.

With particular regard to research activity, based on the demanding measure of new patents created, one would have to conclude that research activity has not been at all successful. On the other hand, in terms of the number of publications produced, available data, comparing publication output by Mauritania to other Arab and developing countries, would indicate that immense improvements were needed. With regard to the usefulness of research carried out in solving actual problems very limited information is available, with the exception of two anecdotal accounts,[209] both concerning innovations in the dairy products industry, carried out through links between specific enterprises and researchers.

At any rate, future policies would do well to address measures aimed at promoting far greater coordination among research institutions, on the one hand, as well as effectively linking these institutions to public and private enterprises with focus on a long list of national challenges and priorities, on the other. With view to avoiding duplication and ensuring optimal outcomes, Mauritania will need to mandate a single institution, endowed with authority and convening power to coordinate national STI policy formulation and oversee its implementation.

### 3.4.4 Morocco's STI Policies, Institutions, and Research Activity

Morocco's STI system witnessed significant developments over the past few years. Nevertheless, scientific research, technology development and innovation are yet to acquire national priority status, in various government as well as civil society circles. This is despite the potential that the country possesses in terms of human capital, regional and international linkages as well as demand by numerous pressing socioeconomic needs.

Gross expenditure on research and development by Morocco as a percentage of GDP was reported at around 0.7%, a figure that approximates the average for upper middle-income economies.[210] Morocco scored higher than average in terms of the number of researchers in R&D than other countries in the Middle East and North Africa with around 860 compared to the MENA average of 613 researchers per million of the population.[211] However, available statistics also indicate that Morocco's high technology exports as a percentage of manufactured products witnessed a decline from around 7.7% in 2010 down to almost 6.6% in 2013, possibly reflecting shortcomings in conversion of growing STI abilities into exportable products. On the other hand, several reports indicate that steps are being taken to attract a series of high-technology enterprises into the country as well as measures targeting creation of a national knowledge economy.

The following paragraphs attempt brief examination of Morocco's STI policy regimes and institutions as well as steps taken to encourage international cooperation in research and to establish links to the country's STI community abroad.

### 3.4.4.1 STI Policies in Morocco

Morocco does not seem to have formulated a comprehensive STI policy document as many other countries did during the latter part of the last century, and the first decade of the present. Rather, it seems that the R&D and innovation activities were guided by 5-year development plans, royal decrees, official pronouncements, as well as individual strategies formulated by various actors on the STI scene, including concerned ministries, principally the Ministry of Higher Education, Professional Training and Scientific Research. Thus a "National Strategy for the Development of Scientific Research toward Horizon 2025" was put forward by this Ministry, including an ambitious action plan for the years 2015–16, with a list of activities and indicators of achievement for each goal. Sectors deemed as most promising recipients of support through research activity are listed in Table 3.12, under headings that relate to their economic performance. One of the major long-term goals of this strategy is for gross expenditure on research and development to reach a level of 3% of GDP by the year 2025.

On the other hand, as indicated in a UNESCO document on "The Science, Technology and Innovation System of the Kingdom of Morocco,"[212] the country's 5-year plans often address capacity building in science, technology and innovation. Thus, the 5-year plan covering the years 2000–04, for example, highlighted research on a selection of socio-economic development priorities including: agriculture and fisheries, health, water, geology and mining, energy, the environment, transport as well as information and telecommunications technologies. This plan

**TABLE 3.12** Sectors Selected for Particular Support by Morocco's "National Strategy for the Development of Scientific Research Toward Horizon 2025"

| Competitive Sectors | Sectors with Good Potential | Fragile Sectors |
|---|---|---|
| • Tourism<br>• Construction and public works<br>• Commerce and services | • Agriculture and fishing<br>• Agroindustries<br>• Artisanal production<br>• Mining | • Textiles |

Adapted from "Strategie Nationale pour le Developement de la Recherche Scientifique an l'Horizon 2025." Published by the Directorate for Scientific Research and Innovation at the Ministry of Higher Education, Professional Training and Scientific Research Downloaded in April 2017 from: http://www.enssup.gov.ma/sites/default/files/PAGES/168/Strategie_nationale_recherche 2025.pdf also stressed the need for effective institutional coordination and collaboration toward a set of socioeconomic objectives.[213]

During its 35th General Conference held in Tunis in 2013, the Federation of Arab Research Councils received official representatives from several Arab countries, including Morocco, whose presentations outlined essential features of a crash program aimed at promoting research and development. An estimated €72 million was allocated to this program which included additional funding for research on national priorities.[214] Morocco's presentation also alluded to plans for implementing another programs during 2013–16, with six tracks, including reviewing measures used for research governance, updating regulatory arrangements as well as development of extensive international research collaboration.

Plans for the establishment and operation of a number of FDI projects, incubation facilities, and technology park initiatives have contributed to a de facto policy with the private sector playing a significant part, whereas concerned government departments contribute toward infrastructure arrangements as well as funding. Government's contributions appear to target research grants as well as scholarships designed for higher studies at home and abroad involving considerable research activity.

Thus, the main trends that seem to characterize STI development plans may feature preoccupation with:

- the country's mineral wealth, which has been confined to its sizeable phosphate deposits, with more recent promise of considerable natural gas resources;
- massive development of renewable energy resources, with wind and solar occupying a central position;
- hosting high-technology production facilities in domains including electronics and avionics products, destined mainly for export.

### 3.4.4.2 STI Institutions

The Ministry of National Education, Higher Education, Professional Training, and Scientific Research is the central government ministry that is most closely linked to STI human capacity development.[215] Two directorates within this Ministry deal directly with S&T matters including the exact and natural sciences, humanities, and social sciences research, as well as technology development, innovation and forecasting.

A "Permanent Inter-Ministerial Committee for Scientific Research and Technological Development" was created in 2001 and is considered the principal S&T policy-making body in Morocco. As its name implies this committee includes representatives of concerned ministries. It is presided over by the Prime Minister. Its scope of responsibilities includes:

- making proposals with regard to promoting scientific research and technological development;
- providing guidelines in order to reap optimal benefits from S&T activities undertaken by various government departments;

- putting forward suggestions for resource allocations to research projects and program areas, in line with national priorities.

Established in 2002, the "Government Authority for Scientific Research" acts as a general secretariat for the above-mentioned Permanent Inter-Ministerial Committee, it is housed within the Ministry and is charged with:

- working out the national S&T research strategy;
- identifying nation-wide priority research projects and programs and proposing related resource allocations, including programs aimed at nurturing the development of human resources in S&T;
- ensuring coordination, financing and evaluation of scientific research activity as well as ensuring effective partnerships and cooperative arrangements between researchers and agents of technological innovation, both national and international;
- development and publishing biannual evaluation reports, with details of progress achieved.

The most recent attempt at securing support for scientific research is the work plan for higher education and research for the period (2013–16). The plan included several measures aimed at tackling human capital development, restructuration of scientific research, developing partnerships, a new vision for the promotion of international cooperation as well as consolidation of innovation infrastructure. Under this plan, support per research and technology development project was provided at levels ranging between €50 and €1000.

The following paragraphs present summary information on institutions active in supporting and undertaking research in Morocco.

### 3.4.4.2.1 The CNRST[216]

Created in 1976, CNRST is a public institution mandated to:

- maintain an inventory of national scientific and technological activities and achievements;
- identify national priority areas in research and technology development on the basis of in-depth studies, taking into account national requirements outlined within development plans;
- provide broad outlines for national research programs and related implementation and monitoring modalities, in close collaboration with implementing institutions;
- encourage and facilitate research through awarding grants, organizing workshops as well as dissemination of research outcomes.

Toward the beginning of the century, CNRST is said to have gone through periods when it was only partly able to fill the above roles. Around 2006, however, it was regarded as the most important actor on the R&D scene in Morocco.[217] NCSRT created the Moroccan Institute for Scientific and

Technological Information (MISTI), during 2003–04 with a mandate focused upon collecting, analyzing, and disseminating scientific and technological information. The following section presents information on research activities sponsored by CNRST.

CNRST is in charge of the selection and funding of research and development activities in Morocco. CNRST publishes annual reports with information on research activities funded and progress toward made their completion. The latest such report made available by CNSRT on its website belongs to the year 2011, indicating that researchers in Morocco were engaged in implementing 168 projects within the priority areas shown in Fig. 3.20. On the other hand, Fig. 3.21 presents a view of the distribution of funds allocated to these projects across expenditure items, indicating that the equipment and materials purchases

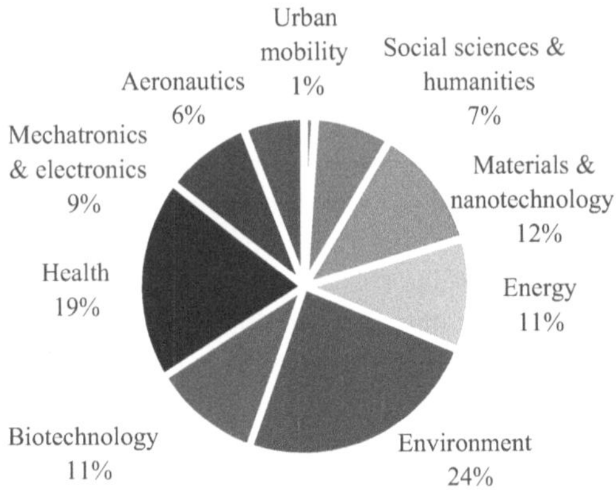


**FIGURE 3.20** Distribution of 168 projects selected for funding by Morocco's National Centre for Scientific and Technological Research.

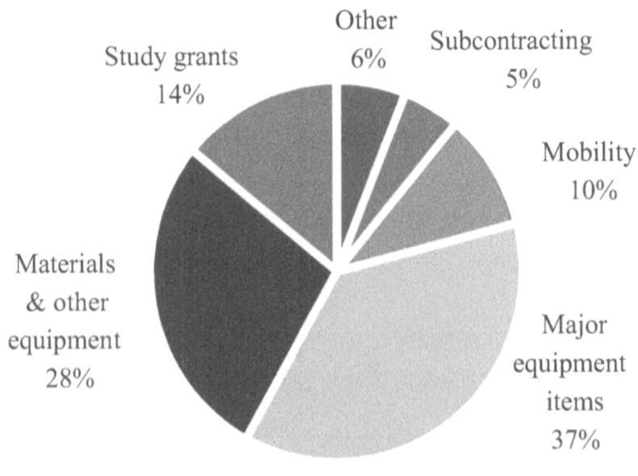


**FIGURE 3.21** Distribution of total funding allocated to National Centre for Scientific and Technological Research projects across expenditure items (2011).

take up a sizeable percentage, amounting to nearly 65% of the total allocations made of around DH268 million, or around US$26.5 million.[218]

### *3.4.4.3 Research Activity at Other Moroccan Institutions*

The majority of centers actively pursuing research in Morocco are engaged in research in the humanities and social sciences. However, several university departments in Morocco undertake research projects in applied science and technology. The following paragraphs provide brief summaries concerning a few such centers, whereas Frame 3.19 presents a list of several other institutions dedicated to research and technical services activities in Morocco.

#### 3.4.4.3.1 Moroccan Association for Research and Development

Created in 1997 MARD, was recognized as a public utility by the Moroccan government in 2001 and charged with promoting innovation and R&D at the national level. Its membership includes representatives of public and private firms, public research institutions and government departments responsible for the management of innovation and research. MARD's principal tasks include the development of Morocco's innovation system, promotion and support of national competitiveness and fostering research development. Achievements listed on the Associations website include:

- a Strategy for building an effective national innovation ecosystem;
- a national innovation indicators survey;
- promoting incubation schemes, with emphasis on innovative businesses based on research outcomes.

**Frame 3.19 Government and Semipublic Research Institutes in Morocco[219]**

- Centre for Development of Renewable Energies
- National Agronomic Research Institute
- National Centre for Forestry Research
- National Centre for Road Studies
- National Fisheries Research Institute
- National Institute for Hygiene
- National Office for Drinking Water
- National Office for Petroleum Exploitation Research
- Pasteur Institute
- Public Testing and Studies Laboratory
- Research Centre for the Study of Phosphates

From "The Science, Technology and Innovation System of the Kingdom of Morocco; " A UNESCO document authored by N. Boshoff. The document is not dated, but appears to have been published post 2007. Downloaded in July 2016 from: http://portal.unesco.org/education/en/files/55536/11998897825Morocco.pdf/Morocco.pdf.

##### 3.4.4.3.2 Research Institute for Solar Energy and New Energies[220]

This institute was created in 2011 by the Ministry of Energy, Mining, Water and Environment, in cooperation with several other national institutions concerned with the energy sector in Morocco. IRESEN is reported as employing around 540 researchers working on 37 projects, with a budget of around US$25 million.[221] Projects undertaken by IRESEN address photovoltaic and thermal solar energy devices as well as means for utilization of biomass resources.

##### 3.4.4.3.3 Hassan II Academy of Science and Technology

This is a nongovernmental body that is essentially responsible for promoting scientific research in areas deemed as constituting national priorities. HAST was created in 1993 but did not hold its first meeting till 2006.

#### *3.4.4.4 Public–Private Sector Research Cooperation*

A report issued by the Arab Thought Foundation in 2013, indicated that, despite several initiatives, cooperation between research centers within universities and the public sector institutions, on the one hand, and production and services institutions within both the private as well as the public sectors, on the other hand, remain quite modest.

The following factors seem to have placed further constraints upon such cooperation:

- demand for research and technical services by the private sector is still quite limited, essentially due to the fact that its activities are confined within domains that enjoy limited sophistication;
- an acute mismatch exists in any case, between research centers' capabilities and the requirements of production and services institutions;
- inflexible bureaucratic procedures add further hindrances.

#### *3.4.4.5 International Research Cooperation*

International cooperation is encouraged by Morocco's government in all aspects related to the country's development priorities. In practice, a considerable proportion of international STI cooperation is carried out with the European Union, through the Framework and Horizon 2020 programs. Moroccan STI institutions have also been reported as implementing collaborative projects, on bilateral bases principally with French and German research and industrial institutions on individual bases. Several reports point to increased collaboration with the United States, South Korea among other countries.

#### *3.4.4.6 Linking to Morocco's Diaspora*

The International Forum of Moroccan Competencies Abroad, a special program referred to as "FINCOME,"[222] was launched by Morocco in 2007, on the basis of a public–private partnership, with view to making use of Moroccan experts

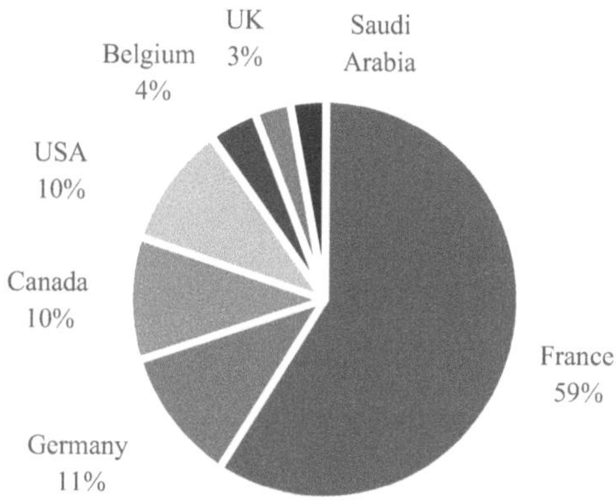


**FIGURE 3.22** Percentage distribution of Moroccan experts who took part in FINCOME 2011 across countries of residence.

working abroad. In 2011, this program was able to make use of the services of only 71 such experts, distributed across their countries of residence as indicated in Fig. 3.22.[223] As the pie chart in this figure indicates, by far the largest proportion of expatriates came from France, followed at some considerable distance by those from Germany, Canada, and the United States.

As Table A.3.3 indicates at least 15 research networks are reported as having been established by various university faculties with view to establish collaboration among researchers working within Moroccan institutions, on the one hand, and with their counterparts abroad, on the other. Research collaboration covered by these networks targets a variety of the applied sciences, engineering and medicine. The fact that such networks have been established by university faculties clearly indicates that these faculties undertake research in the fields of science and technology areas their networks cater for, and might be viewed as compensating for the predominance of the social sciences and humanities as focus areas for a large proportion of national research institutions.

### 3.4.4.7 Innovation Policy and Enterprise Creation in Morocco

Establishing Morocco's industrial and research clusters came about as a result of a broader framework set by the National Plan for Industrial Emergence 2009–15. This National Plan is implemented by the Ministry of Industry, Commerce and New Technologies and the Morocco Innovation Initiative, in collaboration with several other government institutions, including the Ministry of Industry, Commerce and New Technologies, the Ministry of National Education, Higher Education, Professional Training and Scientific Research and the General Confederation of Moroccan Enterprises. The National Plan aims at boosting national competitiveness, through support for SMEs, as well as interventions in aid of creating dynamic local economies. Focus is maintained by the National

Plan on high potential services such as ICTs, biotechnologies, sustainable development, aeronautics and mechatronics, toward which around 80 % of available funds are dedicated.

One of the components of the above-mentioned Innovation Initiative is dedicated to promoting clusters, four of which are reported as having been created in 2010 through joint efforts by concerned ministries. Funding by the Moroccan government in support of the above clusters is reported at an annual €527,000, two-thirds of which is dedicated to "financing collaborative projects."[224]

A report published by Erawatch in 2012[225] indicated that an innovation policy initiative covering the period 2009–14 was formulated under the title "Morocco Innovation Initiative" with provisions including drafting an innovation bill designed to encourage start-ups, spin-offs, and recruit researchers into private sector institutions. Four clusters were reported as having been established and financial instruments designed to facilitate funding of new innovative start-ups as well as research and technology development and innovation (RTDI) projects launched for the benefit of private sector institutions. No detailed information was available on specific outcomes of this initiative with figures concerning funds made available for specific projects or groups thereof. On the other hand, this report underlined the need for this policy initiative to be integrated with other policies in order to improve access to knowledge from various sources, such as FDI, Moroccan diaspora, international cooperation in RDTI. Additionally, the Erawatch report emphasized the need for identifying obstacles that hinder technology and/or knowledge transfer between universities and end-user sectors. Such difficulties often occurring in administrative and financial procedures and legal framework.

### 3.4.4.7.1 Science and Innovation Parks

Morocco was recognized by a United Nations report issued in 2009 as one of only eight African countries to have integrated the development of their science parks into their science and technology policies. Together with Tunisia, Morocco is credited with being one of the most advanced among African countries in terms of deployment of technology parks. The following paragraphs provide a summary outline.

***3.4.4.7.1.1 Morocco's Technopark*** The Technopark has facilities in Casablanca, Rabat and Tangiers. It was set up to help creation and development of businesses in the information and communications and environmental technologies. Morocco's Technopark facilities were established as a result of a public–private partnership and are managed by a single managing company, namely Technopark MITC. Since its creation, this Technopark is reported as having assisted 800 innovative companies in Casablanca and more than 80 companies in Rabat and

20 companies in Tangier. It is presently reported as housing 250 Moroccan start-ups and SMEs, and employing nearly 2000 staff.[226] Technopark opened in Rabat in 2012 and in Tangiers at 2015.

#### *3.4.4.7.1.2 Moroccan Centre for Innovation and Social Entrepreneurship*

This Centre was established in 2012 as a not-for-profit entity dedicated to entrepreneurial and innovative solutions to social challenges facing the country. The MCISE vision statement emphasizes the need for applying innovative ideas in national priority areas. Its programs include a variety of services such as coaching, incubation with seed funding as well as access to crowd funding platforms.

### 3.4.4.7.2 Morocco's Business Incubation Schemes

A report[227] based on a fact-finding mission to both Morocco and Tunisia conducted in 2013 by the Dutch Incubation Association (DIA), under contract with the Dutch Ministry of Foreign Affairs, assessed business incubation activities undertaken by these two countries. The mission's report presented the results of a survey it conducted in Morocco with primary weaknesses and strengths summarized in Frame 3.20 as well as recommendations aimed at tackling obstacles the mission was able to identify.

**Frame 3.20 Strengths, Weaknesses, Opportunities, and Threats Characterizing Business Incubation in Morocco**

| **Strengths** | **Weaknesses** |
|---|---|
| • Large numbers of young people with good ideas;<br>• Availability of multidisciplinary university education;<br>• Adequate subsidies for innovative projects;<br>• A sizeable population of young entrepreneurs in the informal economy. | • Inadequate exploitation of youth's potential;<br>• Entrepreneurship lacking public esteem;<br>• Limited drive toward entrepreneurship among graduates;<br>• Bureaucracy tending to suffocate initiative. |
| **Opportunities** | **Threats** |
| • Morocco's diaspora could provide expertise;<br>• Valuable assistance through government's initiatives;[228]<br>• Incubators will create jobs;<br>• The need to utilize innovative ideas created by universities. | • Limited financing and coordination within the national innovation system;<br>• Lack of entrepreneurship education as well as limited mentoring;<br>• Complex laws and regulations and protracted bureaucratic procedures. |

Recommendations produced at the end of the mission underlined the need for government to institute reforms concerning permits, taxes and resolving insolvency, in short, for greater compliance with World Bank's "Ease of Doing Business" guidelines as well as tackling bureaucracy, which seemed to be a common complaint by both entrepreneurs and enabling institutions.[229] The report furthermore emphasized several measures including the need for:

- collection of data on the national start-up ecosystem as essential bases for evaluating its effectiveness as well as assessing contributions by various actors, including incubators, investors and trainers;
- helping business incubators by simplifying registration procedures;
- providing greater assistance to university incubators, through improved funding for spin-off projects;
- improving the quality and extent of entrepreneurship education, through introduction of entrepreneurship curricula;
- enhancing access to risk capital for start-ups and the availability of start-up credits;
- facilitating market access for newly created companies;
- promoting closer cooperation between various actors within the start-up ecosystem;
- investigating possibilities posed by accelerator programs for early stage growth companies with potential on the international arena.

According to the DIA mission report, both Morocco and Tunisia opted for similar choices in the design of their knowledge infrastructures, including technology transfer units, business incubators and science parks. However, they seem to have different tracks with regard to roll-out or graduation procedures, with the result that Morocco currently has an estimated 13 technology transfer centers and two business incubators. Tunisia, on the other hand, had an estimated 30 business incubators, most of which are linked to institutions of higher education and research centers as well as departments at the Ministry of Industry. Despite these and other differences, both countries were able to modernize their national start-up ecosystem, upgrade high-tech entrepreneurship and create jobs for skilled youth.

#### 3.4.4.8 Concluding Remarks

The public sector in Morocco was reported as funding 90% of all research undertaken in Morocco. Although the semipublic and private sectors covered 9% and 1%, respectively.[230] Public research centers and academic institutions were estimated as spending 88 of all R&D allocations. With rare exceptions, including a few companies active in agro-food, textiles, mining, leather, R&D spending by the private sector has traditionally been quite low. More recently, success in attracting technology-intensive TNCs has marginally improved the private sector's contributions to R&D activity. Nevertheless, this remains relatively small compared to the public sector's contributions which are still in need of much further enhancement.

Current directions taken by measures aimed at revamping Morocco's STI capabilities, generally emphasize support for high-tech industrial activities aimed at the export market, may not be expected to do away with socioeconomic problems witnessed by the country. Overhauling the country's STI systems should in principle contribute toward improving the lot of the majority of Morocco's population, only if STI outcomes will produce tangible and widespread benefits. To this effect STI inputs aimed at upgrading education and infrastructure arrangements in rural Morocco should be accorded a high order of priority.

A draft World Bank report dated in 2009 concluded that, in general, Morocco had "several weaknesses in its innovation system."[231] It added that with the country's limited R&D resources, and knowledge production in general, an important part of any innovations it might be able to introduce would need to depend upon its ability to access, and subsequently adapt and disseminate, global knowledge in order to improve its socioeconomic performance. At 3.72, Morocco's innovation pillar score in 2009 was considerably below the MENA average of 7.57.[232] However, Morocco's innovation scores and related global and regional rankings are likely to have improved since then.[233] Nevertheless, absence of close monitoring and open debate concerning progress achieved remain serious impediments to further gains reflecting into sustainable socioeconomic development.

In summary, Morocco took significant steps over the past few years with a series of nationwide plans to stimulate sectoral and spatial diversification of its economy. The fact that interest is being shown by Morocco in upgrading traditional sectors of its economy, such as its mining and related industries as well as to the acquisition and dissemination of new technologies, particularly ICT and related infrastructures is certainly a healthy sign.

However, structural factors inherent in traditional sectors, such as the small size of the majority of holdings in the agricultural sector, should certainly receive much greater attention by future STI initiatives, at the national policy level and through specific local and territory-wide interventions. Goals for such interventions might well focus on enabling access by small agricultural enterprises to novel technology inputs as well as facilitating moves toward higher value crops.

Although plans have been reported with the aim of mobilizing government and local authorities, their impact on economic activities, job creation, will need to be closely monitored. Perhaps more to the point, creating and enabling communities of change agents beyond traditional government agencies would be essential for reaping sustainable and more inclusive change.

Broader mobilization of both urban and rural populations, utilizing business associations, trade unions, local community actors and educational institutions, should target greater engagement by citizens in knowledge- and innovation-based inclusive development.[234] In particular, such mobilization should facilitate sorely needed improvements in overall business and governance ecologies, engendering trust and wider collaboration.

## 3.4.5 Tunisia's STI System; Its Governance, Policies, and Institutions

Several research institutions were established prior to and immediately following the country's independence in 1956, along lines commonly adopted by their French counterparts. However, they enjoyed far less support than their role models did, in terms of financial and human resources, and symbiotic partnerships among one another. Nor did they establish the mutually beneficial links with private sector enterprises that characterized corresponding institutions in France.

The following section attempts a closer look at Tunisia's STI system, which despite notable achievements would still greatly benefit from coordinated policy making, improved legislation, closer coordination among major actors as well as improved resource allocations.

### *3.4.5.1 Governance of Tunisia's STI System*

A Secretariat of State for Scientific Research and Technology (SERST)[235] was created in 1991 with the aim of promoting research aimed at tackling national development problems. SERT was placed under the authority of the Prime Minister and charged with:

- coordinating related interministerial activities;
- drafting a national research and innovation policy in line with national social and economic priorities;
- monitoring the implementation of this policy through reviewing activities of concerned ministries.

SERST helped finance and lead a number of research programs, including the "National Mobilizing Programmes (PNM)"[236] launched in 1992, targeting a broad range of sectors, including agriculture and fisheries, the environment, natural resources, industrial activity, energy generation, computing and telecommunications, healthcare and pharmaceuticals as well as socioeconomic research. By the time PNM was discontinued in 1998, it had provided support for a total of 600 projects at a total expenditure of around US$20 million. Projects in agriculture, healthcare, and pharmaceuticals were reported as the main beneficiaries with allocations amounting to around 67% of the total.

Other accomplishments of SERST include the design of a program for the valuation of research outcomes (VRR).[237] This program which was also launched in 1992, supported around 90 projects with grants amounting to a total of about US$5 million. Projects within agriculture, information and communications technologies, energy and biotechnology, enjoyed 25%, 21%, 15%, and 11% of the total, respectively.

SERT also plays an important role in enhancing national R&D funding allocation. Thus, in accordance with the goal set by SERT, stipulating that R&D spending would reach 1% of GDP by the end of 2004, R&D allocations were increased by concerned ministries and institutions till this objective was eventually achieved.

In addition to SERT, a Higher Council for Scientific Research and Technology was created in 1992, primarily in order to facilitate interministerial coordination. It was chaired by the Prime Minister and brought together representatives of ministries, labor unions, heads of industry as well as a selection of other institutions with essential input. Frame 3.21 presents a list of additional bodies, essentially national committees and intergovernmental councils created with tasks aimed at promoting scientific research and technology development.

Laws and regulations governing the work of R&D institutions were initially adapted from corresponding French laws and regulations and undertook radical

**Frame 3.21 Committees and Councils Established by Tunisia in Order to Promote Scientific Research and Technology Development**

| | |
|---|---|
| • Le Comité de Haut Niveau pour la Science et la Technologie | • The High-level Committee for Science and Technology |
| • Le Conseil Supérieur de la Recherche Scientifique et de la Technologie | • The High Council of Scientific Research and Technology |
| • Le Conseil Consultatif National de la Recherche Scientifique et de la Technologie | • The National Advisory Council for Scientific Research and Technology |
| • Le Comité National de l'Evaluation des Activités de Recherche Scientifique | • The National Committee for the Evaluation of Scientific Research |
| • Le Comité Technique de la Recherche Scientifique et de la Technologie. | • The Technical Committee for Scientific Research and Technology |
| • Conseil Stratégique de l'Economie Numérique. | • Strategic Council of the Digital Economy |

From "How to harness the National Innovation System in Tunisia" Report by the Technology Centre of the Economic and Social Commission for Western Asia. Published in 2016. Downloaded in April 2017 from: https://www.slideshare.net/MondherKhanfir/report-on-the-tunisian-national-innovation-system?from_action=save.

revision only as late as 1996, with later amendments in 2002 and 2006, in a manner that recognized the special character of national research activities and related fund disbursement modalities.[238]

In order to ensure compliance with STI directives as well as monitor performance, Tunisia took in 1997 the commendable step of establishing a national committee specifically charged with monitoring and evaluation of research activities undertaken by Tunisian institutions. This is the National Committee for the Evaluation of Scientific Research Activities (CNEARS).[239] This committee is vested with authority to refuse the creation of new research facilities and even close others based on their performance. The fact that this Committee may conduct its deliberations and render its proposals independently from the Ministry of Higher Education and Scientific Research is noteworthy. However, there may be a strong need for revising statutes that govern the operations of this Committee, with particular regard to its own governance and operational modalities.

### *3.4.5.2 Tunisia's STI Policy Regime*

The fact that the country does not yet have an integral STI policy document that is formally adopted as basis for strategic action is a major drawback. However, what seems to replace such a policy document in Tunisia is a series of laws and decrees issued by government. A prime example of this is a "guideline law" issued in January 1996, and later amended, in 2000, 2002, and 2006, with the aim of promoting scientific research as well as the acquisition and implementation of technology for the purposes of national development. Frame 3.22, presents the article within this law purporting to set essential goals for the country's science and technology policy.

The latest amendment, in 2006, introduced changes to take into account the fact that recent legislation granted special status to institutions charged with scientific research and technology development with regard to the manner in which they may conclude contracts and disburse funds. As a result, these institutions were required to follow stricter rules and procedures, essentially set by relevant ministries. Another significant amendment to the previous law concerns abolishing a clause that allowed individuals as well as variety of institutions to conduct scientific research and technological development work, thereby restricting such activities specifically to recognized public and private sector organizations.

Apart from the above, Tunisia's virtual STI policy is essentially based on official announcements that generally encourage enterprises and industrial support institutions to take up innovation, technology acquisition and R&D into their strategies and operational plans as well as national 5-year development plans with more or less specific objectives.

Thus, the Tenth Development Plan, covering the period 2002–06, included components favoring moves toward a knowledge economy, whereas the Eleventh Development Plan (2007–11), was essentially conceived with a view

**Frame 3.22 Tunisia's Guideline Law on Scientific Research and Technology**

The law was passed in January 1996 with the aim of guiding scientific research and the acquisition of technology in attaining inclusive development. Article 2 of this law lists the objectives of national scientific research and technology development policy as

- Employing scientific research and technology development in support of economic growth and to ensure that it keeps in pace with international developments.
- Spreading the culture of research and urging innovation and innovation in society with contributions to enriching knowledge in the human, social and basic sciences.
- Supporting the formation of researchers in all fields of knowledge, within institutions of higher education and scientific research as well as public sector research centers.
- Establishing an appropriate framework for innovation and development in various sectors and institutions.
- Promoting cooperation and partnership activities between public institutions for scientific research, institutions of higher education, research and economic institutions and encouraging them to open up toward the national economic, social and cultural environment.
- Exploiting research outcomes and implementing them in response to economic, social and cultural needs in accordance with national priorities.
- Ensuring concordance between the development of scientific knowledge and the preservation of ethics and human values.
- Promoting international cooperation in the fields of scientific research and technology development of common interest in the service of national priorities.

to building a full-fledged knowledge-based economy. Policy reforms called for by this plan were aimed at:

- higher R&D expenditures;
- enhancing the quality of education;
- creation of technology poles;
- wider access to the Internet;
- strengthening the country's intellectual property regime.[240]

With particular regard to information and communication technologies, Tunisia has developed a strategic plan[242] with emphasis on priorities that included:

- high capacity networks;
- security of systems and of networks for critical sectors;
- ICT applications in health, energy, and environment;
- platform development for mobile communications;

- embedded systems, software industry and methodology;
- audio–visual and multimedia for research.

Tunisia's ICT strategic also emphasized the key role of high speed ICT infrastructures to further enhance connectivity and interoperability for research and education platforms in North Africa and with the European Union.

Tunisia may be credited with having made reasonable progress toward all of the above goals, with numerical indicators affirming advances with particular regard to public research spending. Thus, with particular regard to R&D expenditure, Tunisia now leads all other Arab countries, in dedicating the highest proportion of its GDP to R&D expenditure. Thus, at 1% of GDP, Tunisia's R&D expenditure exceeds levels dedicated by other Arab countries, which generally varies between 0.2 and 0.6 of GDP, but still quite low compared to industrialized countries around the world.

### *3.4.5.3 Tunisia's S&T Institutions*

Only three major research institutions were inherited from Tunisia's pre-independence area, namely the Pasteur Institute, the Institute for Veterinary Research (IRVT), the Tunisian Institute for Agricultural Research (INRAT)[243] and the National Institute for Marine Science and Technology (INSTM). Following Tunisia's independence further research institutions were launched with the creation of the Forestry Research Station in 1957, followed by the Atomic Energy Commissariat (CEA)[244] as well as the Water Research Institute (CREGR) in 1959. Two further research facilities were created in the early sixties, namely the Social and Economic Studies Research Centre (CERES)[245] and a center specializing in research on the utilization of saline water in agriculture (CRUESI). This latter center was merged in 1970 with CREGR, giving rise to the present CRGR, which is a multidisciplinary research establishment.

Further institutional entities concerned with scientific research and technological development were also created, with particular emphasis on agricultural research and rural development (Frame 3.23). Thus the National Institute for Research in Rural Engineering, Water and Forestation (INRGREF) was created in 1996, incorporating CRUESI and the National Forestry Research Institute (INRF) with six laboratories specializing in a variety of domains including water management and conservation, forest management, agricultural machinery, soil conservation and the use of renewable energy in agricultural activity.

#### 3.4.5.3.1 The Ministry of Higher Education and Scientific Research

Tunisia's Ministry of Higher Education and Scientific Research (MHESR) was established in 1978. Initially, it included two administrative units, one of which was dedicated to scientific research. Research centers operating under the Ministry's supervision address a variety of research disciplines.

**Frame 3.23 Chronology of the Establishment of Agricultural Research Facilities in Tunisia**

1957—Forestry research station established together with a network of regional stations throughout the country.
1959—First water research center (CREGR) founded.
1962—Saline agriculture (CRUESI) research Centre created with United Nations assistance.
1965—Reforestation institute set up with help from a special United Nations fund.
1966—The National Institute of forestry research is created.
1970—CRUESI and CREGR merged, creating CRGR as a multidisciplinary research institution.
1996—The National Institute of Rural Engineering, Water and Forests (INRGREF),[241] merging INRF and CRGR, and placing it under the authority of the "Institution de la Recherches et l'Enseignement Supérieure Agricoles (IRESA)".

**Frame 3.24 Research Centers Under the Supervision of Tunisia's Ministry of Higher Education and Scientific Research**

- Centre National Universitaire de Documentation Scientifique et Technologique
- Centre d'Etudes et de Recherches Economique et Sociales
- Centre de Recherches et des Technologies de l'Energie
- Institut National de Recherche et d'Analyses Physico-chimique
- Centre de Recherches et des Technologies des Eaux
- Centre de Biotechnologie de Sfax
- Centre National des Sciences et Technologies Nucléaires
- Centre National de Recherche en Sciences des Matériaux

Frame (3.24) presents a list of these centers indicating the wide scope of activities they undertake, spanning a variety of scientific and technological domains. The Ministry supports and supervises a large number of higher education, research and technical services institutions throughout Tunisia. Its website is by far one of the most informative among all others reviewed by this author, with ample statistics on, among other issues, its human resources, budgets, degrees and grants awarded, publications output by researchers active in its institutions, as well as its research laboratories and units.

A total of around 86,000 students graduated from the Ministry's HEIs at the end of the academic year 2009/2010. However, the number appears to have decreased by over 27,100, or around 32% by the end of the academic year 2014/2015. Graduate numbers who qualified in the social sciences, business and law as well as the pure and applied sciences went down by almost 40%, whereas those specializing in engineering went up by over 6%. Indeed, graduates in engineering specializations were the only ones to witness an increase between the above-mentioned academic years, whereas graduates in all other specializations saw reductions of between 50.2% for graduates in education and 13.4% for graduates in health and social services specializations.

Grouped together, graduates in the pure and applied sciences, engineering, as well as specializations pertaining to industry and construction constituted over 41% and 45% of the total number of graduates in the academic years 2009/2010 and 2014/2015, respectively.

It is noteworthy that the average percentages of females qualifying in all specializations during the academic years 2009/2010 and 2014/2015 were around 62% and 67%, respectively. The fact that the percentage of females went up although total numbers of all graduates went down is consistent with the notion that economic difficulties were forcing more male students to start working in order to earn an income before, or rather than, attempting to gain higher qualifications.

The following sections provide information on two further institutions that engage in, or support research and development activity in Tunisia.

#### 3.4.5.3.2 The Institute for Scientific and Technology Research[246]

The National Institute for Scientific and Technical Research was established in 1983,[247] on the basis of an earlier "Institute for Scientific and Technology Research." It is modeled along the lines of France's National Centre for Scientific Research, or "Centre National de Recherche Scientifique (CNRS)," with its mission including:

- the conduct of research programs and participation in the development of scientific and technical research;
- integration of research outcomes within the economic and social fabric;
- nurturing partnerships with industry and promoting industry–university collaboration in research;
- evaluating research outcomes and the impact of their exploitation in the economic sphere.

#### 3.4.5.3.3 The Ministry of Communication Technologies and the Digital Economy

The MCTDE was established in 2004 with a mandate including the formulation of policies designed to promote information and communication technologies and ensuring their implementation with view to consolidating the role of the ICTs in economic and social development. The Ministry is charged with developing national strategies of the ICT sector and ensuring their implementation. This is carried out essentially through, among other means:

- Keeping abreast with the sector's development and striving to exploit the opportunities they offer within the framework of economic and social development plans and proposing associated provisions.
- Contributing to the creation of an environment that encouraging investment in the ICT sector and to the development of relevant skills and expertise.

- Designing programs and projects with viable options to ensure contributions by concerned parties to their implementation.
- Ensuring the continuity of quality ICT infrastructures and services including protection of the security of systems, networks and data in cyber space. With particular regard to e-government and e-commerce applications, the Ministry aims at promoting trust in virtual transactions.

Tunisia's ministry is also responsible for developing international cooperation programs and for action in support of international and regional bodies targeting promotion of ICT capabilities commensurate with national policies and their implementation strategies.

#### 3.4.5.3.4 Research Laboratories and Units Established During 2012–15

Fig. 3.23 presents a view of the growth in the numbers of research laboratories and units established during the period 2012–15, indicating significant growth in the number of laboratories, and remarkable growth in the number of research units established by the Ministry within this period. Thus, the numbers of research laboratories and units witnessed growth of around 17% and 77%, respectively, during the period 2012–15. The numbers of student researchers also witnessed considerable growth, by nearly 43% during the above period, from around 9800 in 2012, to over 14,000 in 2015.

### *3.4.5.4 Research and Development Activity in Tunisia*

Public sector institutions are predominant on Tunisia's R&D scene. A report published in 2009 indicated that the country had 33 research centers with nearly 227 laboratories and over 229 research units covering a variety of specialisations.[248,249] During the 1990s public R&D spending

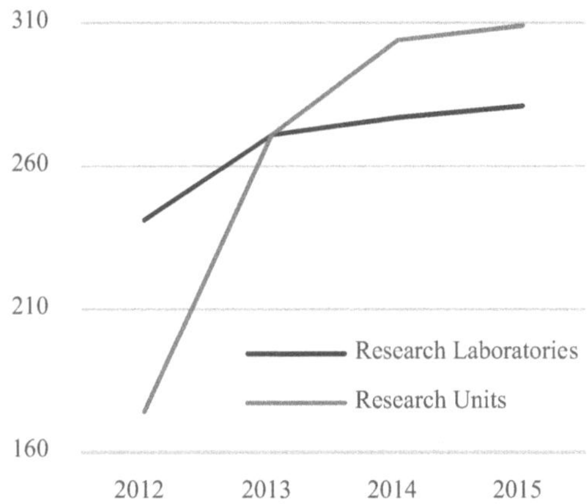


**FIGURE 3.23** Numbers of research laboratories and units operating under the supervision of Ministry of Higher Education and Scientific Research in Tunisia.

was estimated at around 92% of the total, whereas the share of the private sector stood at around 5%, with nongovernmental organizations making up the remainder.[250] R&D activities are undertaken by centers established by a variety of ministries, including the ministries of agriculture, health, industry, and the environment.

The "Federative Research Programme (FRP)," was introduced last decade with the aim of enhancing cooperation between public research organizations and business firms, both private and public, in tackling themes that fell within national development priorities identified following systematic consultation with stakeholders within the research community. Through undertaking research on such themes, this program aims to mobilize skills and promote the creation of research teams and networks in multidisciplinary projects. The current 4-year program is financed by government grant and projects are implemented based on agreements between the General Directorate of Scientific Research and each research team or network. Themes currently earmarked for R&D joint activity are listed in the Frame 3.25.[251]

#### 3.4.5.4.1 Tunisia's Agricultural and Related Research Activities

Agricultural research received considerable attention by the Tunisian government for many decades. The number of researchers in Tunisia's agricultural research centers grew rather rapidly in recent years, with the establishment of four new regional research centers under the country's IRESA. However, agricultural research spending did not increase in order to match inflation rates, despite considerable growth in agricultural output over the past decade or so. As a result, Tunisia's agricultural research is reported as having fallen by almost a half during the decade 2002–12.[252]

In 2015, the Agricultural Science and Technology Indicators (ASTI), an organ of the International Food Policy Research Institute observed that Tunisia's agricultural researchers were among the most highly qualified in West Asia and North Africa. Adding that almost half the researchers with PhD degrees were in

**Frame 3.25 Themes Identified as Focus Areas for Joint R&D Activity**

Renewable energy: solar water heating, development of innovative processes for solar cooling, individual air conditioning using natural gas, development of a wind energy production model, hydrogen power, and development of a new generation of batteries.

Water: enhancing water treatment systems, design and construction of small-scale water treatment stations using innovative methods, rural water purification systems, water desalination using solar energy.

Remote sensing for forests assessment.

Biotechnology: breaking leaves disease in palm trees.

Health (disabilities): learning troubles, cares to poly-handicapped persons, molecular factors for genetic mental retardation in Tunisia.

ICT: e-security, e-government applications.

their 50s and 60s in 2012 and will now have reached retirement age. However, as indicated in Table 3.13 which is based on ASTI data actual expenditure on agricultural research went from around 1% of GDP due to agricultural activity in the year 2000 down to 0.64% in 2012, whereas the number of full-time research equivalents increased from 46 to 66 during the same 2-year period.

Figs. 3.24 and 3.25 present, by way of example, the distribution of research allocations in 2013 across types of institutions and fields of research activity.

Allocations totaling US$1.3 million, a low sum by any standard, were made in 2013 by the Agricultural Research and Higher Education Institution (IRESA)[253] which is dedicated to undertaking and overseeing research and higher education in agriculture. Fig. 3.23 clearly indicates that agricultural research centers

**TABLE 3.13 Tunisia's key agricultural research Indicators 2002–12**

| | 2000 | 2009 | 2012 |
|---|---|---|---|
| Research expenditure (PPP in constant US$ 2005 prices) | 60.8 | 49.4 | 55.9 |
| Overall growth (%) | – | −19 | 13 |
| Researchers full-time equivalents | 348 | 432 | 542 |
| Overall growth (%) | – | 24 | 26 |
| Agricultural research spending as percentage of agricultural GDP | 0.96 | 0.62 | 0.64 |
| Researchers full-time equivalents per 100,000 farmers | 46 | 53 | 66 |

Factsheet published by IFPRI's 'Agricultural Science and Technology Indicators (ASTI) in March 2015. Accessed in April 2017 at: https://www.asti.cgiar.org/pdf/factsheets/Tunisia-Factsheet.pdf.

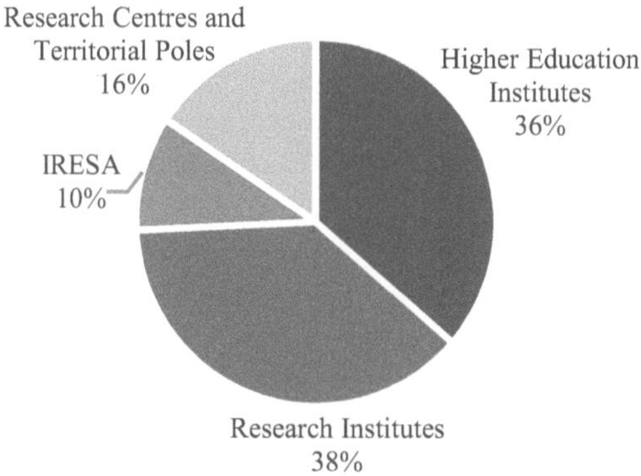


**FIGURE 3.24** Distribution of agricultural research allocations totaling around US$1.3 million by Institution de la Recherches et l'Enseignement Supérieure Agricoles in 2013 across types of institution.

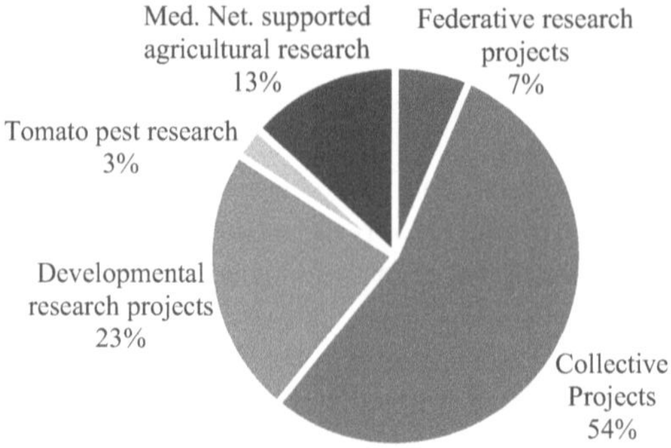


**FIGURE 3.25** Distribution of agricultural research allocations totaling around US$1.3 million by Institution de la Recherches et l'Enseignement Supérieure Agricoles in 2013 across fields of research activity.

and higher education institutions received comparable allocations, amounting to 38% and 36% of the total, respectively.

Figures on expenditure by IRESA on agricultural research are available for the period 2010–13. See Table A.3.4. It is rather remarkable, given the turmoil that Tunisia went through during these years, that total allocations witnessed limited variations throughout this period, between a maximum, equivalent to around US$1.6 million in 2011, and a minimum of around US$1.3 million in 2012. Another observation that may be made with regard to agricultural research allocations by IRESA concerns the relatively large allocations made to research institutes and institutes for higher education relative to those awarded to territorial research poles.

With regard to the distribution of allocations across fields of research activity, indicated in Fig. 3.25, it is noteworthy that, developmental research projects are implemented upon the express request of concerned professional and development organizations on the basis of contractual arrangements.

Furthermore, they are intended primarily to cash in on the results of research outcomes of previous national programs with view to their further adaptation and dissemination for widespread use. The fact that this category of projects received allocations amounting to around 23% of the total is essentially a positive indication, demonstrating a degree of synergy between institutions responsible for agricultural research, on the one hand, and development efforts, on the other.

### 3.4.5.4.2 STI Operation With Its European Neighbors

Tunisia is party to the Barcelona process and was the first Mediterranean country to sign an Association Agreement with the EU in July 2005. Under the terms of this Agreement, the EU and Tunisia committed themselves to cooperation in a range of economic, social, cultural and political domains. Tunisia is also a partner, since 2005, in the European Neighbourhood Policy; an Action Plan, including agreement on priorities and objectives to be attained through cooperation within economic, political, security cultural and scientific arena. A science and technology cooperation agreement was signed in 2003, and entered into force in 2004 allowing improved research collaboration and enhanced participation within the Research Framework Program. As a result, European and Tunisian researchers and research institutions participated within successive Framework Programs. Based on the above a Science and Technology Cooperation Agreement was concluded between the EC and Tunisia with the aim of integrating Tunisia within the European Research Area.

In concluding a rundown on Tunisia's science and technology collaboration with its European neighbors, a document published by the European Union's Research Directorate[255] states that Tunisia's cooperation with its European neighbors with regard to science and technology "is one of the most advanced among the Mediterranean countries and by far one of the most active ones in Euro-Mediterranean cooperation."

Nevertheless, Tunisia came third, behind Egypt and Morocco, among Mediterranean countries taking part in the Framework Program (FP7), covering the years 2007–10. Thematic domains of priority interest in which Tunisia participated included food, agriculture and biotechnologies, the environment and health. As indicated in Frame 3.26, Tunisia's participation in this Framework

**Frame 3.26 Tunisia's Participation in Europe's Framework Program 7, Covering the Years 2007–10[254]**

**Energy:** There were no Tunisian participants in FP7 calls in energy by 2011 and no specific actions targeting Tunisia within the 2011 work program despite the fact that it included activities that were judged by the EU to be of particular interest to Tunisia, including, for example

- Support for off-shore wind power generation;
- Reducing costs of manufacturing PV systems;
- Improving energy storage and efficiency in power plants.

**Environment:** There were five Tunisian partners in five ranked proposals requesting a total EU contribution of around €440.000.

**Food Agriculture and biotechnology**: The first three calls under FP7 included five Tunisian partners in five projects for a total EU contribution of around €657.000.

**Health:** Four projects with Tunisian participation were implemented within this track, with total EU funding of around €1 million. Three of these projects dealt with aspects of leishmaniasis.

**Information and Communication Technologies:** Only two projects with Tunisian institutions. Ten proposals were submitted in the previous call WP2010, including a wide range of areas.

**Nanosciences, nanotechnologies, materials, and new production technologies:** Projects involving Tunisian researchers in this track included activities aimed at "Capture of evaporated water with novel membranes" and "Development of the next generation bioreactor system."

**Socioeconomic sciences and humanities (SSH) and Science in Society (SiS):** No Tunisian partnerships were reported in neither the SSH nor the SiS domains.

Program is somewhat limited. Thus, no applications were made in tracks such as energy, socioeconomic sciences and humanities and only limited participation appeared to have taken place in other tracks, such as agriculture and biotechnology and information and communication technologies. The fact that the bulk of FP7 activities were planned and at least largely implemented prior to the 2010, is noteworthy, indicating that the full potential of collaborative international research programs was not being fully exploited, even before political turmoil took over.

### *3.4.5.5 Institutions Aimed at Promotion Innovation*

Tunisia was a leader among Arab and African countries, over the past two decades, in establishing institutional entities designed, including science parks and techno-poles, with the intention of linking industrial and other business enterprises, on the one hand, and universities and research centers, on the other. Table 3.14 presents a list of such institutions indicating their geographical

**TABLE 3.14 Tunisian Science Parks and Technopoles**

| Institution | Areas of Specialization | Location |
|---|---|---|
| El-Ghazala Technology Park | Principally ICTs | Ariana |
| Pôle de Compétitivité Monastir/ El Fejja | Textile and clothing | Tunis |
| Pôle Industriel et Technologique de Gabès | Renewable energy, water, environment | Gabès |
| BiotechPole Sidi Thabet | Biotechnology for health and pharmaceutical industry applications | Sidi Thabet |
| Technopole Borj Cédria | Renewable energy, water, environment, and biotechnology | Borj Cédria |
| Pôle de compétitivité de Bizerte | Agro-food industries | Bizerte |
| Pôle Elgazala des Technologies de la Communication | Mechanical and electronic engineering, ICTs and multimedia | Ariana |
| Technopole de Sousse | Mechanical and electronic engineering and informatics | Sousse |
| Technopole de Sfax | ICTs & multimedia | Sfax |
| Pôle de Compétitivité de Gafsa | Industrial development | Gafsa |

From Website of the Tunisian Ministry of Industry and Commerce. Accessed in April 2017 at: http://www.tunisieindustrie.nat.tn/fr/doc.asp?mcat=16&mrub=157. *ICT*, information and communication technology.

distribution across various parts of the country as well as specific technology domains of their operations.

One of the earliest examples of the bridging institutions is El-Ghazala technology and communication park, which commenced operations in 2001. El-Ghazala's main objective is to help set up technology companies by providing a variety of resources on site and facilitating links with researchers and other private enterprises. Bringing research outcomes to the market is another El-Ghazala objective. A significant portion of the El-Ghazala activities is dedicated to working with public authorities. El-Ghazala hosts training and research institutions as well as production facilities.

Modalities intended to promote innovative enterprise creation includes a program of incentives in support of researcher mobility between university and enterprises,[256] creation of units within research institutions to explore and promote commercialization of research outcomes, and a one-stop-shop to coordinate incentives and support provided to innovators.

A system of grants or premiums, also referred to as bonuses, was introduced in 1994, under the name 'Prime d'investissement en Recherche et Developpement (PIRD),'[257] with the aim of rewarding researchers and technologists contributing to the:

- development of new products or processes, as well as significant improvements on existing ones;
- design, testing and implementation of technical prototypes;
- acquisition of scientific laboratory equipment necessary for the conduct of R&D projects.

Institutional entities eligible for benefiting from the PIRD initiative are listed as including:

- industrial enterprises;
- enterprises active in agriculture and fishing;
- technical service providers in domains that include computer engineering, health and environment;
- public and private institutions, enterprises and scientific associations engaged in research and technological development projects, judged as contributing to innovations with tangible benefits.

Since 1995, 41 projects are reported as having benefited from this program with funding provided for 38 enterprises.

Research is predominantly undertaken within public institution in Tunisia. There are 33 research centers comprising 139 laboratories and 643 branch research units[258]. Public spending represented 92% of the total in the nineties. The share of enterprises remained modest at 5%. Private sector share is therefore negligible.[259] This raises the question of sustainability and efficiency of the system, notably that large parts of the budget are devoted to salaries (between 50% and 0.60%).

### 3.4.5.6 Concluding Remarks

Institutions charged with overseeing scientific research, technology development and innovation in Tunisia witnessed several changes throughout the past 3 decades. Apart from developments immediately following the country's independence in the mid-1950's, the 1990's and the first decade of the century, may be viewed as a formative period, during which several research institutions were founded or restructured and one that witnessed increased support in terms of financial as well as manpower resources. Key developments that characterized this period included:

- The establishment of the State Secretariat for Research in Science and Technology in the early 1990's with a mandate to oversee scientific research throughout the country.
- The promulgation of a special law recognizing the special nature of scientific research and technology development in 1996 is considered one of the landmarks characterizing the development of STI affairs in Tunisia.
- The creation of a cabinet ministry for scientific research and technology, which came into being twice, first in 2002, and then in 2005, before its responsibilities were relegated each time to specialized departments within the Ministry of Higher Education and Scientific Research.
- The creation of a cabinet ministry charged with promoting information and communication technologies and their role in establishing a digital economy.

Three essential elements may be singled out in characterizing the evolution of Tunisia's science and technology system:

- An understandable drive by Tunisia toward emulating institutional arrangements that existed and was operated more or less successfully for many decades in France, rather than devising its own strategies.
- Agricultural activities, including reforestation and fishing, were awarded special attention throughout.
- Information and communication technologies gained prominence during the late 1990's, with related policy decisions, institutional and infrastructure arrangements continuing to receive considerable attention to date.

Finally, despite successes on several fronts, Tunisia's STI system suffers a number of drawbacks, both structural and functional. Thus, in common with several other developing countries, the lack of a single policy document, ensuring adequate coordination among institutions concerned with building STI capabilities, is often the source of duplication with commensurate loss of resources and limited impact. Additionally, vacillation with regard to monitoring modalities are a cause for concern.

Nevertheless, although Tunisia is one of the Arab countries with the highest level of R&D allocations, measured as a percentage of GDP, many improvements are still needed by its STI system, which still suffers shortcomings common to many other developing countries, and certainly all of its fellow Arab countries in the Middle East and North Africa. Chief among these are inadequate resources,[260] limited cooperation among the variety of actors, as well as an underdeveloped and continually changing institutional landscape.

Yet, it is notable, that despite enormous odds, Tunisia STI system seems to have evaded many of the negative effects of internal political turmoil and security situation created in the aftermath of its uprising. Future development of the Tunisia's STI system will naturally be impacted by a variety of factors. However, several reforms and revisions enacted during the recent past may stand its system in good stead.

## 3.5 THE GULF ARAB COUNTRIES

With their relatively plentiful financial resources, two major sets of issues confront the Arab Gulf countries with regard to STI capacity building. The first concerns creating adequate human capital, whereas the second is to do with implementing STI capabilities for a variety of socioeconomic ends.

Concerning human capital formation, the Arab countries of the Gulf region, took steps to create their own universities as well as a number of research centers and technical services facilities back in the 1960s and 1970s. Although most STI facilities were initially staffed by nonnationals at the time, they now sport predominantly national cadres while still hosting a sizeable portion of staff members from neighboring countries as well as overseas.

A somewhat different situation pertains when it comes to practical applications of scientific and technological knowledge, in areas such as water desalination and treatment, petroleum extraction and transportation as well as the petrochemical industry. Thus, activities in these domains are mostly contracted to private sector firms, which in turn employ sizeable contingents of foreign experts. Little STI capacity building ensues as a result given that these firms are mainly foreign-based with few, if any links to national STI institutions.

In recent years, several Arab countries in the Gulf region took to hosting campuses of prominent universities from the United States, the United Kingdom and France. Although many of the foreign university campuses specialize in nontechnical subjects, a few do in fact provide instruction, presumably involving at least undergraduate research activity in the medical and physical sciences. In-depth analysis is certainly required in order to establish the ultimate impact of foreign university campuses, with particular reference to actual benefits in relation to costs incurred. For the time being, however, the impact on STI capacity building may be assumed as generally positive.

Concerning implementation of scientific and technological knowledge toward socioeconomic developments, pronouncements and plans were made by the Arab countries of the Gulf region, for over 2 decades in the past, with regard to diversifying their economies. However, with the possible exception of the United Arab Emirates, few positive outcomes may be recorded. More recently, several governments in the region, such as Saudi Arabia for example, issued vision documents and associated plans indicating moves toward establishing knowledge economies. However, several such documents lack detail with particular regard to invoking scientific research and novel technologies as bases for future development. The following sections will take a close look at the above as well as other aspects related to development of STI capabilities in the region.

## 3.5.1 Bahrain's STI System; Policies and Institutional Entities

Bahrain's STI system is a small one, in terms of the number of institutions, the people they employ and output they generate. Nevertheless, several government documents, authored mainly during the 2000s indicate much greater interest than in the recent past, in building capacity on several fronts. The following paragraphs include a concise summary of two such attempts, Bahrain's 2030 Vision and Bahrain's National Research and Development Strategy.

As indicated in Chapter II, the Economic Vision 2030 seeks to promote aspects of development directly linked to STI capacity building, including:

- promoting competition through higher productivity;
- enhancing FDI, with associated technology transfer possibilities;
- encouraging knowledge-based and innovative entrepreneurship.

Bahrain's Vision document also seeks to promote quality education as means for realization of national development and diversification, stating that the current education system did not provide requisite skills.

However, as indicated in Chapter II, there is need to devote attention to modalities that will be adopted in attempting to attain the above goals. The lack of short- and medium-term and milestones and benchmarks, is a glaring shortcoming in need of urgent attention.

### *3.5.1.1 Bahrain's R&D Strategy*

Bahrain's first draft of the country's R&D Strategy covering the period 2014–24 was produced under supervision of its Higher Education Council in cooperation with several national institutions.[268] One of the main goals of this Strategy is to increase expenditure on R&D from its current level of around 0.04% of GDP to 1% by the year 2020. [269,270] This is a 25-fold increase that available documents on the Strategy do not appear to explain how it might be fruitfully achieved, in view of difficulties acknowledged by the Strategy document.

Bahrain Research Strategy's SWOT analysis highlights intentions on the part of the country's leadership for improving economic growth through investing in research, the fact that resources may be leveraged and that the country has well-developed human capital. Support for the strategy's implementation is stated as including a strong banking sector, possibilities for regional leadership in selected research areas, an open environment as well as chances for leapfrogging. On the side of weaknesses the Strategy underscores, absence of science agencies, funding infrastructure and a research culture, weakness of linkages to global networks limited access to data and strategic direction. On the other hand, implementing the Strategy was expected to meet threats that included increasing competition from neighboring countries with greater resources and more tellingly that due to other "pressing national priorities, research may not fall on top of priorities list."

Bahrain's research system is considered to be in need of:

- strategic direction and leadership;
- government agencies capable of supporting research institutions and tracking their progress;
- a well-developed scientific infrastructure;
- wider understanding of the societal value of research activity;
- stronger linkages between components of the national research system as well as active collaboration with global knowledge networks.

With particular regard to the needs of university research, the Strategy considers that increased funding, more robust connections to industry and society, administrative capabilities for handling complex externally funded research projects are strongly needed.

The Strategy's vision statement indicates a quest for excellence in "key fields of research of importance to the Kingdom's economy and social welfare." This statement further underscores the need to acquire "broad research capacity needed to access and use global knowledge."

The main goal of Bahrain's R&D Strategy is to contribute to the realization of goals articulated in Bahrain's Vision 2030 document, helping transform the country into a knowledge economy, reducing its dependence on oil revenues, ensuring sustainable social, environmental and cultural well-being, providing citizens with world-class education and enhancing the country's regional and international status.

In order to achieve its goals, Bahrain's R&D Strategy lays down the following five objectives:

- Establishment of a national research governance infrastructure;
- Strengthened university research capacity;
- Consolidated links between national academic institutions and academic research institutions as well as entities concerned with national economic and social priorities;
- Improved public awareness and understanding of research and innovation;
- Having national priorities adequately tackled.

Further key tasks and performance indicators are listed with respect to each one of the above goals. However, in the Strategy's version at the time of writing none of the performance indicators included actual numerical criteria. Thus, with regard to the first objective one of the performance indicators is merely stated: "Increase in available funds for R&D both for academia and private sector."

Similarly, performance indicators associated with the second objectives include entries phrased simply as increase in number of researchers at universities and private enterprises and increase "in the share of students with natural science and engineering majors."

The following three areas are identified by the Strategy document as highest priorities:

- Financial services, Islamic banking, and finance as well as insurance.
- Public health and services with particular reference to services extended to countries in the Gulf region.
- Information and communication technologies.

### *3.5.1.2 STI Institutions in Bahrain*

Institutions of higher education, on the one hand, and incubation facilities, on the other dominate the country's STI scene, with only a small number of research centers and laboratories.

#### 3.5.1.2.1 Institutes of Higher Education

Institutions of higher education in Bahrain date back to the late 1970s with the creation of The University College of Arts, science and Education in 1978. The Arabian Gulf University was established on the initiative of the Gulf Cooperation Council in 1984. The University of Bahrain was formed through a merger of the University College of Bahrain and the Gulf Polytechnic, in 1986. At present, Bahrain's HEIs include four public and 12 private universities, with an overall enrolment of around 30,000 students, with the largest proportion of students specializing Arts, Law and Business Administration.[271] A small proportion of Bahrain's university entrants specialize in science and technology related topics. Institutions of higher learning in medicine were lately established attracting students from Bahrain and other Gulf countries.

With regard to research undertaken at Bahrain's public universities, a review of papers published by staff at the Arabian Gulf University, for example, appears to indicate that the vast majority of research reports were authored in the medical sciences. However, the College of Graduate studies at the same university appears to implement programs that involve postgraduate research in fields including deserts and arid lands, novel farming techniques including hydroponics, geographic information systems and remote sensing as well as biotechnology, environmental management, hydrology and water resources management.

#### 3.5.1.2.2 Scientific Research Centers

With regard to research activity, it would seem that applied science research in Bahrain has had a rather chequered history, while a small number of recently established medical research outfits appear to operate actively. Thus, the Bahrain Centre for Studies and Research (BCSR) was established in 1981 to carry out research activity, emphasizing contractual research projects, as well as providing consultancy services to decision makers in both the public and the private sectors in specializations including economics, the social sciences as well as theoretical and applied studies in areas related to energy, natural resources and the transfer of technology. However, BCSR was dissolved in 2010 by royal decree.[272]

Al-Jawhara Centre for Molecular Medicine (AJC) was established in 2003, at the College of Medicine and Medical Sciences at Arabian Gulf University. Genetics and inherited disorders constitute main areas of research interest at AJC, which also offers courses in Molecular Medicine for medical students and provides clinical as well and molecular diagnostics services. Available facilities enable AJC to carry out research in molecular and cellular biology, biochemistry, genetics, and immunology, mainly for the study of genetic diseases and a range of conditions including diabetes.[273]

The Royal College of Surgeons in Ireland founded the Medical University in Bahrain (RCSI-MUB) in 2004. As part of its research strategy, RCSI-MUB created in 2016 a number research clusters in specializations including cardiovascular and metabolic medicine and oncology.[274]

#### 3.5.1.2.3 Business and Technology Incubators

A report by the World Economic Forum on in the Arab countries[275] characterizes Bahrain, along with Jordan, Lebanon, Oman, Qatar, and the United Arab Emirates, as a haven for entrepreneurial initiatives, while several other GCC countries are classified as "late adopters" and "laggards."[276] This may be due to the relative ease with which credit may be obtained from investors and banking institutions in Bahrain.

Bahrain is one of a few Arab countries to adopt measures aimed at easing the local credit market for start-ups, introducing special programs aimed at encouraging banks to lend SMEs, including exemptions on reserve requirements, credit subsidies and partial credit guarantee schemes. In particular, the Bahrain Development Bank provides a variety of financial services tailored to meet the needs of small and medium businesses in Bahrain, including business finance, Islamic financing to manufacturing, agribusiness, healthcare, and other services sectors.

Several initiatives are being implemented in Bahrain in support of entrepreneurial and innovation activities. Tamkeen, which was established in August 2006, is one such initiative, and part of Bahrain's national reform endeavors, as well as Vision 2030. Tamkeen is essentially aimed at developing Bahrain's private sector with two prime objectives:

- fostering the creation and development of enterprises, and
- providing support to enhance the productivity and growth of enterprises and individuals.

Tamkeen's strategy focuses on quality enhancement, which often involves innovative inputs some of which will include employing novel technologies. Tamkeen programs are claimed as having benefited many thousands of businesses operating in Bahrain, including activities conducted in cooperation with local community actors,[277] NGOs as well as industrial enterprises.

#### *3.5.1.3 Concluding Remarks*

Scientific research in Bahrain is carried out mainly by HEIs, with emphasis on medical research among only a few other domains. Technological development is largely the domain of industrial firms. Innovation and enterprise creation, on the other hand, are recognized as essential strands in the country's development policy, with several measures taken by successive governments, aided by the country's relatively well-developed financial sector, in aid of innovative enterprise creation.

The combination of a broad national vision, emphasizing innovation and entrepreneurial activity and a national research strategy might be considered a likely substitute for a full-fledged innovation policy.[278] However, the STI set up in Bahrain essentially suffers the lack of implementation strategies and work-plans with well-defined phases, clear goals and modalities. This is especially important in the case of the tremendous increase in R&D spending, 25-fold to 1% of GDP by the year 2020, with the limited infrastructural arrangements at hand and specialized human resources in relatively short supply.

Although insufficient detailed information is available on the level of support provided by Bahrain's government for STI-related activities, it would be safe to state that its contributions have essentially been limited. This is particularly the case with regard to government's support for innovation initiatives, which is most certainly secondary to that by private sector organizations, particularly banking institutions. In fact, despite a positive view of Bahrain's enterprise and innovation ecology, the above-mentioned WEF characterizes support by governments in the region, including that of Bahrain, as hardly sufficient to create a healthy financing environment for start-ups.[279]

### 3.5.2 Kuwait's STI System; Policies, Institutions, and Their Activities

Kuwait is one of the first Gulf Arab countries to initiate efforts toward establishing its own national university, the University of Kuwait in 1966 and its Kuwait

Institute for Scientific Research (KISR), in 1967. Despite numerous difficulties they encountered over the past decades, both institutions remain to this day essential components of the country's STI infrastructure.

In 2006, efforts were initiated toward restructuring and advancing research and development in the country. Thus, the Emir of Kuwait commissioned a Kuwait Research Review Panel (KRRP) to examine the organization and performance of R&D in Kuwait, and to produce relevant recommendations, with specific reference to strengthening the performance of KISR.

KRRP was established early in 2007, comprising leading members of Kuwait's R&D community, selected stakeholders, as well as international experts in the management of research and development.

The Panel's findings with regard to the country's R&D scene, were summarized as follows:

- R&D activities in Kuwait tend to be poorly networked, and inadequately linked to national priority areas.[280]
- In the cases where the R&D activities were aligned with national needs, i.e., in the oil sector, they tended to be confined to providing technical services rather than engaging in new knowledge creation or adaptation.
- Evidence indicated that R&D activities suffered poor management and bureaucratic restrictions. Hence, R&D institutions can muster only limited and delayed responses to demand from the production and service sectors. There is also evidence that R&D cadres are insufficiently motivated and suffers from low morale.
- Although transforming Kuwait's current system into something akin to that of a developed country may not be achieved overnight, it would certainly be impossible without a comprehensive policy framework and an appropriate institutional framework.

However, although the panel was originally established with focus on KISR and its performance, there were many reasons for awarding attention to other components of Kuwait's STI system. This, the panel considered to be in the best interest of the final outcome of its deliberations.

Additional tasks for the panel included identifying modalities for strengthening the country's scientific culture, and enhancing coordination among members of Kuwait's R&D community and stronger linkages with external sources of STI knowledge.

The Panel's recommendations are presented in Frame 3.27, with particular regard to

- need for a national body to oversee science, technology and innovation activities;
- organization of R&D activities and the role that KISR departments need to assume in relation to selected thrust areas of special importance to the country's socioeconomic development;

**Frame 3.27 Summary of Recommendations by the Kuwait Research Review Panel**

Regarding the need for a national body to oversee science, technology, and innovation activities, the Panel recommended that a higher body, referred to as the Kuwait Science, Technology and Innovation Council, reporting to the country's highest executive office, should be established to

- ensure effective governance of science and technology activities;
- oversee formulation of relevant policies;
- supervise funding allocations;
- coordinate R&D activities among national stakeholders;
- enhance international alliances;
- facilitate commercialization of products and processes developed through local R&D activities;
- promote a culture that is conducive to taking up science and technology as bases for future national development;
- provide advice to government authorities concerning effective implementation of the above tasks.

Concerning organization of R&D Activities it was recommended that KISR departments in charge of Kuwait's R&D thrust areas, namely: Petroleum, Water, Energy (including alternative energy), Environment, Economics be restructured as centers, each with their own board, comprised of representatives of stakeholders along with international allies/partners, with the objective that they would later become centers of excellence.

In reference to cooperation with external organizations, it was recommended that Kuwait establish international alliances and partnerships arrangements, with international as well as regional partners within each of the R&D areas that KISR is working on.

In relation to commercialization of R&D outcomes, the panel found it essential that gaps in the existing commercialization system of R&D outputs in the country be identified, and alternatives measures worked out to fill these gaps, and to define fresh mechanisms to improve interaction among stakeholders, including international partners.

With respect to directions for future action the panel considered it essential that periodic exercises be undertaken to review emerging challenges and opportunities as well as monitor advances in crucial areas of science, technology and their innovative applications in order to guide future action taking into account available resources as well as likely benefits.

- cooperation with external organizations;
- commercialization of R&D outcomes;
- directions for future action.

The following section provides notes on Kuwait's attempts to formulate a national science and technology policy in the past with indications of salient future steps.

### 3.5.2.1 Kuwait's Science, Technology, and Innovation Policy

KISR had originally been assigned a major role in providing policy input to the Government on STI matters. KISR has, in fact, tried to capture the national S&T needs and objectives through its periodic planning process. Additionally, KISR did attempt to launch a preliminary S&T policy document in 2004, and took steps towards the formulation of a full-fledged national S&T policy.

On the other hand, despite involvement of some external stakeholders in the KISR planning process, the resulting plans have not recently received the broad support and recognition necessary to serve as a robust guide for KISR activities or to serve as a single point of reference for integrating the participation of other STI actors.[281]

In the absence of a comprehensive national STI policy that is accepted and acknowledged by all parties as basis for STI action, and due to KISR's relative weight as principal R&D institution in the country, its research strategies were bound to dominate those of Kuwait as a whole. In principle, however, there are no a priori reasons why KISR's interests should be synonymous with those of the entire country.

Thus the view that the provision of science and technology policy advice be embedded within KISR's goals, whereas perhaps feasible at the time these goals were framed, is no longer justified. Naturally, this should not mean KISR be completely divorced from the process of national decision making on policy matters. It would indeed be essential for KISR to provide input to national policy formulation and should certainly be called upon by the national Council, proposed by the Panel, to provide such input on specific issues in which it has amassed a certain level of expertise.

### 3.5.2.2 Institutions Concerned With STI in Kuwait

Kuwait was foremost among Arab countries in the Gulf region to establish a research and development center. The KISR was established in 1967, in partial fulfillment of the Japan Arabian Oil Company's obligations under an oil concession agreement with Kuwait's government. KISR's original mandate covered applied scientific research in three fields, namely petroleum, desert agriculture and marine biology. It was turned into a "Government Agency" in 1973, operating under a Board of Trustees with its director reporting to the Council of Ministers. Its main objectives were expanded to include applied scientific research, especially related to industry, energy, agriculture, and the national economy as well as provide advice to the government on the country on matters relating to scientific research policies.

#### 3.5.2.2.1 The Kuwait Institute for Scientific Research

As indicated above, KISR was originally established in the late 1960s. However, its status was reviewed in 1981, turning it into a semiautonomous public institution. Although its original scope of work was retained, it was mandated to undertake research in areas such as preservation of the environment; investigation of natural wealth; promotion of energy, water, and food resources; and methods of agriculture exploitation. Additionally, the Institute's mission was broadened, requiring it to undertake research and provide related consultations to both public and private sector institutions in Kuwait, the Arabian Gulf region and the Arab world at large. KISR was also encouraged to engage in collaborative activities with international research institutions.

The Institute has traditionally operated on the basis of a series of 5-year plans. Table 3.15 presents a view of core issues. In particular, the Institute's eighth 5-year plan covers the period 2015–20. During this period KISR pledged to maintain its commitment to:

- Advancing science and technology in Kuwait;
- Promoting growth and local industries;

**TABLE 3.15** Core Issues Addressed by Five-Year Plans Adopted by the Kuwait Institute for Scientific Research (KISR)

| Plan | Years | |
|---|---|---|
| 1 | 1979–84 | Planning of management processes, research priorities and organizational structure. |
| 2 | 1984–89 | Strengthening focus of core technical research programs through strategic recruiting and clear alignment with national development priorities. |
| 3 | 1992–95 | Developing a transitional strategic program to rebuild the Institute's capacity in the aftermath of the Iraqi invasion. |
| 4 | 1995–2000 | Aligned and prioritized national initiatives in collaboration with strategic stakeholders. |
| 5 | 2000–05 | Integrated advanced information technologies in support of core research initiatives and national priorities. |
| 6 | 2005–10 | Building up the capacity of the staff and upgraded facilities. |
| 7 | 2010–15 | Transforming KISR into a leading research and development institution in the region through "growing a regional footprint and building world-class niches." |
| 8 | 2015–20 | Continuing to fulfil goals outlined by the 7th 5-year plan with focus on a regional "footprint" and "world-class niches," aimed at building an international reputation. |

Statement by KISR on its website. Downloaded in February 2017 from: http://www.kisr.edu.kw/en/about-us/history.

- Advancing diversity in energy production and enhancing efficiency of energy utilization;
- Contributing to the country's oil industry;
- Protecting the environment and biodiversity;
- Ensuring that novel infrastructure projects being implemented produce efficient and reliable outcomes;
- Contributing to the country's growing water needs.

In 2015 KISR announced its Innovative & Renewable Energy Program (IREP) as part of its seventh strategic plan. IREP is designed for implementation in three phases, the first of which targets delivery of approximately 2 GW of renewable energy developments in Western Kuwait through construction of several new facilities, including:

- A multitechnology renewable energy park with a capacity of 70 MW;
- A 100 KW photovoltaic test platform to determine the performance and compatibility of a variety of panel and inverter models under local conditions;
- A miniwind farm west of Kuwait City to test small-sized wind turbines, eventually to supply on- and off-grid facilities;
- A solar and wind to hydrogen plant—a 16 KW facility for producing and storing hydrogen as an energy carrier and to be used as a fuel cell.

Frame 3.28 lists a set of the Institute's achievements, which, for the most part, address the country's declared top priorities, namely energy, water and the environment. However, available sources do not provide information on economic returns attained nor any partnerships concluded with public or private sector institutions with view to wider dissemination of the Institute's achievements.

KISR is the principal institution dedicated R&D in Kuwait. Its past as well as more recent achievements require in-depth analysis in order to further enhance its output in both quantitative and qualitative terms.

#### 3.5.2.2.2 The Kuwait Foundation for the Advancement of Sciences[282]

KFAS was established by a 1976 decree as a private nonprofit organization devoted to supporting scientific research. The Foundation's work is overseen by a Board of Directors, chaired by the Amir and is financially supported by Kuwaiti private sector companies, contributing 1% of their net annual profit toward the Foundation's budget.

Principal goals of the Foundation are summarized as promoting scientific development in Kuwait, through supporting scientific projects, the scientific community, and construction of the country's scientific infrastructure.

Based on the above-mentioned KRRP report, an extensive assessment of its past performance as well as consultations with its key stakeholders, KFAS management embarked in 2009 on developing a new strategic plan that would help

**Frame 3.28 Selected Kuwait Institute for Scientific Research Achievements**

Even as far back as the 1980s KISR had successfully established itself as a modern multi-purpose applied R&D institution, with well-equipped laboratories, pilot facilities and a reasonable nucleus of highly qualified research staff including experienced expatriates and growing complement of young Kuwaiti researchers. A regional leader at the time, it pioneered R&D activity in fields that included aquaculture and fisheries, biotechnology, the petroleum industry, water resources and desalination, renewable and alternative energy, as well as environmental protection in urban, marine and arid settings. Notable achievements were made thanks to modern management practices, including both annual as well as strategic planning cycles, with early successes including

- successful culture of some local as well as imported fish species;
- initiation of comprehensive environmental marine and arid zone baseline data;
- development of sea water reverse osmosis desalination technology;
- energy conservation and building codes.

Additionally, income from contract research became tangible in sectors such as oil, energy and power, food, water, building construction and environment. However, direct government funding has continued to constitute the bulk of its budget. Indeed, rebuilding KISR's facilities in record time, following Kuwait's invasion and subsequent Gulf war was, in itself, a considerable achievement.

Several, more recent, achievements are referred to on the Institute's website within a variety of areas, including, in the field of oil extraction, assessment of the controlled water injection system designed to boost oil field output as well as enhance refinery operations. In the field of water resources, KISR is quoted as having designed a commercial pilot wastewater treatment plant, with a capacity of 6 million $m^3$ using a twin-membrane separation system.

the Foundation meet future needs of Kuwait's STI system. A strategy covering the years 2012–16 was drafted with four "Thrust Areas":

- Promotion of a science and technology culture through contributing to the development of science including science education, support for the talented, and creation of an enabling environment;
- Enhancing scientific research capabilities, and aligning R&D capacity in and among Kuwaiti scientific institutions toward national development priorities;
- Support for innovation and providing assistance toward commercialization of R&D outcomes within the framework of an integrated national STI system;
- Participation in the development of the scientific and technological capabilities within private sector institutions, thereby promoting their contributions to building a knowledge economy.

#### 3.5.2.3 *Research and Development Activities in Kuwait*

In addition to research and development activities conducted by KISR, both Kuwait University and the Kuwait Petroleum Company implement extensive research programs on areas that bear direct relevance to the country's development and STI capacity building. The following paragraphs offer a summary of these activities in both institutions.

##### 3.5.2.3.1 Research at Kuwait University

Kuwait University was established in 1966, with four Colleges, of Science, Arts and Education as well as a Women's College. By the academic year 2005/06 it had a total of 14 Colleges, with 82 undergraduate programs. KU remains the only government funded University in the country. In 2014/15, its undergraduate students' population stood at around 36,330, of which around 1930, or 5.3%, were enrolled in graduate courses. Females are well represented among both undergraduate and graduate students populations, with 72.7% and 70.9%, respectively. A total of around 1573 faculty members are employed by the University, of whom 429, or 26% are females.[283]

The University's research activity was reported as having increased sixfold between 1980 and 2005, with the faculties of Science, Medicine and Engineering handling two thirds of the total research load. However, as indicated by Figs. 3.25 and 3.26 research activity undertaken in these domains took up an even larger proportion of research funds, during the period 2000/01–2005/06.

No recent statistics are available on research activities undertaken by Kuwait University. However, the University was estimated as having undertaken around 2500 research projects, at an estimated total cost of nearly US$80 million by the year 2005/06. Annual average expenditure on research, within the 5-year period past 5 years 2000/01–2005/06 came to around US$5.2 million.[284] Since the total annual budget of the University presently stood then at around US$493 million, this means that funding allocated to

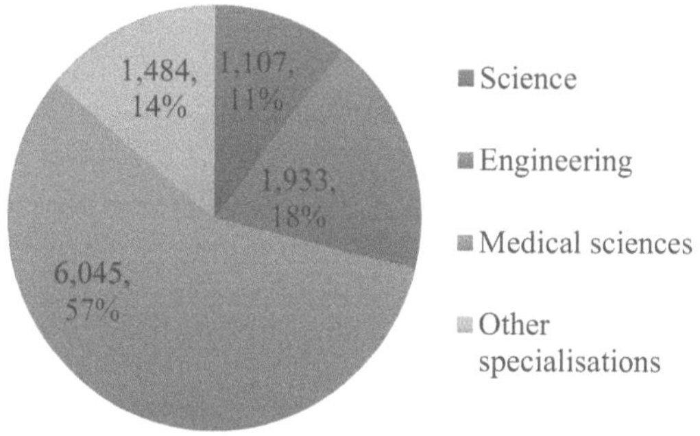


**FIGURE 3.26** Funds allocated to Kuwait University's research projects out of a total of US$10,570 during 2001/02–2005/06.

research activity constitutes a very small percentage, around 1.6%, of the total. However, the fact that only about 67% of research funding gets actually spent means that calls for increased research funding should be coupled to measures that encourage Faculties and Colleges to spend their allocations, including steps to streamline regulations for approval and employment of research assistants. More than 93% of the University's research effort during the above-mentioned period was funded by the University itself, whereas the remaining projects were funded by external sources, mainly KFAS, at around 5% of the total.[285]

### *3.5.2.3.1.1 Obstacles Facing Kuwait University's Research*

Obstacles that hamper Kuwait University's research efforts, in terms of output volume and quality, hence impact include

- Staff preoccupation with teaching leading to limited attention to R&D activity in general;
- Absence of critical mass in any one area of research activity;[286]
- Bureaucracy hindering research cooperation and shunning donations by possible benefactors;
- Appointments made with regard to teaching needs, rather than research standing, at the levels of both faculty and full-time assistants;
- Few privileges offered to senior faculty members engaging in research activity.

In seeking to address the above obstacles, the University's Strategic Plan for the years 2013–17, included a number of goals with numerical targets linked to specific outcomes concerning scientific research. By way of illustration the strategic goal aimed at raising the level of scientific research at the university is set to achieve the objectives and goals presented in Table 3.16, which indicates that numerical targets to be achieved within time limits set by the strategy may be quite unrealistic. The remaining three goals included in the University's Strategic Plan involve acquiring optimal benefits from scientific research output, reaching a high level of international presence in selected research areas, and developing cooperative research ventures.

In concluding this section on Kuwait University's research activity, it is well to emphasize that one of the main difficulties faced by any scheme seeking to enhance the quality of the University's research activities is that ensuing research publications are viewed merely as means toward research promotion. Thus, the performance of research grants programs at Kuwait University is appraised in relation to the number of resulting publications with no reference to quantitative or qualitative impact of research outcomes, in terms of meeting national needs and expectations or some generally recognized development objective.

It would also appear that many of the recommendations made as a result of the above-mentioned KRRP exercise, which was conducted in 2007, and

**TABLE 3.16 Strategic Goals, Objectives and Related Targets[1] Set by Kuwait University's Strategic Plan for the Years 2013–16**

| | | |
|---|---|---|
| **Strategic Goal I.** | | |
| **Raising the Level of scientific Research at Kuwait University** | | |
| **Objective 1.** Increased research productivity | | |
| Target 1 | Target 2 | |
| Increase productivity by 10% annually | Increase the number of research projects by graduate students by 25%. | |
| **Objective 2.** Improving benefits from research funding and enhancing research efficiency | | |
| Target 1 | Target 2 | Target 3 |
| Enhance participation of new faculty by at least 25%. | Increase research allocations by 10%. | reduce time to complete research procedures by 30% |
| **Objective 3.** Greater focus on national priorities | | |
| Target | | |
| Increase the number of research projects targeting priority areas by at least 30%. | | |
| **Objective 4.** Improved criteria for evaluating research output. | | |
| Target 1 | Target 2 | Target 3 |
| Increase number of awards for research excellence by 35%. | Increase research publication in Q1 by 25%. | Increase humanities research publications by 25%. |

Adapted from Kuwait University's Strategic Plan for the years 2013–2016 retrieved in February 2017, from: http://www.planning.kuniv.edu.kw/Strategic_Plan_New_En.aspx

which were, incidentally, in essential agreement with the spirit and the contents of measures proposed by the university's own planning documents, stand true today. In particular, the university's calls for an overriding research policy that binds efforts by all research institutions in pursuit of national priorities and concerns is as essential as ever before.

#### 3.5.2.3.2 The Kuwait Petroleum Company and Its Research Activities[287]

The Kuwait Petroleum Company (KPC) established its own R&D Group in 2012 based on recommendations made back in 2004 by the Company's 2020 strategy as well as a 2005 directive by the Supreme Petroleum Council (SPC), stipulating that KPC establish a research center as an independent for-profit entity, with its own budget and board of directors. Reasoning behind setting up this R&D Group included the need to deal with challenges that are specific to the country's Oil Industry, based on the company's internal capabilities.

Very little information is provided, however, on the Company's website concerning the R&D Group's actual achievements since it was established, whereas its programs and responsibilities are stated as including the design of detailed R&D programs, specifying manpower and financial prerequisites for their implementation and the establishment of R&D labs and required technical equipment.

Preoccupation with technology acquisition and implementation is further reiterated in the Company's 2030 strategy, whose article 5 is intended to "Realise Value from Technology," and calls for pursuing a value-driven approach to research, on identification of as well as development and deployment of appropriate "core and enabling technology solutions through investments, partnerships and technology transfer mechanisms."

### *3.5.2.4 Concluding Remarks*

The performance of Kuwait's STI system has generally been less adequate than many other countries with comparable resources. The low rate of technological capacity-building, disputable quality and meager output may partially be attributed to a variety of factors. Lack of explicit overall national development policies, a small home-market, reliance by the private and even the public sectors on outside sources for scientific and technological inputs as well as fragmented public investment in STI capacity building. Certain deficiencies, however, lie squarely within Kuwait's national system of scientific, technological, and innovation institutions, including:

1. Inadequate infrastructural arrangements, e.g., entities crucial to transforming R&D results into marketable outcomes, such as technology parks and incubators, centers of excellence and research universities to draw, hold on to and multiply talent;
2. Weak linkages and ineffective networking among components of the national STI system coupled to limited collaboration with regional, as well as international, STI systems in the developed and developing countries.

As a result of limited cooperation and coordination between the various components of Kuwait's STI system as well as absence of a formal process for national overall and detailed priority setting, Kuwait's STI system is still inadequately engaged in issues related to the country's main source of wealth. This is especially the case in relation to creating opportunities for employment and enterprise creation linked to the country's principal industries, oil and petrochemicals as well as in diversifying the economy within emerging knowledge-based sectors. Pressing problems caused by climatic conditions, water scarcity may exert even greater stresses upon the country's admittedly immense, but essentially limited, financial reserves.

Opportunities abound for the creation of fresh STI capabilities, and for their effective utilization in various spheres of socioeconomic development in Kuwait, including job and enterprise creation. However, for Kuwait's national STI system to carry out its role, in higher education, research, technology development and innovation, there is need for a radical overhaul of government policies in a variety of spheres and in a manner that takes into full account the need to acquire fresh capabilities and forge stronger linkages with regional and international actors. KISR, in cooperation with public and private sector institutions, including the Kuwait Fund for the Advancement of Science and Kuwait University would need to bear the main responsibility for drafting future STI policies aimed at building the country's future STI system.

In order to bring Kuwait's STI system closer to that enjoyed by most developed, and even some developing, countries, a persistent and an arduous quest has to be initiated and closely followed through. If launched within appropriate policy frameworks and with support and attention at the highest level in government as well as business circles, the chances for success are bound to be excellent.

### 3.5.3 STI in Oman; Policy Frameworks and Principal Institutions

In common with other Arab countries of the Gulf, research and development activities were taken up mainly at universities and selected enterprises concerned with the country's presently dwindling hydrocarbon resources. R&D expenditure was estimated at around 0.1% of GDP, a very low percentage, considering the number goals set for the country's socioeconomic agenda and the urgent need for further diversification of income sources, but in keeping with the order of magnitude pertaining in other GCC countries. The 2011–15 5-year plan targeted national expenditure on R&D at 1% of GDP, however this proved unrealistic due to changes in spending priorities.[288] Attempts were made to redress this situation, through channeling research funding to university departments and centers through The Research Council (TRC), an institution that, since 2009, has funded a number of small research projects in agriculture, fisheries, urban development and road safety, yet TRC contributions remain meagre in view of the list of the challenges posed by economic growth and diversification.[289]

The total number of researchers in full-time equivalent terms was put at around 2000, in a 2014 report by the United Nations Conference on Trade and Development, which also stated that the quality of the research institutions was broadly comparable to that in other Arab Gulf countries.[290]

The following paragraphs will review recent strategic initiatives aimed at improving Oman's STI capabilities, some of the country's principal STI institutions with special focus on activities undertaken by Oman's principal institution, the Sultan Qaboos University.

### *3.5.3.1 Oman's STI Planning*

A literature survey undertaken in 2017 indicated the absence of an integrated science and technology plan, with most STI activity guided by consecutive 5-year plans as well as pronouncements and lists of intended actions by ministries and institutions concerned with STI capacity building.

#### 3.5.3.1.1 Oman's National Innovation Strategy

Nevertheless, formulation of Oman's National Innovation Strategy was initiated in 2014 by The Research Council. It was reported as having been approved by the Council of Ministers in 2017 as complementary to the country's National Research Strategy included as basis for implementation of the prospective national Vision 2040. One of this Strategy's principal aims is to place Oman among the top 20 leaders in innovation by 2040, shifting from a revenue-dependent economy to innovation-based economy.[291]

### *3.5.3.2 Oman's STI Institutions and Their Research Activities*

The national STI scene in Oman is largely dominated by research activities undertaken at the country's universities, on the one hand, and enterprise and innovation activities, promoted by a range of institutions, often in cooperation with the country's principal institutions of higher education. The following paragraphs include, by way of illustration, summary information on research undertaken at the principal institution of higher learning in the country; namely the Sultan Qaboos University (SQU), during 2015, the most recent for which data is available from SQU sources.

The most prominent STI institutions in Oman are the country's higher education institutes, some of which do possess research facilities, as well as linkages to incubation and enterprise support facilities. Two of Oman's main universities, the Sultan Qaboos University and the Arabian Gulf Universities possess departments dedicated to research activities. The following section presents a summary of programs aimed at promoting research and innovation by Oman's Research Council and the Research Deanship at the Sultan Qaboos University and the Research Department concerned with funding and monitoring research activity at the College of Applied Science. A subsequent section will take a closer look at research activity undertaken by the Sultan Qaboos University, on account of the range of issues addressed and their likely outcome with regard to overall STI capacity building.

#### 3.5.3.2.1 The Research Council

Oman appears to lack an institution whose activities are solely dedicated to R&D. In place of such an institution Oman's government created, in 2005, The Research Council (TRC). TRC is mandated with supporting research and innovation activities throughout Oman. Its mission statement emphasizes its role in creating an innovation ecology that responds to "local needs and international trends, fosters social harmony, and leads to creativity and excellence." TRC goals include rendering Oman a regional hub for innovation, and a leader in producing new ideas, products and services, with the largest research capacity in the region and with the means to excel in selected fields of national importance and promote a national culture that is conducive to research and innovation. It is noteworthy that TRC has established a special program with the purpose of promoting innovation at the community level; namely the "Community Innovation Assistance Program."[292]

Table 3.17 presents a list of research programs supported by Oman's Research Council (TRC) during the period 2009–13.[293] As Table 3.17 indicates,

**TABLE 3.17** Research and Development Programs Funded by the Research Council (2009–13)

| Program | Annual budget (US$ million) | Duration (years) | Start date |
|---|---|---|---|
| Open Research Grants Program | 7.8 | Continuous | 2009 |
| Road Safety Research | 1.3 | 5 | 2009 |
| Palm Tree Dubas Bug Research | 1.3 | 5 | 2010 |
| Social Observatory Research | 2.6 | 5 | 2012 |
| Renewable Energy 2013 | – | – | 2013 |
| Nanotechnology Chair at Sultan Qaboos University | 1.6 | 5 | 2011 |
| Undergraduate Research Award Program | 0.52 | 5 | 2013 |
| Oman Eco-House Design | 0.39 | 2 | 2012 |
| The Virtual Science Library (VSL) | 6.76 | 5 | 2012–13 |
| Electronic Research Information System | – | – | 2012 |
| Education Innovation Assistance Program | 0.36 | 3 | 2011 |
| Research Outputs Incentive Program | 0.52 | Continuous | 2012 |
| Animal and Plant Genetic Resource Centre | 18 | 12 | 2012 |
| Institute of Advanced Technologies Integration | – | – | – |

From "Science, Technology and Innovation Policy Review – Oman; " United Nations Conference on Trade and Development (UNCTAD). Call symbol: UNCTAD/DTL/STICT/2014/1. Downloaded in May 2017 at: http://unctad.org/en/PublicationsLibrary/dtlstict2014d1_en.pdf.

the Council planned to fund research and research support activities, at the rate of around US$16 million per year, with several programs running for a period of 5 years with the exception of a program designed to combat an endemic palm tree disease, which is designed to run for 12 years.

Research facilities established by TRC include the Institute of Advanced Technology Integration (IATI), which was established earlier this decade,[294] with the primary objective of helping establish a culture of innovation and entrepreneurship in the fields of research and science. The institute is, thus designed to function as an intermediary between the industry and the business sectors and academia, with focus on scientific research leading to marketable technology outcomes and innovative products. Three key sectors which constitute mainstay of IATI activities; namely water, renewable energy, oil, and gas.

TRC also championed the creation of two further entities to serve research and educational activities, namely:

- the Oman Research and Education Network (OMREN), a high-speed education network, to link scientific research and education in the country[295] and
- Oman's Virtual Science Library.

**The Research Deanship at Sultan Qaboos University (RD-SQU)** was established in order to facilitate research at Sultan Qaboos University which is home to nine centers undertaking research in marine biotechnology, communication and information technology, remote sensing and geographic information systems, earth sciences, earthquake monitoring, water and the environment, oil and gas resources, Omani affairs and the humanities.

The Research Deanship's mission statement underlines its quest for this University to become a foremost research institution in the region.[296] The Deanship's mission includes enhancing a research culture as well as promoting research productivity at SQU, in liaison with TRC as well as the University's directors, through:

- policy formulation, strategic planning and related implementation measures;
- monitoring research activity and enhancing researchers' performance, emphasizing research leading to innovation and entrepreneurship;
- promoting research outcomes at the national and international levels;
- raising external funding in support of SQU research.

The Deanship has the following three departments:

- A Research Department that administers six research grant schemes.
- An Innovation & Entrepreneurship (I&E) Department that is responsible for supporting research innovation, technology transfer, and protection of intellectual property.
- A Publication Department that is responsible for publishing books and scientific journals, administering conference affairs as well linking to society through outreach activities involving visual and print media.

**The essential tasks of the Research Department at the College of Applied Science (CAS)** revolve around promoting scientific research in the Colleges of Applied Sciences with benefits for society and sustainable development at the forefront.[297] The department undertakes activities including management, promotion, support, monitoring, and the implementation of research outcomes. Table 3.18 includes summary information on three projects that the research department lists as focus for its attention in 2017, indicating that allocations to research averaged at around US$200,000, annually. It goes without saying that, this is quite a small sum to dedicate to these projects, even if, as it is most likely to be the case, it does not include salaries and running costs.

#### 3.5.3.2.2 Research Activity at Sultan Qaboos University

The Sultan Qaboos University which was established in the early 1980s, with the first enrolments taking place in 1986, into colleges of Medicine, Engineering, Agriculture, Education and Science. A College of Arts was established in 1987 following which a College of Commerce and Economics, a College of Law and a College of Nursing followed in 1993, 2006 and 2008. SQU is also home to 10 research centers mostly established within the first decade of the present century, to deal with issues including: marine and biotechnology, information and communication technologies, earthquake monitoring, the environment, oil and gas resources, earth sciences, remote sensing and geographic information systems and water resources.

In other words, the University is quite a young entity, but one whose track record indicates a measure of dynamism that appears unmatched by the resources allocated to its research endeavors. Thus, the university's postgraduate enrolment in PhD and MSc programs in 2015–16 totaled 32 and 356 candidates,

**TABLE 3.18** Examples of Research projects Undertaken at the College of Applied Science

| Project | Duration | Funds (US$) |
|---|---|---|
| Building and analyzing an efficient and scalable Arabic information retrieval system based on XML and semantic web technologies | 3 | 200,000 |
| Building a multiagent computer network based on graphic processing unit | 3 | 221,000 |
| Building a face recognition system based on subspace | 2 | 130,000 |

Information adapted from the website of the Research Department of the College of Applied Science. Accessed in May 2017 at: http://ibri.cas.edu.om/pages/Scientific-Research-Department.aspx

respectively. In 2015–16 alone over 2500 workshops and special courses, were held for postgraduate students, consultancy contracts were concluded for a sum totaling the equivalent of over US$4.5 million in the same year, whereas all grants totaled around US$9.4 million. This is equivalent to 0.0013% of the country's GDP in the same year and equal to the cost of building 1.2 km of an express highway, according to a relatively recent UNCTAD review of science, technology and innovation in Oman.[298]

Several comments are worth making on figures published in the Postgraduate Studies and Research Report by Sultan Qaboos University, for the year 2015.[299] First, the number of PhD students, enrolled and graduated in areas related to agricultural and marine specializations is extremely low considering the importance of these areas for the country's future economic growth. Indeed, no PhD students at all were enrolled in programs related to agricultural and marine specializations in 2015, whereas only three out of a total of 18 PhD graduates took up these specializations. A similar situation arises with regard to those graduated and enrolled to gain PhD qualifications in economics. On the other hand, the number of PhD students enrolled in science and engineering specializations combined amounted to 19, out of a total of 32, while taken together, PhD students enrolled in the Arts and Social Sciences as well as Education numbered 13 out of a total of 32.

Second, as indicated in Fig. 3.27, the number of students who, in 2015, were enrolled and graduating with MSc qualifications in education, the arts, the social and political sciences, economics and law by far outnumber those with science and engineering qualifications, including those who took up agricultural and marine specializations. In fact, the percentage of MSc graduates and students in 2015 opting for the former set of specializations constituted 71% of the total.

This, of course, is not to underestimate the importance of qualifications in the former set of specializations; rather it is to indicate that building a knowledge

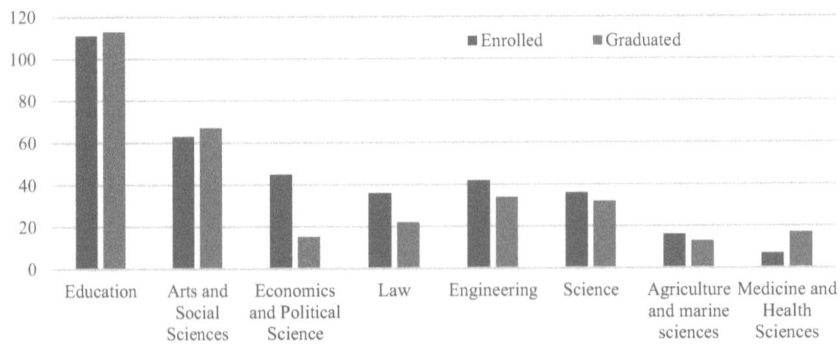


**FIGURE 3.27** Number of MSc students enrolled and graduated from SQU according to area of specialization (2015).

economy would require at least as many experts in the latter qualifications. Those specializing in the agricultural and marine sciences at the MSc level constitute a particularly small percentage considering that Oman sets great store by its agricultural sector and marine resources in advancing its diversification schemes.

Funding dedicated to research projects in domains addressed by various SQU departments appears to originate from two principal sources, internal university allocations and research grants provided The Research Council (TRC). In 2015 only the internal university allocations were made toward research projects in economics the arts, social and political sciences, totaling almost US$100,000, or 7.5% of all internal university allocations to research projects. Figs. 3.28 and 3.29 present the distribution of both internal and TRC grants across research projects carried out by various SQU colleges and centers,

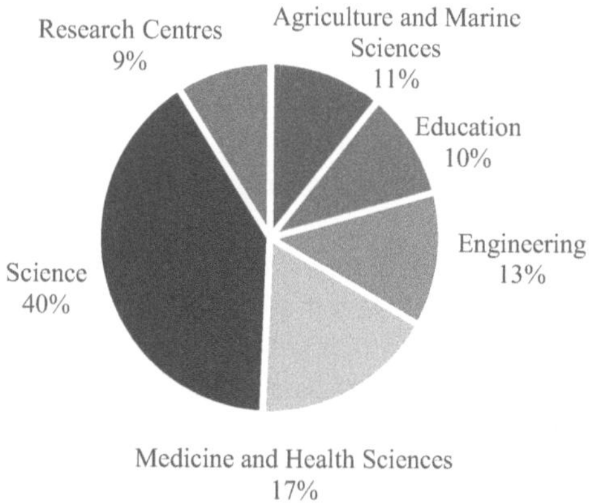


**FIGURE 3.28** Distribution of interned and TRC grants amounting to US$7.9 million across research projects by colleges faculties in 2015.

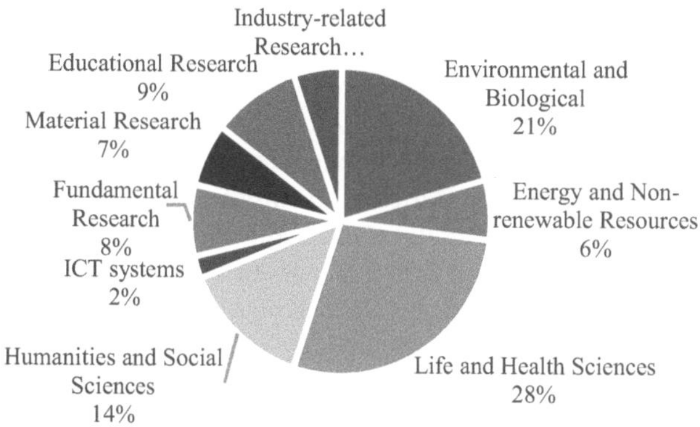


**FIGURE 3.29** Distribution of internal research grants totaling US$1.28 million by themes in 2015.

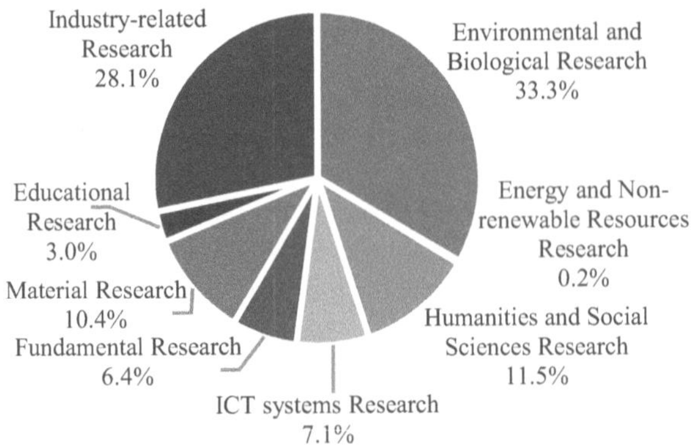


**FIGURE 3.30** Distribution of consultancy contracts totaling US$1.9 million provided by SQU across themes in 2015.

indicating that research projects in science, engineering, medicine and the health sciences as well as agriculture and marine sciences received 81% of all research grants allocated to research in 2015.

Fig. 3.30 provides a view of the distribution of the value of consultancy contracts concluded during 2015 between research facilities in SQU colleges and centers with various entities across research themes. One of the most striking features of this distribution is the miniscule value of research contracts within the domain of energy and renewable resources, amounting to no more than 0.2% of the value of all contracts concluded in 2015. At just over 7%, research consultancies in information and communication technologies and related systems, also amounted to a small proportion of the total. On the other hand, consultancies on environmental and biological as well as industry-related issues captured 61.4% of the total. Incidentally, the proportion of environment related consultancies is consistent with Oman's formal commitment to environmental protection, with laws aimed at ensuring sound environmental developed enforced by the Ministry of Environment and Climate Affairs. On the other hand, it is claimed[300] that despite some successes against polluting enterprises[301] environmental concerns may be losing out to private interests, tourism, real estate and industrial projects.

The absence of research projects and consultancy contracts on issues related to water is noticeable. Although it might be that research classified under themes relating to the environment and renewable sources might be related to water. It would seem that problems faced by Oman, along with all other GCC countries in relation to water resources, their conservation and various modes of utilization and treatment would warrant specific attention, that is not reflected by accounts of research activity at SQU. Indeed, Oman faces water scarcity and periodic droughts. Irrigation-intensive agriculture in the most fertile region, on the Batinah coast, often leads to ground water salinity, whereas overgrazing and desertification are growing problems in the south of the country.

#### *3.5.3.3 Innovation Promoting Institutions*

Most institutional entities dedicated to promoting innovation in Oman date to the beginning of the present decade, apparently through considerable cooperation between universities and the business and banking sectors. The following paragraphs present summary information on three such entities.

**Innovation Park Muscat (IPM)** is an initiative by The Research Council (TRC). It is designed to encourage scientific research and innovation through active collaboration between academic institutions and both local and international private industrial firms.[302] IPM programs highlight building abilities to "utilise knowledge, develop products based on scientific research, and turn ideas into commercial products." The first stage of the IPM initiative was initiated in March 2013. Subsequent stages are supposed to include establishment of institutes with focus on research and innovation in the energy and health sectors, food and biotechnology as well as water and environmental issues.

**The Industrial Innovation Centre (IIC)**[303] was launched in 2010 through TRC cooperation with the Ministry of Commerce and Industry as well as a host of other organizations.[304] According the above-mentioned UNCTAD review, IIC is allocated around US$3.9 million in total, considerably limiting the scope of its industrial innovation activities.[305] This same report relates the low funding levels allocated to IIC, to a general "reluctance or a lack of consensus among Oman's STI policymakers, not least at TRC, to directly support innovation efforts of private firms." This report links the position taken by academics at TRC and related Omani institutions to their "literal interpretation of the principles of the free market," and goes on to state that such an interpretation countered pragmatic approaches adopted by policymakers in several countries, toward ameliorating risks and uncertainties met by technology-based firms. Since, if successful, such approaches would result in innovations with "valuable economic and social benefits."

**The Innovation Hub (I-Hub) Project** is a pilot project intended to nurture a culture of innovation and entrepreneurship, with view to greater economic diversification and job creation. The I-Hub project is conceived as a platform enabling local innovators, to transform their ideas into viable businesses, through providing funding, mentoring as well as possibilities for networking and knowledge sharing. I-Hub consists of innovation assistance programs designed to stimulate innovation within industrial enterprises, within educational systems as well as among individuals and communities. Databases and information services on issues such as prototyping, intellectual property rights, incubating practices, as well as rosters of consultants have been developed within the I-Hub project.

#### *3.5.3.4 Concluding Remarks*

With a rather limited number of small research centers Oman's efforts at STI institution building seem to have favored university-based research facilities, on the one hand, and technology development within business and innovation promotion facilities, on the other. Unlike many of the other countries in the region, many of Oman's STI institutions provide information regarding their activities, with some actually posting copies of their annual reports, with detailed accounts of their funding of various research activities for general viewing.[306] However, in resemblance to most other countries in the region, little information is generally available on outcomes of research projects and innovation ventures, with the exception of papers published and even fewer instances of patent applications made.

Indeed, as far as may be discerned from available sources, current levels of financial support for R&D and innovation in Oman, are not likely to result in considerable progress toward innovation, neither in terms of isolated leaps within priority sectors, nor in the development of a robust national innovation system. In this respect, it is well to recall a contrast evoked by the above-mentioned UNCTAD report on STI in Oman, in which the authors indicate that the Oman's express highway costs approximately the equivalent of US$7.8 million per kilometer, compared to the annual US$52 million per year spent on R&D and innovation programs run by TRC.[307]

### 3.5.4 STI in Qatar

As indicated in the previous chapter, Qatar's Vision 2030 includes several objectives linked to assuming a prominent international role in research and technology development. Objectives linked to building the country's STI capacity are addressed within the Vision's economic development pillar, and include creation of:

- a business climate capable of attracting foreign funds and technologies as well as supporting national enterprises capable of competing worldwide;
- a vigorous oil and gas sector that generates advanced technological innovations, thereby contributing to the advancement of human and economic resources as well as safeguarding the environment through innovations targeting clean energy for Qatar and for the world;
- effective and sophisticated environmental institutions enabling the development, promotion and implementation of environmentally sound technologies.

With specific regard to innovation and creativity, the National Vision emphasizes the need for creating an effective system for funding scientific research shared by the public and private sectors within the country as well as cooperation with specialized international organizations and leading research centers worldwide.

With no indications that Qatar has, or is intending to formulate, a full-fledged STI policy the following paragraphs are aimed at reviewing the country's STI institutions and, wherever possible, outlining their principal activities.

### *3.5.4.1 Institutions Concerned With Promoting STI in Qatar*

Awareness of the importance of building scientific R&D capabilities is reflected in the country's constitution, of which article 24 instructs that the state "promote science, arts, national cultural heritage" and "encourage scientific research."[308]

Action taken to foster national R&D capabilities in Qatar dates back to the early seventies, with research and technical cooperation departments established within various ministries and government agencies.[309] Additionally, when established in 1977, Qatar University was charged with upgrading research skills in selected priority areas, including the environment, water as well as selected branches of engineering. Significant strides toward capacity building in science, technology and innovation were taken later in the nineties, with the establishment of the Qatar Foundation and its Education City. The following paragraphs present a concise summary of the scope of activities of a number of the above STI institutions some of their constituent facilities.

#### 3.5.4.1.1 The Qatar Foundation

Since its establishment in 1995, as a semiprivate chartered, nonprofit organization, the Qatar Foundation for Education, Science and Community Development (QF), has spearheaded Qatar's efforts to establish itself as a regional leader in education, science, and cultural development. One of the Foundations' aims is to support Qatar's transition from a carbon to a knowledge-based economy, through enhancing the country's human potential.

With improving the quality of education, in general, and higher education, in particular, as its principal aims, QF established the Qatar Education City, which was constructed on a 10 $km^2$ campus on the edge of Doha. The City includes campuses of eight satellite universities offering a variety of degree programs. See Table 3.19. A science and technology park, a teaching hospital, a medical research center as well as a Convention Centre. Qatar's Education City is part of Qatar National Vision 2030, and is intended to leverage Qatar's transformation into a knowledge-based economy.

Arguments by QF for hosting satellite branches of well-established institutions of higher education may be summarized by its need to:

- establish the country as a regional research hub;
- cultivate a local pool of well-trained graduates;
- allow young women access to high quality education which they would mostly be unable to have due to cultural restrictions on their travel abroad.

**TABLE 3.19 Satellite Branches of Foreign Universities established Within Qatar Education City**

| | | |
|---|---|---|
| Carnegie Mellon University in Qatar | 2004 | Computer science and business Management. |
| Georgetown University—School of Foreign Service in Qatar | 2005 | Arts and literature, philosophy, international affairs, economics, religious studies |
| HEC Paris in Qatar | 2010 | Business administration |
| Northwestern University in Qatar | 2008 | Media studies and journalism |
| Texas A&M University at Qatar | 2004 | Chemical, electrical, mechanical and petroleum engineering |
| University College London in Qatar | 2012 | Library studies and museology |
| Virginia Commonwealth University in Qatar | 1997 | Design, graphics, fashion and décor |
| Weill Cornell Medicine—Qatar (WCM-Q) | 2002 | Medicine |

However, no unified strategy appears to animate the diversity of satellite university ventures. That is, apart from the intention to modernize the higher educational system and the belief that something is to be gained from the globalization of higher education, both for the universities in question as well as their host countries. Unlike other Arab countries, such as Egypt, Iraq, and Syria which although recipients of external assistance toward developing their systems of higher education during past decades, allowed relevant authorities and concerned ministries a direct role in such efforts, Qatar and perhaps other GCC countries appear to take a hands-off approach.

Possible exceptions to this approach may be confined to allowing freedom of speech and open debate, which are presumably essential values of parent university institutions and possibly quite essential to exploring new ideas as well as bridging cultures. Compromises involved on the part of parent universities might invite deeper questions about forces motivating moves to set up satellite campuses. The question, in essence is why are such universities willing to bypass their foundational values for mere expansion? Proponents of such moves might argue that benefits to students and faculty justify suspension of principles such as free speech and the acceptance of something less than unrestricted academic freedom. Others might charge that as the competition increases for capturing college students, and as private higher education becomes more lucrative, an enhanced global presence makes for financial gains as well as greater prestige.[310]

On the other hand, anecdotal evidence exists with regard to shortcomings that this strategy might have. Thus, claims are made, mainly by staff at national institutions of higher education, to the effect that had expenditures heaped onto "guest" satellite branches of overseas universities been allocated to national facilities, while allowing them to hire high caliber staff, as well as greater freedom in managing their own affairs, then at least similar outcomes might ensue.

Should claims regarding possible neglect of national institutions of higher education as a result of focus on establishing satellite branches of overseas universities be proven, then present policies should lead to creation of multi-tier tracks that favor the privileged, limiting equal opportunities and ultimately undermining future socioeconomic development.

Frame 3.29 presents arguments often made with regard to the pros and cons of satellite branches of overseas universities in the Arab Gulf countries. However, further research is certainly needed in order to arrive at viable cost–benefit analysis of the trend toward hosting satellite branches of outstanding overseas universities, since without such analysis the above arguments would remain essentially speculative. At any rate, it may be noted that Satellite universities in Qatar as well as other Arab Gulf countries appear to employ faculty from both local populations and their parent institutions, in differing proportions. Their student bodies are likewise mixed and variable.

**Frame 3.29 Arguments for and Against Satellite Branches of Overseas Universities in the Arab Gulf Countries**

Reasons as to why Qatar along with other Arab Gulf governments might be keen to host satellites of overseas universities are essentially linked to the latter institutions' reputations for academic excellence. It would seem that Qatar, and presumably its GCC neighbors, consider hosting satellites of prominent universities a mark of prestige worth the considerable investment, in initial construction as well as running costs. Hosting satellite branches of universities with world-class reputations is hence considered a sure sign that the country has become part of an elite global community, reinforcing an innovative self-image while banking on the cultural brand possessed by the parent university as partly their own.

Arguments are also made in favor of the beneficial effects of the attention given by satellite branches of prominent overseas universities to socioeconomic problems that face their host countries. In contrast some argue that although some satellite universities appear to further research agenda related to their host country, many others merely constitute a source of support for postsecondary education within their host countries.

Other arguments emphasize the impact that satellite branches operating on the same turf would have on national institutions of higher learning and research; in effect encouraging national institutions adopt similar methods and evaluation criteria.

#### 3.5.4.1.2 Qatar Science and Technology Park

Formally inaugurated in 2009, the Qatar Science and Technology Park (QSTP)[311] was intended as an international hub for technology-based entrepreneurship and high-tech businesses. Its mission includes:

- support for technology-based entrepreneurs in Qatar through incubation, funding, training, mentorship, and links to selected regional and global institutions involved in innovative ventures;
- accelerating product innovation within the private sector through programs and grants that encourage new product development, innovation, and collaboration with the Qatar Foundation's research institutes;
- attracting tenants focused on development of technology-based products and processes and the commercialization of research outcomes.

QSTP is Qatar's first free-trade zone.[312] As such, it is home to several international technology companies opting to set up facilities in Qatar. Additionally, QSTP houses an incubator for technology-based enterprises. QSTP shares its location with Qatar Foundation's Education City alongside the international universities listed below, thereby facilitating collaboration between tenant companies and university researchers and other staff.

#### 3.5.4.1.3 The Qatar National Research Fund[313]

Qatar National Research Fund (QNRF) was established in 2006 in order to streamline research funding and to ensure adequate responses to the country's research priorities. The QNRF National Priorities Research Program (NPRP) disburses funds with preference to research projects that enjoy the following attributes:

- tackle needs and challenges that local research end-users face;
- result in tangible impacts;
- focus on subjects that are highly promising in terms of commercial and technological potential;
- promote public–private partnerships in Qatar;
- encourage cofunding;
- promote cross-cutting/interdisciplinary approaches;
- stimulate scientific excellence and the advancement of knowledge in Qatar.

NPRP presently provides grants of up to an annual $300,000 per project on competitive bases to research projects selected in accordance with the Qatar National Research Strategy's four pillars, with emphasis on topics in the following domains:

- energy and the environment;
- computer science and information and communications technology;
- health and social sciences;
- arts and humanities.

Satellite universities are eligible for funding by QNRF, with Texas A&M at Qatar, for instance, accumulating total funding of US$236 million during the decade ending 2015 for 289 projects.[314]

#### 3.5.4.1.4 Office of Policy Planning and Evaluation

A Policy Planning and Evaluation (PPE) Office was established by Qatar Foundation with the objective of providing support for research and development in a number of areas, including policy, portfolio analysis, strategic planning, coordination, and evaluation.[315] This office is said to collaborate with national institutions including Qatar National Research Fund, Qatar Science & Technology Park, and the Office of Intellectual Property and Technology Transfer, in order to promote world-class research and innovation with, tangible outcomes. The Office's monitoring and evaluation function extends beyond institutions active within the Qatar Foundation, providing assistance to partner research and university institutions by analyzing their research project portfolios, reviewing their overall research programs, infrastructure arrangements, human capacity and how such attributes compare to other institutions in terms of quality and competitiveness of their research. The principal aim of such efforts does not stop at issuing an evaluation report. Rather, such analyses are designed to shed light on future paths to be taken and how research institutions might learn from others.

### *3.5.4.2 Institutions Undertaking Specialized Research Activities*

In addition to the above institutions, Qatar launched, only during the past few years, a number of research facilities in order to conduct research activities within certain domains, in medicine, information and communication technologies, renewable energy as well as water-related issues. The following sections provide a glimpse of these institutions activities.

#### 3.5.4.2.1 The Sidra Health and Research Centre

Created on the basis of an endowment of nearly US$8 billion, the Sidra Health and Research Centre (SHRC) is the most recent addition to Qatar Foundations panoply of institutions based on novel technology inputs. Utilizing an endowment of nearly US$8 billion,[316] SHC commenced operation in May 2016 with three clinics specializing in dermatology, pediatric general surgery, consultation and obstetrics. SHRC is expected to successively create further facilities as well as a range of educational and research units "through strong partnerships with leading institutions around the world."[317] Its declared mission emphasizes educational and research as well as provision of world-class health care services.

#### 3.5.4.2.2 Qatar Biomedical Research Institute

This is a national research institute established in 2012 within Hamad bin Khalifa University.[318] QBRI is intended to improve and transform healthcare through bringing about innovations in preventive and diagnostic medicine as well as the treatment of diseases affecting populations in Qatar and the region. A QBRI statement underscores its focus on integrative and multidisciplinary approaches allowing insights into the molecular basis of the diseases such as breast cancer, type 2 diabetes as well as neurological disorders such as autism and epilepsy. Other items on its agenda include the development of novel biomarkers, diagnostic and therapeutic strategies to facilitate early diagnosis, treatment and management of the above diseases leading to personalized medicine and healthcare. With the above in mind, QBRI plans to set up and manage three scientific centers of excellence to carry out research on cancer, diabetes and neurological disorders as well as train the next generation of scientists and technicians.

### 3.5.4.2.3 Qatar Computing Research Institute

QCRI was established in 2010 by the Qatar Foundation as a private, nonprofit organization with the essential objective of assisting the country's transformation from a carbon to knowledge-based economy.[319] QCRI operates under the umbrella of Hamad bin Khalifa University and is tasked with conducting world-class multidisciplinary computing research in the several domains, including: Arabic language technologies, cyber security, data analysis, distributed systems, computational science and engineering, social computing, with relevance to the needs of Qatar as well as the Arab region, and the world at large. QCRI research is stated as aligned with the Qatar National Research Strategy, and is carried out in support of strategic priorities underlined in Qatar National Vision 2030.

### 3.5.4.2.4 Qatar Environment and Energy Research Institute

In line with Qatar's National Vision 2030, the Qatar Environment and Energy Research Institute was established in 2011, with essential focus on research in energy and water.[320] With the primary objective of safeguarding the environment QEERI carries out research along four tracks in each of the above two focus areas. Thus, with regard to energy, QEERI carries out research in solar photovoltaics, energy storage, grid integration and energy efficiency.

In relation to water, QEERI research addresses problems in water desalination and treatment, water quality and reuse, aquifer recharge and water conservation. QEERI is also stated as involved in research on global climate change.

### 3.5.4.2.5 Qatar Solar Technologies

Qatar Solar Technologies (QSTec) was established by the Qatar Foundation in 2010 as a joint venture between Qatar Solar, a fully owned QF subsidiary, SolarWorld AG and the Qatar Development Bank. QSTec is stated as having embarked on construction of an 8000 metric ton per year (MTPY) solar grade polysilicon plant in the Ras Laffan Industrial City (RLIC). Exports from this plant are slated to begin in 2017. Ground and rooftop PV installations with total capacity of around one 1 MW of solar power are reported as having been installed at its plant in RLIC.[321]

#### 3.5.4.2.6 Water-Related Research Efforts

Concerning research and technology development efforts aimed at acquiring technologies relevant to water desalination, a review of literature posted by Qatar Environment and Energy Research Institute (QEERI), the country's research institution most closely linked to water resources, is quite informative.

Thus, a list of research highlights posted by QEERI on its website,[322] illustrates the tendency to tackle applied research topics that, although worthy in substance and orientation, still stand at some considerable distance from application. This is most vividly illustrated by one of the projects highlighted in this list. The project in question is aimed at determining an optimal molecular structure for membranes that defy biofouling when used in reverse osmosis desalination plants.[323] Although research at this level is certainly of theoretical and inherent practical importance it is still far removed from immediate or short-term applications, and as such it may not be expected to present immediate or short-term practical benefits, in terms of enhancing the effectiveness and efficiency of Qatar's desalination plants.

The main issue, in effect, is that such research was being conducted in the absence of activities aimed at piloting, testing and eventually manufacturing new membranes compositions for widespread application. In the absence of such "downstream" technology development activities, the most likely beneficiaries of such research in addition to the scientific community at large would be firms that produced membranes, packaged and exported them to Qatar and other countries that rely upon desalination for their water needs.

It would therefore be essential for future technology policies concerned with water desalination and treatment in Qatar, to complement existing applied research in this domain with pilot facilities designed to test and validate research findings prior to their application in water desalination and treatment ventures. Naturally, it might be convenient to start by forging alliances and joint ventures with established international as well as regional actors, with the ultimate objective of creating integrated domestic capabilities.

### *3.5.4.3 Concluding Remarks*

With several facilities established toward the beginning the present decade, it might still be early days to judge the effectiveness and efficiency of Qatar's efforts toward building scientific and technological capabilities.

A promising future may lie ahead, particularly in view of the number of research facilities that have been established, the support they receive and the diversity of areas being addressed. The fact that synergy is being sought among activities undertaken by government ministries, semigovernmental institutions, banks as well as universities and research centers augers well for the future.

On the other hand, however, the fact that an institution of Qatar Foundation's proportions and level of endowments is almost solely in charge of charting the future of the country's STI future has negative as well as its positive ramifications. Criticism of the Foundation's role is often linked to emphasis on superficial issues, in relation to newly established research and educational facilities as well as evaluation of attainments, with much less emphasis on content and methodologies for ensuring quality outcomes.[324]

Unavoidably, several institutions being established are staffed by a larger proportion of expatriates than Qatari specialists.[325] Although the large majority of expatriates may well be trusted to implement high standards and work ethics in discharging their duties, the absence of a sense of ownership could constitute a serious obstacle to establishing long-lasting institutional traditions together and associated quality standards.

A certain degree of tentativeness is apparent in the way that certain scientific research institutions have tended to evolve. An example of this is provided by Qatar University's Scientific and Applied Research Centre (SARC) which was established by the University in 1980 and tasked with conducting basic and applied research emphasizing national development needs and participating in transfer and adaptation of technology for implementation in Qatar.[326] However, in 2005, SARC was renamed as the Environmental Studies Centre (ESC), and its mission was restricted to disciplines associated with Qatar's environmental sustainability.[327]

In conclusion, it would seem that research and development activity in Qatar is in need of radical upheaval before it may participate, let alone perform a leading role, in the creation of a knowledge-based economy. There is almost total absence of critical mass throughout. More importantly, such critical mass is absent where it is needed most, in areas that are closely related to the country's wealth and others that may be invoked in adding value to this wealth. Although expatriate researchers may be able to bridge gaps, temporary commitment and in most cases, slender linkages to the surrounding environment would at best help Qatar attain the "enclave" vision of its intended knowledge economy.

With only a small proportion of the population engaged in research activity[328] the chances for creating research schools of the type that is responsible for widespread innovation in champion knowledge-based economies are very slim indeed, no matter how much financial resources may ultimately be dedicated to the process.

On the issue of "research school" creation, it is worth noting that several fields of research activity appear to suffer imbalance in the relatively large proportion of Qatari and non-Qatari PhD holders, presumably leaders of research in their field, compared to less qualified research students; fresh graduates or Master's degree holders. This is illustrated by the situation in electrical and electronics engineering, in which two Qatari holders of the Bachelor's degree are reported as working in research, led by 11 Qatari and 17 non-Qatari senior researchers. A similar situation pertains in the health sciences, where only one Qatari female researcher is reportedly working in an area in which 6 Qatari and 14 non-Qatari senior researchers are working. Indeed, a rather smaller number of Bachelor's and Master's holders than ordinarily expected appear to be engaged in all research disciplines.

It goes without saying that involving larger numbers of junior Qatari, as well as non-Qatari scientists and engineers in R&D would certainly play a crucial role in favoring the "innovative society" as opposed to an "enclave" version of Qatar's future.

### 3.5.5 Saudi National STI System

Scientific and technological capabilities evolved in a manner that is characteristic of its other Arab neighbors, with universities staffed mainly by non-Saudis from neighboring countries and abroad, with an outburst of public science and technology institution building in the 1970s and 1980s as well as meager allocations to R&D activity. However, Saudi Arabia appears to have escaped thus far, the trend toward hosting satellites of foreign universities.

#### *3.5.5.1 Saudi National Science, Technology, and Innovation Plan*[329]

Saudi Arabia's NSTIP is a strategic document designed to promote the country's transition from an oil to a knowledge-based economy. NSTIP was designed for implementation within four phases with the following principal goals:

- Establishment of a robust national infrastructure for science, technology, and innovation, 2007–14;
- Becoming a leading country in STI:
  - in the region during 2015–19
  - in Asia during 2020–24
- Transforming Saudi Arabia into a knowledge-based economy by 2025–29.

Technology development and innovation within 15 strategic technology areas and domains of application, deemed integral to Saudi Arabia's long-term vision are promoted by NSTIP. The intention is to build scientific and technological capacity in these domains through R&D projects implemented by institutions throughout Saudi Arabia with direct benefits to its socioeconomic development. Funding is provided for these projects on competitive basis, additionally, an evaluation scheme is also in force, involving national as well as international experts.

NSTIP was designed with a number of subsidiary goals, including:

- creation of viable links between key actors including R&D centers, educational and vocational training institutions, firms active in STI-related manufacturing and services activity, investors, innovators, and the media;
- promotion of increased international cooperation;
- raising financial support for its own implementation.

King Abdulaziz City for Science and Technology (KACST) has been assigned primary responsibilities regarding NSTIP implementation and monitoring, chairing its steering committee, coordinating efforts with concerned national institutions, including 17 ministries, 10 universities, and 13 national agencies and commissions as well as publishing evaluation reports in accordance with set performance indicators. In 2014 KACST reported finalization of the first 5-year NSTIP implementation plan (Maarifah 1)[330,331] and outlined measures aimed at implementing the second 5-year plan, designated (Maarifah 2).

KACST designed a number of programs with view to NSTIP's implementation, one of which is the Strategic Technologies Program that is designed with the aim of promoting R&D and related technology acquisition within a number of science and technology specializations and application areas. Information on a number of other programs, funded and managed by KACST is presented in Frame 3.30.

Programs designed by KACST with view to implementing NSTIP have also included the establishment of an incubation scheme, as well as program aimed at setting up Joint Centres of Excellence (JCE's) in partnership with renowned research centers and universities across the world.

The incubation scheme is reported as having hosted 90 clients by January 2015, creating 345 jobs by 2014 and generating around SR11 million, or US$ 2.9 million, during the first half of 2014. Eight clients who eventually graduated from the KACST incubation scheme assumed a market value of around SR150 million, or US$40 million. On the other hand, the JCE's Program, succeeded by 2012 in establishing joint entities in a variety of domains including: genomics, nanotechnology, green nanotechnology, nanomedicine, bio-nanotechnology, nanomaterials for clean energy applications, advanced materials, materials and building systems, petrochemicals, complex engineering systems, decision support, telecom applications, space and aeronautics as well as lunar and near-earth objects. It may be noted that more than one third of the JCE's established by 2012, were dedicated to nanotechnologies and their applications. At any rate all JCE's were established in collaboration with international partners with expertise and strong credentials within the above technology domains.[333]

#### 3.5.5.1.1 Recent Activities Supported by NSTIP

Table 3.20 presents a list of strategic technology areas targeted by NSTIP as well as the number of projects implemented, and funding allocated for R&D in each

**Frame 3.30 KACST Programs Designed in Aid of NSTIP Implementation**

The **Capacity Building Program for Scientific Research and Technology Development** is dedicated to establishing research centers with focus on priority areas, e.g., in oil and gas, environmental management, agricultural production as well as medicine and health.

The **Diversification of STI Funding Sources Program** is assigned responsibilities for establishing a fund for technology development as well as other tasks aimed at securing funding for R&D centers.

The **Technology Transfer and Localization Program** is tasked with developing and implementing strategies in support of innovation and technology development. Initiatives being implemented by this program include an SME development center and technology incubator schemes as well as an industrial city.

The **Technology Innovation Centers program** is charged with promoting university-industry collaboration in applied research, science and engineering education as well as technology transfer.[332]

The **STI Human Resources Program** is assigned responsibilities for establishing creativity and scientific innovation centers, granting higher education scholarships in strategic technology domains, design of curricula highlighting research methods, creativity and innovation in public education and rewarding successful inventors and researchers.

The **Knowledge Society Program,** which is entrusted with a range of initiatives including publication of scientific journals within strategic technology domains, developing national databases for STI indicators and promoting STI awareness within the country's schools and universities.

A number of projects, such as the King Abdullah's **Initiative for Water Desalination using Solar Energy,** are also in place with the aim of ensuring water security through the development of advanced environmentally friendly technological energy solutions.

The **STI Laws Program** is responsible for drafting legislative instruments and IP regulations with impact on the development of STI capabilities for national research programs.

domain during 2014, the most recent year for which data is available. R&D projects in medicine and health topped the list in terms of the number of projects selected for receiving support, in terms of the number of R&D projects (74 or 23.6% of the total), and allocations (around US$35.2 million or 24.8% of the total). On the other hand, projects in three domains, namely, oil and gas, space and aeronautics, as well as building and construction, featured lowest on the list, with only six projects in each domain, or around 2% of the total number of projects. R&D projects within the various technology domains are allocated similar amounts, averaging at around US$1.7 million, with projects in space and aeronautics averaging US$1.94 million on the higher end. Projects in mathematics and physics and building and construction are allocated lowest funding per project, at US$1.24 million and US$1.21 million, respectively.

**TABLE 3.20** Numbers of projects Supported by the National Science, Technology and Innovation Plan's Strategic Technologies Program and Budgetary allocations in 2014

| Science and technology domains and Application areas | Number of Projects | Budget US$ Million[a] | Percentage of Total projects | Percentage of Total budget |
|---|---|---|---|---|
| Medicine and Health | 74 | 35.19 | 23.6 | 24.8 |
| Biotechnology | 39 | 18.45 | 12.5 | 13.0 |
| Information Technology | 23 | 9.45 | 7.3 | 6.7 |
| Advanced Materials | 23 | 11.21 | 7.3 | 7.9 |
| Math and Physics | 23 | 7.62 | 7.3 | 5.4 |
| Energy | 22 | 9.96 | 7.0 | 7.0 |
| Environment | 22 | 10.26 | 7.0 | 7.2 |
| Agricultural | 22 | 10.28 | 7.0 | 7.3 |
| Nanotechnology | 16 | 7.16 | 5.1 | 5.1 |
| Water | 12 | 5.97 | 3.8 | 4.2 |
| Electronics, Communication, and Phonetics | 11 | 5.02 | 3.5 | 3.5 |
| Petrochemicals | 8 | 3.33 | 2.6 | 2.3 |
| Oil and Gas | 6 | 2.81 | 1.9 | 2.0 |
| Space and Aeronautics | 6 | 3.10 | 1.9 | 2.2 |
| Building and Construction | 6 | 1.93 | 1.9 | 1.4 |
| Totals | 313 | 141.76 | 100.0 | 100.0 |

[a]*Converted from values quoted in Saudi Riyals at SR3.75 to US$1.*
From KACST Annual Report 2014. The Annual Report of King Abdulaziz City for Science and Technology, for the year 2014. Downloaded in January 2017 from: https://www.kacst.edu.sa/eng/about/AnnualReports/KACST%202014%20Annual%20Report%20English%20LR.pdf.

A similar picture emerges upon reviewing records of R&D projects and funding allocations within the Strategic Technologies Program since the start of the NSTIP implementation in 2008. Thus, funds totaling around SR3.2 billion, or US$853 million, were provided to cover 1852 projects, was provided to a variety of science and technology institutions, with projects in the healthcare and medical sectors receiving nearly 23% of this total, whereas around 18.9% and 8.4% of the total were allocated to projects in biotechnology, nanotechnology and the environment. The distribution of funding allocated to projects submitted for support in 2012 roughly mirrors that indicated above, with a high proportion, 19.4%, of biotechnology projects selected to receive funding followed by projects in medicine and health, at 16.3%. Although projects in space and aeronautics, petrochemicals, oil and gas and building and construction, ended up with 1.9%, 1.6%, 1.4%, and 0.6%, respectively out of the total number of projects selected.[334]

Evaluation of R&D projects submitted for support within the Strategic Technologies Program involves a number of criteria including innovative content and applicability of outcomes, their likely contribution to knowledge within the proposed field of research, development of the country's scientific and technological capabilities as well as compliance with the strategic orientations of Maarifah 1. It is essential to note these criteria are utilized in judging which R&D projects should be funded and that no indications are available as to whether any analysis is undertaken of research outcome or follow-up activities to determine tangible impact. Although such measures might seldom be followed by R&D funding agencies in other countries, it may well be warranted in the case of Saudi Arabia, its Vision 2030 and NSTIP programs, all of which are supposed to produce specific outcomes by set dates.

### *3.5.5.2 Saudi Institutions Dedicated to Promoting STI*

Several universities in Saudi Arabia undertake research activities with academic as well as practical objectives. However, King Abdulaziz City for Science and Technology (KACST) stands out as principal actors on the STI scene in Saudi Arabia. The following paragraphs will attempt a concise review of KACST and its activities followed by a summary account of major research efforts undertaken by some of Saudi Arabia's universities.

#### 3.5.5.2.1 King Abdulaziz City for Science and Technology

As indicated above, KACST is entrusted with a major role in the implementation of the National Science Technology and Innovation Plan and issues an annual report summarizing progress made and outlining future plans, indicating progress made in setting up specialized research centers and other institutional entities designed for STI capacity building.

Additionally, KACST is also a member in the Saudi Arabia Advanced Research Alliance along with, Saudi Aramco, King Fahd University of Petroleum and Minerals, King Abdullah University of Science and Technology, TAQNIA a government-owned entity, established with the purpose of leveraging research outcomes. The only non-Saudi partner in this alliance is RTI International, a US-based firm that provides research and technical services in association with government agencies, universities, foundations as well as private businesses.

KACST established six institutes and a total of 30 centers and one specialized unit all of which dedicated to STI capacity building in areas deemed of strategic importance for Saudi Arabia. Table 3.21 present a list of the KACST institutes and national research centers.

**TABLE 3.21** Institutes and Constituent National Research Centers established by KACST

| | |
|---|---|
| **Water and Energy Research Institute** | • Desalination and Treatment Technology Centre<br>• Electrical Power Systems Technology Centre<br>• Oil and Gas Technology Centre<br>• Combustion and Plasma Technology Centre<br>• Membrane Technology Centre<br>• Solar Energy Technology Centre |
| **Materials Science Research Institute** | • Nanotechnology Centre<br>• Advanced Materials Centre<br>• Petrochemical Technology Centre<br>• Building and Construction Centre<br>• Applied Physics Centre |
| **Communication and Information Technology Research Institute** | • Electronics and Photonics Technology Centre<br>• Information Technology Security Centre<br>• Computer Technology and Mathematics Centre<br>• Robotics Technologies and Intelligent Systems Centre<br>• Sensor Technology and Defence Systems Centre |
| **Nuclear Sciences Research Institute** | • Nuclear Regulatory Unit<br>• Irradiation Technology Centre<br>• Applied Radiation Technology Centre<br>• Accelerators Technology Centre<br>• Nuclear Technology Centre |
| **Life Science & Environment Research Institute** | • Environmental Technology Centre<br>• Genome Technology Centre<br>• Biotechnology Centre<br>• Agricultural Technology Centre<br>• Stem Cell Technology Centre |
| **Space and Aeronautics Research Institute** | • Remote Sensing Technology Centre<br>• Satellite Technology Centre<br>• Aeronautics Technology Centre<br>• Jet Engine Technology Centre<br>• Astronomy Centre |

Retrieved from King Abdulaziz City for Science and Technology's website in February 2017 from: https://www.kacst.edu.sa/eng/rd/pages/content.aspx?dID=61.

A review of information posted on the KACST website concerning activities undertaken by its national research centers indicates that 109 projects were being implemented toward the end of 2016 and the beginning of 2017, targeting a variety of goals. Tables A.3.5–A.3.10 present lists of these projects. Most projects in these lists are concerned with the acquisition of specific technologies and exposing their staff to new skills. Indeed, literature posted concerning some of the centers declare training of their staff as an essential objective. This is understandable considering that several research centers were established relatively recently. Hence, they may need to amass considerable expertise within their designated fields of activity. This should in no way diminish the value of steps taken to establish such centers. Rather, it is to be taken as underscoring the need for accelerating and intensifying such training.

A number of projects, related to defense and security applications, are directed toward building specific prototypes of turbojet engines, transport aircraft as well as robotic systems. Other projects in the same domain target development of special hardware as well as software for radar systems and various sensors. Some of these projects are being implemented in collaboration with both Saudi and overseas firms and research centers. A considerable number of projects related to water and energy are likewise directed toward the development of specific outcomes, including, for example prototypes for systems that might eventually be commercially produced and marketed or otherwise installed, e.g., to ensure power and water supplies in remote areas. That information posted on such projects do not indicate collaboration with organizations that might take charge of actual production and marketing activities is worth noting.

Additionally, should work on such prototypes be merely initiated based on the feeling that such products would be useful, rather than detailed analysis of needs, backed by strong possibilities for potential funding of production and marketing phases, by concerned authorities or donors, then this would indicate a supply sided approach to R&D activities undertaken, which could diminish their ultimate impact.

Another observation concerning projects being undertaken by KACST research centers concerns the lack of indications of collaboration with other research facilities and university centers, which are known to be handling similar or adjacent research projects within Saudi Arabia. This, however, might be due to such information not being adequately advertised. Should this not be the case then remedies would be needed, with efforts dedicated, toward more intensive networking and wider collaborative ventures.

#### 3.5.5.2.2 Research Facilities at Saudi Arabia's Universities

A number of Saudi universities have managed to set up research centers, units and laboratories to address a variety of issues, with notable focus on water, the environment, oil and gas resources, advanced materials, information and communications technology and petrochemicals. This is particularly the case with regard to the King Abdullah University for Science and Technology (KAUST), the King Fahd University for Petroleum and Minerals (KFUPM). The following subsections provide some further information concerning research facilities established by these two institutions as well as the King Saud University (KSU).

***3.5.5.2.2.1 The Research Institute of the King Abdullah University for Science and Technology*** The Research Institute is part of KAUST, which is a graduate research university distinguished among other Saudi universities by its adoption of a decidedly international outlook, in terms of both its staff and students,[335] whereas work undertaken in its research laboratories is closely linked to the country's priorities. Table 3.22 presents a list of active research centers hosted by KAUST, which is also home to the following academic research facilities:

- Biological and Environmental Science and Engineering Division (BESE);
- Computer, Electrical and Mathematical Science and Engineering Division (CEMSE);
- Physical Science and Engineering Division (PSE).

Additionally KAUST hosts an entrepreneurship program for its graduates, encouraging them to commercially exploit research outcomes.[336]

***3.5.5.2.2.2 The Research Institute of the King Fahd University for Petroleum and Minerals (KFUPM)*** The Research Institute of KFUPM was established with focus on capacity building in selected priority areas. The Institute serves as locus for graduate studies at the university. The institute houses centers specializing in the following research domains:

- Communications and Information Technology;
- Economics and Management;
- Engineering Research;[337]
- Environment and Water;
- Refining and Petrochemicals.

**TABLE 3.22** Research Centers and Academic Research Divisions Established by King Abdullah University for Science and Technology

| KAUST Research Centers |
|---|
| Advanced Membranes and Porous Materials Centre (AMPMC) |
| Ali I. Al-Naimi Petroleum Engineering Research Centre (ANPERC) |
| Computational Bioscience Research Centre (CBRC) |
| Clean Combustion Research Centre (CCRC) |
| KAUST Catalysis Centre (KCC) |
| KAUST Solar Centre (KSC) |
| Red Sea Research Centre (RSRC) |
| Visual Computing Centre (VCC) |
| Water Desalination and Reuse Centre (WDRC) |

**Frame 3.31 Constituent Areas of Activity of the Environment and Water Centre at the KFUPM Research Institute**

**Water**
- Water resources development, planning; management and conservation;
- Irrigation water management;
- Water resources system modeling.

**Environment**
- Environment monitoring and pollution control;
- Environmental modeling;
- Waste treatment and management;
- Environmental chemistry.

**Marine studies**
- Baseline biological and oceanographic data collection;
- Monitoring the marine habitats in the Arabian Gulf and Red Sea;
- Assessing manmade and natural impacts on local marine environments.

From KFUPM website. Accessed in May 2017, at: http://ri.kfupm.edu.sa/cew/.

Frame 3.31 presents by way of illustration areas of activity addressed by the Environment and Water Centre.

An example of research collaboration is provided by an activity reported as under implementation by the Centre for Refining and Petrochemicals (CRP) in 2017, where CRP is reported as collaborating with two other Saudi centers and a US research Centre on the development of novel high-performance supported polyolefin catalysts with support by the KAUST.[338]

***3.5.5.2.2.3 Research Centers of the King Saud University*** Research Centers of KSU were established to undertake research on selected priority issues. Frame 3.32 presents a list of these research centers established by this university, indicating interest in several fields of research that are shared by other higher education and research institutions in the country, essentially, water, energy technologies, the environment as well as healthcare. Additionally, however, this university hosts a range of centers of excellence dedicated to a variety of research domains including information technology, biotechnology, sustainable energy technologies and science and mathematics education. A review of the latter center's research output indicates that several activities were undertaken with view to reforming science and mathematics curricula and reviews of student books dedicated to teaching related topics.

### *3.5.5.3 Concluding Remarks*

Science and technology institutions in Saudi Arabia went through enormous transformations over the past 2 decades, with changes intensifying considerably

**Frame 3.32 Research Institutes at the King Saud Research Institute**

King Abdullah Institute for Research & Consulting Studies;
Prince Sultan Advanced Technology Research Institute;
King Abdullah Institute for Nanotechnology;
Prince Sultan Research Centre for Environment; Water and Desert Institute.

during the past few years. Further transformation are predicted in the wake of the country's Vision 2030.

With particular regard to research collaboration, a review of the activities undertaken by top research institutes and centers reveals that collaboration with "client" industrial and services institutions, active in priority areas, including for example, water desalination, is still minimal. Although KACST appears to encourage research collaboration, a much greater deal of it would be highly desirable with view to avoiding duplication as well as saving precious time and resources.[339]

It is likely, as a result of steps taken to implement Vision 2030, that more concrete cooperation might take place between specialized research centers as well as between these centers and end-users in various domains of application. Institutions concerned with implementing Vision 2030 would be expected to push for such cooperation, particularly in areas where specific numerical goals are to be achieved. However, given that such goals remain relatively few in number and the urgent need to enhance cooperative ventures in domains where no such goals have in fact been set, it would be advisable to establish nationwide coordination across disciplinary and sectoral lines in order to promote even greater collaboration than has hitherto been the case.

It is often claimed that demand patterns by end-user sectors might eventually enforce greater collaboration and coordination. However, in the particular case of Saudi Arabia and most of its Arab neighbors, demand is, more often than not, answered by resorting to overseas firms and institutions. This may partly explain isolationist trends among research facilities which have generally led to overlapping R&D programs and fragmented responses to persist far longer than what might be expected as R&D institutions seek to gather experience and achieve maturity. Hence the urgent need for institutional measures with funding regulations that ensure minimal overlaps and fragmentation.

There is also an urgent need to introduce more effective measures for evaluation and monitoring of institutional performance. Thus, apart from the improvement in publications output over the past few years, which are covered in Chapter IV, activities of several STI facilities are still largely confined to formulating strategies, finalizing infrastructure building, organization of workshops and handing out awards in order to encourage innovation with limited information on tangible advances made within domains set as national priorities.

Research centers active in the social sciences, including economics, business and education, such as King Faisal University's Institute of Research and Consultancy, appear to be active in providing advice to government ministries as well as NGOs, with particular regard to education and family affairs. It is likely that their services would become even more essential as further economic and social transformations take shape.

It is doubtless that years to come will see even more extensive changes take place in Saudi Arabia. Whether forthcoming changes will succeed in overturning overriding models, established and strictly adhered to over many years, is an open question. Given current regional and global developments and their political as well as economic ramifications, tightened spending may well be expected together with distinct changes in the order of research priorities with spending on defense-related research possibly gaining greater weight in the next few years.

## 3.5.6 United Arab Emirate's STI System; Policies and Institutions

The United Arab Emirates' STI system appears to be as open as its economy, with public sector's activities limited to support for universities and research facilities as well as partnerships with financial institutions and enterprise in setting up facilities such as Abu Dhabi's famed Masdar City and enterprise promotion facilities throughout the emirates.

The following subsections provide some information on salient features of the Emirates' STI the policy regime, institutions and their major activities.

### *3.5.6.1 The Emirates STI Policy*

Aimed at preparing the UAE for a postoil world, the STI policy formulated by the UAE and published in 2015,[340] is claimed as "a turning point in the country's march toward progress, economic diversification and prosperity." Thus, this national policy was designed to help:

- place the UAE among the top 10 countries in the world in the Global Innovation Index;
- enhance R&D expenditure three folds by 2021;
- increase the share of knowledge workers to 40% of the national workforce;
- advance rankings of its students in mathematics, science and reading to occupy a position among the top 20 countries by 2021;
- endogenous capacity building within selected STI domains, in support of the country's Vision 2021.

In what must be considered a welcome departure from common practice in several other Arab countries, the UAE established a high-level ministerial committee to oversee both the formulation as well as implementation of its STI policy.[341] This STI Committee is also granted the power to establish teams of experts and officials charged with monitoring the policy's implementation and reporting on annual progress based on a set of indicators.

In addition to overall monitoring the implementation of the country's STI policy and innovation strategy, tasks set for this Committee's activities include:

- monitoring the implementation of initiatives formulated in order to achieve the policy's goals;
- enhance coordination, cooperation and exchange of expertise among federal and local entities;
- engaging the private sector and ensuring that its social and economic contributions stimulate innovation.

Twenty-four focus areas for the United Arab Emirates STI policy were identified on the basis of criteria, including

- present and future needs as well as ability to contribute to tackling national and regional challenges;
- present and future international trends, with view to benefiting from emerging worldwide opportunities;
- the country's capabilities and unique assets, with the goal of having the country assume a global leading position with commensurate returns.

In addition to focus areas concerning basic services, i.e., education, health services, water, energy and food security, planned to receive new science and technology-based inputs, the list of focus areas includes a number of frontier science and technology areas, such as:

- Cubesats[342] and nanosatellites;
- Cybersecurity;
- Semiconductor process development;
- Robotics and artificial intelligence;
- Smart city applications and solutions;
- Internet of things and big data;
- Additive manufacturing (3D printing);
- Arabic digital technology;
- Aerospace advanced materials manufacturing, maintenance, and testing;
- Commercial unmanned aerial vehicles;
- Autonomous vehicles.

It is noteworthy that one of the reasons listed in support of choosing some of the high-tech areas is the fact that their inclusion would attract companies developing autonomous vehicle technologies to the country and enhance the country's transport sector.

A set of five enablers are identified as basis for implementing the UAEs policy, with goals that the UAE would seek to achieve regarding each one of these enablers. Table A.3.11 provides a list of enablers and goals that the UAE policy aims to tackle.

Apart from the intention of a threefold increase in R&D expenditure, a notable feature of the UAE policy is absence of numerical goals, e.g., with regard to the number of full-time researchers employed to promote various R&D activities or numerical indications of expected outcomes resulting from the policy's implementation. Additionally, the absence of a roadmap indicating stages and phases of the policy's implementation and what strategies might be used to attain certain goals are also notable features of the present version of the UAE's policy.

### *3.5.6.2 Innovation and Enterprise Promotion*

The United Arab Emirates' Vision 2021 emphasized a number of goals linked to innovative capacity building, including becoming one of the top 10 countries according to the world in the Global Innovation Index, and rendering the country "among the most innovative countries in the world, owing to advanced and innovative talent, resources, legislations and infrastructure." Essential outlines of innovation strategies are included in official documents seeking such objectives.

The UAE has also put forward a framework for a national innovation strategy in which innovative companies and institutions, innovative government and innovative individuals play essential roles as "innovation champions." Priority sectors set for the UAE innovation strategy are: renewable and clean energy, transportation, technology, education, health, water, and space. The UAE innovation strategy also highlights the need for: an innovation regulatory framework, infrastructural arrangements, investment, incentives, and enabling services as basic conditions in support of its innovation strategy.

Human, financial and technological capital are three essential pillars for any innovation ecosystem. Policies formulated by both Dubai and Abu Dhabi appear to target development of all of three pillars with some particular emphasis on the first. Thus, the UAE is a veritable melting pot with possibilities for tapping the experiences of professionals from a wide range of countries around the world.

The UAE now hosts over 120 institutions of higher education, most are private profit-making facilities encouraged to set up shop in the Emirates in order to cater for the large expatriate community.

On the other hand the UAE government, appears to have taken measures to increase investment in education rather steadily. Budget allocations to education represent more than 20% of the total government's budget, a considerably higher percentage than the 13% benchmark average. Primary, secondary, and tertiary education systems are reported to have been overhauled with several opportunities on offer for vocational training, resulting in the country's education index ranking going up in just 2 years, from 65th in 2011 to 15th in 2013.[343]

On the other hand, support systems for innovation in the form of enterprise incubators and accelerators, science and technology parks have started to emerge. Thus, according to a 2016 report by the Organization for Economic Cooperation and Development (OECD),[344] adequate resources are being committed by the Abu

Dhabi government to support entrepreneurship and SME development. Several institutions are also quoted as running programs specifically targeting enterprise development. A prime example of which is the Khalifa Fund for Enterprise Development (KFED), an agency established in 2007 and charged with entrepreneurship and SME development in Abu Dhabi. KFED was provided with a total capital of around US$545 million. This may well be viewed as a rather generous allocation, considering the small number of likely recipients, essentially Emirate nationals, and preferably resident in Abu Dhabi.

However, the same OECD report indicates that the KFED capital investment has progressively been depleted due to relatively high default rates on its loans. Thus, although adequate resources may be provided to launch an enterprise, success and sustainability will ultimately depend on how such resources are spent. Views reported by the OECD report as those of KFED and other government organizations, indicate that adverse conditions for enterprise creation through current schemes include risk aversion by local populations as well as preference for secure government jobs. More to the point is probably the fact that an estimated 60% of KFED beneficiaries are part time rather than dedicated entrepreneurs and that a dependency culture seems to have pervaded local populations. Reversing such trends would require much greater efforts at many levels, including inculcating an enterprise culture through various stages of the educational process as well as nurturing a sustainable entrepreneurial ecosystem.

Finally with regard to the technological pillar, several initiatives, including, for example the Masdar City and Masdar Institute, constitute attempts at building technological capacity, albeit largely with help from a variety of renowned overseas institutions. Two subsections below offer some information on the objectives and prospective activities of both the Masdar City and Masdar Institution.

### *3.5.6.3 STI Institutions in the UAE*

Promoting STI in the Emirates is distributed across a wide variety of both public and private institutions in the Emirates. With ministries and public authorities at the center of contractual arrangements involving supply of various STI inputs in various sectors. It is highly likely, however, that such arrangements rarely involve local actors within the country's STI institutions.

The Emirates Ministry of Higher Education and Scientific Research performs a central role in promoting scientific research in the Emirates. In order to organize its support for scientific research throughout the Emirates, the Ministry established in March 2008 a National Research Foundation (NRF), whose main objective is to provide research grants on competitive bases to credible researchers and research teams from public and private universities, research centers and even firms. NRF is also mandated to provide advice to the Ministry on matters related to national research activities as well as recommend the creation of research facilities.[345] Research proposals submitted to NRF are subjected to several levels of evaluation involving international peer reviews. Furthermore, the selection process also takes into account evidence that a proposal offers distinct socioeconomic benefits.

#### 3.5.6.3.1 The United Arab Emirates University

The United Arab Emirates University (UAEU) at Al Ain, was founded in 1976 following the Emirates independence as the first public institution of higher education. Current programs cover a wide variety of specializations including science, medicine and health studies, information technology, food and agriculture as well as engineering. UAEU is home to nine research centers undertaking research in a wide variety of areas, including space science and technology, the health sciences, genetic engineering and biotechnology, water, energy and date palm agriculture.[346] UAEU also houses a Science and Innovation Park, about which limited information seems to be available.[347]

A number of prestigious overseas universities have established—or are planning to establish—campuses in the Emirates, in consultation with the Abu Dhabi Education Council (ADEC). Examples include the Sorbonne University of Paris, the INSEAD Centre for Executive Education and Research. Additionally, the Munich Technical University and the University of Bonn are reported as jointly planning to establish college of medicine and the sciences on the Abu Dhabi University campus.

Both directly and indirectly, the above institutions of higher education will help train researchers, technologists, and prospective innovators. However, a review of faculties and colleges established by several institutions seem to indicate a preponderance of courses in business administration, commerce, and related topics, with less attention to science and technology specializations.

#### 3.5.6.3.2 Masdar City

A principal objective of the Masdar City project, which was established by the Emirate of Abu Dhabi in 2006, is to merge R&D, investment and sustainable living, thereby sowing the seeds for an evolving clean energy industry and catering for the transformation toward the fossil fuel free economy of the future. The project is also viewed as a step toward implementing Abu Dhabi's Economic Vision 2030, which is focused on transitioning the emirate to a knowledge-led economy.

Masdar City is essentially a commercially driven renewable energy project coupled to a strategic government initiative. Masdar's mission enshrines enterprise incubation and the establishment of a clean energy industry in the country and around the world.

The Masdar City project has benefited from an initial commitment amounting to US$15billion by the Emirate of Abu Dhabi. Its accomplishments are advertised as including:

- investments in clean energy projects valued at US$8.5 billion globally, around one third of which were undertaken within the Masdar City;
- worldwide renewable energy projects with a combined output amounting to 2.7GW (GW).[348]

The Masdar City activities are claimed as having resulted in the displacement of nearly 2.5 million metric tons of carbon dioxide emissions per year.

Available literature does not indicate how decision making is carried out with regard to various aspects of the Masdar project's running as well as investment activities. However, the fact that Mubadala, an established investment company is a key partner, might indicate that short and medium returns on investment would play an important role in decision making. On the other hand, due to subvention by the Abu Dhabi government and in relation to its Vision 2030 commitments, it is likely that the Masdar City project would also be capable of taking up initiatives that may not be directly linked to such returns.

#### 3.5.6.3.3 Masdar Institute

Masdar Institute is a graduate research university dedicated to renewable energy and sustainable technologies education and research. Research themes taken up by the Institute include water, environment and health, energy systems as well as microsystems and advanced materials, with research projects aimed at tackling issues within the following domains:

- Biomass-derived fuels;
- Smart-grid applications;
- Saltwater desalination;
- Energy-efficiency technologies.

The Masdar Institute's vision statement emphasizes its quest for setting new standards in learning and research, with benefits for Abu Dhabi and the world at the core of its activities. The Institute's activities revolve around interdisciplinary research and collaborative technology development in advanced energy and sustainability. Inculcating an innovative spirit among students in order to "develop technology and enterprises in the region and globally" is declared as a priority.

Masdar Institute's website indicates the relative quality of research output produced by the fact that in 2016 it was ranked among the top 20 academic institutions in Western Asia. However, the article count reported by Nature Index for the duration 1 February 2016–31 January 2017 indicates that the Institute published a total of only 4 articles in journals covered by Nature index.[349] Incidentally, top articles published by the Institute during this period were in the area of new materials involving membranes for water desalination. The same source mentions that no collaborative links existed between the Institute and other research facilities in the country of the region. Rather, all collaborative research activities were confined to institutions in the United States and Italy.[350]

### *3.5.6.4 Incubators and Accelerators*

In addition to the facilities established by several HEIs in the UAE with the aim of promoting innovation and enterprise promotion, such as the above-mentioned Science and Innovation Park at the United Arab Emirates

University, several business incubators and accelerators have been established. Thus, private sector facilities designed to offer space for office and industrial activities, mentoring as well as access to investors have been set up over the past decade. Examples include, impact hub and flat labs. Some of these facilities display special interest in promoting digital enterprises. Additionally, the Dubai government set up a 100% government-owned free zone, the Dubai Silicon Oasis (DSO), 2005 with the Mission to "facilitate and promote modern technology-based industries." In effect, DSO is a technology park, providing integrated living and working facilities community.

### *3.5.6.5 Concluding Remarks*

The United Arab Emirates made considerable progress in reducing its dependence on oil exports through developing tourism, transport, construction, and financial sectors is responsible for several developments in its STI infrastructure and related activities. In seeking leading international positions as tourism, transport and green technology hubs both Dubai and Abu Dhabi, in particular, had to acquire a host of technological capabilities.

However, based on available literature, it is not at all clear to what extent related STI inputs have in fact been acquired and internalized by Emirati institutions, with possibilities for further adaptation and autonomous future development. In fact, it is highly likely that open business models adopted by the Emirates would mean that STI inputs implemented with minimal or no involvement by local actors may not actually be accessible for wider dissemination and future development for reuse in similar settings both at home or abroad.

In concluding this section on STI policies, institutions and activities in the Arab countries of the Gulf region the following remarks are in order.

Disparities are evident with particular regard to skill and technology indicators and the gap that separates the Gulf countries when compared to similarly, even to less endowed countries that have been able to advance rapidly up the development ladder over the past few decades.

All recent development policies and vision documents produced by the Arab Gulf countries emphasize their quest for open knowledge economies and societies in which progress and participatory approaches predominate, whereas traditional values continue to be central in all development activities. To achieve this central goal education is allocated a high order of priority by all policy and vision documents.

Thus, the UAE policy document, for instance, is focused on "transitioning [the country] into a knowledge and innovation based economy independent of oil resources, and on enhancing scientific and technological innovations in its public, private and academic sectors." The UAE policy also recognizes manpower requirements for attaining its objectives by declaring as one of its objectives "seeking to increase the share of knowledge workers to 40% of total workforce."

Additionally, STI policies produced by GCC countries appear to rank education among their top priorities. However, absence of quantitative goals regarding achievements in the educational sector is generally the rule. Among notable exceptions to this general tendency is the UAE, whose STI policy vows to "advance the rank of its students in mathematics, science and reading to become among the 20 highest ranked countries by 2021."

Yet insufficient attention is paid to modalities for building and enhancing local STI capabilities, whether such capabilities are being built and acquired by public or private sector institutions. Additionally, excessive monetization of inputs into all aspects of development brought about by a predominant role for the private sector is likely to limit continued progress in STI capacity building and dissemination.

In essence, in order to attain their stated objectives, the Arab Gulf countries will need to promote local skills and build capabilities for adapting and further developing technologies of particular importance to their specific needs, thereby narrowing the technological gap and enhancing their own inclusive socioeconomic development.

## 3.5.7 Yemen's STI System

Modern higher education in Yemen dates back to the 1970s when the countries two major universities were established in Sana'a and Aden. The first faculties to be established were concerned with literary, religious and legal studies. Science and engineering faculties followed with several of their graduates selected to continue further studies abroad. Resources allocated to research to spare for scientific research and technological development, even when such activities are sorely needed to deal with catastrophic conditions of drought, disease, and malnutrition.

The following paragraphs present an overview of STI policies, institutions, and activities in Yemen, followed by a section aimed at a brief examination of higher education and research systems in Comoros, Djibouti, and Somalia.

### *3.5.7.1 Yemen's Strategy for Higher Education and Scientific Research*

A document intended as basis for reforming higher education by the Ministry of for Higher Education and Scientific Research,[353] included several elements directly aimed at enhancing the country's scientific research capabilities.[354] In discussing scientific research in Yemen, the strategy document, which is intended to cover the period 2006–25, points out quite clearly that scientific research did not receive a great deal of attention at Yemen's universities. The document further stated that national universities lacked a research tradition and culture. Means for carrying out research activity, including equipped laboratories, technical staff, periodicals and other reference material, are also characterized as insufficient. Additionally, the document clearly indicated that lecturers simply did not possess sufficient incentives for carrying out research activities and that publications produced by some members of teaching staff were primarily aimed at gaining promotion, rather than attempting to implement their university's research strategy.[355]

The absence of a national entity dedicated to "promoting, organizing, and monitoring scientific research" activities is quoted by the strategy document as a major deficiency. Accurate information concerning the quantity, quality of research projects undertaken and the areas they target is considered an additional obstacle.

However, the Strategy document acknowledges the importance of pursuing scientific research for two essential purposes:

- expanding and improving "knowledge capabilities" of university teaching staff;
- help resolve socioeconomic problems faced by Yemen.

However, to achieve tangible benefits in these domains the Strategy document emphasizes the need to introduce improved resource allocation and planning.

The small number of students opting for higher study courses is indicated as an added constraint. Thus, the number of students pursuing courses of higher study at the University of Sanaa, at the time the Strategy document was authored, was limited to around 600 students, against nearly 77,000 students registered for BSc courses. Their total number in the entire country was put at around 1500.

Although no reliable figures are available for the number of scientists and technologists undertaking research activity, it is highly likely that their numbers are exceedingly small. Additionally, a 2008 study that relied on government data from 10 Arab countries indicated that women accounted for 40% of researchers in Egypt and Kuwait, 30% in Algeria and Qatar, and 20% in Morocco and Jordan. However, their numbers fell to low levels of between 14% and 4% in Oman, Yemen, and Mauritania.[356]

The ministry's strategy document is openly critical of its own arrangements with regard to supporting scientific research, citing outdated regulations, complex procedures, inadequate coordination and interference in decision making, and calling for a complete overhaul of such arrangements in a manner ensuring transparency and accountability.

Delving further into root causes as to why scientific research fares so poorly in Yemen, the strategy document quotes:

- absence of a clear vision, a well-defined mission as well as strategic planning;
- limited financial resources coupled to constraints on how such resources are ultimately spent;[357]
- inadequate infrastructures, including communications networks and student advisory services;
- lack of regulations aimed at quality assurance and academic accreditation.

In addition to reiterating customary obstacles facing scientific research at the country's universities, the Strategy document underscores difficulties such as limited resources and weak links to the private sector and also the following:

- absence of interaction between students and academic staff as well as limited cooperation between administrative and academic staff;
- lack of transparency in the selection and appointment of academic staff and assistants;
- breaking laws and regulations governing political party activities within academic institutions.

Challenges that scientific research at the country's universities would need to tackle are summarized as including:

- the need to access resources other than what is provided by government institutions;
- stemming the flight of qualified academic staff to seek better opportunities abroad;
- resistance to reform, indeed any attempt at introducing change, to the national system of higher education.

Based on the above, the Strategy document proposed several measures as means for promoting scientific research at the country's universities, including:

- comprehensive review of scientific research activities undertaken at all Yemeni universities;
- allocation of a special budget by the Ministry for promoting research activities, with emphasis on research leading to expanding knowledge boundaries rather than merely producing articles in support of educational processes;
- establishing a National Research Authority (NRA) and entrusting it with formulating a comprehensive scientific research strategy, targeting national priorities as well as securing adequate resources and improving research infrastructures.

The Ministry's Strategy document is undecided about links that the proposed National Research Authority should have to the Ministry of Higher Education and Scientific Research, i.e., whether it should be part of the Ministry itself, or an independent entity while maintaining close links to the Ministry. However, the need to have top Yemeni academics with international experience as well as representatives from universities and the private sector as well as the national industrial production. The document, nevertheless, stressed the need for NRA to maintain strong links to the Ministry, and proposed that its general secretariat is "assumed" by the Ministry and that it is headed by a deputy to the Minister of Higher Education and Scientific Research.

The strategy document also included estimated running costs, covering years till 2025, based on a set of assumptions, which may be supposed to include increasing enrolment, and levels as well as commitments to running scientific research capabilities. Fig. 3.31 summarizes related growth, indicating deficits approaching US$500 million by 2025.

The Strategy document further proposed that universities establish special units mandated to coordinate and monitor research activities undertaken by university staff. These units would be presided over by the deputy chancellor for scientific research.

Additionally, the Strategy document required that NRA establish special committees to take care of priority research domains and address questions relating to financing research activity, renovation of related infrastructures, with strong emphasis on economic development.

### *3.5.7.2 Scientific Research in Yemen*

Measured by its published output in international journals, scientific research undertaken in Yemen appears to be very limited indeed. Thus, Yemeni publications in international journals amounted to around 0.46% of the total number of publications produced by all the Arab countries during 1996–2010. Yet, at around 5.8, the average number of citations these publications received per publication is equal to the average for the entire group of Arab countries, during this period. However, the H-index gained by Yemen on account of the above publications was estimated at around 32, compared to 116 for Egypt and 105 for Saudi Arabia and an average of around 54 for all the Arab countries.

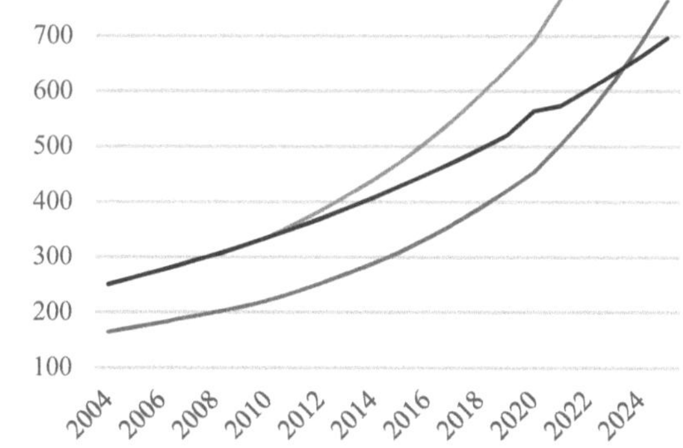


**FIGURE 3.31** Projections for the number of students enrolled in Yemen's public higher education (thousands), running costs and available resources (US$ millions).

Scientific research in Yemen is principally undertaken at the country's universities. With agriculture faculties among the more active in applied research directed toward a variety of crops, animal husbandry and other issues including agricultural pests. As the following paragraph indicates, however, agricultural institutions and their activities, initially beset by considerable shortcomings in the past, went on to suffer debilitating setbacks, as hostilities took over the country.

### 3.5.7.2.1 Agricultural Research in Yemen

A report by a group of experts,[358] titled "A Review of the Agricultural Research System in the Republic of Yemen" was presented in 1993, to the Ministry of Agriculture and Water Resources. The report indicated that although modern agricultural research in Yemen was initiated only two decades previously, it had undergone a significant structural and organizational changes. The Agricultural Research and Extension Authority (AREA) was essentially the major institution dedicated to agricultural research, at the time the above report was written. AREA was essentially a semiautonomous national research facility governed by a board of directors, which was chaired by the Minister of Agriculture and Water Resources and enjoyed a mandate to conduct and supervise agricultural research throughout the entire country. It enjoyed over 200 professional staff members, of which 23, 67, and 83 had PhD and MSc and BSc degrees, respectively, as well as 34 expatriates associated with externally funded projects.

Additionally, faculties of agriculture at Aden and Sanaa universities were also involved in agricultural research. However, only limited information could be retrieved concerning their activities. On the other hand, plenty of press reports were available indicating rampant corruption as well as destruction and looting of office and field facilities established, for example at an agricultural research center in Thimar by the University of Aden's College for Agricultural Sciences, as well as confiscation of field plots established by experimental stations, during 2012–14.[359] Recommendations presented by the report included the need for AREA to

- assume a policy making rather than a managerial role, with an internal management committee to take responsibility of managing day-to-day research and extension services;
- improve linkages to university research units as well as specialized regional and international organizations;
- direct donor funding toward activities identified as a result of comprehensive and approved policy and strategy documents;
- enhance human resource management practices in order to access and retain talented and well-trained staff;
- develop and implement effective monitoring and evaluation system of research activities, in order to ensure their relevance and high quality in the service of overall development;
- make better use of information technologies in the management of research programs as well as the dissemination of technical information on research activities.

All of the above recommendations would naturally need to be embedded within a comprehensive policy document with related implementation strategies and executive plans, including the customary benchmarks and evaluation modalities.

### *3.5.7.3 Initiatives Aimed at Improving Scientific Research in Yemen*

Under the title "World Bank Provides US$13 million to Higher Education and Scientific Research in Yemen;"[360] details were given on the World Bank's official website, in 2010, of a project aimed in support of the country's higher education and scientific research. This project was put forward in response to challenges faced by Yemen's higher education and scientific research and in line with the Ministry's "National Strategy for the Development of Higher Education," which was designed expressly with the aim of addressing these challenges.

According to the above World Bank's report,[361] the higher education sector in Yemen was expanding, with enrolments in growing rapidly due to the

**TABLE 3.23 Components and Objectives of World Bank's 2010 Project in Support of Higher Education and Scientific Research in Yemen**

| Project Components | Objectives | Allocation (US$) |
|---|---|---|
| Quality improvement in university programs | Facilitate the establishment of a sustainable mechanism for channeling financial resources to institutions of higher education, leading to significant qualitative improvement in teaching and learning, in line with the country's social and economic priorities. | 8.2 |
| Quality assurance | Enhance the accountability of institutions of higher education by reinforcing the development procedures for quality assurance and promoting a quality assurance culture. | 1.3 |
| Institutional capacity development | Strengthen capacity for planning and monitoring at the Ministry as well as Yemen's public universities, thereby enhancing trustworthy management and enhancing accountability. | 1.4 |
| Monitoring, evaluation and project management | Manage, monitor and assess activities undertaken under the other three components. | 2.1 |

From "World Bank Provides US$13 Million to Higher Education and Scientific Research in Yemen." Published in April 2010. Downloaded in May 2017, from: http://www.worldbank.org/en/news/press-release/2010/04/25/world-bank-provides-us13-million-to-higher-education-and-scientific-research-in-yemen.

(a) establishment of new private universities and (b) introduction of fee-paying parallel programs within the country's public universities. However, enrolments have concentrated in the social science within specializations that bear limited if any relevance to either local or regional labor needs.

The project was designed for implementation within four components, as indicated in Table 3.23, and was to deliver benefits to all of Yemen's eight public universities, namely the universities of Aden, Amran, Dhamar, Hadramout, Hodeidah, Ibb, Sana'a, and Taiz.

### 3.5.7.4 Concluding Remarks

Formulating national development policies aimed at tackling Yemen's socioeconomic problems has been a constant refrain at many national, regional and international conferences over the past decades. The most recent report on the status of sustainable development in the country[362] included the customary calls for improving access to selected information and communication technologies, remote sensing and higher allocations for scientific research at public universities. On socioeconomic issues, understandable emphasis is placed on reducing poverty, providing wider employment opportunities, access to clean water, better health services, improved educational and vocational training facilities, as well as combatting corruption. Action on all these fronts would require science, technology and innovation inputs and would be greatly facilitated and enriched should viable STI policies be in place.

That numerous similar documents with comparable directives were issued during the past with little or no tangible outcomes, engenders a great deal of skepticism as to their utility in situations where deficiencies abound at all levels of governance. Additionally, the fact that this document, along with many of its predecessors, expresses generally excellent ideas, with no reference to numerical goals, benchmarks and evaluation modalities immensely reduces its utility.

Hopes are often, pinned on help by international development organizations as well as by NGOs, for ensuring that goals such as the ones expressed in the above document are achieved. However, what has certainly gone missing in documents such as the above as well as in initiatives by aid agencies are components that specifically target local as well as national capacity building alongside delivery of relief and development aid. The realization that so much, and no more, may be done by outsiders providing aid for central government authorities seems to have escaped many so far.

Hopes are sometimes entertained on the basis of contributions, essentially in terms of expertise and guidance by expatriate Yemenis, some of whom work as scientists and engineers in the GCC countries as well as other countries of South Asia, Western Europe and the United States. Initiatives along the line of the Palestinian TOKTEN program should be designed and provided with the means for making tangible impacts. There should also be room for ad hoc arrangements with institutions and groups of individuals. However, here again, tangible and long-lasting benefits will materialize only as long as capacity building is targeted at local as well as national levels along with providing scientific and technological knowledge.[363]

## 3.6 SCIENCE, TECHNOLOGY, AND INNOVATION IN COMOROS, DJIBOUTI, AND SOMALIA

All three Arab countries of East Africa have had more than their fair share of political turmoil and lagging socioeconomic development. However, Somalia, in particular stands out as it still faces sporadic outbreaks of violence, that severely interrupt all civilian activity including education at the country's numerous institutes of higher education and whatever research activity might be undertaken on their premises.

All three countries did nevertheless take significant steps over the past few years to build up their STI capabilities. Furthermore, they must be recognized as possessing valuable STI human capital both at home and abroad in neighboring African and Arab countries as well as in Europe.

Research activity in all these three countries was confined to their universities and has often been mediated or supported by external aid agencies and donors. Out of all three countries recent interest in Djibouti's role as a transport hub holds out the promise of future STI-related development in the short and medium terms. An evolving role for Comoros as a tourist destination may also lead to similar developments in the medium term. In Somalia's case, hopes must rest with its educational system to institute much needed reforms, guided by urgent socioeconomic development needs and aided as far as possible by international aid agencies, with capacity building on the ground wherever possible.

The following sections will attempt an overview of STI capacity with emphasis on activities undertaken at the countries' universities.

### 3.6.1 The Comorian STI System

The University of the Comoros (UDC) was created in 2003 as a public institution of higher education with administrative and financial autonomy. UDC includes four faculties, dedicated to Arabic letters and Islamic sciences, the arts and humanities, law and economics, and science and technology. Additionally, UDC has also set up two institutes; namely the University Institute of Technology (IUT), and the Institute for Teacher Education and Research in Education and a school of medicine and public health.

The Agriculture, Fisheries, and Environment Research Institute (INRAPE),[364] is charged with agricultural research but possesses very limited capacity, in terms of human and material resources. Thus the only research activity it was able to carry out in the 1990s, for example, was through a small French-financed project, designed to monitor various farming systems in the country. Further support will certainly be needed if INRAPE is to carry out its mission.[365]

The Ministry of Agriculture, Fisheries, and the Environment is also reported as taking part in surveys and research work, funded by donor and international development agencies with particular regard to agricultural pests as well as compliance with environmental standards with regard to the use of chemical pesticides.

## 3.6.2 Djibouti's STI Scene

Education is considered a priority by Djibouti's government.[366] Expenditure on public education accounted for 4.5% of GDP in 2010. Schooling is free and 7 out of 10 children now attend primary school, although the ratio is higher for boys than for girls.

The University of Djibouti was established as recently as 2006, as a public HEI. The University is officially accredited by Djibouti's Ministry of Higher Education and Research. A strategy document posted by the Ministry[367] indicates that around 8000 students enrolled in higher education course during 2014–15. The University is reported as having established five faculties, providing higher degree courses in

- Law, administration, and economics;
- Science;
- Humanities, languages, and social science;
- Medicine;
- Engineering.

The latter faculty opened in 2013, and plans were announced for setting up a faculty of architecture and a business school in 2015–16.

The above-mentioned strategy document also indicated that plans were being made for the creation of a business incubator at the University of Djibouti.

### *3.6.2.1 Research Activity in Djibouti*

The 2015–19 Strategy Document produced by Djibouti's Ministry of Higher Education and Research underscores the importance of research as a driver of development and provided two examples of research work that led to practical results. The first example concerns geophysical studies which led to the determination of geothermal deposits, and the identification of an exploitable and cost-effective resource. The second example concerns applied research carried out on the use of wind power to launch a sea water desalination project.

National research priorities are listed by the Strategy Document as including:

- Feasibility studies aimed at exploitation of geothermal sources;
- Evaluation of water resources at the a volcanic aquifer in Djibouti;
- Identifying locations of likely oil reserves in the country.

However, the Strategy document further indicates that if research is to produce quality results then it should be adequately supported and financed; as if to indicate that funding and other forms of support were unavailable, or in short supply.

## 3.6.3 Somalia's STI Scene

Many aspects of socioeconomic development came to standstill as political turmoil and armed conflict invaded Somalia's terrains.

Data about the Somalia's education system are scarce and often outdated, particularly with regard to higher education. Somalia's education system effectively collapsed amid political turmoil followed by civil war in the early 1990s, with the majority of qualified teachers and trainers fleeing the country. Rebuilding the country's education system was taken up by actors such as local communities, NGOs and the Somali diaspora. As a result the country witnessed an unprecedented and unregulated proliferation of private universities, with around 40 operating in Mogadishu alone, up from only 15 in 2011. Business administration, health studies, and Islamic studies appear to be most prevalent topics of undergraduate study. Postgraduate programs are rare, but may be found in some of the larger universities, at Mogadishu University for example.

Despite the reopening of Somalia's National University (NUS) in 2014, and the formulation of a draft "National Vision for Higher Education," the role of government is still minimal. Nevertheless, the Ministry of Higher Education and Culture is officially in charge of sectoral education strategies. A European Union aid package has pledged US$60 million investment in the sector over 3 years.

### *3.6.3.1 Research Activity in Somalia*

Scientific research in Somalia is hampered by inadequate resources, including funding, libraries, and access to the internet. Some research activity is undertaken, nevertheless, by research institutions established by NGOs with support by donor organizations. An example of such centers is the Heritage Institute for Policy Studies, a think tank based in Mogadishu, with a very small number of staff members, whose focus is maintained on political and governance issues. The same is true of a project funded jointly by Norwegian Research Council and Somali Stability Fund, which employs a very small group of qualified staff with focus on gender in Somalia's politics. The Mogadishu Centre for Research and Studies, established at the Islamic University in Mogadishu, is funded by a group of businessmen and conducts activities with focus on journalism, surveys, and quantitative research.[368]

In summary, while some of Somalia's universities are reported as undertaking mainly policy-oriented research for donor organizations on consultancy basis, principal obstacles facing research activity include:

- severe lack of funding as well as research facilities;
- limited or no institutional access is available to reference material, electronic books or journals;
- academic staff need to take on many roles in addition to teaching, including administration and management.

Desk research and studies as well as research based on surveys of social or cultural issues is practiced but only to a limited extent, because such research carries special risks of encroaching on political and cultural sensitivities. As a result, researchers are known to have declined offers to conduct such research activities when requested by university authorities, NGOs, or donors, or to withhold publication of their research outcomes.[369]

Research topics commonly addressed, upon donor and NGO requests, include governance, reconciliation, migration, humanitarian issues and health. On the other hand, research on issues in agriculture, economics and environmental issues are accorded far less attention, because they might require higher investments.

On the innovation and enterprise creation front, an innovation hub was established in 2012, with focus on mobile and internet services and incubation accompanied by training with the principal aim of promoting social enterprises. The hub was set up by a South Africa based social enterprise in partnership with a UK-based foundation.

### 3.6.4 Concluding Remarks

Research and development activity in all three countries is at best rudimentary. Table 3.24 presents the number of scientific publications produced by researchers in Comoros, Djibouti, and Somalia over the period 2008–14, with and without international research collaboration. Table 3.24 clearly indicates the low level of scientific output that could find its way to the international academic literature as well as almost total reliance on international research collaboration for producing such output.

While little information is available on STI capacity building ventures by any of the three countries, judging by expenditure on education as a percentage

**TABLE 3.24** Scientific Publications by Researchers in Comoros, Djibouti, and Somalia and International Research Collaboration 2008–14

| | Total Number of Publications | Publications With international Authors | Percentage of Publications With International Authors |
|---|---|---|---|
| Comoros | 18 | 18 | 100 |
| Djibouti | 51 | 45 | 88.2 |
| Somalia | 20 | 20 | 100 |

From "The Science, Technology and Innovation System of the Kingdom of Morocco; " A UNESCO document authored by N. Boshoff. The document is not dated, but appears to have been published post 2007. Downloaded in July 2016 from: http://portal.unesco.org/education/en/files/55536/11998897825Morocco.pdf/Morocco.pdf.

of GDP, it would seem that both Comoros and Djibouti spend amounts higher than the average for all the African countries. Thus, Comoros spent 7.6%, in 2008, and Djibouti spent 4.5%, in 2010, of their public resources on education, whereas the average for all African countries was estimated at 4%.

With regard to expenditure on tertiary education as a percentage of GDP, an indicator that should correlate more closely with STI capacity building, Comoros is credited with having spent an estimated 1.1%, in 2008, whereas Africa's average came to about 0.8%. On the other hand, Djibouti's public expenditure on tertiary education in 2010, amounted to 0.7% of GDP. Expenditure on tertiary education as a percentage of expenditure on all levels of education, by both Comoros and Djibouti, is considerably less than Africa's average, which comes to around 22% – at 14.6 and 16.5, respectively. Incidentally, no figures are available for Somalia on either of the above indicators.

Whatever technological development takes place in all three countries remain confined to trade and industrial initiatives they were meant for, with little or no dissemination into surrounding communities or areas of application. Low literacy levels and language difficulties play a part in low dissemination of science and technology inputs. However, the fact that no policies have been worked out for such dissemination to take place in the first place, should be confronted with a far greater share of the blame.

Table 3.25, which includes gross enrolment ratios in all three levels of education in Comoros, Djibouti, and Somalia, paints a rather dim picture of future prospects, particularly with regard to education in Somalia, and particularly in relation to secondary and tertiary education. That figures for Somalia effectively date back to 2007, whereas those for Comoros and Djibouti are for 2013 and 2012, respectively, should be taken into account. The very fact that no recent figures are available on educational enrolment in Somalia is in itself telling.

Lastly, along with all other Arab countries in Africa and the Middle East, future development of STI capabilities requires vast improvements in statistical abilities for monitoring STI indicators as bases for sound STI policy formulation.

## 3.7 CHAPTER'S CONCLUDING REMARKS

For successful STI systems to emerge and function there is need for sound policies, a nurturing environment, adequate funding and a constant supply of well-trained human capital. Although the latter ingredient is well within reach in almost all of the Arab countries, the first two are seldom so, while adequate financial resources are limited to a few of the Arab countries in the Gulf region.

As a result, STI systems in the Arab countries function far below expectations. Should they continue to operate in a similar fashion then they may not be counted upon to meet the numerous challenges faced by the Arab countries on the economic, social and environmental fronts.

**TABLE 3.25** Gross Enrolment Ratio for Education in Comoros, Djibouti and Somalia in 2012[a]

| | Primary | | | Secondary | | | Tertiary | | |
|---|---|---|---|---|---|---|---|---|---|
| | Male | Female | Total | Male | Female | Total | Male | Female | Total |
| **Comoros** | $105.9^{+1}$ | $99.9^{+1}$ | $103.0^{+1}$ | $62.8^{+1}$ | $65.0^{+1}$ | $63.9^{+1}$ | 10.6 | 9.1 | 9.9 |
| **Djibouti** | 73.1 | 65.9 | 69.5 | 49.4 | 38.1 | 43.8 | $5.9^{-1}$ | $4.0^{-1}$ | $4.9^{-1}$ |
| **Somalia** | $37.6^{-5}$ | $20.8^{-5}$ | $29.2^{-5}$ | $10.1^{-5}$ | $4.6^{-5}$ | $7.4^{-5}$ | – | – | – |

[a]*–n/+n: figures in superscript refer to years before or after 2012.*
From "The Science, Technology and Innovation System of the Kingdom of Morocco; " A UNESCO document authored by N. Boshoff. The document is not dated, but appears to have been published post 2007. Downloaded in July 2016 from: http://portal.unesco.org/education/en/files/55536/11998897825Morocco.pdf/Morocco.pdf.

Many countries in the region did formulate future STI policies. However, there are many cases in which these policies did not go beyond establishing broad guidelines for action, they generally lacked elaborate implementation strategies, nor were they designed with inbuilt measures for evaluation. In any case, very few such policies were effectively adopted as binding documents to which concerned institutions must adhere. The absence of robust sectoral development policies as well as the fact that many national development policies tend to suffer similar shortcomings and are often subject to resource constraints, particularly in the less endowed Arab countries deprive STI policies of frames of reference to operate within.

With the exception of only a few Arab countries, most research activity is conducted within institutions of higher education, which suffer limited resources to start with, and with teaching duties constantly competing with research agenda. At any rate, for many young women and men in several countries of the Middle East and North Africa, doing research at their national university is merely a training opportunity for joining their compatriots in the diaspora.

Significant collaboration does seem to take place between only a few Arab STI institutions and their counterparts abroad, whereas limited collaboration appears to take place among institutions within the region. In compensation, a good deal of informal collaborative activity does take place through university professors and researchers from a variety of Arab countries in the Levant and North Africa working in the Arab countries of the Gulf region.

With rare exception, industry–university STI collaboration is generally absent, resting on ad hoc arrangements rather than regular channels.

## ANNEX TABLES

**TABLE A.3.1** Egypt's Grant Programs, Values, and Durations

| Grant Programs | Maximum Grant Value (LE)[261] | Maximum Duration (Years) |
|---|---|---|
| Basic and Applied Research | 1,000,000 | 3 |
| Young Researchers' Program | 750,000 | 3 |
| Reintegration of expatriate scientists | 1,500,000 | 3 |
| Innovation | 3,000,000 | 3 |
| Research Support and Technology Development | 100,000 | 1 |
| Capacity Building | 5,000,000 | – |

*Continued*

**TABLE A.3.1 Egypt's Grant Programs, Values, and Durations—cont'd**

| Grant Programs | Maximum Grant Value (LE)[261] | Maximum Duration (Years) |
|---|---|---|
| Faculty for Factory Program | 360,000 | 2 |
| Centres of Scientific Excellence | 10,000,000 | 3 |
| Short term fellowships[262] | 38,000 | – |
| Demand Driven Projects | 3,000,000 | 3 |
| Support of Scientific Events | 50,000 | – |
| STDF Incubation Program | – | – |
| Joint Projects for Applied Research | – | 2 |
| Egypt–Japan Cooperation Research Projects | 145,000<br>JPY2,500,000[263] | 2 |
| Egypt–Japan Cooperation Seminars | JPY1,500,000 | – |
| Egypt–France Cooperation[264] | €100,000 or<br>€ 300,000[265] | 3 |
| Egypt–Germany Cooperation (young scientists) | €10,000 | 1[266] |
| Egypt–Germany Cooperation (Joint Research) | €200,000 | 2 |
| **Egypt–United States Cooperation** | | |
| • Junior scientists | US$20,000 | 9 months |
| • Research Planning and Development | US$60,000 | 1 |
| • Collaborative Research | US$200,000 | 3 |
| • Integrated Science Grants | US$500,000 | 3 |
| • Science and Technology Innovation | US$250,000 | 2 |
| AFRICA–EUROPE Cooperation[267] | – | – |

**TABLE A.3.2 Algerian Public and Corporate Research Centers**

| **Public Research Centers** | |
|---|---|
| Centre de Recherché en Astronomie Astrophysique et Géophysique, CRAAG | Astronomy, astrophysics, and geophysics |
| Centre de Développement des Technologies Avancées, CDTA, | Advanced technologies |

**TABLE A.3.2 Algerian Public and Corporate Research Centers—cont'd**

| | |
|---|---|
| **Public Research Centers** | |
| Centre de Développement des Techniques Nucléaires, CDTN | Nuclear technique development |
| Centre de Recherché et d'Exploitation des Matériaux, CREM | Materials |
| Centre de Recherché Scientifique et Technique en Soudage et Contrôle (CSC) | Welding and control |
| Centre de Radio protection et Sûreté, CRS | Radiology |
| Centre de Développement des Énergies Renouvelables, CDER | Renewable energies |
| Centre National des Techniques Spatiales, CNTS | Space technologies |
| Centre de Recherché sur l'Information Scientifique et Technique, CERIST | Science & Technology Information |
| Centre de Recherché Scientifique et Technique en Analyses Physico-Chimiques, CRAPC | Chemistry |
| Centre National de Recherché Appliquée en Parasismique, CGS | Seismic and earthquake studies |
| Centre de Recherché des Industries Agro-Alimentaires Agro-research | Agricultural and food industries |
| Institut Technique de Développement de l'Agronomie Saharienne | Saharan agriculture |
| Commissariat au Développement de l'Agriculture des Régions Sahariennes | Saharan agriculture |
| Centre d'Études et de Recherché Appliquées et de Documentation pour la Pèche et l'Aquaculture | Fishing and aquaculture |
| **Corporate Research Centres** | |
| Centre de Recherche-Développement de SONATRACH | Hydrocarbons |
| Centre de Recherche-Développement ANABIB | |
| Centre de Recherche-Développement SAIDAL | Pharmaceuticals |
| Institut Algérien du Pétrole | Part of SONATRACH |
| Centre d'Études et de Recherché Minière EREM | Mining |
| Centre de Recherché pour la Valorisation des Hydrocarbures et de leurs dérivés (CERHYD) | Part of SONATRACH |

"Mapping research systems in developing countries.; "A UNESCO Country report: The Science and Technology system of Algeria. Accessed in April 2017 at: http://academic.sun.ac.za/crest/unesco/data/Algeria.pdf.

**TABLE A.3.3 Research Networks Established by Moroccan Scientists and Engineers**

| Network | | National Focal Point |
|---|---|---|
| PCQ | Network of Competence Centres for Quality | Ecole Supérieures de Technologie (Casablanca) |
| REMER | National Network for Marine Science and Technologies | Faculté des Sciences (El Jadida) |
| RUPHE | National High Energy Physics Network | Faculté des Sciences Ain Chock (Casablanca) |
| RUSTE | University Network for Space Science and Technology | Ecole Mohammadia d'Ingenieurs (Rabat) |
| MisoBiop | Competence Centre for Soil Microbiology and Plant Biotechnology | Faculté des Sciences (Rabat) |
| STIC | Network of Competence in Information and Communication Sciences & Technologies | Faculté des Sciences (Rabat) |
| PCE | Network of Competence in Economics | Faculté de Droit (Rabat) |
| PCNG | Neuro-genetic Network of Competence | Faculté de Médecine et de Pharmacie (Rabat) |
| RUMEC | University Network in Mechanics | Faculté des Sciences Ain Chock (Casablanca) |
| MACOMS | Network of Competence in Condensed Matter and the Modelling of Systems | Faculté des Sciences (Rabat) |
| REMAT | National Network for the Physical Chemistry of Materials | Faculté des Sciences Ain Chock (Casablanca) |
| PECCA | Network of Competence in Analytical Chemistry and Corrosion Electrochemistry | Faculté des Sciences (Rabat) |
| RELOR | Research Network for Local and Regional Territorial Development | Faculté de Lettres (Rabat) |
| RENAPT | National Network for Theoretical Physics | Faculté des Sciences (Rabat) |
| PC2E | Competence Network for Water and the Environment | Faculté des Sciences Semlalia (Marrakech) |

From "The Science, Technology and Innovation System of the Kingdom of Morocco; " A UNESCO document authored by N. Boshoff. The document is not dated, but appears to have been published post 2007. Downloaded in July 2016 from: http://portal.unesco.org/education/en/files/55536/11998897825Morocco.pdf/Morocco.pdf.

**TABLE A.3.4** Distribution of Allocations to Research Supported by the Agricultural Research and Higher Education Institute in Tunisia (Tunisian Dinars Thousands)

| Institution | Federative Research Projects | Collective Projects | Developmental Research Projects | Tomato Borer Pest | Mediterranean Network in Support of Agricultural Research | Totals |
|---|---|---|---|---|---|---|
| Agricultural Research and Higher Education Institution | 25 | 41 | 55 | – | 97 | 218 |
| Research institutes | | | | | | |
| National Institute for Agricultural Research | 15 | 204 | 108 | – | 32 | 359 |
| National Research Institute for Rural Engineering, Water and Forestry | – | 41 | – | – | 40 | 81 |
| Veterinary Research Institute | – | 40 | – | – | – | 40 |
| Arid Lands Research Institute | – | 17 | 17 | 7.5 | 45 | 86.5 |
| National Marine Sciences and Technologies | – | 129 | – | – | – | 129 |
| Al Zaitouna Institute | – | 60 | 49 | – | 15 | 124 |
| **Total for research institutes** | **15** | **491** | **174** | **7.5** | **132** | **820** |
| **Territorial Research Centers** | | | | | | |
| Territorial Agricultural Research Centre at Daqash | 12 | 95 | 36 | 75 | – | 151 |
| Territorial Agricultural Research Centre at Sidi Bouzeid | – | 66 | 12 | 7.5 | 85.5 | 85.5 |
| Territorial Centre for Biological Research in Agriculture and Horticulture at Shat Meriam | – | 43 | 10 | 15 | – | 68 |
| Territorial Agricultural Research Centre at Baja | 25 | – | – | – | 10 | 35 |
| Territorial Development Research Pole in the semiarid North West | | | | | | |

*Continued*

**TABLE A.3.4** Distribution of Allocations to Research Supported by the Agricultural Research and Higher Education Institute in Tunisia (Tunisian Dinars Thousands)—cont'd

| Institution | Federative Research Projects | Collective Projects | Developmental Research Projects | Tomato Borer Pest | Mediterranean Network in Support of Agricultural Research | Totals |
|---|---|---|---|---|---|---|
| Territorial Development Research Pole in the North East | | | | | | |
| **Total for research centers and territorial poles** | **37** | **204.5** | **58** | **30** | **10** | **339.5** |
| **Higher Education Institutes** | | | | | | |
| National Institute for Agricultural Sciences | 45 | 238 | 69 | 15 | 5 | 372 |
| Higher School for Agro-Food Industries | – | – | – | – | – | – |
| Higher School for Agriculture at Majaz Elbab | – | – | – | – | – | – |
| Higher School for Agriculture at Mater | 18 | – | 12 | – | – | 30 |
| Higher Institute for Fishing and Aquaculture at Benzert | – | – | 40 | – | – | 40 |
| Higher Institute for Agriculture at Shat Meriam | – | 198 | 48 | 7.5 | 43 | 296.5 |
| Higher Institute for Agriculture at Kaf | – | – | 18 | – | – | 18 |
| Higher School for Agriculture at Mugren | – | – | – | – | – | – |
| Higher School for Agriculture at Tubruqa | – | – | 8 | – | – | 8 |
| Higher School for Veterinary Medicine | – | 5 | 18 | – | – | 23 |
| Higher Institute for Preparatory Studies in Biological Science at Sakra | – | – | – | – | – | – |
| **Total for Higher Education Institutes** | **63** | **441** | **213** | **22.5** | **48** | **787.5** |

Information downloaded in April 2017 from the website of L'Institution de la Recherche et de l'Enseignement Supérieur Agricoles (IRESA) at: http://www.iresa.agrinet.tn/index.php/ar/2014-03-05-13-21-41/budget-recherche-ar.html.

**TABLE A.3.5** National Research Centers of the Water and Energy Institute and Their Activities (Saudi Arabia)

| Centers | Projects Being Implemented (January 2017) |
|---|---|
| Water Desalination and Treatment Technology Centre | Design and construction of a pilot mobile desalination plant.<br>Design of water treatment and desalination research units.<br>Technology for agricultural use of desalinated seawater. |
| Electrical Power Systems Technology Centre | Development of a static compensator for voltage regulation.<br>Development of a device to deliver quality power in remote areas.<br>Design of a smart grid controller for PV-grid connected generator.<br>Nanothermal insulation coating for energy efficient glass.<br>Development of an adsorptive cooling system. |
| Oil and Gas Technology Centre | Potential contribution of "shale gas" in Saudi Arabia.<br>Improving oil recovery processes using low salinity water and polymer flooding.<br>Enhancing geodetic expertise at KACST.<br>Application of ground deformation monitoring technologies.<br>Pilot airborne survey of geological and hydrological information. |
| Combustion and Plasma Technology Centre | Development of a multifuel Engine.<br>Design of 2nd generation sodium nickel-chloride batteries.<br>Plasma treatment of fly ash from desalination plants.<br>Development of a plasma fuel reformer.<br>Study of chemical and thermal processes in partial oxidation reactors. |
| Membrane Technology Centre | Synthesis of a zeolite membrane for water treatment.<br>Development of a low-cost ceramic membrane support system.<br>Development of novel water desalination hollow-fibre membrane.<br>Lithium and vanadium recovery from desalination brines. |
| Solar Energy Technology Centre | PV panels production line.Development of ultrahigh concentration photovoltaic solar collectors.<br>Establish reliability of solar collectors.<br>Microinverter.<br>Nanocarbon based supercapacitor pilot unit for energy storage.<br>Development of nanocarbon based water desalination pilot facility. |

Website of the King Abdulaziz City for Science and Technology. Information retrieved in February 2017 from: https://www.kacst.edu.sa/eng/rd/WERI/pages/WERI.aspx.

**TABLE A.3.6 National Research Centers of the Materials Science Research Institute and Their Activities (Saudi Arabia)**

| Centers | Projects Being Implemented (January 2017) |
|---|---|
| Nanotechnology Centre | Hybrid organic-metal halide perovskites for PV applications.<br>Advanced applications of metal-organic frameworks.<br>Acquisition of the SRI-International improved silicon processing technology.<br>Construction of clean-room laboratory facilities.<br>Acquisition of Solid State Lighting (LED) technology.[351]<br>Investigating nanocrystal compositions and structures for specific applications. |
| Advanced Materials Centre | Metal matrix composites reinforced with nanomaterials for improving their mechanical properties |
| Petrochemical Technology Centre | Encapsulation of inorganic materials into polymeric nanoparticles for coating and composite applications.<br>Design and development of zeolite-based 'green catalysts.'<br>Development of nanocatalysts for the production of gasoline additives.<br>Production of low cost modified natural polymers with unique properties.<br>Development of molybdate and vanadate for selective oxidative n-propane dehydrogenation.<br>Synthesis and photovoltaic properties of perylene-based polymers. |
| Building and Construction Centre | Development of transparent ceramics for construction and other applications.<br>Study of optical metamaterial for negative refraction applications.<br>Utilization of plasma technology for the preparation of amorphous and fibrous products using local raw materials. |
| Applied Physics Centre | Development of smart sensors based on metal oxide nanostructures.<br>Study of pion–nucleon interactions; a relativistic quantum mechanical approach.<br>Construction of a modern detector laboratory at KACST.<br>Using image reflection technology in atoms cooling<br>Developing a new method for beam analysis of low-energy storage rings.<br>Acquiring novel technologies in quantum information. |

Website of the King Abdulaziz City for Science and Technology. Information retrieved in February 2017 from: https://www.kacst.edu.sa/eng/rd/WERI/pages/WERI.aspx.

**TABLE A.3.7** National Research Centers of the Communication and Information Technology Research Institute and Their Activities (Saudi Arabia)

| Centers | Projects Being Implemented (January 2017) |
|---|---|
| Electronics and Photonics Technology Centre | Development of software programmed tactical radios.<br>Development of a mobile computing device (K-Tab).<br>Developing platforms to test units of measurement stability and motion sensors.<br>Use radio frequency (RF) to detect moisture content of palm trees (WASEDA).<br>Design of secure data networks for Internet, satellite, phone and fax. communications.<br>Development of a phased-array antenna. |
| Information Technology Security Centre | Building KACST's capabilities in the field of information security.<br>Design of computer-based hospital management system (Smart Hospital).<br>A feasibility study to establish a Cybersecurity laboratory. |
| Computer Technology and Mathematics Centre | Strategy and plans for utilizing open source software for development.<br>Social network modeling and analysis.<br>Advanced AI techniques for breast cancer diagnosis.<br>Medical alert system for epilepsy patients. |
| Robotics Technologies and Intelligent Systems Centre | Development of intelligent algorithm for inspection and maintenance in hostile environments.<br>Developing a robot with sensors as a commercial security platform. |
| Sensor Technology and Defence Systems Centre | Testing and evaluation of a mobile system for measuring radar cross-section of flying objects (Qeyas).<br>Developments in the domain of radar, infrared, and electronic warfare (Anmat).<br>Small radar easily mounted on a vehicle and difficult to be discovered by electronic support devices (Sarab).<br>Digital electronic warfare receiver capable of detecting and analysing radar pulses (Nitaq). |

Website of the King Abdulaziz City for Science and Technology. Information retrieved in February 2017 from: https://www.kacst.edu.sa/eng/rd/WERI/pages/WERI.aspx.

**TABLE A.3.8** National Research Centers of the Nuclear Sciences Research Institute and Examples of Their Activities (Saudi Arabia)

| Center/Unit | Projects Being Implemented (January 2017) |
|---|---|
| Nuclear Regulatory Unit Centre | Ensuring Nuclear and Radiological Safety and Security of the public and KACST workers. |
| Irradiation Technology Centre | Enhancement of the characteristics of polymeric formulations for cable insulation using ionizing radiation and additives.<br>Application of irradiation technology to remove antibiotic residues from food products.<br>Engineering designs for an irradiation building using electron beam accelerator.<br>Potential use of radiation technology for environmental remediation. |
| Applied Radiation Technology Centre | Mutation-induced breeding to improve salinity-, drought-, and disease-resistant crops.<br>Clearing radioactive contamination through osmosis membrane systems. |
| Accelerators Technology Centre | Designing and building a Geiger–Muller survey meter.<br>Designing and building an isotope identifier.<br>Determination of light elements by neutron activation analysis. |
| Nuclear Technology Centre | Design and development of low power research reactor.<br>Chemical and radiological monitoring of the Umm Wu'al phosphate project. |

Website of the King Abdulaziz City for Science and Technology. Information retrieved in February 2017 from: https://www.kacst.edu.sa/eng/rd/WERI/pages/WERI.aspx.

**TABLE A.3.9** National Research Centers of the Life Science and Environment Research Institute and Examples of Their Activities (Saudi Arabia)

| Center | Projects Being Implemented (January 2017) |
|---|---|
| Environmental Technology Centre | Design and development of an air pollution diagnostic and alarm system for Riyadh.<br>Monitoring and control of rock slides on mountain roads using ground-based laser scanning and photogrammetry.<br>Developing an advanced oxidation technologies using synthetic metal nanoparticles. |
| Genome Technology Centre | Identify the genetic bases for diseases in Saudi population using state-of-the-art genome sequencing and bioinformatics.<br>Establish and maintain the necessary infrastructure to perform genetic diagnostic tests for known genetic diseases.<br>Test clinical samples and verify results.<br>Discover genetic mutations that cause genetic diseases. |
| Biotechnology Centre | Developing novel techniques for preservation of mammalian cells in anhydrous state.<br>Study of recombinant protein synthesis on cell membranes. |

**TABLE A.3.9** National Research Centers of the Life Science and Environment Research Institute and Examples of Their Activities (Saudi Arabia)—cont'd

| Center | Projects Being Implemented (January 2017) |
|---|---|
| Agricultural Technology Centre | Red Date Palm weevil project.<br>New design of a greenhouse with novel features for atmospheric and temperature control.<br>Reduction of postharvest spoilage and prolonging the shelf-life of fruit products.<br>Conservation of native Saudi plants exposed to environmental degradation (Rawdat). |
| Stem Cell Technology Centre | Stem cell research with several objectives including the study of genetic diseases. |

Website of the King Abdulaziz City for Science and Technology. Information retrieved in February 2017 from: https://www.kacst.edu.sa/eng/rd/WERI/pages/WERI.aspx.

**TABLE A.3.10** National Research Centers of the Space and Aeronautics Research Institute and Examples of Their Activities (Saudi Arabia)

| Center | Projects Being Implemented (January 2017) |
|---|---|
| Remote Sensing Technology Centre | Establishing stations for the reception of satellite images.<br>Aerial photography using manned aircraft.<br>An early warning system for floods.<br>Setting up satellite-image laboratories for the Ministry of Defence. |
| Satellite Technology Centre | Develop and construct a dual frequency SAR Payload and a high-resolution electro-optical payload (launch planned for 2018).<br>Building ground facilities required for utilising remote sensing satellites.<br>Building a telecommunication satellite in cooperation with Lockheed Martin to serve several end-user organizations.<br>Development of a reliable and cost-effective satellite platform capable of carrying multipurpose missions (Shortest Path to Space).<br>Development of a communication payload to receive Automatic Identification System signals from commercial vessels.<br>Manufacture of PV solar panels. |
| Aeronautics Technology Centre | Construction of multipurpose aircraft capable of carrying a 10 tons payload.[352] |
| Jet Engine Technology Centre | Building a turbojet engine with 3.5 kN thrust.<br>Development of a small turbojet engine (KJ-80 JetCat) to be manufactured locally.<br>Development of a turbojet engine (KJ-200 JetCat Engine) with a 230N thrust for small air vehicles. |
| Astronomy Centre | Crescents observation with the aim of digitizing the lunar crescent database. |

Website of the King Abdulaziz City for Science and Technology. Information retrieved in February 2017 from: https://www.kacst.edu.sa/eng/rd/WERI/pages/WERI.aspx.

**TABLE A.3.11 Policy Enablers and Associated Goals for the United Arab Emirates Science, Technology, and Innovation Policy**

| Policy Enablers | Associated Goals |
|---|---|
| Talent | • Promote excellence in Science, Technology, Engineering and Mathematics (STEM) instruction;<br>• Establish a strong local STEM workforce;<br>• Employ and retain UAE nationals in R<br>• Develop strong local R&D management skills;<br>• Attract and retain the best talents from all over the world;<br>• Ensure knowledge transfer between global and local talents. |
| Investments and incentives | • Provide adequate government funding for basic scientific research;<br>• Increase industry funding of R&D activities;<br>• Ensure availability of risk capital to support entrepreneurial activity;<br>• Encourage the growth of SME's in the fields of Science and Technology. |
| Universities and support institutions | • Strengthen higher education, and basic and applied research in the fields of science and technology;<br>• Equip universities with world-class laboratories and computing facilities;<br>• Expand specialized R&D facilities to support key innovation sectors;<br>• Ensure the availability of supporting institutions for technology transfer and incubating innovation. |
| Regulations and intellectual property protection | • Develop flexible regulations to encourage growth of innovative projects and businesses;<br>• Provide best protection of intellectual property;<br>• Facilitate the import of technology, equipment, and materials critical to research;<br>• Encourage the inflow of scientists, researchers and innovators;<br>• Develop mechanisms and incentives to strengthen individual and corporate innovation;<br>• Encourage innovation, entrepreneurship and reward risk-taking. |
| Partnerships and networks | • Enhance cooperation between universities and private sector in STI;<br>• Make government R&D accessible to the private sector;<br>• Encourage international partnerships with universities and big research institutions;<br>• Establish distinguished expertise in certain unique areas through partnerships. |

Adapted from "Science, Technology & Innovation Policy in the United Arab Emirates; " First Edition, dated November 2015. Posted by the UAE government on the Internet. Downloaded in May 2016 from: http://www.uaeinnovates.gov.ae/docs/default-source/pdfs/science-technology-and-innovation-policy-en.pdf.

## ENDNOTES

1. Later renamed "Science and Technology Policy Research." However, it is often referred to as SPRU in various literature article.
2. Pioneers of science and technology policy initiatives such as C Cooper, G Oldham, R C Desai and F Sagasti, to name but a few, played an important part in promoting essential notions on science and technology policies and their implementation strategies.
3. Several experts were involved in both meetings designed to promote efforts at science and technology policy formulation. Many, such as such as A W Chahid, and A Shihab Eddin, were heads of research establishments in their own countries while others such as A B Zahlan and U Kholi were well known for their contributions on the subject.
4. The UNESCO Report titled "Review of Scientific Research in Iraq." Published in 2013 by the United Nations Education, Scientific and Cultural Organization, Field Office for Iraq in Amman 9, Yacoub Ammari St. Abdoun, PO Box 2270 – Amman 11181 Jordan. Downloaded in May 2017 from: http://www.unesco.org/fileadmin/MULTIMEDIA/FIELD/Iraq/pdf/Review_SR_Report_Eng.pdf.
5. Information from the website of Iraq's Ministry of Higher Education and Scientific Research. Accessed in May 2017 at: http://mohesr.gov.iq/category/%D8%A7%D9%84%D8%A8%D8%AD%D8%AB-%D9%88-%D8%A7%D9%84%D8%AA%D8%B7%D9%88%D9%8A%D8%B1/.
6. "Modernization of the Education Sector in Iraq; Rapid Assessment of the Management of the Education System." Prepared by R Salamé for the UNESCO Office for Iraq in November 2010.
7. France, for example, did not have legislation regulating links between public universities and enterprises until the late 1990's.
8. Document in Arabic provided by Dr. B. Khailany, 30 May 2011; titled "ثحبلل ةيجيترتسا وحن يملعلا."
9. "Iraq Education in Transition: Needs and Challenges;" Published by UNESCO, Paris in 2004.
10. Law no. 16, dated 1963.
11. Its responsibilities concerning both higher education and scientific research were largely established as the result of implementation of Law 40/1988 and its subsequent amendments and the instructions, the most recent of which was promulgated in November 2010.
12. Law 40/1988 entrusting MOHESR with the management of the higher education system throughout Iraq.
13. See endnote 3.
14. The Ministry's website carries pages in both English and Arabic with disparate contents. English pages include information on research centers and directorates on which no further information is included regarding these entities. Website of the Ministry of Higher Education and Scientific Research in the Kurdistan Region of Iraq. http://www.most.gov.iq/.
15. As well as local and regional issues.
16. Critical mass available of expert, support and administrative staff as well as graduate students.
17. See endnote 4.
18. At least by May 2017 when this work was being drafted.
19. The "National Scientific and Technological Requirements and Potential;" was produced by HCST in 1998. Further details on this and other similar studies concerning selected Arab countries may be viewed in a report produced by the Economic and Social Commission for Western Asia (ESCWA). Report produced by the Economic and Social Commission for Western Asia (ESCWA) titled "Science and Technology Policies for the Twenty-First Century, E/ESCW /TECH/1999/4.
20. Additional objectives of this document included identification of national institutions engaged in S&T activities, evaluation of S&T resources, available as well as additional Human capital resources and specialized training requirements, institutional arrangements, and principal equipment items.
21. A total of around 7500 institutions were included in the survey and a response rate of around 70% was achieved.

22. Report produced by the Economic and Social Commission for Western Asia (ESCWA) titled "Science and Technology Policies for the Twenty-First Century, E/ESCW/TECH/1999/4.
23. A previous science and technology policy document by HCST, covered the years 2006–2010.
24. Policy document (Arabic). Downloaded in February 2017 from the HCST website at: http://hcst.gov.jo/sites/default/files/policy-and-stratgy.pdf.
25. In particular, the functions of principal STI institutions were considered and the interactions reviewed.
26. "Defining Scientific Research Priorities in Jordan for the Years 2011–2020;" Higher Council for Science & Technology, General Secretariat. November 2010, Amman – Jordan. Document downloaded in February 2017 from: http://hcst.gov.jo/?q=en/publications.
27. The name given to this category of research topics and issues is rather misleading as it includes mostly subjects and issues normally classified within the domain of the applied sciences.
28. Term used in reference to independent reasoning on polity issues with the aim of arriving at valid but also practical application of religious edicts.
29. Defined as a ruling on a point of Islamic law, concerning a secular or a religious matter, given by recognized religious authority.
30. Verbal records of the teachings of Prophet Muhammad as reported by his companions.
31. Islamic canonical law based on Quran and teachings of Prophet Muhammad.
32. In preparation for the formulation of the national STI Policy for the period 2013–2017, the HCST General Secretariat set up a specialized committee to study and review the status and implementation rates of projects included in the previous 2006–2011 S&T policy document. This committee submitted a report stating that the percentage of completion of projects was about 50%, with no information given on the substance and orientations of implemented activities, the extent of allotted resource utilisation. See "The National Policy and Strategy for Science, Technology and Innovation 2013–2017;" HCST, Amman-Jordan.
33. Website of the Higher Council for Science and Technology at http://hcst.gov.jo.
34. Vision and mission statements of Jordan's Innovation Strategy 2013–2017 document.
35. Prince El Hassan of Jordan is credited with playing a prominent role in setting up HCST and in its later development.
36. Research and development activity within the latter domains is undertaken by the "National Centre for Research and Development," which includes several departments or divisions.
37. The HCST mission statement includes (a) setting the national science and technology policy, strategies, plans and programmes, (b) playing a catalytic role in enhancing and accelerating interaction between the S&T community, production and services sectors, and government agencies, (c) transforming science and technology ideas into commercial enterprises through supporting innovation, entrepreneurship, and technology commercialization.
38. RSS activities and policies are supervised by the Board of Trustees chaired by HRH Prince El Hassan bin Talal. A board of directors is responsible for its operations.
39. A special cooperative project designated "Jordan: Start-up Financing for the Establishment of the National Centre for Innovation (NCI) and Transition to Full Operational Functionality," was designed by HCST for implementation in cooperation between the Jordanian government and the World Bank. A budget of US$2.6 million for establishing NCI was allocated from the Bank's MENA Transition Fund and the European Bank for Reconstruction and Development (EBRD) was contracted to provide assistance.
40. "A Practical Guide to Wealth Creation; Proposals for an Innovation Policy for Jordan;" By David Tee, April 2006.
41. The Philadelphia incubation facilities were established in cooperation with Euro Jordanian Enterprise Development Action (EJADA), a Jordan-based IT services and solutions provider, whose activities cover the Middle East and North Africa. Information retrieved in February 2017 from: http://www.ejada.com/CorporateInformation/Pages/aboutus.aspx.
42. "Study on the National Innovation System in Jordan;" Final Report by U Seidel et al in December 2009. Report produced by the German Agency for Technical Cooperation on behalf of the Federal Ministry for Economic Cooperation and Development.

43. A network of Jordanian innovation centers had been established by the Jordan Enterprise Development Corporation (JEDCO), a governmental organization dedicated to supporting the development of emerging businesses and small and medium-sized enterprises in Jordan. However, the proposal made by GTZ appears to target a more proactive network with strong components aimed at championing national, regional as well as international partnerships.
44. See for example "The National Policy and Strategy for Science, Technology and Innovation. (Arabic)" Downloaded in February 2017 from: http://bluerayjo.com/hcst5/?q=en/publications.
45. An analysis of the strengths, weaknesses, opportunities and threats was conducted for the national STI system in Jordan within efforts aimed at formulation of the National STI Policy and Strategy, 2013–2017. Details of this analysis were not included in the national policy document but were made available to the author by Dr Fawaz Karmi, Assistant Secretary General of the Higher Council for Science and Technology. Some of the weaknesses and threats included above were indicated in the "Status Report on Science, Technology and Innovation activities in Jordan;" Prepared for UNESCO by Dr. Emanuel Hassan. Dated April 13, 2013.
46. Reference is made above to the fact that R&D allocations as a percentage of GDP are still quite low, and have a considerable way to go before they reach the target of 1.5% set by Jordan's National Agenda, 2006–2015. See Table 2.5, above.
47. Virtual absence of venture capital and angel investors is quoted by sources in relation to financial support for start-ups enterprises.
48. With publication of research results at the centre of attention, application of research results is often relegated a much lower order of priority.
49. Protectionist practices by private enterprise often result in the creation of niches and captive markets, obviating the need for constant product and service improvements that they would otherwise need to achieve, through R&D efforts.
50. A number of award schemes have been successfully instituted by various regional organisations, e.g., the Arab Thought Forum and the Mohamad bin Rashid Foundation, in addition to international organisations and corporations, such as the UNESCO – L'Oreal "Women for Science" award scheme. Awards such as these are certainly useful, however, they may not be a substitute for periodic institutional and national reports highlighting citations of research output as well as instances of effective implementation of research output, and tangible benefits achieved as a result. Additionally, the role that incubation facilities and financial support schemes dedicated that innovative ventures, might play in de facto ranking of research and innovation efforts is stymied by the small number of such facilities and schemes and the limited resources at their disposal.
51. "Status Report on Science, Technology and Innovation activities in Jordan;" Prepared for UNESCO by Dr. Emanuel Hassan. Dated April 13, 2013.
52. For institutional funding, rules may be devised taking into account the number of academic staff as well as and students engaged in research activity, the type of research being pursued. Additionally, performance indicators may also be used in order to further refine institutional funding, based for example upon the number of scientific publications in international peer-reviewed journals, citation impact, number of patents registered and granted.
53. Three Task Forces made up of Lebanese experts were involved in drafting this policy document. Contributions were also included from a number of experts, assigned to provide assistance, essentially by UNESCO and ESCWA.
54. Research programmes listed in table (3.6) were intended for implementation within a 5-year period preceded by a preparatory year.
55. However, it does not seem that anything even close to this sum was ever made available for implementing STIP.
56. Not all activities are planned to commence within the first year of the STIP Plan's implementation.
57. These are programmes aimed at stimulating cross-disciplinary, inter-institutional, international and public-private sector STI collaboration.

58. "Charter of ethics and guiding principles of scientific research in Lebanon." A publication of the National Council for Scientific Research, July 15th, 2016. Beirut, Lebanon.
59. All funding figures quoted in this section refer to approved and renewed research projects. Thus, not all projects whose funding is included in the distribution of funding across institutions or groups of disciplines need to have been concluded.
60. The Lebanese Agricultural Research Institute (LARI) was established in 1957 as a governmental organization operating under the supervision of the Minister of Agriculture. LARI has set up eight experimental stations covering 280 ha of agricultural land in areas where subtropical and temperate corps are produced. LARI conducts applied and basic scientific research for the development and advancement of the agricultural sector in Lebanon. In addition, the Institute keeps close ties with Lebanon's farmers and tries to develop research activities aimed at solving their problems. Research issues tackled by LARI include: olive tree propagation, cereal and grain legume development, pasture and forage production, as well as sterilisation as means for biological control of insects. LARI is also entrusted with implementation of a National Research Strategy adopted by both the International Service for National Agricultural Research (ISNAR) and Lebanese Government.
61. CNRS website accessed in July 2016 at: http://www.cnrs.edu.lb/research/geophysicalresearch.html.
62. Berytech's website. Accessed in July 2016 at: http://berytech.org/funding/.
63. Fund I was set up in 2008, amounting to US$6 million, whereas Find II is a US$55 million Venture Capital Fund targeting start-ups and SMEs in the IT and health technology sectors. Berytech's website. Accessed in July 2016 at: http://berytech.org/funding/.
64. Launched by members of the Lebanese-Canadian Information Technology Association (ATILC).
65. Kafalat guaranteed loans benefit from interest rate subsidies financed by the Lebanese treasury and administered by the Central Bank of Lebanon, to mitigate the crowding out effect of high interest rates in Lebanon, induced essentially by public sector borrowing.
66. IRI website. Information accessed in July 2016 at: http://www.iri.org.lb/brief.html.
67. The IRI website includes a list of projects dedicated to: exploitation of renewable energy resources, energy saving, pollution indicators, cleaner production and coastal/marine pollution.
68. IRI website. Information on this activity downloaded on 20th May, 2013.
69. The "Innovation Vouchers" programme was initiated on the basis of €200,000 grant by the European Union in 2012. The initiative appears to be continuing. However, the IRI website does not indicate how many enterprises actually benefited from the above-mentioned €10,000, grant and whether further funds were added in order to extend assistance to a wider range of enterprises. In this connection, the IRI website, merely states that "In February 2012 the CIT was awarded a grant from the European Union for 200,000 Euros for the implementation of the action entitled: 'Innovation Vouchers', with the Presidency of the Council of Ministers as the Contracting Authority." IRI website. Information accessed in March 2017 at: http://www.iri.org.lb/cit.html.
70. Reasons behind partial success and outright failures in making use of Lebanon's STI diaspora in attaining improved STI capacity at home are worthy of further research, through field and disciplinary specific surveys.
71. Website of the United Nations Development Programme (UNDP). Information accessed in March 2017 from: http://www.toktenlebanon.org/.
72. Website of the United Nations Development Programme (UNDP). Information accessed in March 2017 from: http://www.lb.undp.org/content/lebanon/en/home/library/democratic_governance/tokten-lebanon.html.
73. Top-level experts are generally recruited on voluntary bases to share provide expertise on a variety of matters, emphasising policy advice and research activity to Palestinian ministries, universities, research centers, non-governmental organizations (NGO's) as well as private sector enterprises, over periods of between 3 weeks and 3 months.

74. "General Framework of Preparation of the National Development Plan 2014–2016." Ministry of Planning and Administrative Development. Document downloaded in May 2017 from: http://www.ilo.org/wcmsp5/groups/public/---arabstates/---ro-beirut/documents/publication/wcms_236946.pdf.
75. "National Development Plan 2011–2013; Establishing the State, Building our Future." Document downloaded in May 2017 from: http://www.apis.ps/up/1332062906.pdf.
76. "National Agriculture Sector Strategy; Resilience and Development 2014–2016." Document prepared by Palestine's Ministry of Agriculture with technical support from the Food and Agriculture Organisation of the United Nations and financial support from the European Union. Document downloaded in May 2017 from: http://www.apis.ps/up/1417423273.pdf.
77. The above bases for the National Development Plan are adapted from the above-mentioned "General Framework of Preparation of the National Development Plan 2014–2016." by the Ministry of Planning and Administrative Development.
78. "An overview of renewable energy potential in Palestine;" A Juaidi et al. Elsevier; Renewable and Sustainable Energy Reviews 65 (2016) 943. Accessed May 2, 2017 at: https://www.researchgate.net/publication/305845159_An_overview_of_renewable_energy_potential_in_Palestine.
79. Abu Hameda et al. estimate that exploiting solar sources has the potential of reducing 13% of electricity demand. Harnessing wind energy in suitable locations should produce around 7% of current demand, whereas conversion of animal waste into biogas could produce 20% of demand in terms of biogas, whereas processing agricultural residue could replace around 5% of imported diesel fuel. "Renewable energy in the Palestinian Territories: Opportunities and challenges;" T Abu Hamed, H Flamm and M Azraq. Renewable and Sustainable Energy Reviews; Elsevier. Volume 16, Issue 1, January 2012, Pages 1082–1088.
80. A report published on July 1st 2017, by the Israeli news daily Haaretz, related the confiscation of installed solar panels donated by the government of the Netherlands to a humanitarian project to an isolated village of Jubbet al-Dhib east of Bethlehem. Reasons given by the Israeli authorities were that they were built without permits required by Israel to develop in an area of the West Bank designated as under full Israeli military control. The electricity generating equipment, said to be a hybrid power system using both diesel and solar power was donated by the Netherlands toward electrification of the southern Bethlehem region at a cost of around €500,000. "Netherlands outraged after Israel seized Dutch-funded solar panels in West Bank;" The Midle East Monitor; July 2nd, 2017, quoting the Israeli news daily Haaretz's article published on July 1st, 2017.
81. With particular regard to the West Bank, the 1995 Oslo Accords stipulated that 80% of water from a major aquifer, referred to as the "mountain aquifer," would be allocated to Israel, whereas Palestinians would use the rest. The agreement was intended as a 5-year interim measure but has never been revised because both the Palestinian and settler populations in the West Bank nearly doubled. "Water in the West Bank. Nor yet a drop to drink;" The Economist, July 2016. Article accessed in May 2017 at: http://www.economist.com/news/middle-east-and-africa/21702716-palestinians-go-thirsty-despite-sitting-over-underground-ocean-nor-yet-drop.
82. "Water Situation Alarming in Gaza." Article published in November 2016 by the World Bank. Accessed in May 2017 at: http://www.worldbank.org/en/news/feature/2016/11/22/water-situation-alarming-in-gaza.
83. QOU was established in Amman, Jordan, by the Palestinian Liberation Organization (PLO). It started operating in the Palestinian territories in 1991. Information accessed from the QOU website in May 2017 from: http://www.qou.edu/home/indexHomePage.do#.
84. Website of the "Palestine Academy for Science and Technology (PALAST)." Accessed in April 2017 at: http://www.palestineacademy.org/main/.
85. TWAS Newsletter, volume 22, number 4. 2010. Downloaded in April 2017 from: https://twas.org/sites/default/files/attachments/07-sciencepalestine-39-45.pdf.

86. Information obtained from the An-Najah University's website. Accessed in May 2017 at: https://www.najah.edu/en/community/scientific-centers/water-and-environmental-studies-institute/.
87. Information obtained from the An-Najah University's website. Accessed in May 2017 at: https://www.najah.edu/en/community/scientific-centers/energy-research-center/.
88. "Incubators in Palestine;" Information dated September 2015, on the WAMDA website. Retrieved in May 2017 from: https://www.wamda.com/2015/09/palestine-s-top-8-incubators.
89. Information presented in the following sections is, understandably, synoptic in character for two principal reasons. First, very few reliable reports are available on progress made, if any, on various STI fronts during the past few years. Second, a culture that adopts evaluation and accountability in many spheres of social and economic life including STI activity was yet to be come about even before Syria's tragic events unfolded.
90. A series of Sanctions and embargoes were imposed for the past 4 decades against Syrian authorities' for a variety of reasons, linked to Syria's role in regional politics, with emphasis on the purchase of equipment and materials deemed to be of dual, i.e., military as well as civilian, use.
91. See, for example, sections on STI output in terms of scientific publications and patents registered and granted.
92. Information provided through private communications with Dr. B. Mounajed, ex-minister for Communications and Technology, indicates that the ICT Strategy document was ratified by the 'Peoples Assembly' in 2002.
93. The study titled "Future of science and technology in Syria till 2025," was authored in Arabic and distributed by its authors Asef Diab and Marwan Zbibi in 2007 as basis for discussions among concerned experts and institutions. Although this prospective study is not an official as yet, it would be safe to regard its precepts and the measures it proposes as coincident with official thinking on the issues it raises.
94. NSTIP Document issued for final review by HCSR in 2012, in Arabic. National Science, Technology and Innovation Policy document. Made available to the present author by HCSR.
95. Numerical goals adopted by the NSTIP document reiterate goals outlined in a special section dedicated to STI development within the tenth 5-year national development plan.
96. SEBC website. Accessed in June 2016 at: http://www.sebcsyria.com/web2008/index.php.
97. "Development Aid," Document accessed in June 2016 at: https://www.developmentaid.org/#!/organizations/view/263.
98. League of the Arab States; Investment and Business Guide; Arab League Investment Guide Investment Climate and Opportunities. Washington DC. USA International Business Publications, 2011. ISBN 1-4330-0175-6.
99. Most books published in Syria are the Arabic and the majority are authored by Syrians.
100. Legislative decree n°28 of 1980.
101. "Science, Technology and Innovation Indicators of Syria (2008–2009);" A presentation by the Higher Commission for Scientific Research. Downloaded in July 2016 from: http://www.uis.unesco.org/StatisticalCapacityBuilding/Workshop%20Documents/ST%20Workshop%20dox/Damascus-2010/Science,%20Technology%20in%20syria.pptx.
102. No explanation is put forward for this enormous increase in the space of 1 year. However, it is likely that a number of factors might be put forward as an explanation, including issues to do with accounting practices and delayed spending of long earmarked funds.
103. "Diagnostic Study of Syrian Innovation System;" Lena Tsipouri, Samir Elsayyad, Ahmed Elsayad. MEDIBTICAR. Dated 2003 and given call number MED/2007/070–567.
104. Among other issues this report enumerates lack of proficiency in foreign languages and limited or no financial incentives for researchers as particular weaknesses.
105. "Diagnostic Study of Syrian Innovation System;" Lena Tsipouri, Samir Elsayyad, Ahmed Elsayad. MED/2003/070–567.

106. "Study on the National Innovation System in Syria;" issued by the Deutsche Gesellschaft für Internationale Zusammenarbeit (GTZ) on behalf of the Federal Ministry of Economic Cooperation and Development, by Uwe Siedel, Dr. Wolfgang Domröse and Dr. Gerd Meier zu Köcker. This study is dated December 2009. Downloaded in May 2016 from: http://www.arabengineeringindustries.org/uploads/Damascus_Office/2010-02-16_giz_Regional_Seminar_SY/Innovation_Study_Syria.pdf.
107. The survey solicited the views of local experts, who were asked to grade the national innovation system using the above-mentioned determinants. A set of recommendations was put forward on the basis of the survey and subsequent study. Attributes of the various determinants of the national innovation system were evaluated on the basis of interviews with the local experts. Thus, using a scale of 1–4, with lower scores representing absence of an ingredient or attribute of the national innovation system, or its nascent nature, whereas higher scores signified more a mature ingredient or attribute.
108. Projects such as Shabab, Massar. the Syrian Young Entrepreneurs Association and The Syria Trust for Development.
109. The first technopole is reportedly in the feasibility study phase. Higher Institute of Applied Sciences and Technology (HIAST) has requested technical assistance from the United Nations Economic and Social Commission for Western Asia (ESCWA). Information downloaded in May 2016 from: http://www.unido.org/search.html?tx_indexedsearch_pi2%5Baction%5D=search&tx_indexedsearch_pi2%5Bcontroller%5D=Search.
110. According to the GTZ study, Most of these are micro enterprises, by European standards.
111. The Rapid Innovation Action Learning (RIAL) initiative was implemented by the World Bank through missions undertaken by teams of experts to various developing countries around the world. Missions dedicated to Jordan, Lebanon and Syria among other Arab countries were organized in 2010 and 2011.
112. "National Agenda for the Future of Syria;" A Project Document by the United Nations Economic and Social Commission for Western Asia (UN-ESCWA), June 20, 2015.
113. See endnotes 58.
114. Such facilities evolve into more autonomous units, successfully impacting innovation capabilities with their own outreach modalities, clients and modes of operation.
115. The Gordon Memorial College in Sudan and Cairo University in Egypt were established in 1902 and 1908, respectively.
116. This has the status of a ministry of state rather than a full-fledged ministry.
117. Earlier references to the Ministry of Higher Education as the Ministry for Higher Education and Scientific Research, ceased in or around 2006. The website of the Ministry of Higher Education does not refer to responsibilities with regard to scientific research, but indicates its involvement in providing scholarships for higher studies. However, several websites including that of the Federation of Arab Scientific Research Councils seem to retain the earlier appellation for this ministry. Website of the Ministry of Higher Education. Visited in March 2017 at: http://www.mohe-casm.edu.eg/English/Main_menu/Site_Map/Site_Map.jsp.
118. The National Council for Education and Scientific Research (NCESR) was recently created replacing "The Higher Council for Science and Technology." NCESR is essentially a panel of ministers and nongovernmental experts including expatriate scientists, re-elected every 3 years, tasked with identifying national developmental priorities and how they could be supported through effective education as well as scientific research, technology acquisition and development. Incidentally, the Ministry of Scientific Research and the Academy of Scientific Research and Technology do not post direct references to NCESR.
119. These are the Supreme Council for Coordination between Research Centres in the different ministries, The Supreme Council for Research Centres and Institutes and The Supreme Council for Universities.
120. The Ministry's website had no entries under vision, mission, strategy, when visited in March 2017. Website of the Ministry of Higher Education. Visited in March 2017 at: http://www.msr.gov.eg/.

121. List of STDF tasks was downloaded in March 2017 from: http://www.stdf.org.eg/doc%20center/STDF%20Funding%20Mechanisms.pdf.
122. Information on STDF activities and their impact was accessed in March 2017 from: http://www.stdf.org.eg/index.php/features/about-stdf.
123. Little could be found in relevant literature concerning achievements made during this decade, apart from cooperation programmes concluded with selected governments in the European Union and with the United States in support of research and scholarship grants. See section on "Egypt's Research Funding."
124. The remaining nine pillars address: Economic Development, Energy, Transparency and Efficiency of Government Institutions, Social Justice, Health, Education and Training, Culture, Environment and Urban Development.
125. Data used for current values of the various indicators were extracted from WEF's, Global Competitiveness Report for the year 2013 and the WIPO's Global Innovation Index report for the year 2014.
126. Such as the Oslo and Frascati manuals.
127. That is, including government, private sector, and civil society from domestic and international resources as a percentage of GDP.
128. Attributed by the Third Pillar's strategy document to "decades of inherited negligence, whether in the terms of instilling the culture of innovation in youth, or in fostering talents in schools, or in creative appreciation and motivation."
129. Studies carried out in 2007 with the intention of reforming Egypt's science and technology system identified the following areas as national priorities: solar and wind energy. water management, including desalination, efficient irrigation and new groundwater sources, food and agriculture, with emphasis on new cash crops and improved production of wheat, rice and aquaculture, health, with focus on Hepatitis C, cancer and obesity, ICT, space technology and the social sciences. Document posted by OECD. Accessed in March 2017 at: https://www.oecd.org/sti/outlook/e-outlook/sticountryprofiles/egypt.htm.
130. "Science, Technology and Innovation for Sustainable Development, 2013 ECOSOC Annual Ministerial Review; ' Regional preparatory meeting for Western Asia. Joint publication by UNDESA, UN-ESCWA and UNESCO. November 26th, 2012.
131. The Council is Chaired by the Minister of State for Scientific Research, with the President of the Academy acting as Deputy Chairman. Other members of the Council include university presidents nominated by the Minister of Higher Education, presidents of national research centers and 10 selected scientists or experts appointed by the Prime Minister. Website of the InterAcademy Partnership, accessed in March 2017 at: http://www.interacademies.net/academies/byregion/africa/egypt/13207.aspx.
132. "Science and innovation in Egypt;" By M Bond, H Maram, A Soliman and R Khattab. Joint publication by a number of organisations including the British Council, the Royal Society and the Organisation of Islamic Cooperation. Downloaded in March 2017 from: https://royalsociety.org/~/media/policy/projects/atlas-islamic-world/atlas-egypt-report.pdf.
133. Information retrieved from the website of the Zewail City of Science and Technology. Accessed in March 2017 at: https://zewailcity.edu.eg/main/content.php?lang=en&alias=about_zewail_city.
134. The Altmetric score for a research output provides an indicator of the amount of attention that it has received.
135. See the website of Egypt's Observatory for Science, Technology and Innovation. Viewed in March 2017 at: http://www.estio.eg.net/.
136. Previously named "Mubarak City for Science and Technology."
137. Developed with the Italian National Agency for New Technologies, Energy and Sustainable Economic Development. As reported in the "Science and innovation in Egypt Science and innovation in Egypt." "Science and innovation in Egypt Science and innovation in Egypt;" by M Bond et al. Published under Creative Commons by several organisations including the British Council, the Royal Society, and the Organisation of the Islamic Cooperation. ISBN: 978-0-85403-952-4. Downloaded in March 2017 from: https://royalsociety.org/~/media/policy/projects/atlas-islamic-world/atlas-egypt-report.pdf.

138. No indications were available at the time of writing this as to the actual number of employees at SVC facilities.
139. "Egypt looks to Suez for ICT expansion;" Article published online by Oxford Business Group in December 2015. Viewed in March 2017 at: http://www.oxfordbusinessgroup.com/news/egypt-looks-suez-ict-expansion.
140. "Smart Village Eyes to Launch New Subsidiary in Damietta Within 2014;" Article by N Eid and S El-Feki, published by Amwal AlGhad in December 2013. Accessed in March 2017 at: http://en.amwalalghad.com/investment-news/technology-news/24003-smart-village-eyes-to-launch-new-subsidiary-in-damietta-within-2014.html.
141. An OECD report on innovation in Egypt indicates that gross expenditure on research in Egypt stood at 0.21 of the country's GDP. OECD report on innovation in Egypt based on the "OECD Science, Technology and Industry Outlook 2012," published online in "Innovation Policy Platform." Report accessed in March 2017 at: https://www.innovationpolicyplatform.org/content/egypt.
142. A statement by the President of the Academy of Scientific Research and Technology (ASRT) in Egypt, at a meeting organised by BioVision, a roaming international science forum held in 2012 indicated that research funding reached 0.4%. The ASRT President went on to state that this sum was still inadequate and that research spending should ideally be at least 1%.
143. "Science and innovation in Egypt;" M Yahia in Nature Middle East. Published online 30 April 2012. Downloaded in March 2017 from: http://www.natureasia.com/en/nmiddleeast/article/10.1038/nmiddleeast.2012.63.
144. Funding from this source is stated as having dwindled over the past few years. Report by the Deutsches Wissenschaftszentrum Kairo. Accessed in March 2017 at http://www.dwz-kairo.de/research-funding-0.
145. "Egypt's Science and Technology Parks Outlook: City for Scientific Research and Technology Applications;" Y R Abdel-Fattah, et al.; Supreme Council for Research Centres and Institutes, The Ministry of Scientific Research. http://www.wtanet.org/ds_imgs/sub04/wtr6/WTR02020604BP.pdf.
146. See endnotes 145.
147. Article by M K Al Sayed, Published by Al Shorouk Newspaper fifth March, 2017. Article downloaded in March 2017 from: https://www.shorouknews.com/columns/view.aspx?cdate=05032017&id=748f5fd2-618d-452e-98eb-2db2a624d57c.
148. This percentage was enshrined in Egypt's constitution.
149. Several research centers that had to cease operations were financed by external sources, mainly western government agencies and international NGOs, thereby drawing criticism as instruments foreign interference. This was particularly the case with regard to research centers concerned with social policy issues.
150. Egypt's 'country profile' from the "OECD Science, Technology and Industry Outlook 2012," published online by "Innovation Policy Platform." Profile accessed in March 2017 at: https://www.innovationpolicyplatform.org/content/egypt.
151. "Innovation System Landscape in Sudan;" Published by UN-ESCWA in November 2016. Downloaded in May 2017 from: https://www.unescwa.org/sites/www.unescwa.org/files/page_attachments/sudan-ntto-innovation-landscape-study-en.pdf.
152. Information obtained from the Website of the Ministry of Higher Education and Scientific Research. Website accessed in May 2017 at: http://mohe.gov.sd/index.php/ar/pages/details/49/48.
153. Other functions carried out by the Council concern setting minimal qualifications for the appointment of lecturers and researchers as well as establishing rules for their promotion.
154. See endnotes 151.
155. SAS website. Information retrieved in May 2017 from the at: http://www.sas.edu.sd/earth.php.
156. See endnotes 151.
157. Information retrieved in May 2017 from the SAS website at: http://www.nctr.sd/partner/
158. Website of the African Portal on Agriculture. Accessed in May 2017 at: http://www.erails.net/SD/ncr-dic/sudan-ncr/.

159. Vision statement posted on the website of the Agricultural Research Corporation. Accessed in May 2017 at: http://www.arcsudan.sd.
160. Website of the Agricultural Research Corporation. Accessed in May 2017 at: http://www.arcsudan.sd/centers.html.
161. See endnotes 151.
162. LRRC is based on the Veterinary Research Corporation (VRC) which was established as early as 1913.
163. The Industrial Research and Consultancy Centre was established in 1965 with assistance by UNIDO. IRCC hosts a UNESCO Chair established in 2006 to promote the transfer of industrial technologies.
164. As reported on the Industrial Research and Consultancy Centre's website. Website of the Industrial Research and Consultancy Centre. Accessed in May 2017, at: http://www.ircc.gov.sd/journal.html. However, the latest issue of the journal which appears to be published on annual basis dates to 2010.
165. "Assessment of Science and Technology Indicators in Sudan;" S Mohamed Nour, Maastricht University. School of Business and Economics. Paper prepared for the Eighth GLOBELICS International Conference held in November 2010, under the title: "Making Innovation Work for Society: Linking, Leveraging and Learning" at the University of Malaya, Kuala Lumpur, Malaysia. Paper downloaded in May 2017 from: http://umconference.um.edu.my/upload/43-1/papers/290%20SamiaSattiOsmanMohamedNour(1).pdf.
166. This paper was based on a review of reports by United Nations organisations such as UNDP, including values of S&T indicators for Sudan as well as other countries around the world, as well as 25 face-to face interviews with the government policy-makers and the academics in public and private universities. Interviews were designed with the aim of exploring the main factors that hinder or promote S&T development in Sudan.
167. Information retrieved from the website of "Africa City for Technology," in May 2017 at: http://www.act.sd/en/projects.
168. This must be considered an improvement on quoted for the year 2005 by the World Bank, indicating 168 researchers mere 34 technicians, per million of the country's population. World Bank's World Development Indicators, February 2016.
169. Report, titled "Algeria to open research sector to foreign scientists;" published by SciDev, and accessed in June 2017 at: http://www.scidev.net/global/r-d/news/algeria-to-open-research-sector-to-foreign-scientists.html.
170. See endnotes 169.
171. Loi n° 98-11 du 22 Août 1998 d'Orientation et de Programme à Projection Quinquennale sur la Recherche Scientifique et le Développement technologique (1998–2002)
172. Abdelkader DJEFLAT, "Building Knowledge Economies for job creation, increased competitiveness, and balanced development: Individual country overviews with data on STI systems in the Arab countries. Report produced for the High Level International Conference, Organized by the Tunisian Government and ISESCO, In collaboration with the World Bank, Carthage, December 1–3, 2009.
173. Algeria to open research sector to foreign scientists;" published by SciDev. The report was accessed in June 2017 at: http://www.scidev.net/global/r-d/news/algeria-to-open-research-sector-to-foreign-scientists.html.
174. The "Fond National de la Recherche (FNR)."
175. See endnotes 172.
176. Civil society institutions, such as the "Maghtech Network" and the "Algerian Association for Technology Transfer (A2T2)," were involved.
177. Acronym is derived from the French "Conseil National Economique et Social."
178. Information presented on the website of the United Nations Education, Science and Culture Organisation (UNESCO). Accessed in April 2017 at: http://www.unesco.org/new/en/communication-and-information/portals-and-platforms/goap/access-by-region/arab-states/algeria/.

179. SONATRACH, stands for "Société nationale pour la recherche, la production, le transport, la transformation et la commercialisation des hydrocarbures"
180. UNESCO report titled "Mapping research systems in developing countries Country report: The Science and Technology system of Algeria." Accessed in April 2017 at: http://academic.sun.ac.za/crest/unesco/data/Algeria.pdf.
181. This acronym derives from the French "Direction Générale de la Recherche Scientifique et du Développement Technologique."
182. The acronym is derived from the French "Conseil National de la Recherche Scientifiques et du Développement Technologique."
183. The acronym is derived from the French "Comité National d'Évaluation des Projets de Recherche Universitaire."
184. The acronym is derived from the French "Agence Nationale de Valorisation de la Recherche et du Développement Technologique."
185. The acronym is derived from the French "Centre d'Etudes et de Services Technologiques de l'Industrie des Matériaux de Construction."
186. The acronym is derived from the French "Centre National des Technologies et du Consulting"
187. "L'Entreprise Algerienne et l'Innovation: un d'Analyse;" I Mancer (Université Akli Mohend Oulhaj de Bouira). Paper published by Les Cahiers du Cread n°98–99/2011–2012. Downloaded in April 2017 from: http://dspace.cread.dz:8080/bitstream/CREAD/138/1/2.L%E2%80%99entreprise%20alg%C3%A9rienne%20et%20l%E2%80%99innovation.pdf.
188. The European Union's TEMPUS initiative. The report was downloaded in April 2017 from: http://eacea.ec.europa.eu/tempus/participating_countries/overview/libya_overview_of_hes_final.pdf.
189. "Scientific Research in Libya: The Role of the New Generation of Researchers;" O A Tashani, Libyan Journal of Medicine. 2009; 4(4): 129–130. Published online December 1st, 2009 (http://dx.doi.org/10.4176/090716). Downloaded in April 2017 from: https://www.ncbi.nlm.nih.gov/pmc/articles/PMC3066747/.
190. See relevant sections in parts dedicated to discussion of research publications and patent applications.
191. Several others among the high human development index group of countries managed to dedicate considerably more than Libya to education.
192. "SMEs in Libya's Reconstruction: Preparing for a Post-Conflict Economy;" OECD Report, issued in 2016. http://dx.doi.org/10.1787/9789264264205-3-en. Executive summary accessed in April 2017 at:http://www.keepeek.com/Digital-Asset-Management/oecd/development/smes-in-libya-s-reconstruction/executive-summary_9789264264205-3-en#.WQSfMIiGPIU#page1.
193. Presentation by InfoDev titled "Establishing a Libyan Incubator for Technology and Innovation;" Seidel Uwe. 16th November, 2007. Retrieved in April 2017 from: https://www.infodev.org/infodev-files/resource/idi/document/Status%20of%20establishing%20Libyan%20incubator,%20Seidel%20Uwe.pdf.
194. "SMEs in Libya's Reconstruction: Preparing for a Post-Conflict Economy;" OECD Report, issued in 2016. http://dx.doi.org/10.1787/9789264264205-3-en. Executive summary accessed in April 2017 at:http://www.keepeek.com/Digital-Asset-Management/oecd/development/smes-in-libya-s-reconstruction/executive-summary_9789264264205-3-en#.WQSfMIiGPIU#page1.
195. "Libya 2030 Vision Committee holds first meeting;" S Zapita, 25th May, 2013. Article accessed in April 2017 at: https://www.libyaherald.com/2013/05/25/libya-2030-vision-committee-holds-first-meeting/.
196. "SMEs in Libya's Reconstruction: Preparing for a Post-Conflict Economy;" OECD Report, issued in 2016. http://dx.doi.org/10.1787/9789264264205-3-en. Executive summary accessed in April 2017 at: http://www.keepeek.com/Digital-Asset-Management/oecd/development/smes-in-libya-s-reconstruction/executive-summary_9789264264205-3-en#.WQSfMIiGPIU#page1.

197. See table annexed to the section on tertiary education.
198. "The Science, Technology and Innovation Policy (STIP) Review;" The United Nations Conference on Trade and Development (UNCTAD). Downloaded in May 2017 from: http://unctad.org/en/Docs/dtlstict20096_en.pdf.
199. The UNCTAD Report ("The Science, Technology and Innovation Policy (STIP) Review;" The United Nations Conference on Trade and Development (UNCTAD). Downloaded in May 2017 from: http://unctad.org/en/Docs/dtlstict20096_en.pdf) uses the term STI strategy, instead of STI policy. The latter terminology is used here instead, whereas the term "strategy" is preserved to refer to the range of modalities adopted in order to achieve specific policy objectives.
200. Including the Ministry of Health, the Ministry of Agriculture and Livestock, the Ministry of Fisheries, the Ministry of Islamic Affairs and Traditional Education, the Ministry of Culture and the Ministry of Civil Service and the Modernization of the Administration.
201. Acronym derived from the French institution's name "Institut Mauritanien de Recherche Océanographiques et des Pêches."
202. Acronym derived from the French institution's name "Centre National de Recherche Agronomique et du Développement Agricole."
203. Acronym derived from the French institution's name "Centre National de l'Élevage et des Recherches Vétérinaires."
204. See endnotes 198.
205. "Mauritania Agricultural Science and Technology Indicators;" Stads G–J, et al. (ASTI) Country Brief No. 15. Washington, D.C. International Food Policy Research Institute (IFPRI) and The Hague: Netherlands: International Service for National Agricultural Research (ISNAR). 2004.
206. See endnotes 198.
207. Figure retrieved from database published on database posted by NationMaster. NationMaster website. Accessed in May 2017 at: http://www.nationmaster.com/country-info/stats/Education/Scientific-and-technical-journal-articles.
208. Figure retrieved from the World Bank's database accessed in May 2017 at: http://data.worldbank.org/indicator/IP.JRN.ARTC.SC.
209. See endnotes 198.
210. "The UNESCO Science Report: Toward 2,030." ISBN 978-92-3-100129-1. Downloaded in May 2017 from: http://unesdoc.unesco.org/images/0023/002354/235406e.pdf.
211. World Bank's World Development Indicators. Downloaded from: http://databank.worldbank.org/data/reports.aspx?source=world-development-indicators#.
212. "The Science, Technology and Innovation System of the Kingdom of Morocco;" A UNESCO document authored by N. Boshoff. The document is not dated, but appears to have been published post 2007. Downloaded in July 2016 from: http://portal.unesco.org/education/en/files/55536/11998897825Morocco.pdf/Morocco.pdf.
213. Internet searches aimed at tracing the "national R&D strategy for the next 20 years," referred to in the above-mentioned UNESCO document did not yield any results.
214. Reported by an official website of the Moroccan Government. Official website of the Moroccan Government. Accessed in April 2017 at: http://www.maroc.ma/ar/السيدة-سمية-بنخلدون-تستعرض-بتونس-تجربة-المغرب-وتوجهاته-الاستراتيجية-للنهوض-بالبحث-العلمي/أخبار.
215. Several other ministries are naturally concerned with S&T capacity building are the ministries of Industry, Trade, Mines, Energy, Health and Agriculture.
216. Acronym derived from the French: "Centre National pour la Recherche Scientifique et Technique."
217. See endnotes 212.
218. Report made available by CNSRT on its website belongs to the year 2011. Accessed in April 2017 at: http://www.cnrst.ma/images/CNRST/PDF/PPR-Statistiques_relatives_aux_186_projets_retenus.pdf.

219. Only centers which could be traced on the Internet, using their English or French names, are included in this list.
220. The UNESCO document of "The Science, Technology and Innovation System of the Kingdom of Morocco;" A UNESCO document authored by N. Boshoff. The document is not dated, but appears to have been published post 2007. Downloaded in July 2016 from: http://portal.unesco.org/education/en/files/55536/11998897825Morocco.pdf/Morocco.pdf, mentions a National Centre for Energy and Nuclear Science and Technology (NCENST) that was established in 1986 a public institution with a principal role to develop national capabilities in energy research, including nuclear energy research. However, no recent information was found on this Centre.
221. Information posted on the IRESEN website. Accessed in April 2017 at: http://www.iresen.org/?lang=en%2F#recent-projects.
222. Acronym derived from the French appellation of the programme: "Forum International des Compétences Marocaines Résidant à l'Étranger."
223. Data from CNRST report for the year 2011. CNRST report for the year 2011. Downloaded in April 2017 from: http://www.cnrst.ma/index.php/fr/publications/rapports-d-activites.
224. "Cooperation opportunities in the Mediterranean Basin;" Publication by Sophia Antipolis and others. Downloaded in April 2017 from: http://www.eurosfaire.prd.fr/7pc/documents/1345711827_final_version_v2.pdf.
225. "Morocco Country Reports, 2012;" The Erawatch Network, by Ilyas Azzioui. Downloaded in April 2017 from: https://rio.jrc.ec.europa.eu/en/q/Morocco.
226. Information from Technopark's website. Accessed in April 2017 at: http://technopark.ma/en/presentation/.
227. "Fact Finding Mission Start-up Ecosystems Morocco & Tunisia;" Report by DIA - Dutch Incubation Association, i.c.w. contract partner Huygens-XC b.v. Delft. The Netherlands. Report dated January 2014. Accessed in April 2017 at: http://dutchincubator.nl/wp-content/uploads/sites/5/REPORT-Start-up-Ecosystems-Morocco-and-Tunisia.pdf.
228. The report quotes reference to the 'Grand Projets de Maroc; '
229. The report highlights 'soft corruption' practices sustained by bureaucratic procedures, involving payments or benefits provided by businesses to government officials in return for speeding up procedures.
230. "Le système national d'innovation au Maroc (National innovation system in *Morocco*.)" B *Cugusi*, Centro Studi di Politica Internazionale. Working paper, January 2008.
231. See endnotes 172.
232. Knowledge Assessment Methodology Database. World Bank Institute. 2008. World Bank. Washington DC. World Bank. Washington DC. 2008.
233. Morocco's Global Innovation Index in 2016 allowed it a ranking in 72nd position globally, seventh among its income group of countries and 12th within the MENA region. Its global ranking is a decided impro,vement upon its ranking as 84th worldwide, only 2 years ago in 2014.
234. Report titled "Transforming Arab Economies: Traveling the Knowledge and Innovation Road;" Issued in 2013 by the Centre for European Integration, the European Development Bank and the World Bank. WB 82736v2. Downloaded in April 2017 from: http://documents.worldbank.org/curated/en/664441468059651205/pdf/827360ESW0v10P00Bo x 379869B00PUBLIC0.pdf.
235. The acronym 'SERST' is derived from the French "Secrétariat d'Etat à la Recherche Scientifique et à la Technologie."
236. The acronym 'PNM' is derived from the French 'Programmes Nationaux Mobilisateurs.'
237. The acronym 'VRR' is derived from the French 'Valorisation des Resultats de la Recherché.'
238. In essence, this revision was aimed at changing the character of scientific and technological establishments from 'public administration institutions' to 'public scientific and technological institutions.'

239. The acronym ‘CNEARS’ is derived from the French ‘Comité National de l’Evaluation des Activités de la Recherche Scientifique.’
240. See endnotes 172.
241. The Institut National de Recherche en Génie Rural, Eaux et Forêts (INGREF) employs presently 45 permanent researchers, 26 technicians, 11 administrative staff and 334 qualified farming workers. See website created by System National d’Information sur l’Eau at: http://193.95.21.66/fr/content/institut-national-de-recherches-en-g-nie-rural-eaux-et-ts.
242. “EU-Tunisia Scientific and Technological Cooperation Agreement;” Draft Road Map for 2010–2011. Downloaded in April 2017 from: https://ec.europa.eu/research/iscp/pdf/policy/tunisia_roadmap_2010-11.pdf.
243. Institution de la Recherche et de l’Enseignement Supérieur Agricoles. Established in 1893. Website of Tunisian Institute for Agricultural Research (INRAT). Accessed in April 2017 at: http://www.inrat.agrinet.tn/.
244. Commissariat à l’énergie atomique.
245. Website of the “Centre d’Etude et de Recherches Economique et Sociale (CERES).” Accessed in April 2017 at: http://www.ceres.rnrt.tn/.
246. Acronym derived from the French ‘Institut National de Recherche Scientifique et Technique.’
247. A European, Comission’s document “Le Systems National de Recherche en Tunisie.” Projet ESTIME Contrat n°510696 Appel d’offre FP6-2002- INCO- COMultilatRTD/SSA-5. Downloaded in April 2017 from: http://www.estime.ird.fr/IMG/pdf/Tunisie_estime_1.pdf.
248. The Arab Knowledge Report.
249. Document published by H. Al Feel on the website of the Tunisian Institute for International Relations, 23rd June 2010. Titled in Arabic “ثحبلا ةموظنم روّطت لحارم سنوت في يملعلا.” Accessed in April 2017 at: https://tunisitri.wordpress.com/2010/06/23/محبياشلا -رهاط-دمحم-يباقنلا-لاقم-ىلع-در/.
250. In the report referred to under Abdelkader DJEFLAT, “Building Knowledge Economies for job creation, increased competitiveness, and balanced development: Individual country overviews with data on STI systems in the Arab countries. Report produced for the High Level International Conference, Organized by the Tunisian Government and ISESCO, In collaboration with the World Bank, Carthage, December 1-3, 2009, the following ministry is quoted as the source of figures on R&D expenditure: “Ministère de le Recherché Scientifique, de la Technologie et du Développement des Compétences.” With information stated as falling under “Le système de Recherche Scientifique et de Développement Technologique en Tunisie.”
251. “Erawatch; A European Commission Support Measure Document;” Last updated in 2009. Kindly provided by Dr Hatem Mhenni. Available at: http://cordis.europa.eu/erawatch/.
252. As reported by Agricultural Science and Technology Indicators (ASTI), an initiative of the Consultative Group on Agricultural Research (CGIAR), led by the International Food Policy Research Institute. Report by “Agricultural Science and Technology Indicators (ASTI),” produced by the Consultative Group on Agricultural Research (CGIAR), led by the International Food Policy Research Institute. Information accessed in April 2017 at: https://www.asti.cgiar.org/tunisia.
253. ‘The Agricultural Research and Higher Education Institution’ is a public institution endowed with financial autonomy. It was created in 1990. The acronym IRESA is derived from the French name of the institution: ‘L’Institution de la Recherche et de l’Enseignement Supérieur Agricoles.’
254. Information in this frame is derived from a report by the European Union. “EU-Tunisia Scientific and Technological Cooperation Agreement;” Draft Road Map 2010–2011 July, 2010. Downloaded in April 2017 from: https://ec.europa.eu/research/iscp/pdf/policy/tunisia_roadmap_2010-11.pdf.
255. See endnotes 254.
256. This is the Ministry of Industry’s ‘Upgrading Plan’ or ‘Plan de Mise à Niveau (PMN)’ launched in 1995, essentially to support industrial enterprises through embedding researchers within their ranks for set periods.

257. Prime d'investissement en Recherche et Developpement (PIRD). Information published on the website of the Ministry of Industry and Commerce. Accessed in April 2017 at: http://www.tunisieindustrie.gov.tn/pird.html.
258. See endnotes 248.
259. Ministère de le recherché scientifique, de la Technologie et du Développement des compétences, Le système de recherche Scientifique et de Développement Technologique en Tunisie, July 2006.
260. Note for instance the level of resources allocated to certain aspects of agricultural research.
261. Unless another currency is indicated.
262. Fellowships should last for 3 weeks to 4 months.
263. Amount allocated to each project. The total number of concurrent projects is not stated.
264. Sums indicated here are for joint research projects. The programme includes funding of other activities, including visiting professors.
265. Funding allocated per project. The total number of concurrent projects is not stated.
266. This amount is allocated per year with the grant presumably available for more than 1 year, possibly for the duration of a Masters or PhD studentship.
267. Grant values and durations arranged on project-by-project basis.
268. A workshop was held on February 25, 2014 to discuss this draft, which was reportedly produced in collaboration with Stanford Research Institute International. Website of the Ministry of Education. Accessed in May 2017 at: http://moedu.gov.bh/hec/news.aspx?id=2015&group_key=news&keywords=Research.
269. "Science, Technology and Innovation for Development; Innovation in the Arab region;" The United Nations Economic and Social Commission for Western Asia's (ESCWA) Committee on Technology for Development. 12th January 2017. Call symbol E/ESCWA/TDD/2017/IG.1/5 (Part II). Downloaded in May 2017 from: https://www.unescwa.org/sites/www.unescwa.org/files/events/files/mlmh_lbtkr_fy_lmntq_lrby.pdf.
270. "National Research Strategy, 2014–2024; Creating a smart Bahrain, based on knowledge and innovation." The Higher Education Council. Downloaded in May 2017 from: http://gulfuniversity.edu.bh/media/document/Bahrain%20National%20Research%20Strategy%20-%20Summary.pdf.
271. Website of the Kingdom University (KU). Accessed in May 2017 at: http://www.ku.edu.bh/wp-content/uploads/2016/09/President-Speech-5-How-Bahrains-Higher-Education-Has-Developed-Over-the-Years-English-Only.pdf.
272. No reasons for dissolving the centre were available at the time of writing this chapter.
273. Website of the Arabian Gulf University's. Accessed in May 2017 at: www.agu.edu.bh.
274. Website of Medical University in Bahrain. Accessed in May 2017 at: http://www.rcsi-mub.com/index.jsp?n = 2211&p = 2210.
275. "Accelerating Entrepreneurship in the Arab World;" Report by the World Economic Forum in partnership with Booze & Company, October 2011. Downloaded in May 2017 from: http://www3.weforum.org/docs/WEF_YGL_AcceleratingEntrepreneurshipArabWorld_Report_2011.pdf.
276. Thus, Saudi Arabia is grouped with 'late adopters' and Kuwait with 'laggards.'
277. Information retrieved from Tamkeen's website in May 2017 at: https://www.tamkeen.bh/Tamkeen-Hosts-Community-Partnership-Event-to-Meet-with-Local-NGOs.
278. See, for example, the Report titled "Innovation Policy for Inclusive Sustainable Development in the Arab Region;" United Nations Economic and Social Commission for Western Asia (UN-ESCWA), 2016.
279. Page 15 of the Report titled "Accelerating Entrepreneurship in the Arab World;" Report by the World Economic Forum in partnership with Booze & Company, October 2011. Downloaded in May 2017 from: http://www3.weforum.org/docs/WEF_YGL_AcceleratingEntrepreneurshipArabWorld_Report_2011.pdf.
280. The level of collaborative research was found to be low, with little R&D done on issues in the petroleum sector and high value-added petroleum derivatives.

281. Technopolis Report, submitted to the Panel in February 2007. Made available through private communication with the author.
282. Based on material presented on the Foundation's website. Website of the Kuwait Foundation for the Advancement of Science. Accessed in February 2017 from: http://www.kfas.org/.
283. At just over 23, the student to faculty staff ratio is significantly higher than the average in developed countries around the world.
284. This is an average taken over the previous 6 years, during which a four-fold increase was witnessed in research spending, from around US$1.93 million to US8.1 million. Relevant part of the report by Bishara and Kadhimi.
285. 34 projects were awarded funding during the period 2001/02–2005/06 as a result of collaboration with these two institutions, amounting to around 5% of research grants during this period.
286. This is reflected in the small values of research grants allocated to individual projects.
287. The Kuwait Oil Company. Website accessed in February 2017 at: https://www.kockw.com/sites/EN/Pages/Profile/Research/Research.aspx.
288. Website of Oman's Research Council. Accessed in May 2017 at: https://www.trc.gov.om/trcweb/.
289. Research Department at the College of Applied Science. Information retrieved in May 2017 from: http://ibri.cas.edu.om/pages/Scientific-Research-Department.aspx.
290. "Science, Technology and Innovation Policy Review - Oman;" United Nations Conference on Trade and Development (UNCTAD). Call symbol: UNCTAD/DTL/STICT/2014/1. Downloaded in May 2017 at: http://unctad.org/en/PublicationsLibrary/dtlstict2014d1_en.pdf.
291. According to the Strategy Project's manager Dr Sharifa Al Harthiyah, of The Research Council. However, attempts to obtain a copy of the Strategy, from this website on the same date were unsuccessful. News report on the Strategy project. Accessed on 21st May 2017 at: https://www.trc.gov.om/trcweb/topics/media/news/6531.
292. Information regarding this programme was retrieved from The Research Council's website in May 2017 at: https://home.trc.gov.om/tabid/151/language/en-US/Default.aspx.
293. See endnotes 290.
294. IATA was established on the basis of a study carried out in cooperation with the Helmholtz Association of German Research Centres, cooperative activities are maintained with centers belonging to this association. Website of the Institute of Advanced Technology Integration (IATI), accessed in May 2017 at: https://www.trc.gov.om/trcweb/node/38.
295. Information on the "Oman Research and Education Network (OMREN)" was retrieved in May 2017 from: https://www.trc.gov.om/trcweb/node/41.
296. Information on the Research Deanship at Sultan Qaboos University was retrieved in May 2017 from: https://www.squ.edu.om/dor/and https://www.squ.edu.om/squresearch/.
297. Website of the Research Department at the College of Applied Science. Accessed in May 2017 at: http://ibri.cas.edu.om/pages/Scientific-Research-Department.aspx.
298. Clearly road construction costs vary according to location, terrain quality, number of lanes as well as detailed specifications. It is also undeniable, that road construction will almost always yield solid outcomes for many years ahead, which may not be seen as case for allocating resources to research activity. Nevertheless, the analogy here is intended as a rough illustration of the level of financial support being made available to scientific research. The figure of around US8 million per kilometre is quoted in "Science, Technology and Innovation Policy Review - Oman;" United Nations Conference on Trade and Development (UNCTAD). Call symbol: UNCTAD/DTL/STICT/2014/1. Downloaded in May 2017 at: http://unctad.org/en/PublicationsLibrary/dtlstict2014d1_en.pdf.
299. Report posted on the website of the Deanship of Research at the Sultan Qaboos University. Report downloaded in May 2017 from the website of the Deanship of Research at the Sultan Qaboos University at: http://www.squ.edu.om/dor/Research-Department/AnnualReports.
300. "Country Report; Oman;" Bertelsmann Stiftung's Transformation Index (BTI) 2016. Downloaded in May 2017 from: https://www.bti-project.org/fileadmin/files/BTI/Downloads/Reports/2016/pdf/BTI_2016_Oman.pdf.

301. Such as the temporary shutting-down of a fish oil and powder producing factory in Jaalan Bani Bu Ali in January 2015.
302. The Research Council's website. Information retrieved in May 2017 from: https://www.trc.gov.om/trcweb/node/36.
303. The Industrial Innovation Centre is located in the Knowledge Oasis within the Muscat industrial district.
304. By 2013, IIC had been engaged in 29 projects of which 10 made significant progress toward of prototyping and market testing, however, a variety of reasons appear to have hindered full commercialization of some of these projects. See endnotes 290.
305. The Research Council's website. Information retrieved in May 2017 from: https://home.trc.gov.om/tabid/194/language/en-US/Default.aspx.
306. See endnotes 299.
307. "Science, Technology and Innovation Policy Review - Oman;" United Nations Conference on Trade and Development (UNCTAD). Downloaded in May 2017 at: http://unctad.org/en/PublicationsLibrary/dtlstict2014d1_en.pdf.
308. Additionally, awards were set up to honour outstanding achievements in research in a variety of areas including social science, economics, exact and applied sciences, in addition to art, literature and Islamic studies.
309. "Scientific Research and Development in the State of Qatar;" by the Technical Cooperation Department in the General Secretariat of the Council of Ministers. Report dated September 2008. (*An Internet search for this report in Feb 2017 was unsuccessful.*).
310. At least in the case of satellite university campuses established in Qatar, the Qatar Foundation is reported as having paid for all buildings, overhead, and staff salaries, whereas the universities receive the student fees. "In Qatar's 'Education City,' U.S. Colleges Build Atop a Gusher." *The Chronicle of Higher Education*, April 22, 2005. Retrieved in February 2017 from: http://rlp.hds.harvard.edu/faq/education-city.
311. Press reports indicated that an investment amounting to more than $800 million by Qatar Foundation and a group of associated institutions was dedicated to establishing QSTP. Which had been operating as a QF research hub since 2004. "Qatar Foundation pumps $800 million into R&D hub;" by V. Menon, dated 23rd February 2009 in "Arabian Business." Downloaded in February 2017 from:http://www.arabianbusiness.com/qatar-foundation-pumps-800-million-into-r-d-hub-65118.html.
312. A news article published in 2015 by "Doha News" concerning a joint Qatari-Spanish joint venture aimed at establishing a free zone in a coastal location in Qatar, indicates that QSTP operates one of three free zones, allowing its tenants 100% ownership, hiring expatriate employees, although remaining exempt from paying corporate taxes. "Qatari-Spanish JV win QR1.6bn contract to build new economic zone;" by L. Walker, 14th July 2015. Retrieved in February 2017 from: https://dohanews.co/qatari-spanish-jv-win-qr1-6bn-contract-to-build-new-economic-zone/.
313. Vision and Mission statements by the Qatar National Research Fund (QNRF), retrieved in February 2017 from: http://www.qnrf.org/en-us/About-Us/Vision-Mission.
314. Report posted on the website of the Texas A&M University at Qatar. Retrieved in February 2017 from: http://www.qatar.tamu.edu/research/current-research/.
315. "Furthering A Vision For Research;" Interview with F. van den Boom, Published in July 21, 2016. Downloaded in January 2017 from: http://www.qf.org.qa/content/the-foundation/issue-91/furthering-a-vision-for-research.
316. "Sidra's Qatar staff told to expect more delays before hospital opens;" P Kovessy, Doha News 18th June, 2015. Retrieved in February 2017 from: https://dohanews.co/sidras-qatar-staff-told-to-expect-more-delays-before-hospital-opens/.
317. Sidra Health Centre's statement on its website. Downloaded in February 2017 from: http://www.sidra.org/about/.
318. Qatar Biomedical Research Institute (QBRI). Information retrieved in February 2017 from: http://www.qbri.org.qa/about-us/overview.

319. Information on Qatar Computing Research Institute (QCRI) was retrieved in February 2017 from: http://qcri.com/about-qcri.
320. Information on Qatar Environment and Energy Research Institute (QEERI). Retrieved in February 2017 from: https://www.qeeri.org.qa/en/about-us/overview.
321. Information on Qatar Solar Technologies (QSTec) was retrieved in February 2017 from: http://www.qstec.com/about/who-we-are.
322. Website of the Qatar Environment and Energy Research Institute (QEERI). Accessed in February 2017 from: http://www.qeeri.org.qa/en/research-highlights?topic=All&sort_by=created&sort_order=DESC&sorting=created%7CDESC.
323. See, for example, a list of successful research projects posted in April 2016 on the QEERI website, which included a paper titled "Charge and Size Selective Ion Sieving Desalination Membranes;" published by the American Chemical Society's ACS Nano, by scientists at the Institute. The research utilised relatively advanced research methods, including scanning electron microscopy (SEM), and transmission electron microscopy (TEM), in order to investigate the link between the molecular structure of a particular desalination membrane and its resistance to bio-fouling. Although inherently quite important in relation to the chemical constitution, design and future manufacture of future membrane materials, such research would hold benefits for membrane manufacturers rather than more immediate gains in terms of the cost or effectiveness of on-going desalination operations underway in Qatar. An account of the publication was retrieved in February 2017. Information on Qatar Environment and Energy Research Institute (QEERI). Retrieved in February 2017, from: http://www.qeeri.org.qa/en/research-highlights?topic=All&sort_by=created&sort_order=DESC&sorting=created%7CDESC. See also Paper's abstract viewed in February 2017, on the American Chemical Society's website at: http://pubs.acs.org/doi/abs/10.1021/acsnano.6b00181.
324. The Sidra Health Centre's 'About' statement on its website, downloaded in February 2017, for example extolls the role of the Chairperson as well as the fact that a highly renowned architect was responsible for the Centre's building design and a separate outpatient clinic. Before going on to indicate the number of beds offered and the maladies it might be able to deal with.
325. The workforce of the high technology Sidra health centre, for example, is made up of 75% of expatriates. Statement by the Qatar Foundation's website. Statement retrieved in February 2017 from: http://www.qf.org.qa/news/sidra-medical-and-research-center-restructures.
326. Other tasks that SARC was required to carry out included providing advice to industrial establishments, participating in economic feasibility studies of industrial projects, research on

    - agricultural practices;
    - marine and desert flora and fauna;
    - alternative sources of water resources, desalination and conservation;
    - alternative energy resources.

327. See statement on the history of Qatar's Scientific and Applied Research Centre (SARC). Statement posted on Qatar University's website, retrieved in February 2017 from: http://www.qu.edu.qa/offices/research/esc/history.php.
328. The UNDP Human Development Report for the year 2007–08 lists considerable information on technology diffusion, including the number of researchers per million of a country's population. No data is available for Qatar, nor indeed many other Gulf and Arab countries. However, figures given for other countries range widely, with, for example, around 7800 for Finland and almost 1949 for Portugal and 640 for Cyprus. Figures provided in the Report issued by the Council of Ministers on "Scientific Research and Development in the State of Qatar; ' point to a proportion of the order of 0.1%, counting only Qatari researchers in science and engineering subjects, and to a little more than twice this percentage when non-Qatari researchers are taken into account. It is by no means certain that the researchers counted thus are all fully dedicated to research activity. Indeed, because more than half of the researchers, both Qatari and non-Qatari, work in institutions of higher education, it is

highly likely that they would have considerable teaching loads preventing them from being fully dedicated to their research duties. The low proportion of technicians among personnel engaged in research activity is certainly not a great source of help in such a situation.

329. Article by Abdulaziz M Al-Swailem, published on-line by the United States National Centre for Biotechnology Information. Downloaded in January 2017 from: https://www.ncbi.nlm.nih.gov/pmc/articles/PMC4075461/, Presentation made by President of the King Abdulaziz City for Science and Technology (Mohammed Al Suwaiyel), on 26th January, 2015, published on-line by the Global Competitiveness Forum, Saudi Arabia. Downloaded in January 2017 from: www.gcf.org.sa/en/.../MOHAMMED%20AL%20SUWAIYEL%2026th%20am.pptx, Information from the KACST website, retrieved in January 2017 from: http://maarifah.kacst.edu.sa/NSTIPWCPortal/, and The Annual Report of King Abdulaziz City for Science and Technology, for the year 2014. Downloaded in January 2017 from: https://www.kacst.edu.sa/eng/about/AnnualReports/KACST%202014%20Annual%20Report%20English%20LR.pdf.
330. This programmes designation 'Maarifah' comes from the word for 'knowledge' in Arabic.
331. The Annual Report of King Abdulaziz City for Science and Technology, for the year 2014. Downloaded in January 2017 from: https://www.kacst.edu.sa/eng/about/AnnualReports/KACST%202014%20Annual%20Report%20English%20LR.pdf.
332. A number of technology innovation centers are reported as having been established since 2011 at various Saudi universities including KSU, KFUPM, KAU, KAUST to deal with technology acquisition in areas such as radio frequency and photonics for the e-Society, carbon capture and sequestration, personalized medicine, solid-state lighting and geo-informatics.
333. Examples of collaborating institutions include The United States National Aeronautics and Space Administration (NASA), International Business Machines (IBM), The University of California at Los Angeles (UCLA), the University of Cambridge and the University of Oxford in the UK. Presentation made by President of the King Abdulaziz City for Science and Technology (Mohammed Al Suwaiyel), on 26th January, 2015, published on-line by the Global Competitiveness Forum, Saudi Arabia. Downloaded in January 2017 from: www.gcf.org.sa/en/.../MOHAMMED%20AL%20SUWAIYEL%2026th%20am.pptx.
334. NSTIP Annual Report issued by KACST in 2012.
352. Project implemented in collaboration with the Ukrainian firm 'Antonov' and TAQNIA Aeronautics.
351. Implemented in collaboration with King Abdullah University for Science and Technology (KAUST) and the University of California at Santa Barbara.
335. The bulk of KAUST staff and student body is drawn from a variety of countries around the world, rendering it a truly international school of graduate studies.
336. This is part of the University's competitive 'Entrepreneur Accelerator' programmes referred to as 'TAQADAM.' Its aims is to develop early stage entrepreneurial concepts into new high-potential start-ups. Early stage start-ups catered for by this programme in 2017 targeted 28 student- and faculty-led start-up teams, chosen from 13 Saudi universities, including KAUST, with interest in starting enterprises based on a wide range of technologies and industries including healthcare, bioinformatics, new materials, the internet of things as well as software development.
337. Essentially electrical, hydrological and mechanical research is carried out at this centre.
338. Research centers with which CRP collaborated with included: the Centre of Research Excellence in Petroleum Refining & Petrochemicals, established by the Ministry of Higher Education, and another newly established centre dedicated to Transformative Research in Petrochemicals and Polymers as well as the California Institute of Technology, USA. Information retrieved from the KFUPM website in February 2017. Accessed at: http://ri.kfupm.edu.sa/crp/polymers-plastics/.
339. Incidentally, it would seem that cooperation between research institutions and likely 'clients' is pursued most effectively in relation to defence-related research, e.g., in relation to the development of sensors as well as communications and surveillance devices.

340. This first edition of the policy document was published in November 2015, following its approval by the UAE President. Policy document published in November 2015. Downloaded in May 2017 from: https://government.ae/-/media/Science-Technology-and-Innovation-Policy.ashx.
341. The STI Committee was established in November 2014. Its membership includes the ministers for Cabinet Affairs, Higher Education and Scientific Research, Economy, Education, health, energy and financial Affairs, Telecommunications Regulatory Authority Chairman of UAE Space Agency.
342. Miniaturized satellites used for space research, constructed out of several cubic modules of units. CubeSats have a low mass of not more than 1.33 kg per unit and often use commercial off-the-shelf (COTS) electronics components.
343. Chapter 6 of the WIPO Report on the Global Innovation Index. Downloaded in June 2017 from: http://www.wipo.int/edocs/pubdocs/en/wipo_pub_gii_2014-chapter6.pdf.
344. "Entrepreneurship, SMEs and local development in Abu Dhabi; Boosting the entrepreneurial ecosystem." OECD Report downloaded in June 2017 from: https://www.oecd.org/countries/unitedarabemirates/AbuDhabi-Ecosystem-Final-Web.pdf.
345. Statement by the NRF director. Accessed in June 2017 at: http://www.nrf.ae/EFRMA2015.aspx.
346. The United Arab Emirates University is reported as having filed around 55 invention patents, with around 20 patents granted to the university by 2014. The United Arab Emirates University also reports strong research partnerships on oil and gas, water, health care, agricultural productivity, environmental protection as well as traffic and traffic safety issues. Information obtained from the UAEU website in May 2017 at: http://www.uaeu.ac.ae/en/sip/.
347. Information presented in Information obtained from the UAEU website in May 2017 at: http://www.uaeu.ac.ae/en/sip/. However, press reports indicate that the UAEU Science and Innovation Park was inaugurated as recently as November 2016, which might explain the dearth of information concerning its activities on the UAEU website. See also See the press report published by the Nation titled "Mohammed bin Rashid launches UAE University Science and Innovation Park;" Dated 22 November 2016. Downloaded in June 2017, from: http://www.thenational.ae/uae/education/mohammed-bin-rashid-launches-uae-university-science-and-innovation-park.
348. Some which had not become operational, in 2017.
349. Information retrieved from the Masdar Institute's website in May 2017 at: https://www.masdar.ac.ae/.
350. The Nature Index website. Information retrieved in June 2017 from: https://www.natureindex.com/institution-outputs/united-arab-emirates/masdar-institute-of-science-and-technology/513906c134d6b65e6a00082f.
353. Source: Yemen's Higher Education and Scientific Research Strategy document. Downloaded in May 2017 from the website of the Ministry of Higher Education and Scientific Research at: http://www.yemen.gov.ye/portal/Portals/14/استراتيجية%20التعليم%20العالي.pdf.
354. The strategy document is billed as covering the years 2006–13. However, as its budgetary projections would suggest, a much longer time span, reaching the year 2025. See Figure (3.31).
355. On the face of things, this statement is somewhat inexplicable, because promotion is often used by universities around the world as a legitimate incentive in promoting university research output. On the other hand, it is not clear why university teaching staff are not required to direct their publications in a manner that concurs with the university's strategy, in order for these publications to help their promotion prospects.
356. The 2013 ECOSOC Annual Ministerial Review Regional preparatory meeting for Western Asia "Science, Technology and Innovation for Sustainable Development" BACKGROUND NOTE 26 November 2012. Hashemite Auditorium in the Royal Scientific Society ESCWA Technology Centre El Hassan Science City Amman, Jordan.
357. The Strategy document indicates that around 70% of available financial funding is dedicated to staff salaries and student grants.

358. Report was sponsored by the Arab Organisation for Agricultural Development (AOAD) and the International Service for National Agricultural Research (ISNAR), with support by the United Nations Development Programme (UNDP). "A Review of the Agricultural Research System in the Republic of Yemen." Report presented in 1993, to the Ministry of Agriculture and Water Resources. Sponsored by the Arab Organisation for Agricultural Development (AOAD) and the International Service for National Agricultural Research (ISNAR), with support by the United Nations Development Programme (UNDP).
359. "Systematic destruction of Yemen's agricultural infrastructure;" posted in April 2014, on a website (The Green Dream الحلم الأخضر) dedicated to Yemen's environmental affairs. Website accessed in May 2017 at: http://holmakhdar.org/news/investigations/item/269-2014-04-02-22-27-30.html.
360. "World Bank Provides US$13 Million to Higher Education and Scientific Research in Yemen." Published in April 2010. Downloaded in May 2017, from: http://www.worldbank.org/en/news/press-release/2010/04/25/world-bank-provides-us13-million-to-higher-education-and-scientific-research-in-yemen.
361. See endnotes 360.
362. Report published by the United Nations Economic Commission for Western Asia (UN-ESCWA) in 2015, as background document to the Arab Sustainable Development Report. Downloaded in May 2017 from: http://css.escwa.org.lb/SDPD/3572/YemenAr.pdf.
363. Thus, judging by the example of the American Association of Yemeni Students and Professionals, the main purpose of such organisations appears to target improving the lot of their own communities abroad rather than extending a hand to their country of origin. Statements on the website of the American Association of Yemeni Students and Professionals. Website accessed in May 2017 at: http://www.aaysp.org/index.html.
364. Acronym from the French "Institut National de Recherche pour l', Agriculture, la Pêche et l'Environnement."
365. "Pilot Agricultural Services Project." World bank Report No. 15174-COM. Completed in December 1996. Downloaded in May 2017 from: http://documents.worldbank.org/curated/en/607051468770051538/pdf/multi-page.pdf.
366. See endnotes 212.
367. "Document de Strategie Enseignement Superieur et Recherche; 2015–2019." Posted by Djibouti's Ministry of Higher Education and Research. Downloaded in May 2017 from: http://www.mensur.gov.dj/Textes/Strategies/STRATEGIE%20ENSEIGNEMENT%20SUPERIEUR%20ET%20RECHERCHE.pdf.
368. "Research, knowledge and education-based systems in Somalia;" FA Mursal. Downloaded in May 2017 from: http://www.inasp.info/uploads/filer_public/70/97/70970d71-3157-496b-a765-df1ffeb1fed4/somali_regions_countrystudy.pdf.
369. See endnotes 368.

Chapter 4

# Research Output by the Arab Countries

## Chapter Outline




Science, Technology, Innovation, and Development in the Arab Countries
http://dx.doi.org/10.1016/B978-0-12-812577-9.00004-0

The volume of research output produced by a given country, as expressed by the number of research publications in international refereed journals, as well as patents applied for and registered, constitutes a reliable measure of research activity. While the Arab countries' research output currently makes up only a small proportion of the global yield of research publications and an even smaller proportion of patents, a review of orientations and quality of such output, as measured by commonly used indices, is informative in many ways.

The present chapter is divided into two distinct parts. The first is concerned with research articles published in refereed journals by the various Arab countries while the second considers the intensity of patenting activity across countries and fields of applications.

## 4.1 RESEARCH PUBLICATIONS

Both in terms of their numbers as well as orientations across domains of science and technology, research publications provide invaluable information concerning scientific activity and the extent to which this activity might help attain national development goals. Drafting novel policy regimes for the Arab countries should therefore greatly benefit from both quantitative and qualitative analyses of research publications.

## 4.1.1 Introduction

Evaluating the extent and quality of research activity undertaken by various countries and institutions witnessed significant advances over the past few decades. Procedures developed based on citations made of them in subsequent literature articles now play a prominent role in ranking countries, institutions, and individual scientists, as well as their respective contributions to a global body of scientific literature. Applying such procedures to research output by Arab research institutions is invaluable for future efforts aimed at overhauling their capacity for research, at the national, institutional, and individual levels.

Similarly, looking into patterns of coauthorship should help in drafting optimal policy directives and in the design of collaborative ventures with particular reference to interinstitutional as well as regional and international cooperation.

The following sections will look at publications produced by researchers in the Arab countries during the period 1996–2015 in comparison to researchers in selected neighboring countries and around the world. In addition to comparing overall numbers of research documents published by the Arab countries to those produced by colleagues in the region and worldwide, the following sections will also consider the distribution of research publications across various disciplines. Efforts aimed at impact evaluation are also considered.

## 4.1.2 Research Publications by the Arab Countries: An International Perspective

Based on records of scientific research publications collected within a database dedicated to monitoring research publications,[1] Figs. 4.1 and 4.2 present a view of the annual research publications by all countries around the world for which

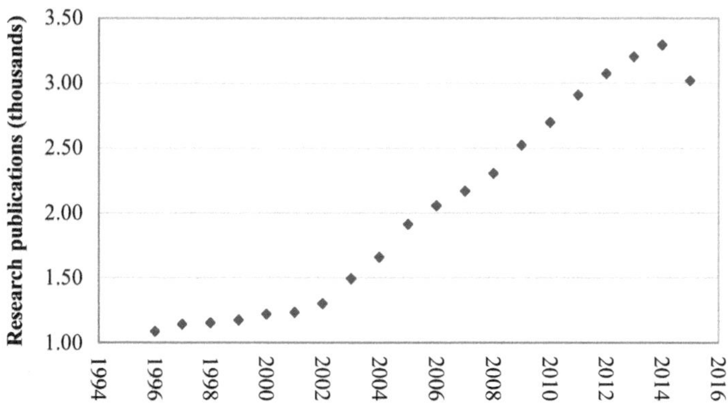


**FIGURE 4.1** Growth of research publications by all countries during 1996–2015. http://www.scimagojr.com/countryrank.php?year=1996.

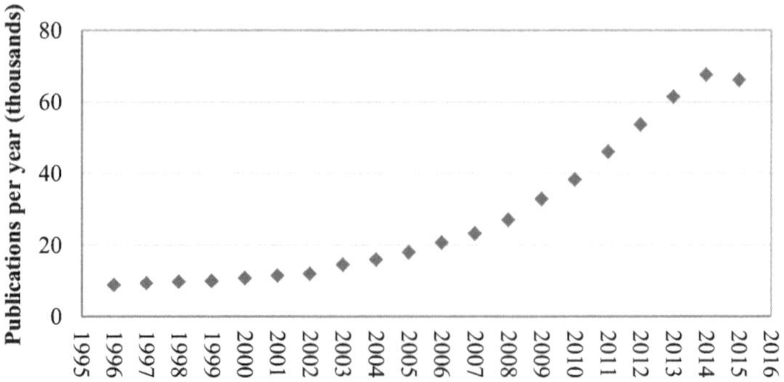


**FIGURE 4.2** Evolution of research publications by all Arab countries during 1996–2015. http://www.scimagojr.com/countryrank.php?year=1996.

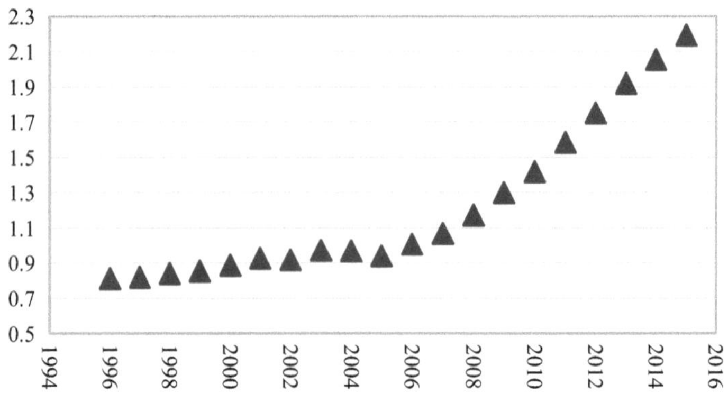


**FIGURE 4.3** Percentage of AC publications out of publications by all countries. http://www.scimagojr.com/countryrank.php?year=1996.

data is available, and for all the Arab countries, during the period 1996–2015. According to this database, researchers in the Arab countries published a total of over 556,000 documents during this period, representing a mere 1.37% of the total number of research documents published worldwide.[2]

Fig. 4.1 indicates that worldwide research output reported for the period 1996–2015 went essentially through four phases.[3] In the first, between the years 1996–2000, research output recorded by the database increased at a somewhat modest rate of around 33,000 publications per year. This rate went up considerably during 2001–10, assuming an annual average of around 170,000 publications, before settling down to around 130,000 publications per year during the years 2011–14, and finally experiencing a downturn between 2014 and 2015.[4]

By contrast, Fig. 4.2 shows that the number of research publications made by all the Arab countries combined was in the vicinity of 9000 publications around 1996. It then grew at a rate of 500 publications per year for the next 6 years before starting to climb by a much higher annual average of more than 2000 publications during the period 2002–07 and then at more than 7000 publications a year during 2009–14.[5] Thus, publications by the Arab countries seem to have increased at a much higher rate relative to world publications, particularly during this period. This trend is made clear by Fig. 4.3, which presents the variation of the percentage of research publications by all the Arab countries relative to worldwide publications. This figure essentially indicates that research publications by the Arab countries grew from a little over 0.8% in 1996 to around 2.2% of worldwide publications in 2015.[6]

Additionally, during the period 2002–13, annual averages for publications by the Arab countries ranged between a minimum value at 10.6% for the years 2003–04, and a maximum of 21.2% for the years 2008–09 with an overall average for the years 2003–13 at 16.1%. Corresponding values for worldwide publications were 4.2% for the years 2012–13 and 15.6% for the years 2004–05, with an overall average for the years 2003–13 at 8.6%. Similarly, the compound annual growth rate for publications produced by the Arab countries during 2009–14 stood at around 15.6%, in contrast to 5.5% for publications produced by all countries worldwide. Overall, publications by the Arab countries grew by 325% during the period 2003–13, while those by all countries around the world grew 115% during the same period.

#### *4.1.2.1 Arab Countries' Research Publications Compared to Combined Output of Iran and Turkey*

Comparing, to start with, the total number of research publications produced during the period 1996–15 by the Arab countries combined to the number of publications produced by Iran and Turkey taken together, it appears that, combined, the latter two countries managed to produce a total of over 786,000 research publications while the Arab countries combined produced less than 552,000.[7] That is to say, the total number of publications produced by the Arab countries over the period 1996–15 is less than 72% of what Iran and Turkey managed to produce during the same period. This percentage must naturally be looked at with the differences that exist between populations and resources possessed by the Arab countries and their two closest neighbors, in mind.[8]

Fig. 4.4 presents a view of the evolution of the share of the Arab countries taken together as well as both Turkey and Iran combined, out of all publications made worldwide during the years 1996–16. This figure clearly indicates that the combined performance by both Turkey and Iran was superior to that of the Arab countries taken together. A point to bear in mind is that this has not always been the case. Rather, the share of the Arab countries out of worldwide publications was superior to that of both Iran and Turkey combined until the first few years of the present century. However, while the percentage of publications by all the Arab countries increased only modestly between the years 1996–05, before following a steep upward path around 2006, that achieved by Turkey and Iran combined began an upward swing at the turn of the century. At any rate, this figure demonstrates that the percentage of research publications produced annually by Iran and Turkey during the period 2006–12 relative to worldwide research publications exceeded that for Arab countries' publications by around 0.75%. This difference appears to have narrowed down to around 0.3% by 2015.

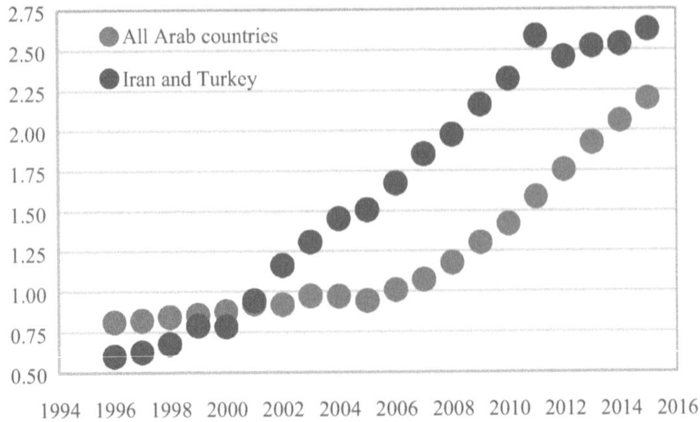


**FIGURE 4.4** Share of worldwide research publications by all the Arab countries compared to the combined share of Iran and Turkey during 1996–2015.

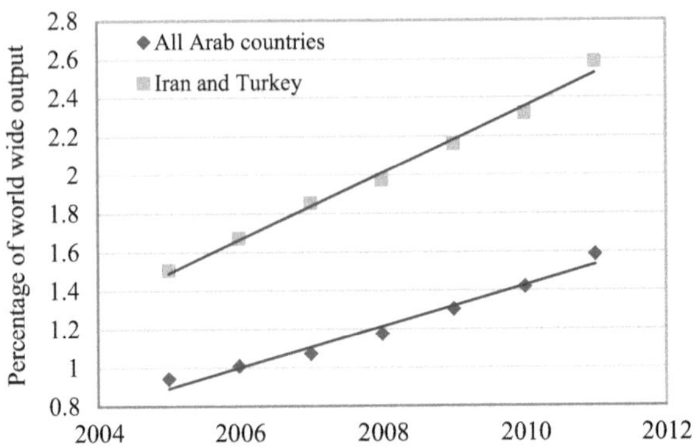


**FIGURE 4.5** Evolving share of scientific publications by all the Arab countries worldwide compared to the combined share of Iran and Turkey.

Additionally, the rate at which research publications by Turkey and Iran grew exceeded the corresponding rate for the Arab countries combined. This is most apparent during the period 2005–11, which witnessed almost linear growth for both groups. At around 0.107, the coefficient of the variable term in the linear equation describing the rate of increase of the Arab countries' share of worldwide publications during this period is considerably lower than the corresponding coefficient for the combined share of Iran and Turkey, which stood at 0.173. See Fig. 4.5.

### *4.1.2.2 Research Output by Selected Groups of Arab Countries*

It is tempting to group the Arab countries within five categories based on their research output, as portrayed by annual publications within journals covered by the Elsevier database during the past 5–6 years. At the very top of these categories, it would be fair to place Saudi Arabia and Egypt, with their annual publications since 2010 close to or exceeding 10,000 research documents annually. Second would be a group of four Arab countries, including the Maghreb countries, Algeria, Tunisia, and Morocco, as well as the United Arab Emirates, with between 4000 and 6000 publications, annually. Third, one might group Iraq, Jordan, Lebanon, Kuwait, Oman, and Qatar, with annual publications of 1000 to 4000 documents. Fourth, it would be logical to have Bahrain, Palestine, Sudan, Syria, and Yemen, within a penultimate group of Arab countries whose research publications fall between 100 and 500 documents per year. Lastly, one might group Mauritania, Comoros, Djibouti, and Somalia, whose research publications have fallen below 50 per year for the period covered by available statistics. Indeed, publications by Somalia and Comoros often fall below 20 per year while those by Mauritania, the most prolific of these four African Arab countries hover between 20 and 30 annually.

It is also possible to classify the Arab countries in relation to their research publications output around the middle of the present decade, into three categories, neglecting output produced by the least prolific countries, i.e., Mauritania, Comoros, Djibouti, and Somalia, and amalgamating the second and third categories in the above classification. Within this scheme, Egypt and Saudi Arabia would constitute sole members of the first category of most prolific producers of research publications. The second category would be populated by the North African Arab countries of Algeria, Morocco, and Tunisia, and the Gulf Arab countries of Kuwait, Oman, Qatar, and the United Arab Emirates, as well as three countries of the Levant, namely Iraq, Jordan, and Lebanon. The third category would then include all the remaining Arab countries, namely Bahrain, Libya, Palestine, Syria, Sudan, and Yemen, which seem to operate below their potential as they produce a significant number of publications, while they possess the capacity for considerably larger contributions.

The above grouping should be useful in guiding future national and regional policies and collaborative initiatives aimed at promoting future research efforts with substance and impact meriting their publication in reputed international journals.

The following paragraphs will take a closer look at published research output by Egypt and Saudi Arabia in comparison to one another and in relation to research output published by two of the Arab countries' closest neighbors, namely Iran and Turkey. Subsequent paragraphs are dedicated to a review of research publications for less publication-prolific subgroups.

#### *4.1.2.3 Research Publications by Egypt and Saudi Arabia*

Figs. 4.6 and 4.7 provide a closer view of the growth of publications made by researchers in the two most prolific Arab countries, namely Egypt and Saudi

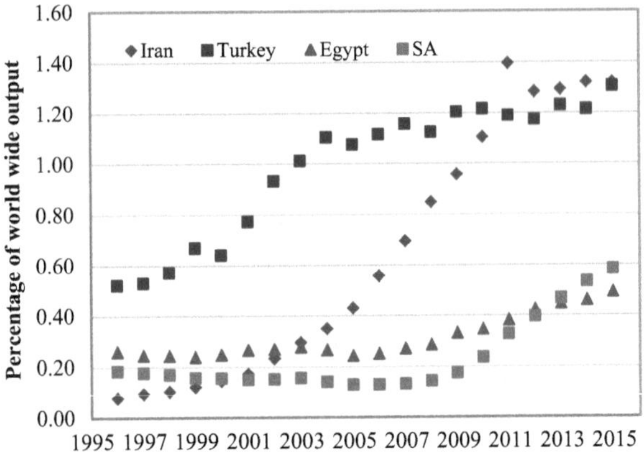


**FIGURE 4.6** Evolving world share of scientific publications by Egypt, Saudi Arabia, Iran, and Turkey.

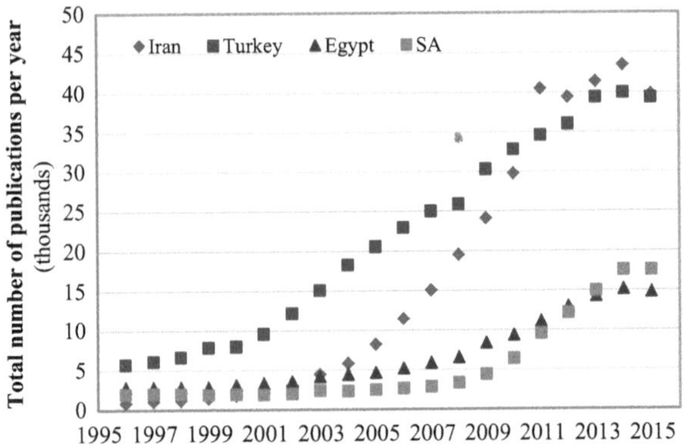


**FIGURE 4.7** Growth in all publications by researchers in Iran, Turkey, Egypt, and Saudi Arabia over the period 1996–2015.

Arabia, in comparison to those made by both Iran and Turkey, two of their closest non-Arab Middle East neighbors, during 1996–2015.[10] Both figures are in broad agreement with the essential conclusions of a report published in 2011 by Thomson Reuters, some of which are summarized in Frame 4.1.[11] Thus, both figures clearly indicate that the combined publications output by Iran and Turkey is far superior to that produced by both Egypt and Saudi Arabia.

Additionally, Turkey's share of world output more than doubled between 1996 and 2015, while Iran's share grew at an even higher rate, by more than 16-fold from around 0.08% in 2000 to 1.32% in 2015. On the other hand, as shown in Fig. 4.7, Egypt and Saudi Arabia, displayed a much flatter

**Frame 4.1 Report by Thomson Reuters on Changing Landscape of Arabian, Persian, and Turkish Research**

A report published in 2011 by Thomson Reuters (TR) on the Changing Landscape of Arabian, Persian, and Turkish Research indicated that despite the low rate of publications in various scientific disciplines by contributors in the region, many countries did witness considerable growth over the decade. According to this report's findings Turkey's share of world output increased almost threefold from 0.7% in 2000 to 1.9% in 2009 while Iran's share grew at an even higher rate, from less than 0.2% to 1.3% during the same decade.

In comparison, Egypt, Saudi Arabia, and Jordan, which led all other Arab countries in terms of research publications in refereed international journals, displayed a much flatter trajectory throughout. Nevertheless, all three countries have in fact increased the numbers of their refereed publications during the decade covered by the survey.

The TR report also indicated that research publications even by institutions in the less prolific Arab countries of the Middle East have increased during the past decade and have generally continued on an upward path.

Countries that produced the least number of publications exhibited substantial output growth. Thus, Iraq, Jordan, Qatar, and Yemen more than doubled their (admittedly miniscule) share of world output between 2000 and 2009, indicating their potential in future scientific research activity. However, the United Arab Emirates, Kuwait, and Lebanon achieved the highest rates of growth within the group of less prolific Arab countries, while Tunisia and its two neighbors, Algeria and Morocco, increased their annual rate of publications from less than 1000 in 1996 to over 5000 in 2014. Among the group of least productive countries, Bahrain had the lowest rate of growth while the Sudan and Syria had the highest.[9]

trajectory throughout. And while they did in fact increase their share of worldwide publications considerably since 2005, they still produced less than half the number of research documents published by either Iran or Turkey by 2015.[12]

#### 4.1.2.3.1 Controversy Surrounding Saudi Arabia's Publication Output and University Ratings

The number of research publications and the citations they secured over the past 5–6 years has qualified some of Saudi institutions of higher education for much higher world rankings than they had only a few years earlier. Indeed, Saudi Arabia gained the second highest rating by Nature Index in Western Asia as the number of publications produced and the citations they received soared during the recent past.[13] While Saudi Arabia's enhanced research publications output and citation rankings might be attributed to policies allowing generous allocations for graduates on study missions abroad, controversy arose as it was linked to attempts by some Saudi universities to enlist prominent scientists abroad as affiliates. Thus, in 2011, an article in Science claimed that top scientists in a variety of disciplines were offered generous allowances in return for declaring their affiliation to Saudi universities as research fellows or visiting professors. In explaining their moves, Saudi universities claimed that such arrangements allowed Saudi researchers to become heavily involved in research collaboration with top ranking scientists from highly regarded institutions, thereby enhancing their skills and promoting improved international collaboration in scientific research.

As described by the above-mentioned article, see Frame 4.2, policies designed to enhance the ranking and status of Saudi universities appear to be skewed in favor of institutional and national prestige, with few if any apparent measures aimed at capturing the full benefits of collaborative research to attain lasting quality improvements in research and higher education. This said, however, it may be argued that enhanced institutional status attained through

**Frame 4.2 Saudi Arabia's Initiative to Enhance Ratings of Its Major Universities**

In 2011, Science published an article titled "Saudi Universities Offer Cash in Exchange for Academic Prestige." The article was generally critical of King Abdulaziz University's (KAU) aggressive efforts aimed at improving its international ranking by paying highly cited scientists to declare themselves as KAU affiliates. KAU is reported as responding that its goal was essentially to "encourage and enhance its multidisciplinary research" as well as initiate "collaborations with other leading institutions around the globe." As a result of this program around 130 highly cited researchers from various countries around the world listed KAU as a secondary affiliation on Thomson Reuters' highly cited researcher database. That this figure is four times higher than that attained by Harvard University, with the next highest number of secondary affiliations of 32, is underlined by critics of the KAU scheme. It has also been alleged that the KAU scheme, which was initiated in 2010 by its math department, was later emulated by other university departments and research institutions in Saudi Arabia.

On the other hand, the Science report appears to indicate that some of the scientists who joined the Saudi program were never required to even visit Saudi institutions, adding a worrying angle to the entire episode.

In conclusion, it is possible that policies designed to enhance rankings of Saudi Arabia's universities were essentially flawed since they favored quick wins in institutional and national prestige over tangible and long-lasting improvements in quality of research and higher education. Further analysis leading to subsequent revision of these policies is essential in order to find out how improved status is to ultimately impact actual quality, how contracts with top scientists might positively reflect on actual research activity in terms of dynamic responses to actual needs, and ultimately national development.

collaboration with well-known researchers might ultimately reflect in future quality improvements, possibly through attracting further talented teaching staff and researchers. Additionally, any contacts that Saudi researchers might have with top scientists from distinguished institutions abroad should positively reflect on their research activity, in terms of exposure, dynamism, and ultimately quality.

While further analysis would certainly be needed to determine interim and end results of Saudi Arabia's strategies aimed at enhancing the rankings of its higher education and research institutions, both the above press reports as well as anecdotal evidence seem to point out that such strategies might have limited impact in terms of ultimate and long lasting quality improvement. At any rate, it seems more likely than not that lower oil prices, coupled to enhanced defense expenditure, may cause the Saudi government to devote lower levels of expenditure to such initiatives, possibly even rendering the entire scheme unsustainable.

### *4.1.2.4 Research Publications by Less Prolific Arab Countries*

Fig. 4.8 presents patterns of growth in research publications by four groups of Arab countries in decreasing order of overall publications productivity during 1996–2015. These are namely three Maghreb countries, four Gulf countries, three countries in the Levant, as well as six of the least prolific countries, within all subject categories covered by Elsevier's database.

Fig. 4.8 shows Tunisia leading its two neighbors, Algeria and Morocco, with the rate of growth of its publications rising more steeply from beginnings inferior to that of Morocco and close to that of Algeria in 1996 to exceed these two countries by 1500–2000 publications per year throughout the period 2010–14. Morocco's decline from a position of leadership during 1996–2002 to one in which it distinctly trails behind its two neighbors, during 2007–15 is notable. It is also to be noted that Algeria, a country that enjoys almost four times Tunisia's

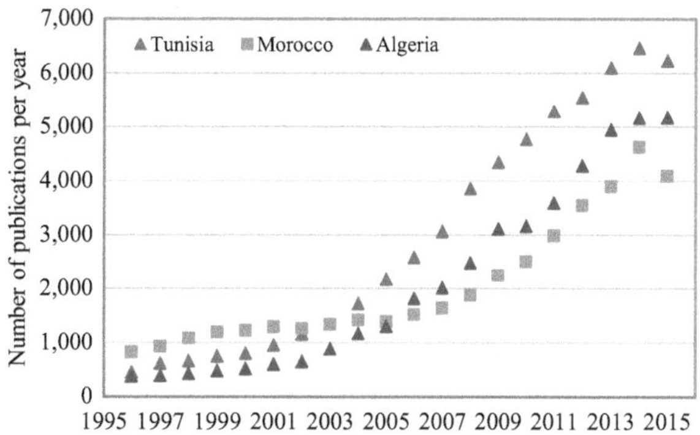


**FIGURE 4.8** Growth of publications by researchers in Algeria, Morocco, and Tunisia over the period 1996–2015.

gross nominal national product and population, trails behind its neighbor in terms of research publications.

Fig. 4.9 indicates that Qatar, the United Arab Emirates, and Oman made considerable gains in their research output, while Kuwait managed modest progress. Thus, during the period 2003–14, the United Arab Emirates and Qatar managed to increase their rate of annual publications by nearly 5- and 17-fold, respectively, Oman managed a 3.5-fold rise in the numbers of its research publications, while Kuwait barely managed a 2-fold increase.

Fig. 4.10 depicts growth of research publications produced by Iraq, Jordan, and Lebanon during 1996–2015. The remarkable recovery made by Iraq, from low publication levels during 1995–2003 to levels approaching Lebanon's output, is notable.[14] Even more notable is the fact that Lebanon and Jordan, two countries in possession of fewer financial and manpower resources managed to maintain relatively high levels of research activity, with their published output taking an upward path starting around the years 2001–03.

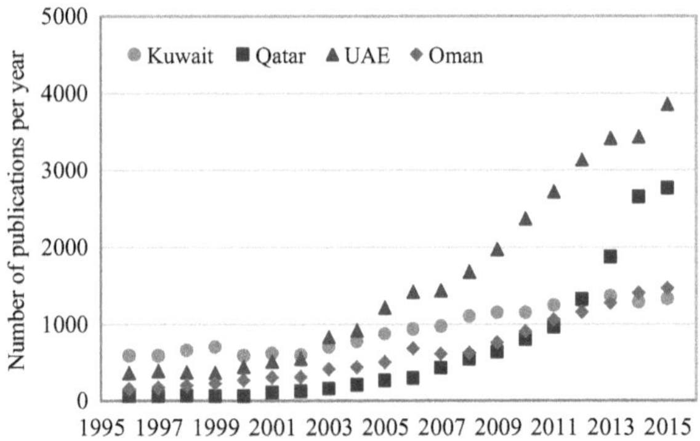


**FIGURE 4.9** Growth in all publications by researchers in Jordan, Lebanon, Kuwait, Qatar, and the United Arab Emirates, 1996–2015.

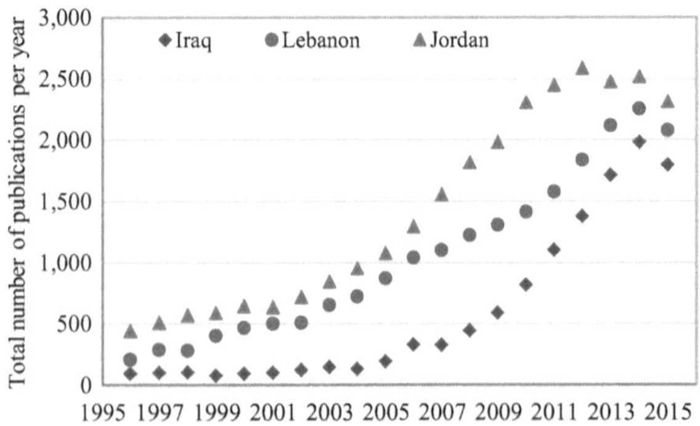


**FIGURE 4.10** Growth in publications by researchers in Syria and Iraq over the period 1996–2015 SC-10.

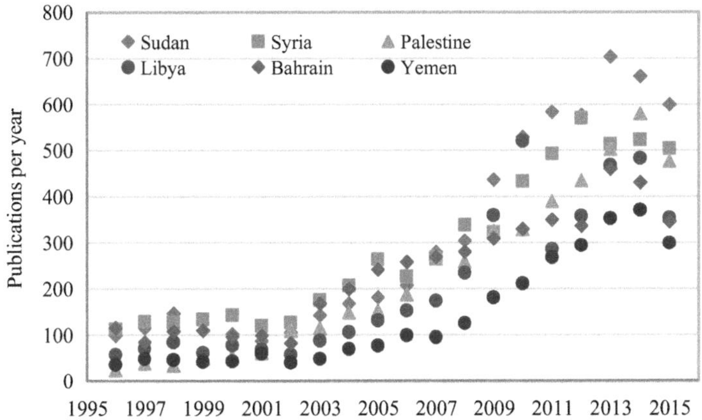


**FIGURE 4.11** Research publications by the least prolific Arab countries, 1996–2015.

Fig. 4.11 portrays the evolution of research publications in six Arab countries of the Middle East and North Africa, namely Bahrain, Libya, Palestine, Sudan, Syria, and Yemen, indicating that even these countries managed to increase their research output since the late 1990s, from levels of around 100 to between 400 and 700 research publications per annum. Significant improvement in publications of these countries relative to the rates during the mid-1990s appears to have been made across all disciplines since 2004–05, reaching maximum levels around 2013–14. Notably, Syria trails behind its closest neighbors, producing one-half to one-fifth of the annual publications produced by Lebanon and Jordan for much of the period covered by this chart. In effect, the level of scientific research output published by Syrian institutions throughout the period covered by Fig. 4.11 is disappointing, as it falls below 600 published documents at the highest point attained in the year 2011.

### *4.1.2.5 Comparing Research Publications in 2014 to Those Produced in 1996*

Fig. 4.12 illustrates the extent to which various Arab countries managed to enhance their research publications output over the years 1996–2014.[15] The bar chart in this figure compares the ratios of publications produced in 2014, by 18 Arab countries, four of their neighbors, namely Cyprus, Iran, Israel, and Turkey, as well as all countries around the world, to those produced in 1996. This bar chart clearly indicates that, with the exception of Kuwait, all the Arab countries increased their research publications between 1996 and 2014 to a larger extent than the world's average. Qatar, Palestine, Iraq, Tunisia, and Algeria made the most remarkable gains, increasing their research publications output by ratios of around 42, 25, 21, and 15, respectively, over the period 1996–2014.

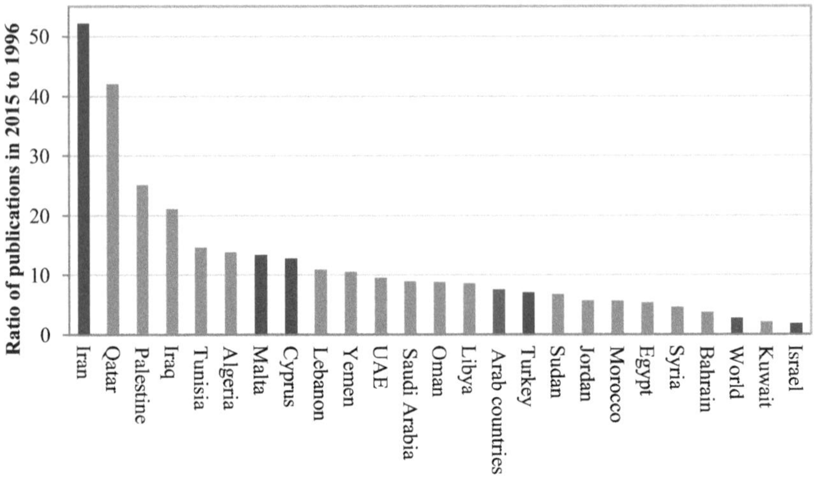


**FIGURE 4.12** Ratio of research publications produced by the Arab countries and four of their ME neighbors in 2014 to those produced in 1996.

It is possible to credit the meteoric rise in research output exhibited by Qatar to policies implemented over the past decade or so aimed at hosting overseas branches of world-class academic institutions as well as the establishment of the Qatar Foundation (QF), one of whose primary objectives is promoting research activity. The same may be true, though perhaps to a lesser extent, in the case of the United Arab Emirates. It is also possible to argue that gains made by Saudi Arabia, as it enhanced its publications output nearly 10-fold between 1996 and 2014, might also be due to policies encouraging widely cited authors to become affiliated to its universities as well as devoting further resources to higher education both at home and abroad. See Section 4.1.2.3.1.

On the other hand, several prolific producers of research publications, such as Egypt, among the Arab countries, and Turkey and Israel in particular, among their Middle East neighbors, made far more modest gains in the intervening years. The argument that some of these countries have achieved their potential in terms of institutional capacities and resources allocated to research activity may not be valid, however. This must particularly be the case in a country the size of Egypt; with its rising population level and wide range of socioeconomic development issues, a considerable proportion of which would certainly require intensive research activity resulting in increased publications. Notable among the other countries that could not manage significant gains in the volume of their research publications are both Syria and Kuwait. At any rate, none of the Arab countries matched the enormous increase in the number of research publications achieved by Iran, which managed to enhance its output by around 52-fold between 1996 and 2015.

#### 4.1.2.6 *World Rankings of the Arab Countries Based on Their Research Publications*

A country's world ranking based on the volume of its research publications within refereed international journals reflects the relative maturity of its scientific institutions. Based on the latest Elsevier world rankings for the year 2015,[16] Saudi Arabia and Egypt ranked 33rd and 36th, respectively, while Algeria, Iraq, Jordan, Kuwait, Lebanon, Morocco, Oman, Qatar, Tunisia, and the United Arab Emirates ranked between the 51st and 81st positions. Bahrain, Libya, Palestine, Sudan, Syria, and Yemen ranked between the 98th and 122nd positions, while the lowest ranking Arab countries, including Comoros, Djibouti, Mauritania, and Somalia, ranked in the 209th, 196th, 181st and 195th positions, respectively.

Fig. 4.13A–D presents records of world rankings for four groups of Arab countries in terms of the number of research documents they published in international refereed journals over the period 1996–2015. Rather than grouping countries within categories delineated by their present or average rankings, Fig. 4.13A–D reflects the extent to which rankings varied over the period 1996–2015. This should highlight the impact of policies implemented over this period with a view to enhancing scientific research output.

Thus, the first group, for which information is presented in Fig. 4.13D includes four countries, namely Egypt, Jordan, Morocco, and Syria, that witnessed an average variance in their rankings estimated at only five points. The second group, in Fig. 4.13B, includes Lebanon, Oman, Tunisia, and the United Arab Emirates, whose rankings varied by an average of around 18 points. The third and more populous group of countries, whose rankings varied considerably, with an average variance of 38 points, is portrayed in Fig. 4.13C. This group includes Algeria, Bahrain, Kuwait, Libya, Saudi Arabia, Sudan, and Yemen. The fourth group, portrayed in Fig. 4.13D, includes Iraq, Palestine, and Qatar, all of which witnessed large upward changes in their world rankings during the period 1996–2015, with the average variance of observed changes amounting to around 226 points.

As Fig. 4.13A indicates, Egypt's worldwide rankings averaged at around the 40th position during the period 1996–2015, varying between the lowest rank as the 43rd in 2008 to the highest at 36th in 2015. Morocco's rankings varied within an even narrower range, between the 54th and 60th positions, giving this country an average ranking of 56 during the same period. Jordan's rankings varied between the 59th and 65th positions, granting the country an average ranking of 61st.

On the other hand, Syria's rankings were the lowest and the most variable; giving this country an average ranking in the 100th position worldwide while varying between the 96th and 107th positions.

Fig. 4.13B displays changes in rankings for Lebanon, Oman, Tunisia, and the United Arab Emirates, which managed to drive their ranking up by more than 10 positions over the period 1996–2015. Thus, Lebanon's rank took off from the 83rd position in 1996 to reach 67th in 2015, spending the years 1999–2015 within a narrow range between the 70th and 66th positions. Oman's ranking,

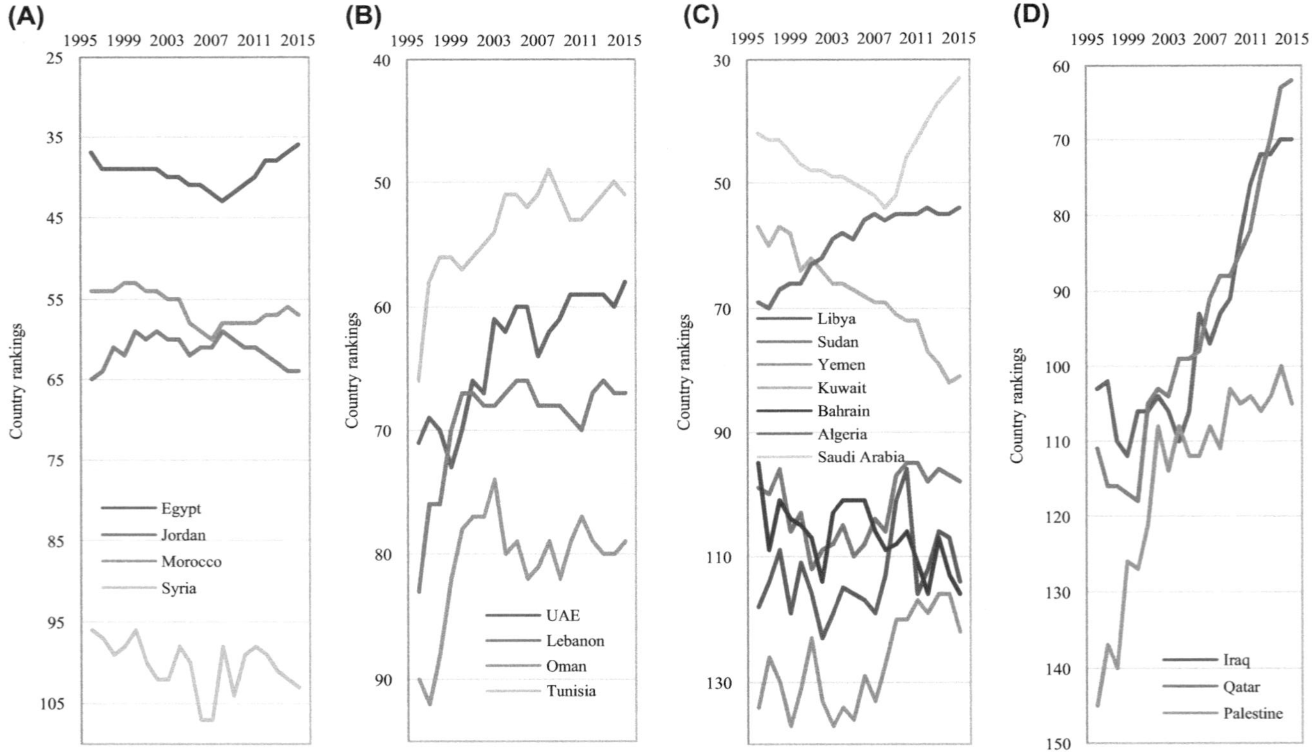


**FIGURE 4.13** Changes in world rankings for (A) four Arab countries which experienced small changes, (B) four Arab countries which experienced significant changes, (C) seven Arab countries which experienced considerable changes, and (D) three Arab countries which experienced large changes in their rankings in terms of research publications during 1996–2015.

on the other hand, went up from the 90th in 1996 up to the 74th in 2003 before settling in at around the 80th position for most of the past decade.

Fig. 4.13C portrays how rankings changed for seven Arab countries that differed somewhat widely in terms of locations and resource levels yet managed considerable upward changes in their rankings. These countries may be grouped within two rather distinct sets. The first encompassing Algeria, Kuwait, and Saudi Arabia, which enjoyed relatively high rankings, as they occupied positions above the 60th rank for most of the period covered by this figure. While Bahrain, Libya, Sudan, and Yemen attained much lower rankings, averaging 107th, 114th, 102nd, and 127th positions, respectively. Kuwait's declining rank is the most notable feature of this figure. This admittedly small country managed to attain a world ranking of 57 in 1996 before dropping down to the 82nd position by 2014.

Libya's rankings appear to be the most variable, having remained close to or below the 110th position throughout the period 1996–2008, before climbing up to the 96th position in 2010 and then plunging back below the 110th position in 2011, and going further down to the 114th position in 2015. These latter rankings might well be linked to the violence and political turmoil suffered by this country over the first half of the present decade. Sudan's rankings display a rather variable path as well, with this country near or below the 100th position for much of the period 1996–2008 before assuming rankings between the 95th and 98th positions during 2009–15. Saudi Arabia's world rankings over the period 1996–2015 declined from the 42nd position in 1996 down to the 54th position in 2008, before rising once more to the 33rd position ahead of all the other Arab countries.

Several features of Fig. 4.13D are worthy of comment, including the rise in Qatar's ranking from the 118th position it occupied in 2000, before managing to move up to the 62nd rank in 2015. Similar progress appears to have taken place in Iraq's case, with this country's ranking beginning to escalate in 2006, following an initial period at rankings within the range 102nd to 112th during the years 1996–2005, to reach its current ranking as the 70th among the world's countries. The path taken by Palestine's rankings during 1996–2015 is also noteworthy for the steep upward trend it took throughout 1996–2015, starting at a rank as 145th in 1996 and ending up on a relatively flat path between the 100th and 105th positions during the years 2011–15.

Linking changes in world rankings to specific policy and institutional changes is necessarily a complex affair. However, as indicated above, in the case of Saudi Arabia, it is unlikely that policies designed to enhance country rankings based on enlisting widely cited researchers with prolific research output as affiliates would be sustainable unless closely linked to aggressive capacity building measures at the national and institutional levels. Furthermore, as indicated in relation to the rapid rise of Qatar's and the United Emirates' rankings, policies aimed at hosting overseas campuses of distinguished foreign universities would need to be closely linked to national and regional priorities and to roles by local actors for their impact to be sustainable.

### 4.1.3 Citations and Impact of Publications by Researchers in the Arab Countries

The extent to which research publications are cited generally indicates their quality and relevance to issues of current interest to researchers in a given field of the scientific enterprise and practitioners in various application areas. Figs. 4.14 and 4.15[17] present a view of the growth of annual citations to papers published by researchers in the Arab countries and worldwide, respectively. These figures indicate that citations of articles, reviews, and conference papers produced by researchers in the Arab countries, as well as worldwide, took an upward path from their levels in the late 1990s, peaking around 2010–11 in the case of Arab country publications before rapidly declining to lower levels during the period 2012–15. Worldwide citations, on the other hand, experienced a maximum around 2004–05 before declining more gradually to minimum values in 2015.[18]

The dip taken by the annual count of citations to the Arab country and worldwide publications around 2011 and 2004, respectively, might be explained by the fact that citations will always trail behind the publication of a given scientific document. That is, a period of a few years will normally elapse before fresh

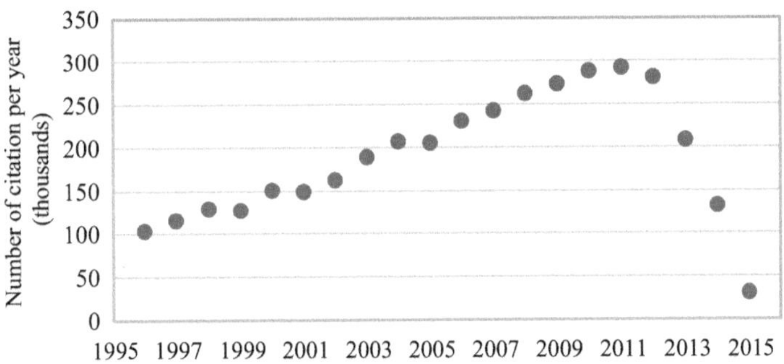


**FIGURE 4.14** Evolution of citations to publications by the Arab countries during 1996–2015 (thousands).

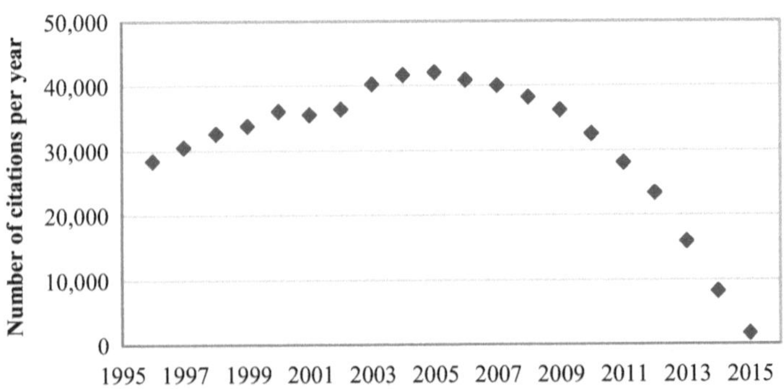


**FIGURE 4.15** Evolution of citations to publications by all countries around the world (thousands).

research articles include citations to previous research publications. However, explaining the earlier climax, and the more gradual downward trend, experienced by citations to worldwide publications is somewhat more difficult based on available information. Further research is needed to explain trends in citations to publications by researchers around the world and from the Arab countries.

The evolution of the annual rate of citations per document published by institutions in the Arab countries follows a pattern similar to that exhibited by worldwide publications. Fig. 4.16 presents trends followed by the average number of citations per research publication by Arab researchers and researchers around the world. Both trends take similar paths best described by a quadratic curve,[19] with maximum values attained in both cases over the period 1998–2003, falling down to only a few citations per document in recent years.

Two further attributes of the evolution of research publications by the Arab countries are worth looking into. The first is the percentage of citations to these publications out of all citations made to worldwide output. The second, addressed below, concerns self-citations by authors from the Arab countries. As shown in Fig. 4.17, the percentage of citations to Arab publications increased quite

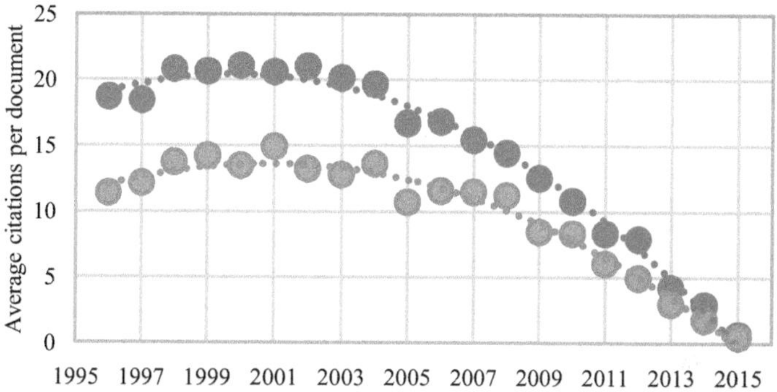


**FIGURE 4.16** Average number of citations per research publication for all countries over the period 1996–2015.

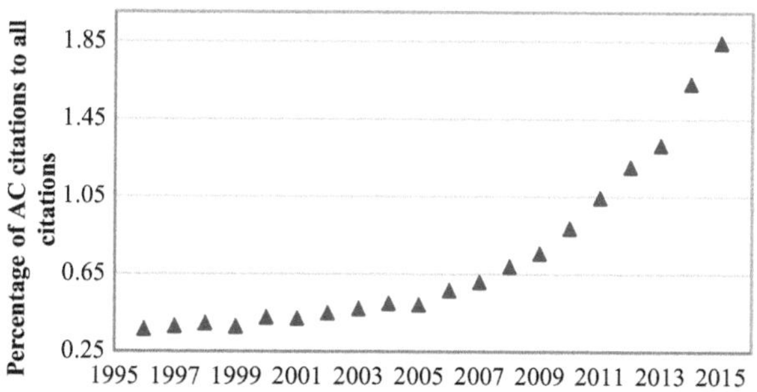


**FIGURE 4.17** Percentage of citations of AC publications to citations of publications by all countries.

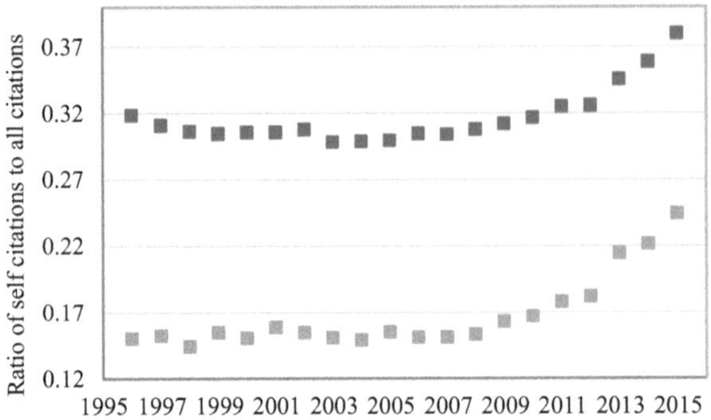


**FIGURE 4.18** Ratio of self-citations to all citations for Arab country and worldwide research publications.

noticeably from low levels of around 0.3% at the start of the period covered by the Elsevier database to levels approaching 2% by 2015. This indicates rising levels of recognition of research publications put out by the Arab countries.

Fig. 4.18 provides a view of the variation with time of the ratio of instances of self-citation to all citations made by published documents by the Arab countries and worldwide, throughout the period 1996–2015. As shown in Fig. 4.18 the ratio of self-citations in research documents to all citations made of publications by the Arab countries remained constant for a considerable period, essentially from 1996 till 2007, before embarking on an upward trend. The ratio of self-citations in research documents to all citations made of publications by the Arab countries took off from levels around 0.15 over the period 1996–2008 to around 0.25 in 2015. The worldwide trend, presented in Fig. 4.18, is somewhat similar, with the ratio of self-citations to all citations assuming values around 0.31 during the years 1996–2008, before rising steadily thereafter to around 0.4 toward the year 2015.

Fig. 4.18 seems to indicate that self-citation is considerably less common in research publications by the Arab countries than it is the case for worldwide publications. The rate of self-citations to all citations of research publications worldwide was consistently higher than the corresponding ratio for research publications produced by researchers in the Arab countries. During the period 1996–2009, this ratio was lower in the case of publications by Arab country researchers than for researchers worldwide by a factor of around 2. The difference narrowed slowly thereafter to reach a factor of around 1.5 by the year 2015. This is to say that the tendency for self-citation appears to be on the rise in Arab country publications. In fact it seems to have increased during the period 2011–15 at an overall rate that is over 20% higher than the corresponding rate for research publications worldwide.[20]

Somewhat counterintuitively, self-citations might signify a rather positive attribute in the evolution of research capabilities, as they could highlight persistence of research efforts by authors as well as possible establishment of research schools, rather than short-lived commitment to publishing research output, generally characteristic of early stages in the evolution of research capabilities. In other words, that the ratio of self-citations to all citations is increasing for Arab country publications may be viewed as an indication of continuity of research efforts and a tendency to build on past research activity rather than a means of self-promotion. Indeed, a plot of this ratio for all countries against country rankings reveals that the ratio of self-citations to all citations tends to increase exponentially as country rankings rise.[21]

#### *4.1.3.1 H-Index Values for Research Publications From the Arab Countries*

Measuring research quality is fraught with numerous difficulties. Nevertheless, several tools have been developed for the purpose and one such is the H-index, which measures citations made to research publications by peers in the same or adjacent disciplines.[22] Table 4.1 presents values of this index as well as corresponding world rankings for 14 Arab countries and 5 neighboring countries in the Middle East for comparison purposes. Data presented in this table cover a fairly long period stretching between 1996 and 2013. Rankings presented in the first column of the table are derived from data pertaining to 141 countries, for which citation data are available.

**TABLE 4.1** H-index Values and Country Rankings for Arab Countries as Well as Five Neighboring Countries in the Middle East

| Country | H-Index Values 1996–2015 (1996–2013) | H-Index Rankings[23] 1996–2015 (1996–2013) | Rankings Based on Publications[24] |
|---|---|---|---|
| Algeria | 106 (89) | 85 (81) | 55 |
| Bahrain | 55 (43) | 138 (122) | 110 |
| Comoros | 13 (–) | 213 (–) | 208 |
| Djibouti | 18 (–) | 203 (–) | 195 |
| Egypt | 184 (148) | 51 (49) | 42 |
| Iraq | 59 (–) | 131 (–) | 85 |
| Jordan | 112 (92) | 81 (76) | 66 |
| Kuwait | 108 (92) | 84 (76) | 70 |
| Lebanon | 138 (108) | 63 (67) | 68 |
| Libya | 51 (–) | 145 (–) | 113 |
| Mauritania | 32 (–) | 177 (–) | 173 |
| Morocco | 129 (92) | 71 (67) | 56 |

**TABLE 4.1 H-index Values and Country Rankings for Arab Countries as Well as Five Neighboring Countries in the Middle East—cont'd**

| Country | H-Index Values 1996–2015 (1996–2013) | H-Index Rankings[23] 1996–2015 (1996–2013) | Rankings Based on Publications[24] |
|---|---|---|---|
| Oman | 91 (74) | 95 (93) | 80 |
| Palestine | 60 (–) | 129 (–) | 111 |
| Qatar | 86 (60) | 98 (105) | 77 |
| Saudi Arabia | 195 (144) | 43 (52) | 44 |
| Sudan | 70 (58) | 120 (109) | 99 |
| Syria | 81 (–) | 102 (–) | 101 |
| Tunisia | 123 (97) | 75 (73) | 53 |
| UAE | 130 (100) | 70 (71) | 61 |
| Yemen | 50 (42) | 149 (125) | 121 |
| **Neighbouring Countries in the Middle East** | | | |
| Cyprus | 127 (100) | 74 (71) | 71 |
| Iran | 199 (158) | 42 (44) | 22 |
| Israel | 536 (456) | 16 (15) | 23 |
| Malta | 83 (70) | 99 (90) | |
| Turkey | 296 (237) | 37 (36) | 20 |

Based on information from the Elsevier database. Downloaded in October 2016 from: http://www.scimagojr.com/countryrank.php?year=1996 and The INSEAD report "Global Innovation Index 2015." Downloaded in May 2016 from: https://www.globalinnovationindex.org/userfiles/file/reportpdf/GII-2015-v5.pdf.

H-index values for the Arab countries range between 50 for Yemen and 195 for Saudi Arabia. With an H-index of 184, Egypt is in second position of the Arab publications impact league. Apart from Malta, whose H-index is 83, scientific publications by researchers in the Arab countries' closest neighbors listed in this table possess impact factors exceeding those of most of the Arab countries. Thus, H-index values for Israel, Turkey, and Iran were computed at 536, 296, and 199, respectively.

The bar chart in Fig. 4.19 provides further support for the notion that while research publications output by the Arab countries has been on the rise over the past decade or so, the impact resulting from this output still has some way to go before reaching levels enjoyed by many other countries, including some of their

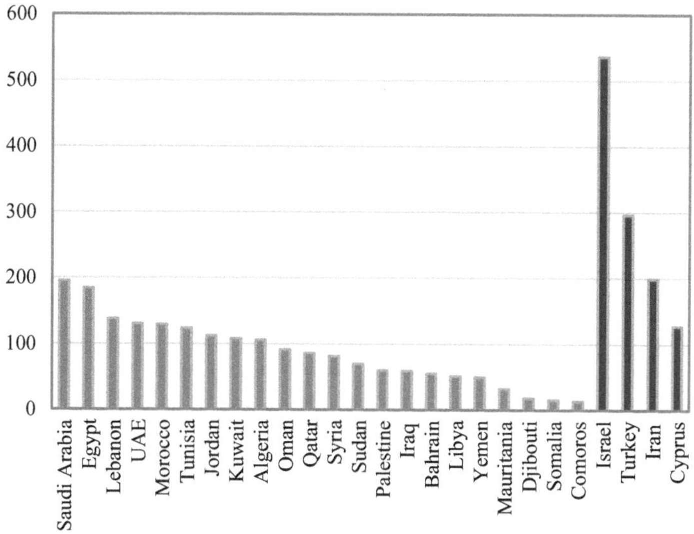


**FIGURE 4.19** H-index values for all the Arab countries as well as four of their closest neighbors.

closest neighbors. Thus, only Saudi Arabia and Egypt possess H-index values that are close to those of Iran, while the rest of the Arab countries possessed inferior values of the index; and both Iran and Turkey still have some way to go before attaining the world's average. Israel and Turkey, on the other hand, stand at the forefront of all countries in the Middle East and North Africa with values for the H-index for their publications during 1996–2015, at 536 and 296, respectively.

With the exception of only six Arab countries, namely Saudi Arabia, Egypt, Lebanon, the United Arab Emirates, Morocco, and Tunisia, all the rest fall within the lower two-thirds of the world's countries in terms of rankings based on the value of their H-index values. Three of their neighboring countries, namely Israel, Iran, and Turkey achieved the 16th, 42nd, and 37th positions worldwide, respectively. That is all three countries made it to the top 20% of all countries in terms of their H-index rankings.

On the other hand, H-index rankings of almost all the Arab countries seem to have improved, albeit to a much lesser extent than those of their three closest neighbors, Israel, Turkey, and Iran.[25] This is evident upon comparison of H-index values computed from publications and citation data covering the period 1996–2013 to corresponding values extracted from data pertaining to the period 1996–2015. Thus, eight Arab countries, namely Egypt, Jordan, Lebanon, Morocco, Qatar, Saudi Arabia, Tunisia, and the United Arab Emirates, seem to have improved their H-index ratings by increments equal to or greater than 20 points, with Saudi Arabia, Morocco, and Egypt securing the highest gains at 51, 37, and 36 points, respectively.

These gains are in agreement with previous results published by the above-mentioned Thomson Reuters Web of Knowledge (TRWK) report,[11] which considered the relative citation impact of three Arab countries in the Middle East, namely Egypt, Jordan, and Saudi Arabia, as well as Iran and Turkey.[26] Comparing H-index values attained by these countries over 5-year overlapping periods between 1990 and 2009, this report stated that the average citation impact of a great deal of the research conducted in the region remained below the world's average, set at 1.0. However, as this report indicated, the trend is upward toward increasing impact throughout.[27] According to this report, relative citation impact of the above five countries fell between 0.44 and 0.52, with Egypt on top, trailed by Turkey and Iran, then Saudi Arabia and Jordan, with the latter two countries in fourth and fifth positions, respectively. In summary, based on citation impact information, the TRWK report concluded that publications produced by Egypt, Saudi Arabia, and Jordan, made considerably more impact around 2009 than they did at the start of the 1990s.

### 4.1.4 Research Capacity Building and Disciplines Targeted by Arab Researchers' Publications

Commonly available references seldom address details of national capacity building within specific research areas targeted by researchers. National surveys conducted in preparation for the formulation of science, technology, and innovation policies, underline broad areas in which various research groups in a given country may be active. Directives subsequently issued within national policy documents often earmark certain branches of science and technology as priority areas. However, for a variety of reasons research publications need not always conform to survey results nor to policy stipulations. One of the reasons for such departures is that researchers, particularly those who have recently finished their training in developed country institutions, may continue to work and publish within disciplines of their initial training, which may bear little or no relevance to national priorities, as stipulated by national policy documents. The fact that promotion in some universities may be linked to research publications while the fields in which such research is to take place are seldom clearly linked to national priorities may reinforce such tendencies.

The following section summarizes the results of the above-mentioned TRWK report published in 2011.[28] This will be followed by a brief analysis of publications cataloged within two databases, Nature Index and SCImagojr, established by Nature Research and Elsevier, respectively.

#### *4.1.4.1 Research Disciplines Highlighted by a Recent Thomson Reuters' Report*

On the basis of a survey of research publications in a wide variety of international and refereed journals, the above-mentioned TRWK report concluded that engineering and agricultural sciences were among fields in which Arab scientists contributed their largest world share of publications in 2005–09. Numbers of research publications from Arab countries in the Middle East in the latter domain, i.e., agricultural sciences, were also found to have risen, with growth ranking among the top highest in recent global output of research papers. The fields of clinical medicine, chemistry, mathematics, computer science, and especially microbiology also exhibited substantial growth since the previous 5-year period.[29]

Even in the social sciences, which still ranked lowest in terms of interest by researchers compared to other research fields, publications exhibited the highest rate of growth within the countries of the Middle East surveyed by the TRWK report.[30] However, the fact that publications in the social sciences took off from a comparatively low starting point means that the social sciences remain an underrepresented field of research activity in the Arab countries.

Fig. 4.20, which is based on the above TRWK report, provides information on disciplines tackled by research publications made during the period 2005–09 by three Arab countries, namely Egypt, Jordan, and Saudi Arabia, which count among the more publication-prolific Arab countries, as compared to both Iran and Turkey. Table 4.2 reiterates the information provided in the above figure but with the breakdown of publications across research areas for Egypt, Jordan, and Saudi Arabia.

One of the most striking features of both Table 4.2 and Fig. 4.20 is the small percentage of publications in the fields of medicine and the environment and computer sciences. Indeed, the three Arab countries' combined share in world publications made in medicine is less than one-tenth of that attained by Iran and Turkey combined.

On the other hand, the relatively large research contributions made as a share of world publications in pharmacology is out of proportion with the attention devoted to publications in medicine. With particular regard to

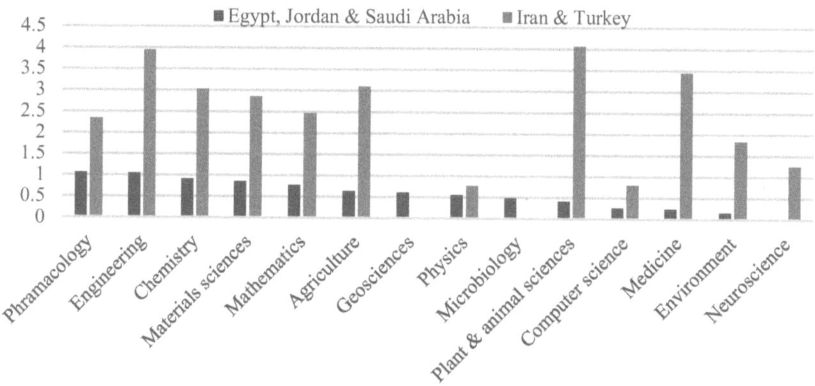


**FIGURE 4.20** Shares of worldwide publications during 2005–09 by three Arab countries compared to those by Iran and Turkey within 10 research disciplines (%). *Adapted from table (2) in the TRWK 2011 Report.*

**TABLE 4.2** Shares of Worldwide Publications by Three Arab Countries Compared to Those by Iran and Turkey Within 10 Research Disciplines in Which the Individual Countries Are Best Represented (%)

| | Mathe-matics | Physics | Chemistry | Materials Sciences | Agricul-ture | Plant & Animal Sciences | Environ-ment | Geosci-ences | Medicine | Pharmacy[a] | Micro-biology | Neuro-science | Engineering | Computer Sciences |
|---|---|---|---|---|---|---|---|---|---|---|---|---|---|---|
| Egypt | 0.31 | 0.40 | 0.64 | 0.66 | 0.48 | 0.32 | – | 0.34 | – | 0.71 | 0.35 | – | 0.57 | – |
| Jordan | 0.13 | — | 0.11 | – | 0.15 | 0.09 | 0.16 | 0.10 | 0.07 | 0.12 | – | – | 0.15 | 0.11 |
| Saudi Arabia | 0.32 | 0.14 | 0.15 | 0.19 | – | – | – | 0.16 | 0.26 | 0.22 | 0.13 | – | 0.31 | 0.15 |
| Iran | 1.16 | 0.76 | 1.68 | 1.19 | 1.19 | 0.93 | – | – | 0.60 | 1.05 | – | – | 1.71 | 0.79 |
| Turkey | 1.3 | – | 1.34 | 1.67 | 2.87 | 2.17 | 1.82 | – | 2.84 | 1.29 | – | 1.25 | 2.22 | – |

[a]Publications grouped under this category are likely to include research results dealing with "Pharmacology".
Adapted from table (2) in "Global Research Report, Exploring the Changing Landscape of Arabian, Persian and Turkish Research", J. Adams et al.; February 2011. Thomson Reuters Science. ISBN:1-904431-27-5.

publications in the former domain, it may be remarked that Egypt, together with several other Arab countries, appears to be targeted by several prominent pharmaceutical multinational firms for drug trials. Some of these firms advertise their links to university departments and invite researchers to avail themselves of their services for clinical testing of recently developed drugs. See Frame 4.3. Hence, it is likely that research results are reported in publications jointly authored with staff at university hospitals and researchers within associated pharmacology faculties, thereby enhancing the number of pharmacology-related research publications.[31]

Computer sciences do not feature among the disciplines within which Egypt contributed sufficiently to count among the top 10 disciplines in which it is best represented in terms of its share of research publications across the world. Similarly, environmental sciences were not among the top 10 disciplines addressed neither by Egypt nor Saudi Arabia.

**Frame 4.3 International Clinical Trials Conducted in the Arab Countries**

The past few decades witnessed a considerable shift in locations targeted for clinical drug trials financed by transnational pharmaceutical companies to low- and middle-income countries, including several countries in the Middle East and North Africa.

A recent report lists a number of factors that rendered Egypt in particular, one of the most popular destinations for offshore testing of new drugs. In addition to Egypt's viable research infrastructure, its rapidly growing and essentially treatment-naive population, its lower research costs, and inadequate regulation for protecting patients' rights have rendered the country a haven for TNC-sponsored clinical trials.[32]

The website of CenterWatch a firm whose declared mission is to provide "clinical trials information for both clinical research professionals and patients"[33] lists several Arab countries as conducting a total of 334 international clinical trials.

**Arab countries listed as hosting mainly TNC-sponsored clinical trials**

| Country | Algeria | Egypt | Jordan | Kuwait | Lebanon | Morocco | Oman | Qatar | Saudi Arabia | Tunisia |
|---|---|---|---|---|---|---|---|---|---|---|
| Number of trials | 14 | 57 | 12 | 8 | 110 | 6 | 1 | 17 | 48 | 61 |

From CenterWatch website. Information downloaded in December 2016 from: http://www.centerwatch.com/

It may be noted, however, that while promoting medical and pharmacological research and perhaps contributing to public health in host countries, inadequate regulations aimed at protecting patients' rights may render such trials unethical. An additional issue that bears emphasis is that some of the drugs being tested are extremely expensive rendering them unaffordable by the largely uninsured populations of countries hosting the clinical trials.

Among the basic sciences, chemistry appears to have received greater attention than that accorded to physics. Additionally mathematics seems to have been well represented within published research papers during the period 2005–09.

The combined share of the three Arab countries included in Fig. 4.20 out of world publications made in engineering is one of the highest among all other disciplines. Yet, this share amounts to one-fourth the share of both Iran and Turkey in engineering publications worldwide. Egypt's highest shares in the world's research publications were made in five disciplines, namely pharmacology, materials sciences, chemistry, engineering, and microbiology, at 0.71%, 0.66%, 0.64%, 0.57%, and 0.35%, respectively, while Jordan's highest five shares were made in engineering, at 0.15%, mathematics and environment both at 0.13%, pharmacology at 0.12%, and computer science as well as chemistry, at a share of 0.11% of the world's total in these two disciplines. Saudi Arabia's five top disciplines spanned mathematics, engineering, medicine, pharmacology, and materials sciences at 0.32, 0.31, 0.26, 0.22, and 0.19, respectively. By contrast, both Iran and Turkey have engineering among their top five disciplines in which both countries attained shares in world publications estimated at 2.22% and 1.71%, respectively. Both countries also attained high shares of the world's publications in agriculture at 1.19% and 2.87%, respectively, and in medicine, attaining 0.66% and 2.84%, respectively. However, neither country had the geosciences among the top 10 disciplines with their best shares among the world's research publications. On the other hand, the top 10 disciplines in which Egypt, Jordan and Saudi Arabia made their best percentage contributions did not include the neurosciences, which appeared among Turkey's list only, with a relatively sizeable contribution at 1.25% of all publications.

#### *4.1.4.2 Research Publications Tracked by the Nature Index Database*

A more recent view of the distribution of research publications across disciplines is provided by data from a Nature Index (NI), a dynamic database, set up by Nature Research, an arm of the Journal Nature, with the view to tracking research articles published in a selected group of 68 peer-reviewed science journals.[35]

While the journals monitored since 2013 by "Nature Index" represent less than 1% of all journals covering the natural sciences as compiled by the Thomson Reuters' Web of Science, they do account for close to 30% of total citations to all journals in the natural sciences.

Contents of the database include reference to authors' national and institutional affiliations, thereby providing means for closer monitoring of high-quality research output within four broad fields of scientific research, namely chemistry, earth and environmental sciences, life sciences, and physical sciences. Data provided by this database also enables comparison of output by institutions and countries across the world or in a particular region, within the above-mentioned areas.

Table A.4.9 presents the results of interrogating the contents of the NI database for 10 Arab countries that managed to publish more than 10 articles each in the journals surveyed by Nature Research during 2015.[36] Data presented in this table and the accompanying figures clearly indicate that Saudi Arabia occupies a leading position in terms of the total number of articles published in the science journals covered by the database. Egypt follows with Morocco, Qatar and the United Arab Emirates, Lebanon, Tunisia, Jordan, Algeria, and Iraq trailed Saudi Arabia in this same order.

It may be noted that the term "Article Counts" in Table A.4.9 refers to the number of articles produced by a country, with an article count equaling 1.0 assigned if one or more authors of the research article are from that country, regardless of how many researchers from other countries are listed as coauthors. Thus, numbers presented in the second column of this table in fact constitute a reasonable measure of scientific research activity undertaken by authors from a given country but do not mean that research efforts were entirely carried out by institutions within that country.

On the other hand, the term "Fractional Counts" (FC) listed in the third column of this table refers to the count of articles published by a given country taking into account the percentage of authors from institutions within that country contributing to the article in question. An important point to bear in mind here is that in calculating FC, all authors are considered to have contributed equally to the article in question. Clearly, the maximum value that a combined FC that any article can have is 1.0.

Table 4.3 lists the numbers of research articles to which at least one scientist from one of the 18 Arab countries contributed and was eventually published in journals covered by Nature Index. The numbers of research articles including contributions by scientists from institutions in four neighboring countries, namely Cyprus, Iran, Israel, and Turkey, in these same journals are also included for comparison.

Table 4.3 clearly indicates the low level of research documents that the Arab countries managed to publish in the above set of 68 highly regarded journals over the period May 1, 2015 till April 30, 2016. The number of publications produced by all of the Arab countries during 2015—around 950 articles—fell below what Israel alone managed to publish during that period within the same set of journals—over 1190 articles. Five countries, namely Saudi Arabia, Egypt, Qatar, Morocco, and the United Arab Emirates published nearly 76% of the total number of articles produced by all the Arab countries. As indicated above, Saudi Arabia followed by Egypt top the list of the Arab countries, in terms of the total number of publications made during the above-mentioned period. In fact, Saudi Arabia and Egypt produced around 60% of all publications made by all of the Arab countries, with scientists in the former producing around 66% of all publications made by Arab countries in the Middle East, while those in Egypt produced 53% of the publications made by all Arab countries in Africa.

**TABLE 4.3** Research Publications Produced by Arab Countries During May 1, 2015–April 30, 2016[34] Within Four Broad Areas

| Country | Article Count | Fractional Count | Weighted Fractional Count[51] |
|---|---|---|---|
| **Arab Countries in the Middle East** | | | |
| Saudi Arabia | 450 | 103.84 | 101.09 |
| Qatar | 68 | 11.49 | 8.29 |
| UAE | 74 | 16.26 | 11.86 |
| Lebanon | 28 | 6.45 | 3.99 |
| Iraq | 13 | 1.04 | 0.89 |
| Jordan | 12 | 1.51 | 1.40 |
| Kuwait | 11 | 2.76 | 2.68 |
| Oman | 5 | 0.87 | 0.87 |
| Palestine | 4 | 0.29 | 0.29 |
| Yemen | 2 | 0.11 | 0.11 |
| Syria | 1 | 0.05 | 0.05 |
| **Arab Countries in North Africa** | | | |
| Egypt | 136 | 13.19 | 10.01 |
| Morocco | 81 | 7.79 | 6.92 |
| Tunisia | 22 | 3.54 | 3.01 |
| Algeria | 16 | 2.19 | 1.91 |
| Sudan | 2 | 0.11 | 0.11 |
| Libya | 1 | 0.04 | 0.04 |
| Mauritania | 1 | 0.07 | 0.07 |
| **Other Neighboring Countries** | | | |
| Cyprus | 67 | 12.86 | 12.38 |
| Israel | 1192 | 582.50 | 514.75 |
| Turkey | 337 | 72.55 | 59.72 |
| Iran | 184 | 87.92 | 68.88 |

Data downloaded in July 2016 from Nature Index at: http://www.natureindex.com/country-outputs/generate/All/Africa/All/n_article and Data downloaded in July 2016 from Nature Index at: http://www.natureindex.com/country-outputs/generate/All/Western%20Asia/All/n_article.

#### 4.1.4.3 Research Capacity Building by the Arab Countries Within Four Broad Research Areas

The following paragraphs will take a closer look at publications by Arab researchers in the 10 Arab countries which managed to publish more than 10 articles in journals covered by Nature Research during 2015. The essential intention is to arrive at possible indications regarding capacity building in general as well as areas in which some of these countries have managed to establish solid capabilities. The fact that articles published by Saudi Arabia in the above journals amount to more than what was published by all the other Arab countries combined would favor dealing with this country separately. Hence, research publications by the countries listed in Table A.4.9 will be considered firstly with the exception of Saudi Arabia, following which this country's research publications will be similarly looked at. The pie chart in Fig. 4.21 presents the distribution of research publications made during 2015, within the above-mentioned four broad disciplines across the Arab countries on which information is presented in Table A.4.9, with the exception of Saudi Arabia.[37] The share of the two small Gulf countries represented in this figure, namely Qatar and the United Arab Emirates is slightly higher than that of Egypt. On the other hand, Egypt's output is a little higher than that of Tunisia, Morocco, and Algeria, combined. The fact that Morocco was the largest producer of research publications among this latter group of countries is also clearly indicated. Only three countries of the Levant, Iraq, Lebanon, and Jordan, with significant contributions to journals monitored by Nature Research, published a little less than what the United Arab Emirates did during 2015. When resource levels, populations, and related factors are taken into account, it would seem that at least Lebanon might count among countries that performed above expectations.

Fig. 4.22 presents a view of the Fractional Counts for papers published by nine Arab countries during 2015. This figure reveals significant improvements in the positions of four countries, namely Lebanon, Algeria, the United Arab Emirates, and Tunisia, in terms of participation by their scientists in papers

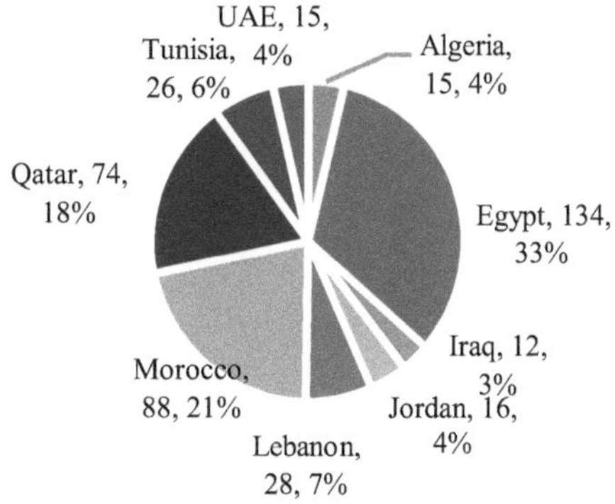


**FIGURE 4.21** Papers published by nine Arab countries during 2015 in 68 high-ranking science journals.

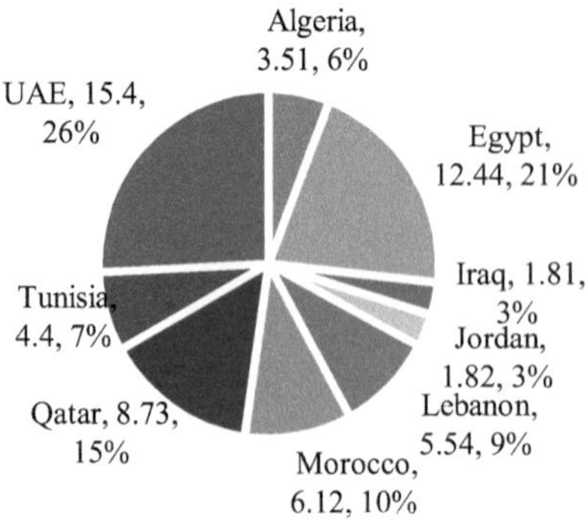


**FIGURE 4.22** Fractional article counts of papers published by nine Arab countries during 2015 in 68 high-ranking science journals.

authored in cooperation with scientists from other countries. Thus, scientists from Lebanon increased their proportional contribution to research publications made by all the countries represented in this figure from 7% by article counts to 9% by fractional article counts.

Algerian researchers enhanced their standing somewhat similarly, as their fractional article count was 6%, compared to an article count of 4%. Even more impressively, the United Arab Emirates fractional article count was estimated at 26%, while it attained only 4% of the total article count by all nine Arab countries represented in Fig. 4.21. Tunisia's position may also be said to have improved, albeit by 1% from an article count of 6% to a fractional article count of 7%.

By contrast, researchers from countries, such as Egypt, Morocco, and Qatar, are shown to have made fewer contributions to the total fractional article count, with the former two countries losing 8% and the latter only 1%, of the contributions they made as measured by article counts.

The pie chart in Fig. 4.23 presents the distribution of over 1000 publications by the Arab countries on which information is presented in Table A.4.9 with the

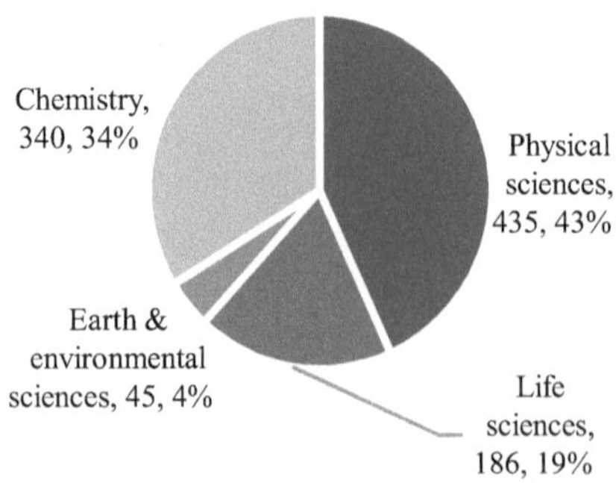


**FIGURE 4.23** Article counts of research publications by nine Arab countries in four broad areas of specialization (2015).

exception of Saudi Arabia across four broad areas of specializations, namely chemistry, earth and environmental sciences, life sciences, and the physical sciences. This figure clearly indicates a high proportion of research and ensuing publication, activity in chemistry and the physical sciences. Thus, publications in subjects within these two admittedly broad disciplines accounted for 77% of the total. On the other hand, publications in earth and environmental sciences accounted for a meager 4% of the total, while the life sciences were represented by around one-fifth of the research publications.

This limited level of activity in the earth and environmental sciences, on the one hand, and the life sciences, on the other, ought to be tackled by future research strategies and related funding allocations. This is especially important since the Arab countries are known to be facing various environmental problems, including, for example desertification, deforestation, and water scarcity, while the oil and mineral wealth owned by some of their prolific performers in research publications would be expected to cultivate greater interest in the earth and environmental sciences.[38] Equally important is the need for cultivating an improved climate for research in the life sciences.

The pie chart in Fig. 4.24 presents a similar view to that of Fig. 4.23, only that the numbers of papers published and their percentages of the total produced by all nine countries represented within both figures[39] refer to fractional article counts rather than mere article counts. Comparison of these two figures allows further insight into the relative contributions made by researchers in these countries to articles published within the four broad areas of specialization by the entire group. Thus, it may be readily seen in the case of publications in the physical sciences that fractional article counts constitute a higher percentage of total contributions to research publications than they did in the case of simple article counts. This should indicate that researchers in the physical sciences featured more prominently in related publications than their colleagues contributing to publications in other specializations. The same is true in the case of research publications in the earth and environmental sciences, with the percentage of fractional article counts twice double that of the simple article counts,

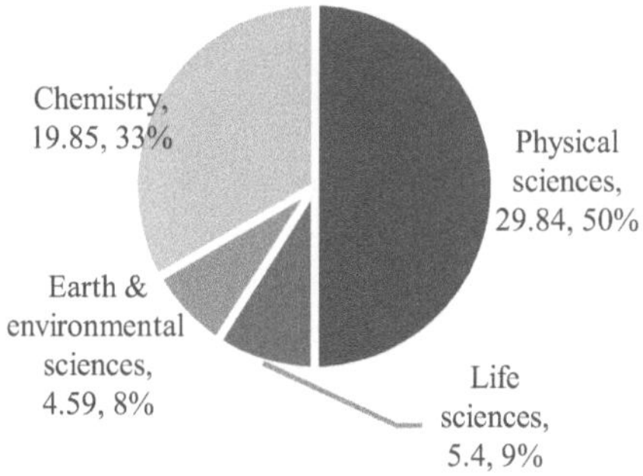


**FIGURE 4.24** Fractional article counts of research publications by nine Arab countries in four broad areas of specialization (2015).

indicating a higher involvement by researchers from the nine Arab countries included in the present analysis in related research activity. The remaining two areas of specialization, namely chemistry and the life sciences, naturally, end up losing out, with the percentage of article counts in the latter specialization dropping down to less than half its value when measured by fractional article counts, as compared to simple article counts.

### *4.1.4.4 Publications by Researchers in Saudi Arabia Across Research Domains*

To start along the lines followed above for the remaining nine Arab countries in comparing figures for article and fractional article counts, Saudi Arabia's researchers seem to have achieved a higher capacity compared to other Arab countries, as indicated by publications in the journals covered by Nature Research. Thus, at 0.23, the ratio of fractional article counts to article counts for publications by Saudi researchers is the second highest exhibited by the Arab countries for which data is presented in Table 4.3.[40] It is exceeded by Kuwait, which scored a ratio of 0.25 and equaled by Lebanon at 0.23, while the United Arab Emirates took third position at 0.22. Interestingly, Palestine, Yemen, and Syria come to the bottom of the list with values for this ratio at 0.07, 0.06, and 0.05, respectively. On the other hand, three of the Arab countries' closest neighbors, namely Iran, Israel, and Turkey, possessed corresponding ratios estimated at 0.48, 0.49, and 0.22.

Turning now to the distribution of Saudi publications across research domains, the pie chart in Fig. 4.25 presents information on articles contributed by Saudi researchers during the year 2015, to journals monitored by Nature Research, within the four groups of broad specializations considered for the nine Arab countries addressed above.

This pie chart clearly indicates the preponderance of research articles in chemistry and the physical sciences as opposed to both the life sciences as well as earth and environmental sciences. Thus, the former two areas take up 78% of all publications, with publications in fields allied to chemistry alone constituting

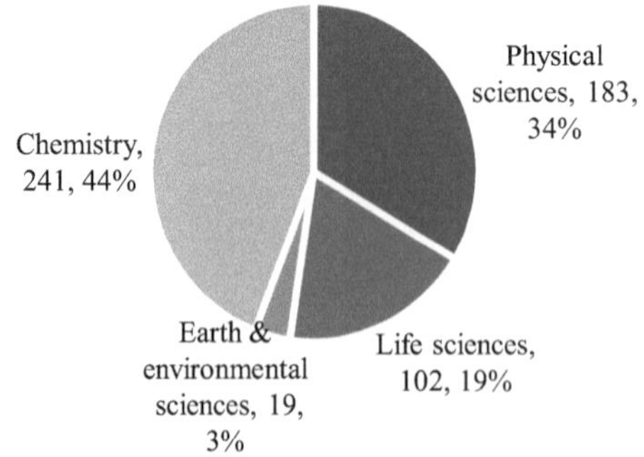


**FIGURE 4.25** Article counts of publications in four broad areas of specialization by Saudi researchers (2015).

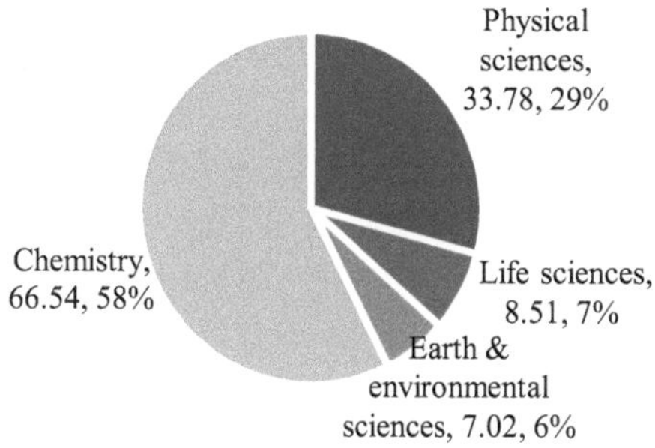


**FIGURE 4.26** Fractional article counts of publications in four broad areas of specialization by Saudi researchers (2015).

twice the number of publications in the latter two areas, i.e., the life sciences as well as the earth and environmental sciences.

The predominance of contributions by Saudi authors to articles in fields related to chemistry is reaffirmed by the pie chart in Fig. 4.26 with information on fractional article counts. In this pie chart, the proportion of articles dedicated to chemical specializations is enhanced even further, to constitute around 67% of all articles while the fractional article counts in the physical and life sciences are diminished to 36% of the total, rather than 53% of the total according to simple article counts. In line with the above observation, concerning publications by the nine Arab countries which contributed more than 10 articles in the journals monitored by Nature Research during 2015; the percentage of fractional article counts in disciplines related to earth and environmental sciences is higher than the corresponding percentage for simple article counts. This indicates that somewhat higher levels of endogenous capability had attained in the earth and environmental sciences.

Further analysis of research activity across the above research areas as well as individual disciplines within these areas is essential to guide future policy and subsequent interventions to consolidate STI capacity in research disciplines where some progress was made, to build and further advance similar capabilities in disciplines in need of further development. Naturally, such analysis would need to be carried out at country levels taking into account both national and local community needs while maintaining optimal links to research institutions in neighboring and overseas countries.

## 4.1.5 Scientific Research Publications in Arab Journals

Universities in several Arab countries publish their own periodicals featuring research papers contributed by their own staff as well as research results produced by workers in other universities and research centers in their country and the region.[41] A total of 141 journals were published in 2015 by institutions in the Arab countries of the Middle East and North Africa. See Table 4.4. Data was downloaded from the Elsevier's SCImagojr database on August 12, 2016 from: http://www.scimagojr.com/countryrank.php.

**TABLE 4.4** Journals Published by Universities and Other Institutions in the Arab Countries

| | All Journals | Open Access Journals | Average SJR Ratings | Ratings' Variance | H-index | H-index Variance |
|---|---|---|---|---|---|---|
| Bahrain | 4 | – | 0.27 | 0.017 | 17.25 | 181.7 |
| Egypt | 67 | 46 | 0.547 | 0.346 | 11.94 | 40.56 |
| Iraq | 2 | 1 | 0.134 | 0 | 2.5 | 0.25 |
| Jordan | 11 | 4 | 0.177 | 0.005 | 8.8 | 49.79 |
| Kuwait | 7 | 5 | 0.173 | 0.004 | 6.29 | 19.35 |
| Lebanon | 3 | – | 0.153 | 0.002 | 10.70 | 40.22 |
| Morocco | 4 | 2 | 0.176 | 0.007 | 6.5 | 48.75 |
| Oman | 3 | 3 | 0.248 | 0.005 | 8.33 | 6.22 |
| Qatar | 2 | 1 | 0.118 | 0 | 3.5 | 0.25 |
| Saudi Arabia | 11 | 2 | 0.208 | 0.015 | 9.73 | 85.83 |
| Sudan | 1 | – | 0.25 | 0 | 5 | 0 |
| Tunisia | 2 | – | 0.127 | 0 | 10 | 4 |
| UAE | 24 | 6 | 0.28 | 0.079 | 7.46 | 54.5 |
| **Arab countries** | **141** | **70** | **0.220** | **0.036** | **8.308** | **40.878** |
| **Three Closest Neighbors** | | | | | | |
| Iran | 138 | 82 | 0.277 | 0.024 | 8.69 | 82.66 |
| Israel | 27 | 2 | 0.261 | 0.046 | 9.26 | 85.82 |
| Turkey | 180 | 144 | 0.201 | 0.017 | 8.59 | 56.11 |
| ACs & neighbors | 486 | 298 | 0.225 | 0.035 | 8.41 | 47.25 |
| **Selected Country Groups** | | | | | | |
| Western Europe | 12,034 | 1330 | 0.692 | 1.63 | 28.54 | 1458.0 |
| Eastern Europe | 1601 | 423 | 0.229 | 0.03 | 8.74 | 74.84 |
| North America | 11,896 | 625 | 0.583 | 1.661 | 23.28 | 1811.27 |

Data downloaded from the Elsevier's SCImagojr database on August 12, 2016 from: http://www.scimagojr.com/countryrank.php.

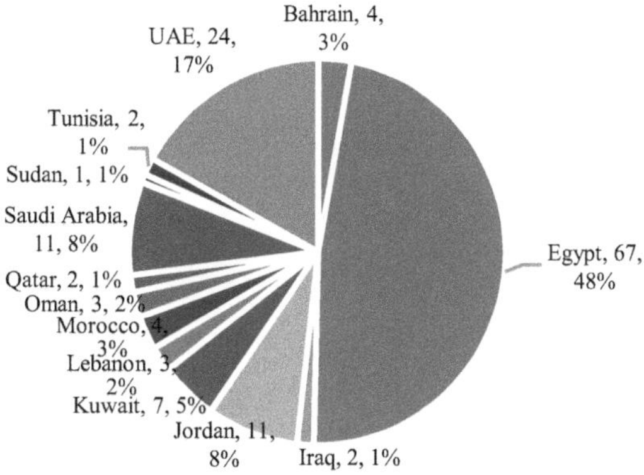


**FIGURE 4.27** Scientific journals published by the Arab countries in 2015.

Fig. 4.27 presents a view of the distribution of journals published in the region across the Arab countries, indicating that Egyptian institutions produced nearly half of all journals published by the Arab countries in 2015. Egypt is followed by the United Arab Emirates, which publishes 17% of the total, then by both Jordan and Saudi Arabia, publishing 8% each. The remaining countries publish much smaller percentages of the total. The fact that Saudi Arabian institutions published only 8% of the Arab countries' total journals is especially notable as it stands at odds with its much superior showing in terms of publications attributed to its researchers in international refereed journals.

On the other hand, the number of journals published by Egypt's institutions seems more in line with its long established position as a leading producer of research publications. See for example Fig. 4.13A, above.

As indicated in Table 4.4, the Arab countries published a considerably smaller number of journals than that published by only two of their closest neighbors, namely Iran and Turkey. Between them, these two countries published 318 journals, as opposed to only 141 journals published by all the Arab countries taken together. The fact that half of the journals published by the Arab countries and 71% of the journals published by Iran and Turkey enjoy open access is noteworthy. The corresponding percentages for countries in Western and Eastern Europe as well as North America are much smaller at 11.1%, 26.4%, and 5.3%, respectively.

Other noteworthy aspects of the list presented in Table 4.4 include the absence of several Arab countries, such as Libya, Syria, and Yemen as well as the relatively small numbers of journals produced by institutions in others, such as Morocco, Tunisia, Sudan, Iraq, Oman, and Qatar.

Judging the quality of the journals published by the Arab countries is essentially fraught with difficulties. However, if commonly used citation-based indicators, such as the H-index and the SJR indicator[42] are adopted, the latter indicator would place journals produced in Egypt on average at the top of the list of all journals produced by institutions in all countries of the region,

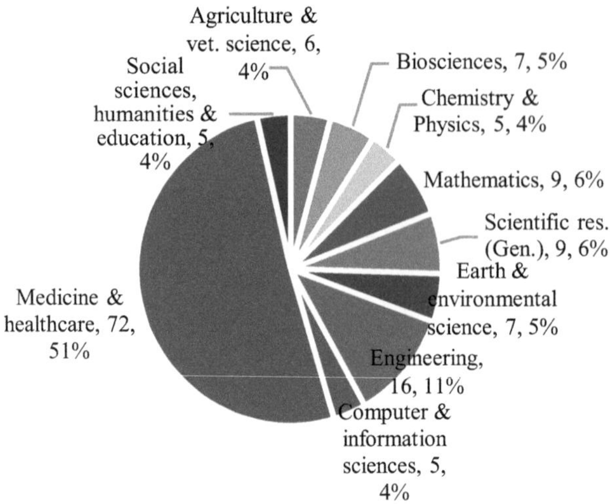


**FIGURE 4.28** Disciplines tackled by Journals published by the Arab country institutions.

including Iran, Israel, and Turkey.[43] This rating calls for some circumspection, however, for reasons including that it would place Egypt's journals ahead of well-established journals published in Western Europe and North America as well as the fact that the degree of variance relating to H-index ratings of Egypt's journals is highest among all ratings observed for other countries in the region. On the other hand, as indicated in Table 4.4, average H-index ratings of journals produced by institutions in Western Europe and North America exceed those of all journals published in the region, including those published by Iran, Israel, and Turkey.

The pie chart in Fig. 4.28 presents the distribution of Arab science journals across disciplines and areas of specialization they address. Journals dedicated to medicine and healthcare are at the very forefront with percentages of the total equal to 51%. They are followed by journals dedicated to engineering, at 11%, general scientific research as well as mathematics at 6% each and the bio- as well as the earth and environmental sciences with equal percentages of the total at 5%. Comparison of the above distribution to that found for journals produced by Iran, Israel, and Turkey, presented in Fig. 4.29, reveals that a similarly high percentage of their total output of scientific journals is dedicated to medicine and healthcare, around 54%. Journals published by Iranian, Israeli, and Turkish institutions in social sciences and humanities constitute 7% of the total. This is a considerably higher percentage than the corresponding percentage for the Arab countries, with the latter amounting to only 4% of the total number of journals published by these countries in 2015. Additionally, Turkey appears to be the only country in the Middle East and North African region with journals dedicated to education. Further information is presented in the annex Table A.4.1 on the distribution of journals published by countries in the Middle East and North African Arab countries, Iran, Israel, and Turkey across areas of specialization and individual countries.

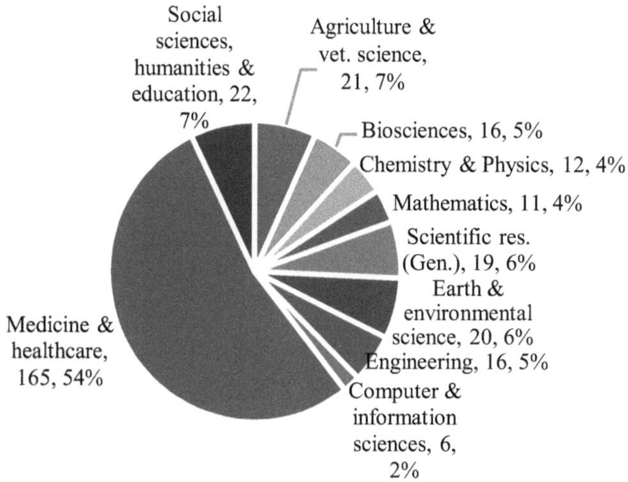


**FIGURE 4.29** Disciplines tackled by Journals published by Iran, Israel, and Turkey.

However, only around 2% of the total number of journals produced by university institutions in Iran, Israel, and Turkey are dedicated to computer and information sciences, against 4% in the case of university institutions in the Arab countries.

All journals listed in the SCImago database are associated with specific universities and other institutions concerned with research, and are generally published by national entities, including the issuing institution and local publishing firms. However, presumably, to improve the dissemination of their research output some institutions appear to have entrusted production and hosting of their journals to well-known international publishing houses. Examples of such institutions include Cairo University's Journal of Advanced Research and its Bulletin of Faculty of Pharmacy,[44] which is published under open access arrangements by Elsevier, an international publishing house. Another example is provided by the Association of Arab Universities, based in Bahrain, which handed production and hosting of its Journal for Basic and Applied Sciences, also under the open access regime to this same publisher, also as an open access journal.[45]

Table 4.5 presents the numbers of Saudi research publications produced during 2003–12 in areas earmarked as "strategic technologies," by a 2013 report[46] together with the numbers of publications produced globally in corresponding technologies as well as the ratios of Saudi to global publications in various fields. Table 4.5 clearly indicates inordinate emphasis on advanced materials and nanotechnology in comparison to other admittedly strategic research domains such as water, oil and gas, petrochemicals. Even when looked at in comparison to global publications at 0.3 and 0.4, Saudi publications in these two fields make up higher ratios compared to publications in fields such as information technology, biotechnology, medicine and healthcare, electronics, and agriculture. Such emphasis would certainly merit further investigation, with

**TABLE 4.5 Summary of Saudi Research Output and Share of Global Output During 2003–12 in "Strategic Technologies"**

| Strategic Technology Areas | Saudi Research Output | Global Research Output | Saudi Share of Global Output |
|---|---|---|---|
| Advanced materials | 2597 | 805,788 | 0.3 |
| Medicine and healthcare | 1796 | 1,183,228 | 0.2 |
| Energy | 1716 | 520,456 | 0.3 |
| Biotechnology | 1607 | 829,130 | 0.2 |
| Nanotechnology | 1478 | 375,574 | 0.4 |
| Electronics, communications, and photonics | 833 | 578,838 | 0.1 |
| Oil and gas | 738 | 151,194 | 0.5 |
| Information technology | 580 | 236,215 | 0.2 |
| Petrochemicals | 536 | 100,574 | 0.5 |
| Agriculture | 406 | 264,457 | 0.2 |
| Environmental technology | 292 | 109,331 | 0.3 |
| Water | 249 | 27,403 | 0.9 |
| Space and aeronautics | 47 | 29,445 | 0.2 |

From "Research Performance Indicators; The 3rd Report;" King Abdulaziz City for Science and Technology and Reuters Thomson, 2013. Downloaded in February 2017 from: http://publications.kacst.edu.sa/SystemFiles/Books_Pdf/PDF_635391133237974950.pdf.

the objective of establishing whether it is in fact due to mere academic interest on the part of researchers schooled in these fields or the product of policy directives ultimately linked to practical applications, industrial or otherwise.

### 4.1.6 Scientific Collaboration Reflected by Research Publications

Records of research publications may provide useful information on international collaboration in scientific research as well as illuminate paths for future policy regimes and cooperative ventures. Table A.4.2 in the annex presents statistics on international scientific collaboration undertaken by researchers in 17 Arab countries with coworkers in 28 countries around the world. Figures presented in this table were computed on the basis of weighted fractional counts (WFC) of articles jointly authored by scientists in a given Arab country with their colleagues in the countries listed in the first column. Disciplines addressed by such research cooperation span four broad areas of specializations: chemistry, earth and environmental sciences, life sciences, and physical sciences.

Figures presented in Table A.4.2 indicate that for most of the Arab countries, collaboration with the United States tops collaboration with all other countries around the world.[47] Thus, most Arab countries listed in this table had levels of collaboration with authors and institutions in United States exceeding 20%. Indeed, with the exception of Mauritania, Algeria, and Yemen most other Arab countries[48] included in this table had the United States at the very top of all countries with whose scientists and institutions counterparts in the Arab countries actively cooperated in publishing scientific research articles throughout 2015.

The United States is followed by the UK among the top 10 collaborating countries with scientists and institutions in the Arab countries. Germany and Italy occupied third and fourth ranks, respectively, while Canada and France shared fifth rank in the list of top 10 overseas countries to cooperate with scientists and institutions from the Arab countries. See Table 4.6.

Notably, collaboration between French scientists and their counterparts in the Arab countries is not confined to countries which France held in the past as colonies or under mandate, in North Africa and the Levant, such as Tunisia, Algeria, and Morocco, on the one hand, as well as Syria and Lebanon, on the other. Thus, France is among the top 10 collaborating countries for scientists

**TABLE 4.6** Number of Instances Where an Overseas Country Occupied a Position Among the Top 10 Collaborators With Three or More Arab Countries in 2015

| Country | Number of Instances |
|---|---|
| USA | 14 |
| UK | 12 |
| Germany | 11 |
| Italy | 10 |
| Canada | 9 |
| France | 9 |
| Spain | 7 |
| Switzerland | 7 |
| Russia | 6 |
| China | 6 |
| Japan | 4 |
| Netherlands | 3 |

and institutions in Saudi Arabia, Egypt, Kuwait, and the United Arab Emirates. Spain and Switzerland follow with seven instances each among the top 10 countries whose scientists collaborate with counterparts in the Arab countries. The positions occupied by both Russia and China in this table are also noteworthy, as both counted among the top 10 collaborators for six of the Arab countries listed in annex Table A.4.2.

Table A.4.2 also indicates that collaboration among the two top performing Arab countries, in terms of the total number of publications, namely Saudi Arabia and Egypt is predominant. One the other hand, significant collaboration appears to take place between Saudi Arabia and four less prolific countries, namely Jordan, Algeria, Oman, and Yemen.

Table 4.7 presents information concerning international scientific research collaboration undertaken by four of the Arab countries' closest neighbors in the Middle East, namely Cyprus, Iran, Israel, and Turkey, together with 16 countries around the world, as monitored by "Nature Research."[49] A similar pattern pertains in collaborating toward joint research publications to that encountered in the case of the Arab countries; with the United States at the top of the 10 countries with whose scientists, Cypriot, Iranian, Israeli, and Turkish scientists collaborated. Several countries in Europe follow the United States among the top 10 collaborating countries in decreasing order. Somewhat predictably, Israeli

**TABLE 4.7** Top 10 Collaborating Countries Across the World With Four Neighboring Countries (Percentages)

| | Cyprus | Iran | Israel | Turkey |
|---|---|---|---|---|
| USA | 25.85 | 19.58 | 32.99 | 22.79 |
| UK | 4.03 | 4.84 | 6.61 | 6.74 |
| Switzerland | 6.84 | 4.17 | 3.5 | 3.73 |
| Italy | 8.15 | 7.22 | 3.6 | 5.57 |
| Germany | 11.74 | 10.51 | 14.05 | |
| France | 4.76 | | 5.31 | 5.26 |
| China | | 4.63 | 3.71 | 5.7 |
| Spain | 3.00 | 4.26 | | 2.64 |
| Belgium | 2.9 | 4.93 | | |
| Canada | | | 2.73 | 10.12 |
| Netherlands | | | 2.1 | 2.65 |
| Russia | 3.03 | | | 2.54 |
| Greece | 4.64 | | | |

**TABLE 4.7** Top 10 Collaborating Countries Across the World With Four Neighboring Countries (Percentages)—cont'd

| | Cyprus | Iran | Israel | Turkey |
|---|---|---|---|---|
| Australia | | 3.38 | | |
| Japan | | | 2.23 | |
| Austria | | 3.44 | | |
| Averages | 7.49 | 6.70 | 7.68 | 6.77 |

Percentages presented in this table are based on WFC indicating mutual collaboration in scientific publications monitored by "Nature Research."
From Nature Research website and databases. Figures downloaded in July 2016 from: http://www.natureindex.com/.

scientists appear to cooperate most with their counterparts in the United States, with Iranian scientists' collaboration with their United States counterparts the lowest, but still far from insignificant, at around 20%. Average percentage contributions by Cypriot, Iranian, Israeli, and Turkish scientists to collaborative publications with counterpart countries around the world,[50] are relatively close, falling between 6.7% and 7.7%, with Israeli scientists' contributions foremost followed by Cyprus, Turkey, and then Iran.

Annex Tables A.4.3–A.4.6 provide detailed information on collaboration by four groups of countries as measured by average percentages of WFC they attained in scientific publications with their top 10 partners. Thus, Table A.4.3 presents, by way of illustration, a view of collaboration, as measured by weighted fractional counts, between the United States, Canada, as well as four countries in Europe with one another as well as with a few other countries around the world, including China and Russia. Tables A.4.4 and A.4.5, on the other hand, present similar data for seven Arab countries as well as their neighbors, while Table A.4.6 attempts a similar characterization of scientific collaboration for an additional 26 countries, developed, developing, and in transition.

Naturally, the level of contributions to coauthored journals by a given institution or country is influenced by a variety of factors, including relative maturity of cooperating institutions, policies they adopt, expertise accumulated through past cooperation, as well as resources available to respective partners.

Table A.4.4 presents WFC data on jointly published articles within areas covered by "Nature Research," by scientists from the Arab countries in regard to articles published with counterparts in the developed countries, in comparison with corresponding contributions made to articles coauthored solely by developed country scientists. Comparison of WFC data presented in Table A.4.4 to those included in Table A.4.3 reveals at least one significant difference. Thus, average contributions by scientists working in Arab universities and research centers to articles published through collaboration with scientists in developed countries ranged from a low value of around 12.6% for Egypt to a high value of 34.8% for Tunisia, with an average standard deviation of 7.51%. In the case of articles coauthored in collaboration among scientists working in institutions in developed countries, average contributions ranged from a low value of 46.2% for Spain to a high value of 54.8% for Germany, with an average standard deviation of 2.87%.

It would therefore seem that collaboration among scientists in countries with more mature institutions in the developed countries takes place on the basis of higher percentage contributions to coauthored articles by respective partners as well as a narrower range of deviation from one country to another. On the other hand, collaboration between Arab country researchers with counterparts in the developed countries is characterized by lower percentages of contributions by the former to coauthored articles, with larger values of standard deviation.

Table A.4.5 presents similar data for four countries neighboring the Arab countries in the Middle East, namely Cyprus, Iran, Israel, and Turkey. Figures presented in this table indicate lower levels of contributions to coauthored articles by scientists in Cyprus and Turkey, 19.4% and 22.4%, respectively, with an average of contributions to articles coauthored by scientists from all four countries with their counterparts in several developed countries amounting to 28.4% and a rather large average deviation of around 9.3%.

In the absence of detailed and up-to-date information on actual research spending in most developing countries, it is tempting to explore the link between levels of national contributions to coauthored articles reporting joint scientific research and national economic output as measured by gross domestic product. This approach must naturally be taken as tentative and provisional. Indeed it is merely aimed at highlighting wider disparities, whenever they arise. Fig. 4.30 presents a plot of the levels of contributions to coauthored research reports, as measured by the average of the percentage of weighted fractional counts of articles coauthored with the top 10 collaborating countries against their GDP per capita. This figure utilizes data for 42 countries—including six Arab countries,[52] four of their closest neighbors in the Middle East as well as several countries in Europe, Latin America, and Asia—to their respective gross domestic product per capita.

At 0.4423, the modest value of the coefficient of determination $R^2$ may be expected as a measure of correlation between weighted fractional counts and GDP per capita. This is essentially due to the complex nature of the relationship between these two variables. A host of measures and supporting arrangements would be needed for a given country to establish research teams capable of forging regional and international partnerships and engaging in effective research collaboration. At any rate it may be expected that countries would proceed to set up such research teams and subsequently engage in scientific cooperation to an extent that corresponds to resources at their disposal.

However, what Fig. 4.30 shows in effect is that four out of the six Arab countries, namely Egypt, Morocco, Saudi Arabia, and the United Arab Emirates, perform far below expectation, as compared to other countries with similar resource

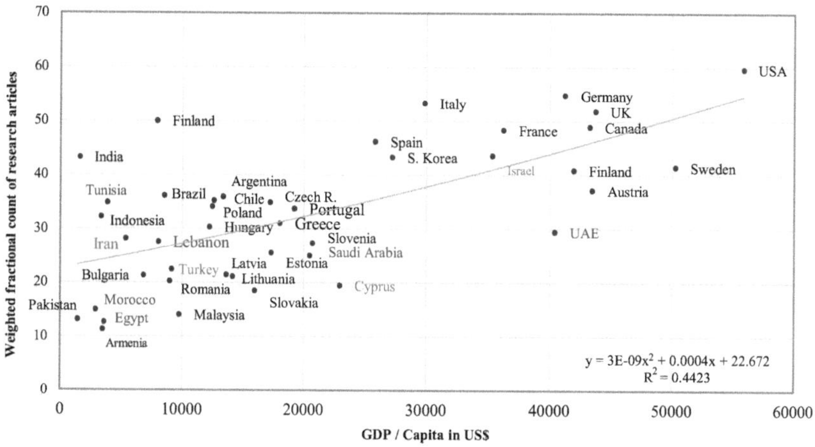


**FIGURE 4.30** Variation of weighted fractional count (WFC) with GDP/capita for selected countries.

levels. On the other hand, both Lebanon and Tunisia fall above the trend line, while Egypt, Morocco, Saudi Arabia, and the United Arab Emirates fall below that line. Incidentally, both Turkey and Cyprus also seem to lag behind somewhat, while both Iran and Israel occupy positions above the trend line.

### 4.1.7 Arab Research Publications: Concluding Remarks

Taken collectively, an admittedly risky endeavor on numerous grounds, the Arab countries produce less research output than their financial and manpower resources would warrant. On the other hand, gains made by a few of the Arab countries in the recent past fall considerably below their resource levels and future development needs.

Additionally, recent progress made by some Arab countries, such as Saudi Arabia, in terms of research publications is unlikely to be sustainable given that country's initiatives aim to improve institutional ratings rather than tangible outcomes. In addition to this, the fact that the cost of schemes seeking affiliation by well-cited scientists across the world is rather high while tangible and long-lasting benefits may be minimal. While in the case of countries such as Egypt, which stands at the forefront of other Arab countries in terms of research publications, output has been relatively stagnant, given potential resource levels and development needs.

The need is all too evident for fresh policy regimes that are more responsive to urgent socioeconomic development priorities, and with greater emphasis on networking various national policy regimes and attention to social and environmental challenges.

Science, technology, and innovation policies produced by several Arab countries over the past few decades have invariably underscored the importance of scientific research and pledged increasing support for research and development institutions as well as various entities dedicated to innovation. However, as indicated in Chapter 6, the Arab countries still trail behind many other developing countries in terms of both financial as well as human resources dedicated to scientific research, technology development, and innovation activities.

## 4.2 PATENTING ACTIVITY BY THE ARAB COUNTRIES

The number of patents filed by, and granted to, a given country or institution, by widely recognized international patent offices and, to a lesser extent, national patent offices is regarded as an important indicator of its innovative activity and ability to translate scientific knowledge into viable products and processes.

The following sections present a review of patent records published by the United States Patent and Trademark Office (USPTO) as well as the European Patent Office (EOP) covering many years during the past few decades. Subsequent sections present information based on data provided by a database, developed and maintained by the Organisation for Economic Cooperation and Development (OECD), concerning areas of technology, disciplines, and application sectors within which patents, to these and other Arab countries were in fact developed.

### 4.2.1 Introduction

In general, the Arab countries are not known for their intense patenting activity. To the contrary, patents granted to residents, and not necessarily nationals, of the Arab countries during the 1990s for instance were very few and far between. Mediocre patenting by the Arab countries is essentially due to limited industrial activity as well as inadequate links between research institutions and various production and services sectors, which in any case utilize imported, off-the-shelf STI inputs for most of their operations.

As further elaborated below, all sources seem to indicate that despite recent improvements, patenting activity in the Arab countries as recorded by USPTO and EPO remains inferior to some of their closest neighbors as well as to world averages.[53]

### 4.2.2 Patents Granted by the United States Patent and Trademark Office

Annex Table A.4.8 presents data on patents granted by USPTO to nationals and residents in the Arab countries, five of their closest neighbors namely Cyprus, Iran, Israel, Malta, and Turkey, as well as the world at large. The numbers of patents granted before 2002 until 2015 and in both 2002 and 2015 as well as corresponding ratios are listed in this table. Table 4.8, on the other hand, presents a comparative view of patenting activity undertaken by the more patent-prolific Arab countries, as well as their above-mentioned neighboring countries, together with collective data on patents granted to citizens and residents of the Arab countries and their neighbors.

**TABLE 4.8** Number of USPTO Patents Granted to Selected Arab and Neighboring Countries Compared to Numbers of Patents Granted Across the World and Timeframes

| | Pre-2002 | | 2015 | | Total Through 2015 | | Ratios of | |
|---|---|---|---|---|---|---|---|---|
| | Number of Patents | Percentage of World Total | Number of Patents | Percentage of World Total | Number of Patents | Percentage of World Total | Patents Granted 2015 to 2002 | Patents including 2015/Patents Before 2002 |
| Saudi Arabia | 168 | 0.0066 | 364 | 0.1117 | 1513 | 0.0264 | 36.4 | 9.01 |
| Kuwait | 59 | 0.0023 | 64 | 0.0196 | 448 | 0.0078 | 8 | 7.59 |
| Egypt | 54 | 0.0021 | 30 | 0.0092 | 275 | 0.0048 | 6 | 5.09 |
| UAE | 29 | 0.0011 | 56 | 0.0172 | 254 | 0.0044 | 9.33 | 8.76 |
| All Arab countries | 428 | 0.0168 | 570 | 0.1749 | 2890 | 0.0503 | 15.83 | 6.75 |
| Israel | 8886 | 0.3486 | 3804 | 1.1669 | 35900 | 0.6255 | 3.43 | 4.04 |
| Turkey | 75 | 0.0029 | 136 | 0.0417 | 748 | 0.0130 | 7.56 | 9.97 |
| Five closest neighbors | 9030 | 0.3543 | 3988 | 1.2234 | 37008 | 0.6448 | 3.54 | 4.10 |

Table 4.8 also compares patenting performance of selected countries in the region to world activity as well as across timeframes for which data is available, showing that up to and including 2015, citizens and residents of the Arab countries were granted a mere 2890 patents in all by USPTO. This represents a mere 0.05% of a total of around 5.7 million patents granted worldwide, while USPTO patents granted to their closest neighbor, Israel amounted to just under 36,000, that is more than 12 times the figure for the Arab countries combined, and in excess of 0.6% of the global total.

The situation before the year 2002 was even more gloomy, with the total number of USPTO patents granted to citizens and residents of the Arab countries up to that year a mere 428 patents, constituting around 0.017% of the world's total.[54] In other words the number of patents granted by USPTO to citizens and residents of the Arab countries increased almost sevenfold by the end of 2015, due mainly to higher patenting activity in Saudi Arabia, the United Arab Emirates, Kuwait, and Egypt, while their closest non-Arab neighbors managed a fourfold improvement on their pre-2002 patent total.

As indicated in the pie chart of Fig. 4.31, nationals and citizens of Saudi Arabia received around 57% of all USPTO patents granted to all Arab countries until and including 2015. Furthermore, it appears that Saudi Arabia together with the UAE, Kuwait, Oman, Bahrain, and Qatar account for close to 80% of all USPTO patents granted to nationals and citizens of the Arab countries.

The bar chart in Fig. 4.32 offers a more detailed view of patenting activity by some of the more patent-prolific Arab countries, their closest neighbors, and the world at large. Fig. 4.32 allows comparison of the shares of world patents granted to the above countries and groups of countries at all times to those granted before 2002 as well as the ratio of patents granted during 2015 to those granted in 2002. This figure indicates that the ratio of all patents granted to citizens and residents in Saudi Arabia, until and including 2015, amounted to nine times the number of patents granted before the year 2002. It also indicates that patents granted during the year 2015 amounted to over 36 times those granted in 2002. Incidentally, as indicated in Fig. 4.33, the

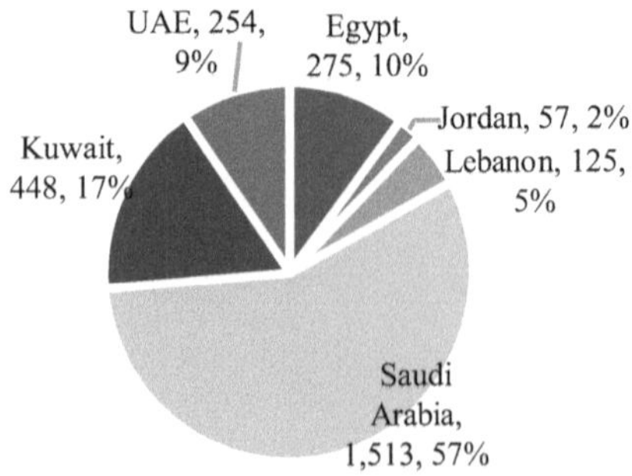


**FIGURE 4.31** Distribution of USPTO patents granted to eight Arab countries (before 2002 and through 2015).

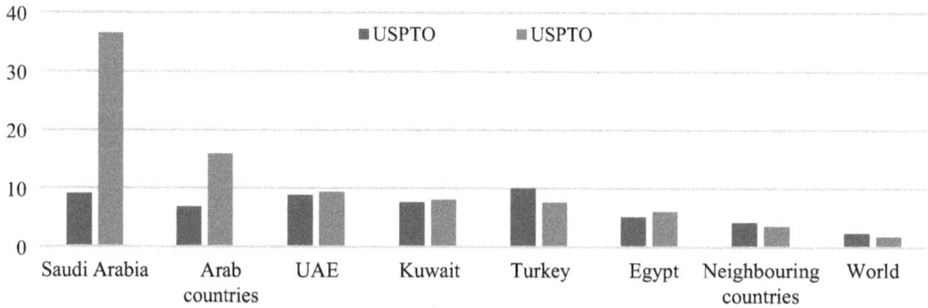


**FIGURE 4.32** Ratio of patents granted by USPTO to selected countries and country groups: (1) all patents relative to those granted before 2002 and (2) 2015 patents to those granted in 2002.

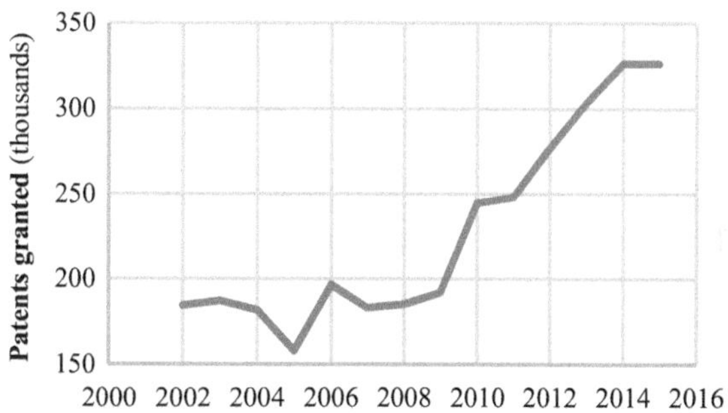


**FIGURE 4.33** Growth of patents granted by USPTO worldwide (2002–15).

number of patents granted by the United States Patents and Trademark Office has undergone considerable growth throughout the second half of the present decade.

That the Arab countries' closest neighbors made more modest gains relative to their previous performance over the same period is more than compensated by the fact that these countries were granted a total of around 21, 31, 7, and 12 times the number of patents granted before 2002, in 2002, during 2015, and at all times, respectively.

While patenting activity from the Arab countries is undeniably on the rise as indicated in Fig. 4.34, the rates of growth are clearly not in their favor. In the sense that the rate of patenting activity in the Arab countries would never bring up the number of patents filed or granted annually any time soon even up to levels comparable to corresponding levels attained by their neighboring countries and world averages.[55]

As indicated in Fig. 4.33 the path followed by USPTO patents granted worldwide took a much steeper upward path between the years 2008–15, than that observed during 2002–08.[56] Similar patterns are followed by USPTO patents granted to both the Arab countries and their closest neighbors, with growth in patents granted to both groups of countries following an almost linear and steep upward paths during the period 2008–15. Fig. 4.34 illustrates this observation,

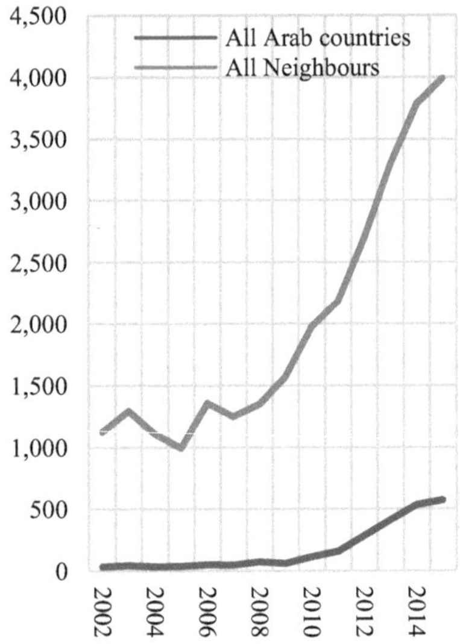


**FIGURE 4.34** US patents granted to nationals and residents of the Arab and neighboring countries (2002–15).

additionally indicating a much lower rate of growth of patents granted to the Arab countries as a whole compared to those granted to their closest neighbors. The rate of growth in the former countries is in fact around one-fifth of that exhibited by the latter, which is essentially due to USPTO patents granted to Israel. Despite the predominant share that Israel has traditionally had, it would seem that Turkey, in particular, gained considerable grounds during 2008–15, bringing its contributions to worldwide USPTO patents up to 0.013%, from an initial pre-2002 share of 0.003.

Fig. 4.35 presents a view of the growth of patents granted to the four most patent-prolific among the Arab countries during 2002–15, namely Saudi Arabia, Egypt, Kuwait, and the United Arab Emirates. This figure underlines the high rate of patenting activity by citizens and residents in Saudi Arabia compared to other Arab countries as well as the fact that this rate increased sharply between 2010 and 2015, reaching an average of around 76 new patents added to the total annually.

### 4.2.3 Patents Filed With the European Patent Office

A somewhat similar picture emerges upon reviewing patents awarded by the EPO to citizens and residents of the Arab countries, as presented in the annex Table A.4.10. The Arab countries combined were granted a total of 360 patents during the years 2006–15, corresponding to 0.06% of a total of nearly 615,000 patents granted worldwide. As in the case of patents granted by USPTO, Saudi Arabia, with 62% of patents, is by far ahead of all Arab countries in terms of all

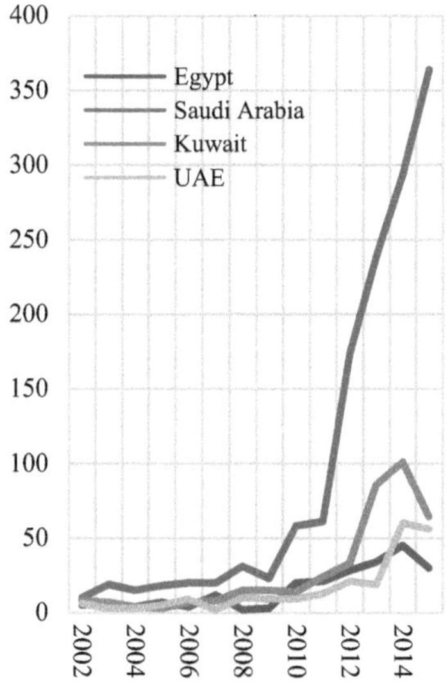


**FIGURE 4.35** US patents granted to nationals and residents in selected Arab countries (2002–15).

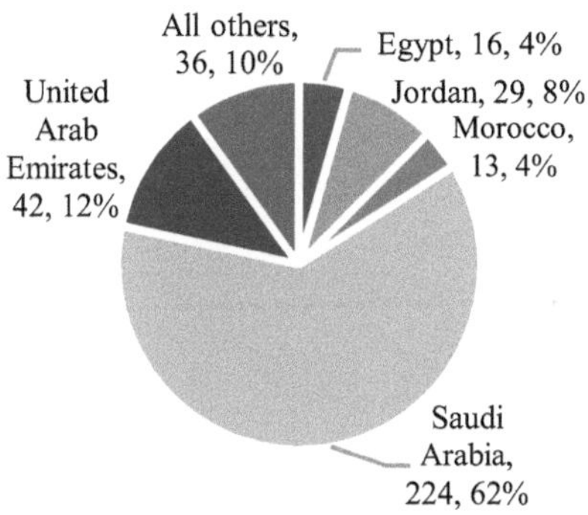


**FIGURE 4.36** Distribution of patents granted by the EPO to Arab countries during 2006–15.

patents granted during the 10-year period 2006–15. See Fig. 4.36. Israel, on the other hand, dominates the Arab countries' five closest neighbors, totaling just under 70% of the total number of patents granted by EPO to all five countries during 2006–15.

Fig. 4.37 presents a view of patenting activity in Saudi Arabia and the United Arab Emirates, indicating increasing numbers of patents granted by EPO to Saudi Arabia at a rate exceeding an additional five patents per year, during the period 2010–15. This is approximately one-fifth the corresponding rate of growth for EPO patents granted to Israel, in contrast to the much flatter path taken by growth of UAE patents over the same period.

In fact, as indicated by Table 4.9 Saudi Arabia managed to increase its share of patents granted by EPO sevenfold between 2006 and 2015, from 0.01% to 0.07%. Even so, this country's share was still minimal in comparison to Israel, and even Turkey, which exceeded Saudi Arabia's share with rates of growth of around 14- and 4-fold, respectively.

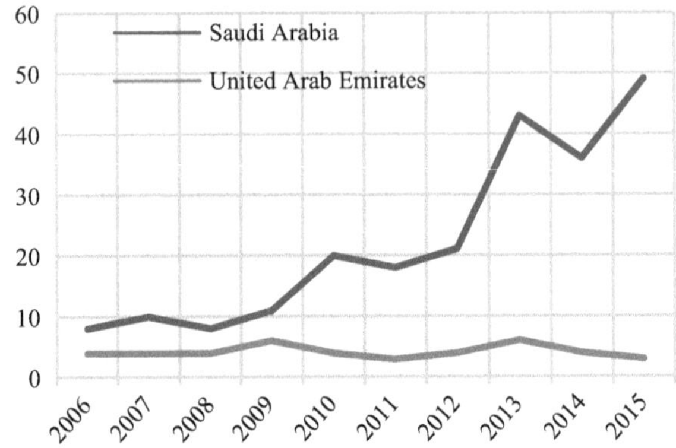


**FIGURE 4.37** EPO patents granted to Saudi Arabia and the United Arab Emirates.

**TABLE 4.9** EPO Patents Granted to Saudi Arabia, Turkey, and Israel as well as all Arab and Five Neighboring Countries; Numbers and Comparisons

| | Patents Granted During | | | | | |
|---|---|---|---|---|---|---|
| | 2006 | | 2015 | | 2006–15 | |
| | Patents Granted | Percentage of Total for 2006[a] | Patents Granted | Percentage of Total for 2015[b] | Patents Granted | Percentage of Total During 2006–15[c] |
| Saudi Arabia | 8 | 0.01 | 49 | 0.07 | 224 | 0.04 |
| Turkey | 31 | 0.05 | 218 | 0.32 | 994 | 0.16 |
| Israel | 258 | 0.41 | 397 | 0.58 | 3086 | 0.50 |
| All Arab countries | 13 | 0.02 | 62 | 0.09 | 360 | 0.06 |
| Closest neighbors | 312 | 0.50 | 648 | 0.95 | 4435 | 0.72 |

[a] *Total number of EPO patents granted in 2006: 62,777.*
[b] *Total patents granted in 2015: 68,421.*
[c] *Total patents granted during 2006–15: 614,850.*

The weak position of other Arab countries, regarding EPO patents is underscored by Table 4.9. This table indicates that their share in all EPO-granted patents increased from 0.02% in 2006 to represent only 0.09% in 2015, including Saudi Arabia's share, which represented 0.07 of all EPO patents granted in 2015. In effect, the number of EPO patents granted to Israel by far exceeds the number of EPO patents granted to all the Arab countries as well as their closest neighbors.

### 4.2.4 Distribution of Patents From the Arab Countries Across Patent Classes[57]

Looking at the distribution of world patents across disciplines, technologies and areas of application should help characterize directions taken by invention activities at the national, regional, and international levels. A reliable resource on patent information with the possibility for examining such a distribution is the OECD statistical database.[58]

Thus, OECD statistics are available for a group of six technology-based classes of patents, namely "ICT," "Biotechnology," "Nanotechnology," "Selected environment-related technologies," "Medical technologies," and "Pharmaceuticals," as well as another group of eight additional categories. This latter group includes patents within eight classes, referred to as the International Patent Classes A–H, covering a wide range of disciplines and application areas: "Chemistry and metallurgy," "Physics," "Electricity," "Mechanical engineering," "Textiles and paper," "Fixed construction," "Performing operations, transporting," and "Human necessities." For brevity, these two groups of patent classes will henceforth be referred to as Group I and Group II of patent classes. In essence, distinctions between the patent classes comprised by these two groups and the subgroups that their constituent classes may relate to areas where the patent in question might find most widespread applications. On the other hand, it is doubtless that many of the patents within classes in both groups are likely to involve varying degrees of novel technology content as well as potential for implementation within common areas of application.

According to the OECD database, the Arab countries filed a total of just over 4360 patents in all patent classes during the period 1999–2013, around 1650, or 38% of this total, fell within Group I of patent classes, while the remaining 2710 patents, or 62% of the total were dedicated to Group II. The total number of patents filed by the Arab countries during this period corresponded to around 0.12%, 1.37%, and 7.75%, respectively, of all patents filed worldwide, in the non-OECD countries and in a small group of the Arab countries' closest neighbors, namely Cyprus, Iran, Israel, Malta, and Turkey. Taking into account the populations of the various Arab countries and resources available to them, these percentages indicate woefully inadequate performance.

Fig. 4.38 compares patents in the above 14 classes of patents filed by the Arab countries as a whole to those filed by two groups of countries; its five immediate neighbors, Cyprus, Iran, Israel, Malta, and Turkey, on the one

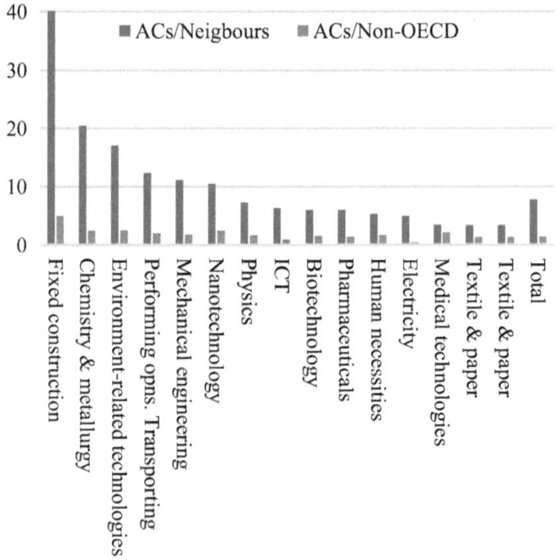


**FIGURE 4.38** Percentage of Arab country patents filed within various patent classes during 1999–2013 relative to those filed by neighboring countries and non-OECD countries (%).

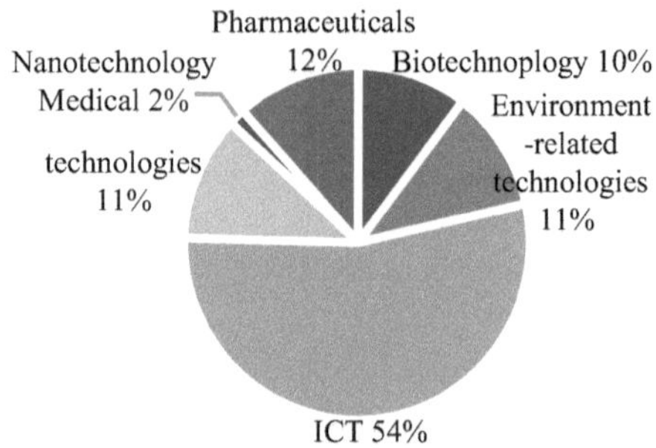


**FIGURE 4.39** Distribution of world PCT patents across patent classes in Group I, including 1,554,820 patents (1999–2013).

hand, and almost 100 non-OECD countries around the world, on the other. This figure presents the percentages of patents filed by the Arab countries within a particular class of patents out of those filed by all members in each group of countries. Fig. 4.38 clearly indicates that the Arab countries acquired a reasonable positions compared to their neighbors as well as non-OECD countries in patent classes such as "Fixed construction", "Chemistry and metallurgy," and "Performing operations." On the other hand, the Arab countries underperformed, compared to both groups of countries, and in comparison to all other countries of the world, with regard to patent classes such as "ICT," "Pharmaceuticals," "Medical technologies," and "Human necessities."

The percentages of patents filed by the Arab countries relative to its closest neighbors, to other non-OECD countries as well as to the world at large were estimated at 6.2%, 0.89%, and 0.094%, respectively.

Patents filed within the class dedicated to "Selected environment-based technologies" occupied a middle positing, totaling over 300 patents throughout the period 1999–2013; equivalent to 0.18% of world patents filed within the same class and over 16% of patents filed by the Arab countries' closest neighbors.

The following paragraphs will compare patents filed by the Arab countries within specific classes to those filed by countries around the world. A closer look will also be taken at growth patterns followed by patents in selected classes. This should provide indications of the Arab countries' relative strengths in relation to inventive capacity as well as possible directions for future policy trajectories targeting future investment and growth possibilities.

Based on data from the OECD database, Fig. 4.39 presents the average distribution of more than 1.5 million patents filed by countries around the world in accordance with the Patent Convention Treaty (PCT), during the period 1999–2013, across Group I of patent classes specializing in selected novel technologies.[59]

The pie chart in this figure indicates a predominant share for the "ICT" class of patents followed by the "Medical technologies" and "Pharmaceuticals" classes, with a comparable share for the "Biotechnology" and patents within the "Environment-related technologies" class. The corresponding pattern for patents filed by the Arab countries altogether is presented in Fig. 4.40. This figure indicates that patents filed by the Arab countries follow a pattern rather similar to that exhibited by countries around the world. Yet, two significant differences may be noted. A considerably larger share of patents was filed by the Arab countries within environment-related technologies, 19%, as opposed to 11%, while a somewhat smaller percentage of patents filed by the Arab countries addressed ICT patents, 48%, in contrast to 52%.[60]

A rather similar comparison is presented by Figs. 4.41 and 4.42, with regard to patents filed within the Group II of IPC patent classes, i.e., those dedicated

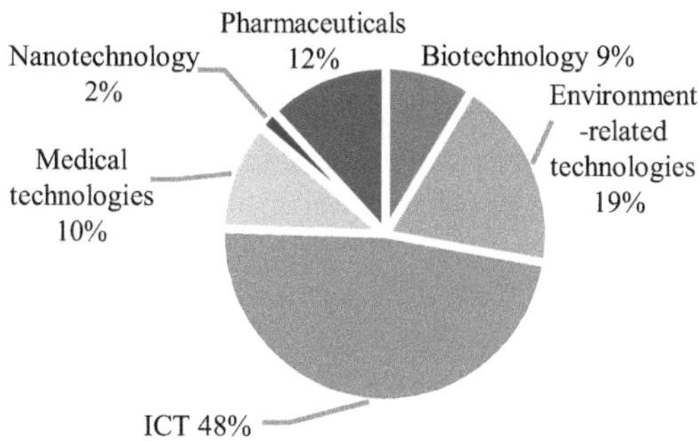


**FIGURE 4.40** Distribution of Arab country PCT patent across patent classes in Group I, including 1,652 patents throughout 1999–2013.

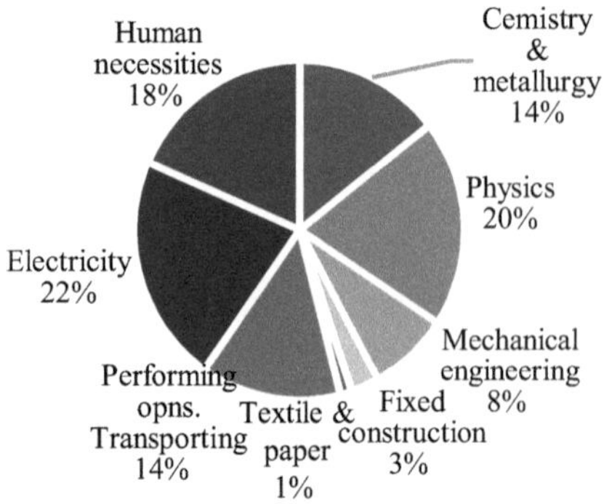


**FIGURE 4.41** Distribution of world PCT patents across patent classes in Group II patent, including 2,179,938 patents (1999–2013).

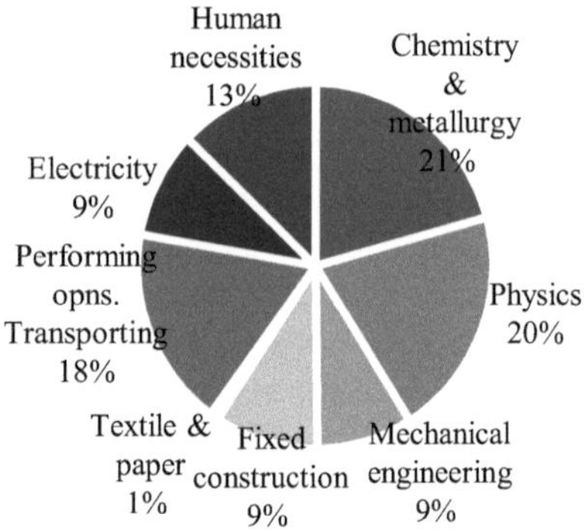


**FIGURE 4.42** Distribution of Arab country PCT patents across classes in Group II patent, including 2709 patents (1999–2013).

to selected disciplines and areas of application. Thus, a more divergent pattern is exhibited here, with considerably larger shares of patents filed within "Chemistry and metallurgy," "Fixed construction," and "Performing operations, transporting," by the Arab countries relative to the countries around the world. On the other hand, the Arab countries attained a considerably smaller share in the "Electricity" and "Human necessities" patent classes. A smaller number of patents were filed from the Arab countries in the "Textile and paper" class, constituting only 0.44% of the Arab countries' total as opposed to the world's 0.65%.

More patents were filed from the Arab countries as opposed to the world's averages in three domains, namely "Chemistry and metallurgy," "Environment-related technologies," and "Fixed construction-related" patents. Percentages of patens filed out of the total in the Arab countries within these domains were 60%, 50%, and 250% higher than corresponding percentages for the entire world. On the other hand, patents in domains such as medical technologies, pharmaceuticals, biotechnology, nanotechnology, and physics are reassuringly similar, with differences amounting to no more than 1%–2%, between patents filed in the Arab countries and those filed worldwide.

With regard to positive signals due to higher patenting activity in environment-related technologies, in-depth analysis would certainly be needed to establish to what extent such patenting activity might be geared to utilizing and assimilating inventions aimed at tackling water treatment methodologies as well as a variety of environmentally sound production technologies.

#### *4.2.4.1 Sectoral Aspects of Patenting Activity by the Arab Countries*

Low levels of patenting in classes pertinent to medical services and healthcare may be noted. Thus, the combined percentages of patents filed by the Arab countries in the two classes that would normally be expected to impact health services, namely "Medical technologies" and "Pharmaceuticals", relative to their closest neighbors the non-OECD countries and the world's averages are estimated at only 6%, 4%, and 0.2%. It may be noted that the Arab countries performed considerably worse with regard to patents related to "Medical technologies," in comparison to pharmaceuticals, vis-à-vis both their closest neighbors as well as the non-OECD countries taken as a whole. However, in comparison to all countries around the world the Arab countries' performance in both classes of patents is quite close at around 0.1% for total patents in both classes filed throughout the period 1999–2013. In fact, as indicated in Fig. 4.43 the percentages of patents filed in almost all classes[61] by the Arab countries relative to world averages fall within a rather narrow range, between 0.08% and 0.12%.

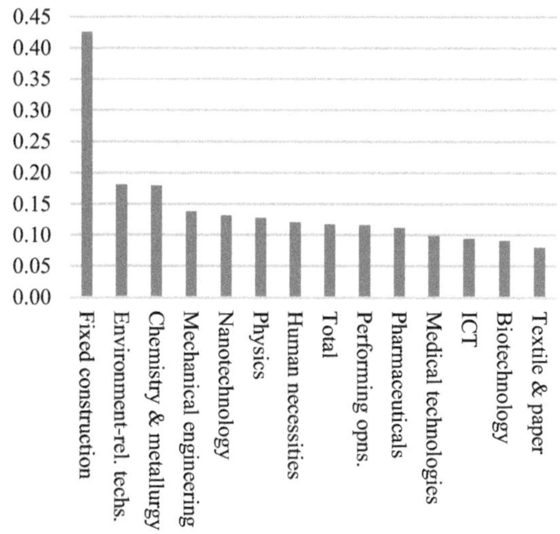


**FIGURE 4.43** Percentages of patents filed by the Arab countries relative to those filed worldwide within 14 classes of patents.

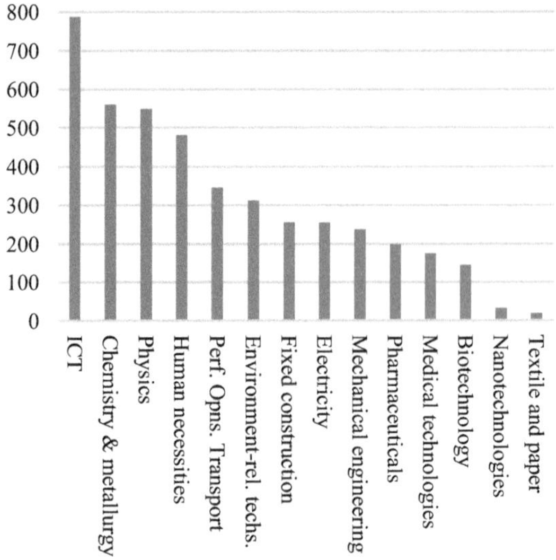


**FIGURE 4.44** Patents filed by the Arab countries within 14 patent classes during (1999–2013).

Fig. 4.44 presents the distribution of the number of patents filed by the Arab countries across 14 patent classes within Groups I and II, i.e., the six technology-based classes of patents as well as the eight classes dedicated to various disciplines and application areas. This figure clearly shows the predominance of patents filed within the "ICT" class followed by the "Chemistry and Metallurgy," "Physics," and "Human Necessities" classes. Patents filed within classes termed "Medical technologies" and "Pharmaceuticals," occupied an intermediate position, with a total of around 374 patents; around 0.1% of patents filed worldwide within these two classes.

On the other hand, patents with applications in the industrial sectors, such as those filed within the "Mechanical engineering" and "Electricity" classes, totaled around 500, constituting around 0.07% of patents filed within these two classes worldwide. By comparison, the construction sector, for which patents would be included mainly within the patent class dedicated to "Fixed construction," received around 260 patents in all. At 0.42% and 40% of patents filed in this same class across the world, and by citizens and residents in their closest neighbors, respectively, this is the highest percentage of Arab country patents in any class of patents for which data is available, possibly indicating a relatively high level of infrastructure building.

Patents filed under the class "Performing operations, transporting" which include patents that target the transport sector amounted to nearly 450 patents throughout the 15-year-period covered by OECD statistics, which corresponds to about 0.16% of worldwide patents in this class. The smallest number of patents filed by the Arab countries was in the "Textile and paper" and "Nanotechnology" classes. At 19 and 32 patents, respectively, the Arab countries contributed around 0.11% toward worldwide patents filed in each one of these two classes.

### 4.2.4.2 *Distribution of Filed Patents Within Patent Classes Across Selected Arab Countries*

Nearly 75% of the total number of patents filed from the Arab countries during 1999–2013, originated from only four countries, namely Egypt, Saudi Arabia, the United Arab Emirates, and Morocco. See Fig. 4.45. The remaining five countries from which patents were recorded by the OECD database contributed between 1% in the case of Kuwait and 4% in the cases of both Jordan and Tunisia.

Figs. 4.46 and 4.47 present a view of the distribution of patents across technologies, disciplines, and areas of application for the four most patent prolific Arab countries. As shown by Figs. 4.46 and 4.47, patents from Saudi Arabia were most numerous in the "Chemistry and metallurgy," "ICT," and "Physics" classes, while Egypt and the United Arab Emirates contributed a considerable number of patents in the "ICT" class, with the former taking the lead in patents within the patent class dedicated to "Human necessities."

With particular regard to patents filed under "Environment-related technologies," all four countries represented in Figs. 4.46 and 4.47 appear to have made significant contributions, while Saudi Arabia took the lead with around 110 patents.

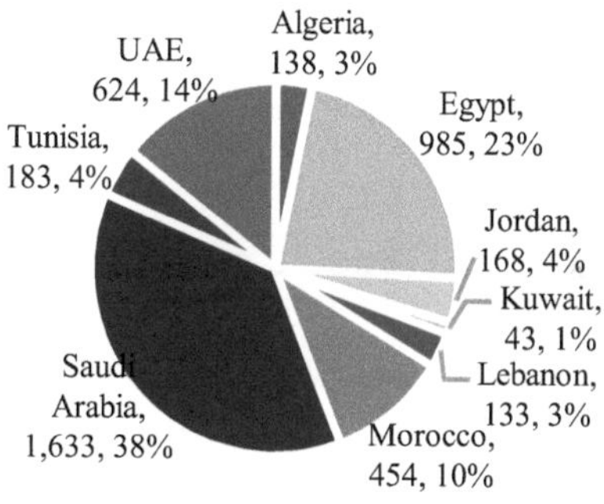


**FIGURE 4.45** Distribution of patents included in the OECD database a cross the Arab countries they were filed from during 1999–2013.

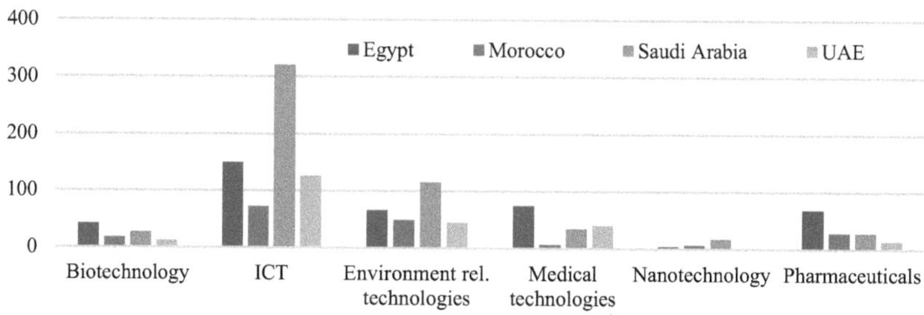


**FIGURE 4.46** Patents filed from four Arab countries across technology-related patent classes.

As indicated above, negligibly small numbers of patents were filed from all Arab countries in the "Nanotechnology," on the one hand, and "Textile and paper" classes, with patents from Saudi Arabia taking the lead in the "Nanotechnology" class.

### 4.2.4.3 Growth of Patents Filed From the Arab Countries Within Specific Patent Classes

The number of patents filed by citizens and residents of the Arab countries within several classes witnessed considerable growth relative to their humble beginnings toward the start of the present century. Thus, taken together patents filed early on during the period 1999–2013 by the Arab countries altogether never managed to exceed 10 patents annually in any of the classes included in the OECD database. Fig. 4.48 presents an overview of the growth of patents filed during this period from all the Arab countries for which data was available.

Fig. 4.49 presents ratios of the number of patents filed in 2013 to those filed in 1999 within Groups I and II of patent classes, described above, by the Arab countries taken together as well as their closest neighbors, non-OECD countries as well as the countries of the world, taken altogether. Larger growth is observed in the case of Group I of patent classes. Thus, the average of the ratios of patents

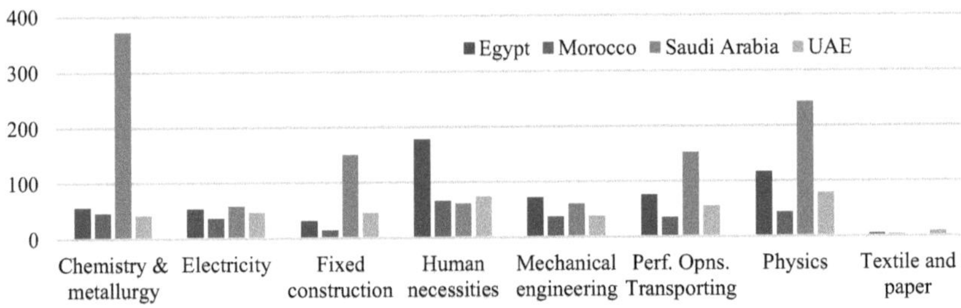


**FIGURE 4.47** Patents filed from four Arab countries across disciplines and areas of application patent classes.

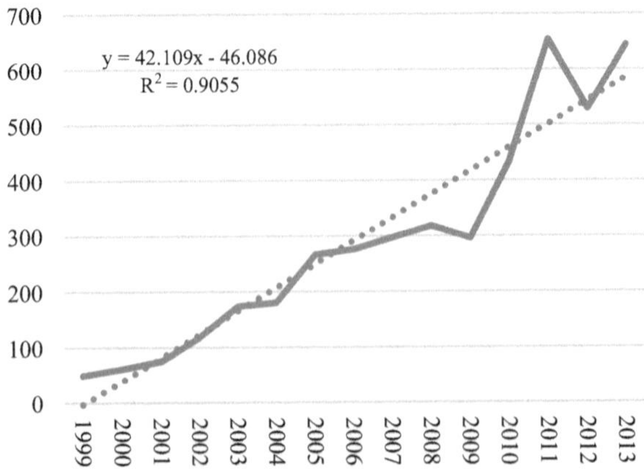


**FIGURE 4.48** Patents filed by citizens and residents of the Arab countries according to OECD records on PCT patents during 1999–2013.

filed within Group I patent classes in 2013 against those filed in 1999 is estimated at 21, while the corresponding average within the remaining eight Group II patent classes, dedicated to particular disciplines and areas of application, was around 15. See Table 4.10, with rates of growth for patents filed within a few patent classes by way of illustration.

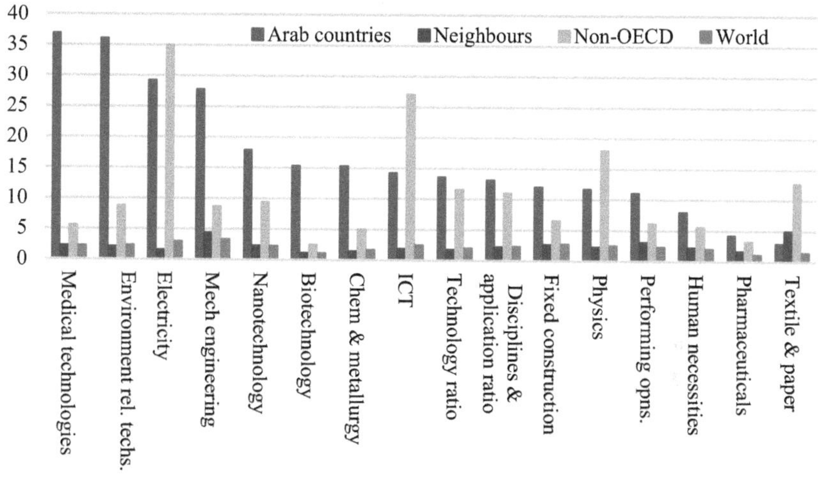


**FIGURE 4.49** Ratio of patents filed in 2013 to those filed in 1999 by citizens and residents of the Arab countries, five of their closest neighbors, non-OECD countries, and the world.

**TABLE 4.10 Comparing the Rates of Growth[a] of Patents Filed From All Arab Countries Within Selected Patent Classes Over Two 5-Year-Periods During 1999–2013 and for the Entire Period 1999–2013**

| Patent Class | 1999–2004 | 2008–13 | 1999–2013 |
|---|---|---|---|
| Chemistry and metallurgy | 1.8 | 16.7 | 6.6 |
| Information and communications technology | 4.6 | 14.2 | 7.7 |
| Physics | 3.7 | 9.8 | 5.2 |
| Performing operations | 2.5 | 6.6 | 3.0 |
| Environment-related technologies | 2.7 | 6.3 | 3.6 |
| Mechanical engineering | 2.5 | 4.7 | 2.7 |
| Electricity | 1.6 | 3.5 | 2.4 |
| Biotechnology | 0.5 | 1.9 | 0.9 |
| Pharmaceuticals | 0.9 | 0.7 | 1.0 |
| Medical technologies | 2.3 | 0.2 | 1.5 |

[a]*As measured by coefficients of the variable term in linear equations that most accurately describe growth of the number of patents filed during the periods indicated.*

#### 4.2.4.3.1 Overall Growth in Patenting Activity

On average, patents filed from the Arab countries in all 14 classes multiplied over 17-fold, during the 15-year period 1999–2013, while those filed by their neighbors, non-OECD countries, and the world at large increased 2-, 3-, and 11-fold, during the same period.

A noteworthy feature of Fig. 4.49 is that the highest growth in patents filed from the Arab countries occurred within two classes of technology-based patents, namely "Medical technologies" and "Environment-related technologies," with the number of patents in these two classes multiplying 37- and 36-fold, respectively, between 1999 and 2013. Patents filed in the ICT class registered considerable growth, by a factor of around 14. However, despite considerable growth witnessed by ICT patents in the Arab countries, the total number of these patents filed throughout the period 1999–2013 was around 0.9% of patents filed by non-OECD countries around the world, and around 0.09 of all patents filed in this class worldwide. Patents in the "Nanotechnology" and "Biotechnology" classes witnessed even more impressive growth; by factors estimated at 18 and 15, representing 0.13% and 0.09% of world patents in these two classes, respectively.

Within Group II patents filed from the Arab countries, two classes namely "Electricity" and "Mechanical engineering" witnessed massive growth by factors of 29 and 28. It is also noteworthy that although impressive growth was evident in patents filed from the Arab countries within the "Electricity," "ICT," and "Physics" patent classes, this growth was exceeded by non-OECD countries, by factors of 1.2, 1.9, and 1.5, respectively.

#### 4.2.4.3.2 Annual Growth Patterns

Figs. 4.50–4.53 present patterns followed by annual growth of patents filed from the Arab countries in all 14 classes addressed by the OECD database. In common with observations made in relation to growth of research publications

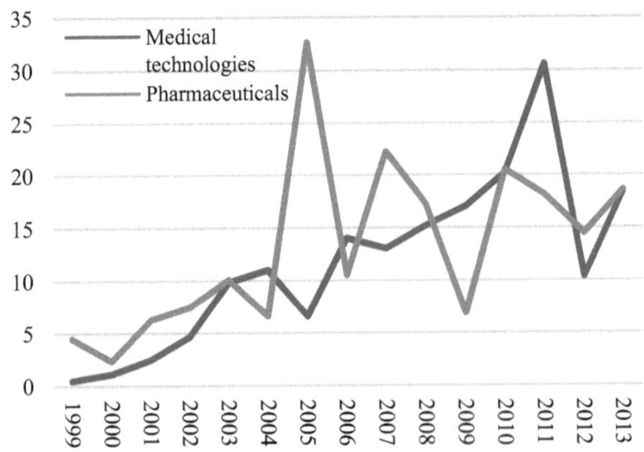


**FIGURE 4.50** Growth of patents filed by the Arab countries in medical technologies and pharmaceutical (1999–2013).

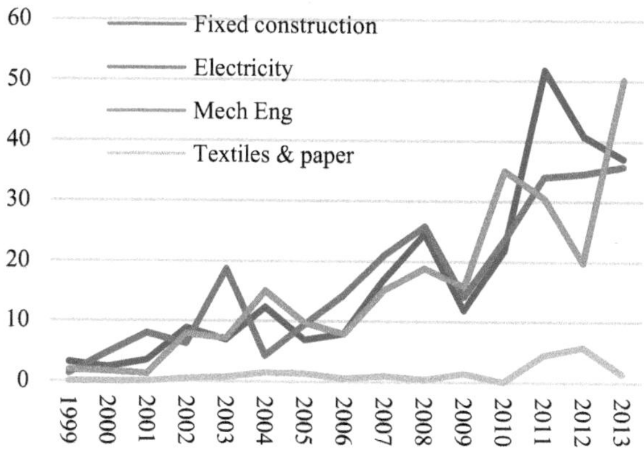


**FIGURE 4.51** Growth of patents filed by the Arab countries in selected fields of engineering and industry (1999–2013).

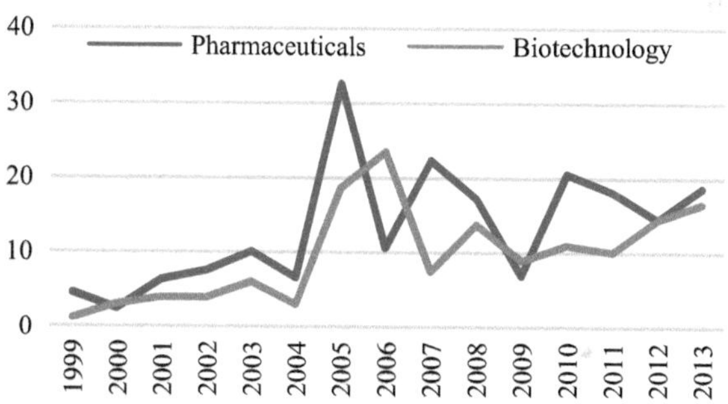


**FIGURE 4.52** Growth of pharmaceuticals and biotechnology patents filed from the Arab countries during 1999–2013.

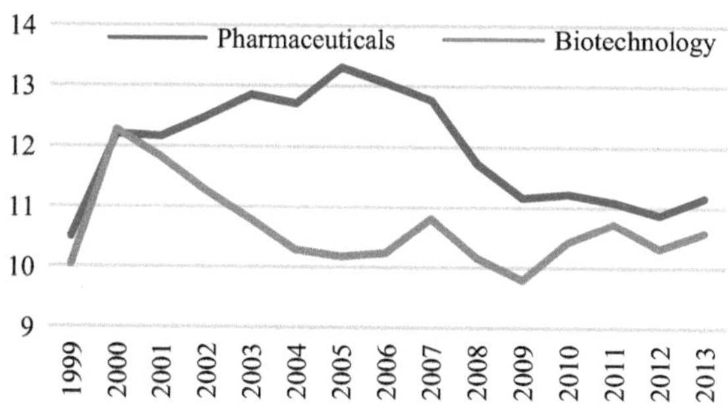


**FIGURE 4.53** Growth of patents filed by all countries in biotechnology and pharmaceuticals during 1999–2013 (thousands).

and tertiary enrolment, higher rates of growth are exhibited by patents filed within patent classes such as "ICT," "Chemistry and metallurgy," and "Physics," compared to the first few years of the present century. This table indicates that the rate of growth of patents in the "Chemistry and metallurgy" class was the highest during the last 5 years of the periods for which data was available from the OECD database, followed by patents in the "ICT" and "Physics" patent classes.

While generally on an upward path, growth of patents filed from the Arab countries appears to take a rather sporadic path in a number of classes. An obvious example of such growth is presented by patents filed in the "Medical technologies" class. See Fig. 4.50. Another aspect that is characteristic of patents filed from the Arab countries in both the classes of "Medical technologies" and "Pharmaceuticals" is the low rate of their growth during the years 2008–13, compared to all other patent classes.

Fig. 4.51 depicts growth in the number of patents filed from the Arab countries in engineering and industry-related classes during 1999–2013. This figure indicates higher rates of growth for patents in the "Mechanical engineering," "Electricity," and "Fixed construction" classes of patents during the last 5 years of that period. By contrast, patents within the "Textiles and paper" class witnessed minimal growth.

Despite their meager numbers and sporadic character of growth observed in most patent classes, patents filed from the Arab countries generally exhibited higher rates of growth throughout the period 1999–2013, but particularly during 2008–13, compared to world trends. This is particularly obvious upon comparing growth rates in patent classes such as "Biotechnology" and "Pharmaceuticals," in which worldwide patents grew at an almost constant rate, particularly during the period 2009–13, while patents filed by the Arab countries multiplied several folds, as shown in Figs. 4.52 and 4.53.

Fig. 4.54 shows that the largest growth rates in Group I technology patents filed from the Arab countries were exhibited in the "ICT" class of patents, which, as indicated above, grew by more than threefold within the period

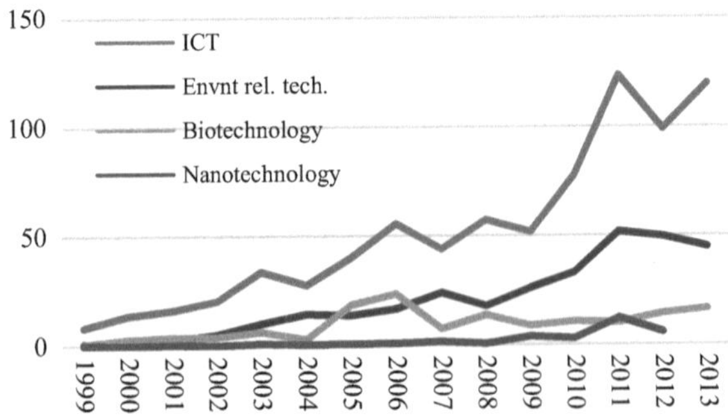


**FIGURE 4.54** Growth of patents filed by the Arab countries within four broad technology areas during 2004–13.

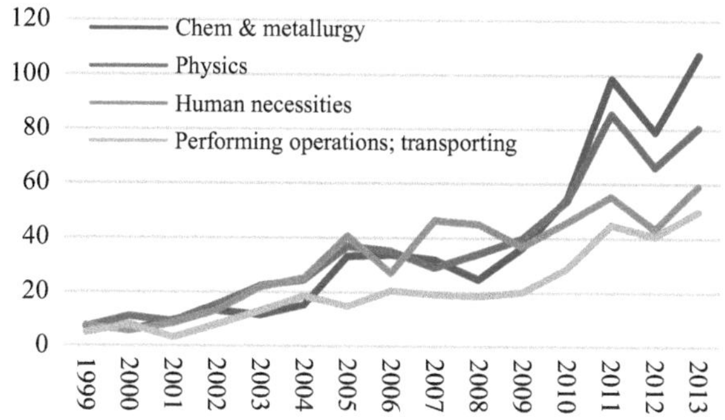


**FIGURE 4.55** Growth of patents filed by the Arab countries in selected disciplines and application domains (1999–2013).

2008–13. This figure also shows that patents in "Environment-related technologies" increased more than twofold during the period 2008–13.

As Fig. 4.55 indicates, high rates of growth were witnessed by patents in the "Physics" and "Chemistry and metallurgy" classes, during the period 2008–13. Patents within these two classes multiplied three- to fourfold, while patents in the other two classes depicted in this figure, namely "Human necessities" and "Performing operations" increased by factors of 1–2.

### 4.2.5 Concluding Remarks on Patenting Activity by the Arab Countries

To summarize, the numbers of patents filed from the Arab countries grew in all patent classes, with some stark distinctions between some. Yet, the numbers of patents filed by, or granted to, citizens and residents of the Arab countries constituted a miniscule percentage of patents filed worldwide and were far exceeded by the number of patents filed from countries in their neighborhood, in sharp contrast to the Arab countries' populations and the resources that at least some of these countries command.

With regard to patenting patterns and their distribution across technologies and application areas, certain similarities are evident. Thus, in keeping with world trends, patents in the ICT class soared while those in nanotechnology witnessed exceedingly limited growth.

Additionally, while patenting activity by citizens and residents of the Arab countries did witness significant improvement during the past decade or so the following points must be borne in mind:

- Patenting activity was confined largely to a few Arab, mainly oil and gas dependent, economies.
- While patents granted to, and filed by, the majority of the Arab countries multiplied several folds from exceedingly low levels toward the end of last century, their numbers are still far inferior to those characteristic of countries in their vicinity as well as regional and world averages.

- Low numbers of patents granted by USPTO and EPO to several Arab countries should not obscure the fact that patents are also being filed with and granted by national patent offices in these countries.
- In the absence of more detailed information on the fields of application as well as actual holders of patents granted to the Arab countries, it would be difficult to establish in what manner patenting activity by the Arab countries might impact sustainable development and national economic growth.[62]

## 4.3 CONCLUDING REMARKS

Whether in terms of quantity, quality, or orientation, output produced by the Arab countries' science, technology, and innovation systems, be it in terms of published scientific reports or filed and granted patents, leaves a great deal to be desired.

In principle, enhancing scientific output by the Arab countries is contingent upon a number of institutional as well as resource-related factors. Many more researchers are needed and higher resource levels must be dedicated to their activities. Initiatives aimed at tackling specific development problems must be designed with provisions for national, regional, and international networking.

Removing regulatory contradictions that hinder closer and more persistent involvement in research activity by teaching staff at Arab universities is another priority. Thus, while many universities in the Arab countries impose scientific publications as an essential condition for promoting staff, faculty members are frequently overloaded with teaching duties, leaving them little time for research activity. On the other hand, faculty members in some countries often welcome extra teaching duties as a source of much-needed supplementary income.

With particular reference to applied scientific research, it is essential to allocate sufficient resources to extension services and multidisciplinary research, thereby creating channels to actual demand, would need to be drastically increased.

The above issues must be taken up within future policy systems designed to promote wider and more profound contributions by scientific research toward national development.

However, most policies formulated by the Arab countries have tended to adopt supply-side approaches, emphasizing issues directly related to science and technology, rather than holistic and multifaceted approaches taking into account demand considerations as well as interactions among a multitude of factors that eventually impact STI capacity building.

An illustration of the above is provided by the mixed results attained by Arab institutions of higher education attempting to link staff promotion to publications in refereed international journals. Directives made to this effect, in emulation of policies adopted by many institutes of higher learning across the world, were seldom accompanied by recruiting teaching and research assistants to reduce teaching load per staff member, with the result that little time is available for staff to engage in high-quality research.

Joint funding that takes into account urgent development needs of the various Arab countries as well as resources available to them is essential for improved networking. The European Union's Framework Programme offers a useful model for analysis and future emulation. Thus, EU funding comes in the form of project-based funding provided on competitive bases, in addition to what most European countries provide their own research institutions.

At a more fundamental level, steps must be taken to promote the creation of learning societies. And, toward this end, research projects would need to be included as mandatory activities at all levels of education.

Few synergies exist between scientific institutions in the Arab countries, at both the national and regional levels. Joint research activity involving Arab scientific research institutions working in the same or similar fields is extremely rare. It is essential to see this scarcity of research cooperation as an outcome of limited cooperation in various other areas, including industry, agriculture, and environmental protection, among others. Indeed, the success of efforts, such as those of the European Framework Programme, must be viewed from a perspective that takes into account collaboration in a variety of economic activities.

While inter-Arab research cooperation is admittedly quite low, more funding is becoming available through international schemes. Thus, France, Italy, Spain, Germany, and—to a lesser extent—the UK have been active in funding research through bilateral programs in the Middle-East region and the Maghreb. Additionally, institutions with national and regional mandates such as Qatar Foundation and the Islamic Development Bank may dedicate resources aimed at filling the gap.

Despite an inadequate record of inter-Arab collaboration in the past, steps must be taken to promote further regional collaboration at many levels and in various directions, placing common environmental and socioeconomic challenges at the top of national lists of priorities.

Several indications point to the notion that funding is not in fact an insurmountable obstacle to wider research collaboration, and indeed to research activity in general, whether collaborative or otherwise. Rather, it is the weaknesses in institutional mechanisms dedicated to research and its funding that seems to impede research activity, particularly when involving collaborative efforts. Evidence, gathered mostly in field visits made to the Arab countries, by the author indicates that challenges lie mostly in building capacity to efficiently and effectively manage and spend research allocations.

## ANNEX TABLES

**TABLE A.4.1 Journals Published in 2015 by Institutions in the Arab Countries and Three of Their**

| | Agriculture and Veterinary Science | Biosciences | Chemistry and Physics | Mathematics | Scientific Research (Gen.) | Earth and Environmental Sciences |
|---|---|---|---|---|---|---|
| Bahrain | | | | 1 | | 1 |
| Egypt | 2 | 4 | 4 | 5 | 4 | 2 |
| Iraq | 1 | | | | | |
| Jordan | 1 | 2 | | | 1 | |
| Kuwait | | | | 1 | 1 | |
| Lebanon | | | | | | |
| Morocco | | | | | 1 | 2 |
| Oman | | | | | | |
| Qatar | | | | | | |
| Saudi Arabia | | | | | 2 | 2 |
| Sudan | | | | | | |
| Tunisia | | | | | | |
| UAE | 2 | 1 | 1 | 2 | | |
| **Total ACs** | **6** | **7** | **5** | **9** | **9** | **7** |
| Iran | 5 | 7 | 4 | 7 | 7 | 7 |
| Israel | 3 | | 1 | | 4 | 3 |
| Turkey | 13 | 9 | 7 | 4 | 8 | 10 |
| **Total neighbors** | **21** | **16** | **12** | **11** | **19** | **20** |
| **Grand total** | **27** | **23** | **17** | **20** | **27** | **27** |

Closest Neighbors; Distribution Across Countries and Areas of Specialization

| Engineering | Computer and Information Science | Medicine and Healthcare | Social Sciences and Education | Unspecified | Totals |
|---|---|---|---|---|---|
| | | 2 | | | 4 |
| 7 | 2 | 37 | | | 67 |
| | | 1 | | | 2 |
| 2 | 2 | 2 | 1 | | 11 |
| 2 | | 2 | 1 | | 7 |
| | | 2 | 1 | | 3 |
| | | | 1 | | 4 |
| 1 | | 2 | | | 3 |
| | | 2 | | | 2 |
| 1 | | 5 | 1 | | 11 |
| | | 1 | | | 1 |
| | | 2 | | | 2 |
| 3 | 1 | 14 | | | 24 |
| **16** | **5** | **72** | **5** | | **141** |
| 7 | 5 | 82 | 3 | 4 | 138 |
| 1 | 1 | 6 | 3 | 5 | 27 |
| 8 | | 77 | 16 | 28 | 180 |
| **16** | **6** | **165** | **22** | **37** | **345** |
| **32** | **11** | **237** | **28** | **37** | **486** |

**TABLE A.4.2** Top 10 Collaborating Countries Across the World With 17 Arab Countries

| | Saudi Arabia | Egypt | Morocco | Qatar | UAE | Lebanon | Tunisia | Jordan |
|---|---|---|---|---|---|---|---|---|
| USA (14) | 31.24 | 24.7 | 15.56 | 28.19 | 29.55 | 27.5 | 5.05 | 37.51 |
| UK (12) | 7.53 | 5.96 | 8.87 | 8.62 | 4.89 | 7.95 | 6.82 | 2.08 |
| Germany (11) | 6.32 | 6.27 | 11.49 | 4.22 | 5.8 | 8.65 | 4.62 | 14.68 |
| Italy (10) | 2.67 | 4.4 | 5.77 | 6.49 | 3.7 | | 2.68 | 2.94 |
| Canada (9) | 5.15 | 6.03 | 3.12 | 3.42 | 4.67 | 2.63 | 5.71 | 9.35 |
| France (9) | 2.72 | 3.46 | 11.62 | | 12.87 | 21.85 | 55.33 | |
| Switzerland (7) | 2.33 | | 5.43 | 3.49 | 2.3 | | | |
| Spain (7) | 3.61 | | 4.12 | 2.97 | | 2.28 | 1.98 | |
| Russia (6) | | 3.53 | 3.25 | | | 3.21 | | |
| China (5) | 10.56 | 5.14 | | 8.13 | | | | 4.22 |
| Japan (4) | 3.67 | 5.91 | 3.19 | | | 3.13 | | |
| Netherlands (3) | | | | | 2.83 | 2.23 | | 4.29 |
| Sweden (2) | | | | | 2.28 | | | 2.05 |
| Romania (2) | | | | | | | | |
| South Korea (2) | | | | 4.64 | 3.61 | | | |
| Pakistan (2) | | | | | | | | |
| India (2) | | | | | | 5.13 | | |
| Austria (2) | | | | 2.95 | | | 1.41 | |
| Belgium (2) | | | | | | | | |
| Portugal (1) | | | | | | | | |
| Poland (1) | | | | | | | | 2.27 |
| Ireland (1) | | | | | | | | |
| Czech Republic (1) | | | | | | | | |
| Hungary (1) | | | | | | | | |
| Australia (1) | | | | | | | | |
| Brazil (1) | | | | | | | | |

(Percentages)

| Algeria | Kuwait | Iraq | Oman | Palestine | Syria | Libya | Mauritania | Yemen |
|---|---|---|---|---|---|---|---|---|
| | 34.7 | 6.75 | 26.71 | 21.07 | 35.42 | 53.85 | | |
| 9.21 | 3.53 | 40.38 | | 5.09 | | | | |
| 6.66 | | | 5.88 | 2.91 | | | | |
| | 14.94 | 4.14 | 2.39 | | | | | |
| | 1.7 | | | | | | | |
| 51.66 | 2.66 | | | | | | 90 | |
| 4.96 | | 4.92 | 2.28 | | | | | |
| 0.8 | | 6.63 | | | | | | |
| | | 6.03 | 2.24 | 17.21 | | | | |
| 8.13 | | | 3.06 | | | | | |
| | | | | | | | | |
| | | | | | | | | |
| | | | | | | | | |
| 1.65 | 10.65 | | | | | | | |
| | | | | | | | | |
| | | 4.05 | | | | | | 8.33?? |
| | | | 23.05 | | | | | |
| | | | | | | | | |
| | | | | 5.33 | | | 10 | |
| | | | 7.48 | | | | | |
| | | | | | | | | |
| 0.8 | | | | | | | | |
| | 3.55 | | | | | | | |
| | | | | | | | | |
| | | 10.72 | | | | | | |
| | | | 3.28 | | | | | |

*Continued*

**TABLE A.4.2** Top 10 Collaborating Countries Across the World With 17 Arab Countries

| | Saudi Arabia | Egypt | Morocco | Qatar | UAE | Lebanon | Tunisia | Jordan |
|---|---|---|---|---|---|---|---|---|
| Singapore (1) | | | | | | | | |
| Slovenia (1) | | | | | | | 1.6 | |
| **ACs** | | | | | | | | |
| Saudi Arabia (6) | | 4.35 | | | | | | 2.98 |
| UAE (3) | | | | | | | 1.53 | |
| Qatar (1) | | | | | | | | |
| Egypt (2) | | | | | | | | |
| Algeria (1) | | | | | | | | |
| Iraq (1) | | | | | | | | |
| Oman (1) | | | | | | | | |
| Yemen (1) | | | | | | | | |
| **Selected Neighboring Countries** | | | | | | | | |
| Israel (1) | | | | | | | | |
| Turkey (1) | | | | | | | | |
| Cyprus (1) | | | | | | | | |
| Rest of the world | 24.18 | 30.25 | 27.58 | 26.88 | 23.51 | 15.44 | 13.27 | 17.62 |

Percentages presented in this table are based on WFC indicating mutual collaboration in scientific publications monitored by "Nature Research."
Bahrain is excluded from this table for the lack of detailed data. However, NI records indicate Bahrain did in fact collaborate with Taiwan in authoring one article in the life sciences.
From Databases on the Nature Research website. Data downloaded in 2016 from: http://www.natureindex.com/.

(Percentages)—cont'd

| Algeria | Kuwait | Iraq | Oman | Palestine | Syria | Libya | Mauritania | Yemen |
|---|---|---|---|---|---|---|---|---|
| | | | | | | 30.77 | | |
| | | | | | | | | |
| | | | | | | | | |
| 6.82 | | | 5.24 | | 31.25 | | | 46.15 |
| | | 4.23 | | | 8.33 | | | |
| | | | | | | 7.69 | | |
| | 3.22 | | | | | 7.69 | | |
| | | | | | | | | 53.83 |
| | | | | | | | | 8.33 |
| | | | | 1.94 | | | | |
| 4.43 | | | | | | | | |
| | | | | | | | | |
| | | | | 45.66 | | | | |
| | 2.54 | | | | | | | |
| | | 2.92 | | | | | | |
| 4.98 | 11.96 | 9.22 | 18.38 | 0.79 | 25.00 | | | |

**TABLE A.4.3** Collaboration Among Seven Developed Countries With Their Top 10 Partners in

| | United States | China | Japan | Netherlands | United Kingdom | Germany | France |
|---|---|---|---|---|---|---|---|
| UK | 42.2 | 47.6 | 54.2 | 57.3 | – | 47.8 | 52.6 |
| Germany | 43.1 | 55.2 | 53.8 | 58.0 | 52.2 | – | 54.4 |
| Italy | 39.9 | 56.6 | 55.5 | 61.2 | 50.3 | 48.4 | 50.3 |
| Canada | 39.3 | 49.2 | 50.4 | 54.6 | 46.3 | 44.3 | 54.0 |
| France | 41.8 | 47.9 | 49.0 | 51.7 | 47.4 | 45.6 | – |
| Switzerland | 38.5 | 56.0 | 55.2 | 51.0 | 45.2 | 45.1 | 44.7 |
| Spain | 37.5 | 45.6 | 48.8 | 53.5 | 44.7 | 42.2 | 47.8 |
| Averages %ages | 40.3 | 51.2 | 52.4 | 55.3 | 47.7 | 45.6 | 50.6 |

[a]*As measured by WFC of jointly published articles within areas covered by "Nature Research."*

**TABLE A.4.4** Percentages of Collaboration Between Seven Arab Countries With Their Top 10

| | United States | United Kingdom | Germany | France | Italy | Spain | Canada | China | Japan | Switzerland | Australia |
|---|---|---|---|---|---|---|---|---|---|---|---|
| Saudi Arabia | 17.2 | 31.8 | 16.8 | 24.6 | 25.4 | 21.7 | 22.6 | 26.7 | 31.1 | – | – |
| Egypt | 7.4 | 15.0 | 11.1 | 13.1 | 8.8 | 10.2 | 18.8 | 11.3 | 10.9 | – | – |
| Morocco | 5.5 | 8.9 | 6.4 | 26.8 | 6.8 | 32.8 | 11.9 | – | 28.6 | 8.7 | – |
| Qatar | 15.2 | 15.8 | 15.9 | 28.4 | 8.4 | – | 27.6 | 2.0 | – | 10.9 | 38.9 |
| UAE | 21.9 | 24.6 | 32.2 | 30.4 | 30.5 | 65.5 | – | – | – | 17.1 | 43.6 |
| Lebanon | 15.5 | 16.9 | 13.6 | 22.9 | 54.0 | 43.5 | – | 27.8 | 3.5 | – | – |
| Tunisia | 27.8 | 28.8 | 25.2 | 23.9 | 37.7 | 41.1 | 52.7 | – | – | – | – |
| Average %ages | 15.8 | 20.3 | 17.3 | 24.3 | 24.5 | 35.8 | 26.7 | 17.0 | 18.5 | 18.4 | 20.3 |

[a]*As measured by WFC of jointly published articles within areas covered by "Nature Research."*

**TABLE A.4.5** Percentages of Collaboration Between Cyprus, Iran, Israel, and Turkey With Their Top

| | United States | United Kingdom | Germany | Italy | Switzerland | France | China | Spain |
|---|---|---|---|---|---|---|---|---|
| Cyprus | 8.4 | 18.3 | 38.6 | 14.0 | 9.3 | 21.8 | – | 27.7 |
| Iran | 11.8 | 34.6 | 13.5 | 31.3 | 15.4 | – | 39.8 | 22.1 |
| Israel | 37.5 | 42.6 | 42.9 | 39.1 | 42.4 | 45.0 | 42.3 | – |
| Turkey | 14.5 | 21.6 | 16.5 | 19.3 | 19.0 | 29.5 | 21.3 | 33.0 |
| Average %ages | 18.1 | 29.3 | 27.9 | 23.4 | 21.5 | 32.1 | 34.5 | 27.6 |

[a]*As measured by WFC of jointly published articles within areas covered by "Nature Research."*

Publishing Scientific Articles.[a] (During the Period May 1, 2015 and April 30, 2016)

| Italy | Switzerland | Spain | Australia | Russia | Canada | Sweden | Average WFC |
|---|---|---|---|---|---|---|---|
| 49.7 | 54.7 | 55.3 | 56.9 | | – | – | 51.83 |
| 51.6 | 54.9 | 57.8 | – | 67.1 | – | – | 54.81 |
| – | 53.4 | 56.3 | – | 61.3 | – | – | 53.32 |
| 44.0 | 55.6 | – | 51.6 | – | – | – | 48.93 |
| 49.6 | 52.3 | 52.3 | – | – | 45.8 | – | 48.34 |
| 46.6 | – | 58.5 | – | – | 44.4 | – | 48.52 |
| 43.6 | 41.4 | – | – | – | a | 56.9 | 46.2 |
| 47.5 | 52.1 | 56.0 | 54.3 | 64.2 | 45.1 | 56.9 | 50.3 |

Partners in Publishing Scientific Articles.[a] (During the Period May 1, 2015 and April 30, 2016)

| South Korea | Netherlands | Russia | Sweden | Saudi Arabia | Taiwan | India | Slovenia | UAE | Austria | Average WFC |
|---|---|---|---|---|---|---|---|---|---|---|
| – | – | – | – | – | 31.9 | – | – | – | – | 24.98 |
| – | – | – | – | 19.6 | – | – | – | – | – | 12.62 |
| – | – | 12.9 | – | – | – | – | – | – | – | 14.93 |
| 26.8 | – | – | – | – | – | – | – | – | – | 18.99 |
| 14.5 | – | – | 14.7 | – | – | – | – | – | – | 29.5 |
| – | 34.4 | – | – | – | – | 43.3 | – | – | – | 27.54 |
| – | – | – | – | – | – | – | 59.5 | 12.8 | 38.9 | 34.84 |
| 20.7 | 34.4 | 12.9 | 14.7 | 19.6 | 31.9 | 43.3 | 59.5 | 12.8 | 38.9 | 23.3 |

10 Partners in Publishing Scientific Articles.[a] (During the Period May 1, 2015 and April 30, 2016)

| Netherlands | Belgium | Russia | Japan | Greece | Austria | Australia | Canada | Average WFC |
|---|---|---|---|---|---|---|---|---|
| – | 6.6 | 7.4 | –00 | 42.3 | – | – | – | 19.44 |
| – | 38.4 | – | – | – | 41.8 | 33.0 | – | 28.17 |
| 52.9 | – | – | 47.1 | – | – | – | 43.7 | 43.55 |
| 31.7 | – | 17.5 | – | – | – | – | – | 22.39 |
| 42.3 | 22.5 | 12.5 | 47.1 | 42.3 | 41.8 | 33.0 | 43.7 | 23.8 |

**TABLE A.4.6** Percentages of Collaboration Between 26 Countries Around the World With and April 30, 2016)

| | United States | China | Japan | Netherlands | United Kingdom | Germany | France | Italy | Switzerland | Spain | Australia | Russia | Canada | Sweden |
|---|---|---|---|---|---|---|---|---|---|---|---|---|---|---|
| Latvia | 0.51 | | | 12.9 | 11.5 | 40.1 | 16.4 | | 12.1 | 24.1 | | 38.3 | | |
| India | 33.0 | 43.8 | 46.1 | | 43.3 | 37.1 | 46.8 | 40.0 | 48.0 | 47.8 | 46.5 | | | |
| Pakistan | 13.0 | 16.6 | 20.3 | | 25.3 | 7.6 | 15.3 | 10.3 | 6.8 | | | 8.9 | | |
| Portugal | 21.8 | 35.8 | | 35.1 | 34.1 | 27.2 | 35.9 | 24.9 | 33.4 | 41.6 | | | | |
| Brazil | 25.7 | | | 43.6 | 34.2 | 33.3 | 38.2 | 31.1 | 29.1 | 46.4 | | 35.4 | 44.2 | |
| Hungary | 23.6 | | 30.1 | | 32.7 | 26.5 | 27.5 | 26.9 | 26.3 | 34.4 | | | | 34.6 |
| Sweden | 37.2 | 40.6 | | 39.0 | 40.6 | 38.9 | 45.9 | 34.3 | 45.5 | 43.0 | | | | |
| Poland | 24.3 | | 44.9 | | 35.7 | 30.7 | 36.4 | 22.9 | 29.1 | 42.7 | | 36.0 | | 37.8 |
| Greece | 20.3 | | | 33.4 | 33.4 | 27.9 | 36.9 | 23.2 | 24.7 | 41.7 | | 31.7 | | |
| Chile | 27.2 | | 33.4 | | 29.8 | 32.2 | 37.7 | 31.1 | 41.9 | 32.4 | | | | |
| Argentina | 39.9 | | | | 27.9 | 24.4 | 36.2 | 31.6 | 35.9 | 33.4 | | | 30.7 | |
| Malaysia | 11.5 | 20.1 | 21.4 | | 13.7 | 12.4 | 8.0 | 5.5 | 8.7 | 19.9 | 19.1 | | | |
| Armenia | 3.7 | | 10.2 | | 12.3 | 6.1 | 7.2 | 22.9 | 5.6 | 15.3 | | 14.4 | 15.2 | |
| Indonesia | 31.3 | | 34.4 | 40.6 | 28.6 | 23.3 | | | | | 31.0 | | | |
| Romania | 19.6 | | 18.2 | 28.6 | 14.2 | 13.1 | 25.8 | 17.1 | 12.2 | 36.9 | | 16.7 | | |
| China | 46.2 | | 47.3 | | 52.3 | 44.8 | 52.2 | | 44.3 | 54.3 | 57.9 | | 50.7 | |
| Bulgaria | 10.7 | | | | 25.8 | 16.3 | 24.0 | 15.8 | 20.8 | 25.1 | | 27.6 | | |
| Lithuania | 7.9 | | | | 21.8 | 16.9 | 17.9 | 5.3 | 21.5 | 31.6 | | 18.1 | | 55.9 |
| Slovakia | 10.4 | | 18.5 | | 11.7 | 16.1 | 15.0 | 11.9 | 12.9 | 32.5 | | 20.9 | | |
| Czech Republic | 41.2 | | 30.0 | | 37.8 | 33.5 | 27.0 | 27.9 | 22.7 | 50.9 | | 34.4 | 43.2 | |
| Slovenia | 21.1 | | 27.3 | 28.1 | 17.2 | 19.4 | 29.7 | 16.4 | 38.5 | | | 25.1 | | |
| South Korea | 36.8 | 43.4 | 45.8 | | 44.5 | 40.0 | 42.2 | 38.5 | 42.2 | | 52.0 | | 47.4 | |
| Finland | 29.2 | | 33.2 | 41.6 | 35.6 | 38.0 | 45.0 | 32.9 | 48.8 | | | 56.9 | | 47.9 |
| United States | | 53.8 | 61.9 | | 57.8 | 57.0 | 58.2 | 60.1 | 61.5 | | 62.9 | | 60.7 | |
| Austria | 34.0 | | 30.7 | 46.7 | 37.0 | 36.0 | 35.2 | 31.9 | 35.9 | 49.0 | | | 35.6 | |
| Estonia | 13.6 | | | 28.9 | 18.4 | 25.4 | | 33.0 | 14.5 | | | 19.8 | | 38.9 |
| Average %ages | 23.3 | 36.3 | 32.7 | 34.4 | 29.9 | 27.9 | 31.7 | 25.9 | 28.9 | 37.0 | 44.9 | 27.4 | 41.0 | 43.0 |

[a]*As measured by WFC of jointly published articles within areas covered by "Nature Research."*

Their Top 10 Partners in Publishing Scientific Articles.[a] (During the Period May 1, 2015

| | Portugal | India | Saudi Arabia | Brazil | Austria | Denmark | Chile | Belgium | Singapore | South Korea | Malaysia | Finland | Poland | Czech Republic | Average WFC |
|---|---|---|---|---|---|---|---|---|---|---|---|---|---|---|---|
| | 25.5 | 33.0 | | | | | | | | | | | | | 21.4 |
| | | | | | | | | | | | | | | | 43.2 |
| | | | 7.1 | | | | | | | | | | | | 13.1 |
| | | | | 47.6 | | | | | | | | | | | 33.7 |
| | | | | | | | | | | | | | | | 36.1 |
| | | | | | 40.7 | | | | | | | | | | 30.3 |
| | | | | | | 50.3 | | | | | | | | | 41.5 |
| | | | | | | | | | | | | | | | 34.1 |
| | | | | | | | | 36.8 | | | | | | | 31.0 |
| | | | | 38.5 | | | | 54.7 | | | | | | | 35.9 |
| | | | | 49.0 | | | 42.6 | | | | | | | | 35.2 |
| | | | | | | | | | | | | | | | 14.0 |
| | | | | | | | | | | | | | | | 11.3 |
| | | | | 38.4 | | | | | 37.1 | 15.7 | 41.9 | | | | 32.2 |
| | | | | | | | | | | | | | | | 20.2 |
| | | | | | | | | | 49.3 | | | | | | 49.9 |
| | | 31.9 | | | | | | 15.4 | | | | | | | 21.3 |
| | | | | | | | | 14.5 | | | | | | | 21.1 |
| | | | | | | | | | | | | | | 35.3 | 18.5 |
| | | | | | | | | | | | | | | | 34.9 |
| | | | | | | | | | | | | | 50.2 | | 27.3 |
| | | | | | | | | | | | | | | | 43.3 |
| | | | | | | | | | | | | | | | 40.9 |
| | | | | | | | | | | 63.2 | | | | | 59.7 |
| | | | | | | | | | | | | | | | 37.2 |
| | | 21.1 | | | | | | | | | | 41.7 | | | 25.5 |
| | 25.5 | 28.7 | 7.1 | 43.4 | 40.7 | 50.3 | 42.6 | 30.4 | 43.2 | 39.5 | 41.9 | 41.7 | 50.2 | 35.3 | 31.3 |

**TABLE A.4.7A** Publications and Citations by Arab Countries Throughout the Period 1996–2015

| Year | All Published Documents | Citable Documents | Citations | Self-citations | Ratio of Self-citations to All Citations | Ratio of Self-citations to All Citable Documents | Ratio of All Citations to All Citable Documents | Average Citations per Document |
|---|---|---|---|---|---|---|---|---|
| 2015 | 66,031 | 61,568 | 30,028 | 7336 | 0.244 | 0.119 | 0.488 | 0.43 |
| 2014 | 67,549 | 63,851 | 131,723 | 29,144 | 0.221 | 0.456 | 2.063 | 1.71 |
| 2013 | 61,431 | 58,121 | 207,484 | 44,446 | 0.214 | 0.765 | 3.570 | 2.94 |
| 2012 | 53,740 | 51,050 | 280,505 | 50,986 | 0.182 | 0.999 | 5.495 | 4.89 |
| 2011 | 45,912 | 43,966 | 291,192 | 51,819 | 0.178 | 1.179 | 6.623 | 6.00 |
| 2010 | 38,162 | 36,446 | 287,159 | 47,854 | 0.167 | 1.313 | 7.879 | 8.29 |
| 2009 | 32,687 | 31,430 | 272,433 | 44,284 | 0.163 | 1.409 | 8.668 | 8.38 |
| 2008 | 26,968 | 25,658 | 261,278 | 39,995 | 0.153 | 1.559 | 10.183 | 11.18 |
| 2007 | 23,109 | 22,259 | 241,390 | 36,444 | 0.151 | 1.637 | 10.845 | 11.44 |
| 2006 | 20,759 | 19,863 | 229,726 | 34,688 | 0.151 | 1.746 | 11.566 | 11.52 |
| 2005 | 17,925 | 17,301 | 204,503 | 31,682 | 0.155 | 1.831 | 11.820 | 10.65 |
| 2004 | 15,987 | 15,552 | 206,451 | 30,755 | 0.149 | 1.978 | 13.275 | 13.56 |
| 2003 | 14,462 | 14,042 | 188,138 | 28,323 | 0.151 | 2.017 | 13.398 | 12.74 |

| | | | | | | | | |
|---|---|---|---|---|---|---|---|---|
| 2002 | 11,900 | 11,596 | 161,975 | 25,025 | 0.154 | 2.158 | 13.968 | 13.17 |
| 2001 | 11,173 | 10,926 | 144,403 | 23,045 | 0.158 | 2.110 | 13.336 | 14.88 |
| 2000 | 10,546 | 10,344 | 143,551 | 21,934 | 0.150 | 2.130 | 14.187 | 13.43 |
| 1999 | 9979 | 9778 | 126,943 | 19,626 | 0.155 | 2.007 | 12.983 | 14.16 |
| 1998 | 9667 | 9509 | 128,223 | 18,433 | 0.144 | 1.939 | 13.488 | 13.68 |
| 1997 | 9335 | 9210 | 115,209 | 17,506 | 0.152 | 1.901 | 12.509 | 12.10 |
| 1996 | 8761 | 8668 | 102,673 | 15,399 | 0.150 | 1.777 | 11.845 | 11.34 |
| Totals | 556,083 | 531,138 | 3,754,987 | 618,724 | | | | |

Figures included in this table were downloaded on August 12, 2016 from: http://www.scimagojr.com/countryrank.php.

**TABLE A.4.7B** Publications and Citations by All Countries

| Year | All Published Documents | Citable Documents | Citations | Self-citations | Ratio of Self-citations to All Citations | Ratio of Self-citations to All Citable Documents | Ratio of All Citations to All Citable Documents | Average Citations per Document |
|---|---|---|---|---|---|---|---|---|
| 2015 | 3,011,688 | 2,721,140 | 1,630,270 | 618,236 | 0.379 | 0.227 | 0.599 | 0.62 |
| 2014 | 3,287,871 | 3,017,852 | 8,090,662 | 2,896,745 | 0.358 | 0.960 | 2.681 | 2.87 |
| 2013 | 3,199,114 | 2,915,515 | 15,804,234 | 5,456,683 | 0.345 | 1.872 | 5.421 | 4.15 |
| 2012 | 3,069,510 | 2,791,552 | 23,330,056 | 7,594,177 | 0.326 | 2.720 | 8.357 | 7.86 |
| 2011 | 2,902,198 | 2,667,563 | 27,991,810 | 9,087,957 | 0.325 | 3.407 | 10.493 | 8.27 |
| 2010 | 2,691,321 | 2,472,102 | 32,503,412 | 10,277,423 | 0.316 | 4.157 | 13.148 | 10.74 |
| 2009 | 2,516,676 | 2,322,007 | 36,171,936 | 11,275,265 | 0.312 | 4.856 | 15.578 | 12.48 |
| 2008 | 2,300,501 | 2,125,672 | 38,099,872 | 11,715,049 | 0.307 | 5.511 | 17.924 | 14.37 |
| 2007 | 2,161,289 | 2,006,255 | 39,910,676 | 12,117,744 | 0.304 | 6.040 | 19.893 | 15.39 |
| 2006 | 2,051,840 | 1,910,321 | 40,733,378 | 12,392,428 | 0.304 | 6.487 | 21.323 | 16.77 |
| 2005 | 1,908,002 | 1,789,592 | 41,923,961 | 12,544,660 | 0.299 | 7.010 | 23.427 | 16.65 |
| 2004 | 1,650,713 | 1,564,015 | 41,579,636 | 12,408,100 | 0.298 | 7.933 | 26.585 | 19.61 |
| 2003 | 1,488,355 | 1,404,246 | 40,130,803 | 11,954,127 | 0.298 | 8.513 | 28.578 | 20.06 |

| | | | | | | | | |
|---|---|---|---|---|---|---|---|---|
| 2002 | 1,296,840 | 1,246,568 | 36,287,639 | 11,153,082 | 0.307 | 8.947 | 29.110 | 20.96 |
| 2001 | 1,230,658 | 1,186,057 | 35,443,682 | 10,826,556 | 0.305 | 9.128 | 29.884 | 20.54 |
| 2000 | 1,216,389 | 1,177,505 | 36,031,507 | 10,809,482 | 0.300 | 9.180 | 30.600 | 21.03 |
| 1999 | 1,169,016 | 1,133,626 | 33,695,966 | 10,260,454 | 0.305 | 9.051 | 29.724 | 20.56 |
| 1998 | 1,148,532 | 1,119,643 | 32,573,349 | 9,971,382 | 0.306 | 8.906 | 29.093 | 20.75 |
| 1997 | 1,138,308 | 1,116,886 | 30,407,787 | 9,451,514 | 0.311 | 8.462 | 27.226 | 18.36 |
| 1996 | 1,080,721 | 1,059,831 | 28,350,261 | 9,030,124 | 0.319 | 8.520 | 26.750 | 18.63 |
| Totals | 40,519,542 | 37,747,948 | 620,690,897 | 191,841,188 | – | – | – | |

Figures included in this table were downloaded on August 12, 2016 from: http://www.scimagojr.com/countryrank.php.

**TABLE A.4.8** Patents Granted by the United States Patent and Trademark Office (USPTO) to Arab Countries, Five of Their Closest Neighbors, and Worldwide Before 2002, During 2002, and 2015 and Corresponding Ratios

| | Number of Patents Granted | | | | Ratio of Patents Granted | |
|---|---|---|---|---|---|---|
| | Pre-2002 | 2002 | 2015 | All Years Till 2015 | All years Till 2015 to All Patents Before 2002 | In the Year 2015 to Those Granted in 2012 |
| **Arab Countries** | | | | | | |
| Algeria | 3 | 0 | 1 | 7 | 2.33 | – |
| Bahrain | 5 | 0 | 1 | 13 | 2.60 | – |
| Egypt | 54 | 5 | 30 | 275 | 5.09 | 6.00 |
| Iraq | 9 | 0 | 1 | 11 | 1.22 | – |
| Jordan | 16 | 1 | 11 | 57 | 3.56 | 11.00 |
| Kuwait | 59 | 8 | 64 | 448 | 7.59 | 8.00 |
| Lebanon | 31 | 2 | 16 | 125 | 4.03 | 8.00 |
| Morocco | 28 | 0 | 4 | 55 | 1.96 | – |
| Oman | 1 | 0 | 5 | 27 | 27.00 | – |
| Palestine | 0 | 0 | 1 | 1 | – | – |
| Qatar | 1 | 0 | 8 | 35 | 35.00 | – |
| Saudi Arabia | 168 | 10 | 364 | 1513 | 9.01 | 36.40 |

| | | | | | | |
|---|---|---|---|---|---|---|
| Syria | 11 | 2 | 2 | 23 | 2.09 | – |
| Tunisia | 11 | 1 | 6 | 43 | 3.91 | 6.00 |
| UAE | 29 | 6 | 56 | 254 | 8.76 | 9.33 |
| Yemen | 2 | 1 | 0 | 3 | 1.50 | – |
| Total | 428 | 36 | 570 | 2890 | 6.75 | 15.83 |
| **Arab Countries' Closest Neighbors** | | | | | | |
| Cyprus | 24 | 1 | 10 | 87 | 3.63 | 10.00 |
| Iran | 32 | 0 | 31 | 199 | 6.22 | |
| Israel | 8886 | 1108 | 3804 | 35,900 | 4.04 | 3.43 |
| Malta | 13 | 1 | 7 | 74 | 5.69 | 7.00 |
| Turkey | 75 | 18 | 136 | 748 | 9.97 | 7.56 |
| Total | 9030 | 1128 | 3988 | 37,008 | 4.10 | 3.54 |
| World's total | 2,548,916 | 184,424 | 325,979 | 5,739,851 | 2.25 | 1.77 |

**TABLE A.4.9** Arab Countries That Published in Excess of 10 Papers in Highly Regarded Journals Tracked by Nature's Index During 2015

| Country/Discipline | Article Count | Fractional Count | Country/Discipline | Article Count | Fractional Count |
|---|---|---|---|---|---|
| Saudi Arabia | | | Lebanon | | |
| Chemistry | 241 | 66.54 | Chemistry | 7 | 0.97 |
| Earth & environmental sciences | 19 | 7.02 | Earth & environmental sciences | – | – |
| Life sciences | 102 | 8.51 | Life sciences | 10 | 0.5 |
| Physical sciences | 183 | 33.78 | Physical sciences | 11 | 3.98 |
| Total | 545 | 115.85 | Total | 28 | 5.45 |
| Egypt | | | Tunisia | | |
| Chemistry | 39 | 3.98 | Chemistry | 6 | 1.51 |
| Earth & environmental sciences | 6 | 0.44 | Earth & environmental sciences | 4 | 0.48 |
| Life sciences | 23 | 1.66 | Life sciences | 9 | 0.5 |
| Physical sciences | 66 | 6.36 | Physical sciences | 7 | 1.91 |
| Total | 134 | 12.44 | Total | 26 | 4.40 |
| Morocco | | | Jordan | | |
| Chemistry | 5 | 1.73 | Chemistry | 5 | 0.91 |
| Earth & environmental sciences | 3 | 0.5 | Earth & environmental sciences | 1 | 0.17 |

| | | | | | |
|---|---|---|---|---|---|
| Life sciences | 4 | 0.23 | Life sciences | 7 | 0.37 |
| Physical sciences | 76 | 3.66 | Physical sciences | 3 | 0.37 |
| Total | 88 | 6.12 | Total | 16 | 1.82 |
| Qatar | | | Algeria | | |
| Chemistry | 17 | 3.92 | Chemistry | 4 | 0.76 |
| Earth & environmental sciences | – | – | Earth & environmental sciences | 4 | 0.85 |
| Life sciences | 13 | 0.37 | Life sciences | 2 | 0.08 |
| Physical sciences | 44 | 4.44 | Physical sciences | 5 | 1.82 |
| Total | 74 | 8.73 | Total | 15 | 3.51 |
| UAE | | | Iraq | | |
| Chemistry | 11 | 5.62 | Chemistry | 5 | 0.45 |
| Earth & environmental sciences | 8 | 2.15 | Earth & environmental sciences | – | – |
| Life sciences | 14 | 1.55 | Life sciences | 2 | 0.14 |
| Physical sciences | 35 | 6.08 | Physical sciences | 5 | 1.22 |
| Total | 68 | 15.35 | Total | 12 | 2.81 |

Article count (AC) is the count of articles produced by an institution or country. An article count equaling 1.0 is assigned if one or more authors of the research article are from that institution or country, regardless of how many coauthors there are from other institutions or countries.
FC is the count of articles published by a given country or institution that takes into account the percentage of authors from that institution (or country) and the number of affiliated institutions per article. For calculating FC, all authors are considered to have contributed equally to the article. The maximum combined FC for any article is 1.0.

**TABLE A.4.10** Patents Granted by the European Patent Office to Nationals and Residents in Arab Countries and Their Closest Neighbors

| Country | 2006 | 2007 | 2008 | 2009 | 2010 | 2011 | 2012 | 2013 | 2014 | 2015 | Total |
|---|---|---|---|---|---|---|---|---|---|---|---|
| Egypt | 0 | 0 | 0 | 1 | 2 | 2 | 2 | 3 | 4 | 2 | 16 |
| Jordan | 0 | 2 | 5 | 1 | 5 | 12 | 1 | 3 | 0 | 0 | 29 |
| Kuwait | 0 | 0 | 0 | 0 | 1 | 1 | 0 | 1 | 1 | 3 | 7 |
| Lebanon | 0 | 2 | 2 | 1 | 0 | 0 | 1 | 2 | 2 | 0 | 10 |
| Morocco | 1 | 2 | 1 | 0 | 0 | 2 | 4 | 0 | 3 | 0 | 13 |
| Saudi Arabia | 8 | 10 | 8 | 11 | 20 | 18 | 21 | 43 | 36 | 49 | 224 |
| Tunisia | 0 | 2 | 0 | 2 | 0 | 1 | 0 | 0 | 1 | 3 | 9 |
| United Arab Emirates | 4 | 4 | 4 | 6 | 4 | 3 | 4 | 6 | 4 | 3 | 42 |
| Other Arab countries[a] | 0 | 1 | 2 | 2 | 0 | 0 | 0 | 2 | 1 | 2 | 10 |
| Totals (Arab countries) | 13 | 23 | 22 | 24 | 32 | 39 | 33 | 60 | 52 | 62 | 360 |

| Closest Five Neighbors | | | | | | | | | | | |
|---|---|---|---|---|---|---|---|---|---|---|---|
| Cyprus | 16 | 15 | 16 | 25 | 18 | 21 | 13 | 17 | 26 | 14 | 181 |
| Iran | 1 | 2 | 0 | 1 | 4 | 2 | 1 | 0 | 2 | 0 | 13 |
| Israel | 258 | 192 | 299 | 228 | 279 | 279 | 335 | 420 | 399 | 397 | 3086 |
| Malta | 6 | 6 | 10 | 12 | 17 | 25 | 23 | 22 | 21 | 19 | 161 |
| Turkey | 31 | 39 | 49 | 59 | 89 | 95 | 134 | 144 | 136 | 218 | 994 |
| Totals (AC neighbors) | 312 | 254 | 374 | 325 | 407 | 422 | 506 | 603 | 584 | 648 | 4435 |
| Patents granted to all countries | 62,777 | 54,700 | 59,800 | 51,952 | 58,117 | 62,108 | 65,655 | 66,707 | 64,613 | 68,421 | 614,850 |

[a]*These countries include Algeria, Bahrain, Iraq, Oman, and Qatar with only very few patents granted by EPO throughout.*
The European Patent Office. Data downloaded in November 2016 from: https://www.epo.org/about-us/annual-reports-statistics/statistics.html.

# ENDNOTES

1. The SCImagojr Journal & Country Rank is a portal established by Elsevier with the aim of collecting, collating, and disseminating data on publications within a wide range of scientific and technological domains. Indicators posted on the SCImagojr website are derived from information contained in the Scopus database (Elsevier B.V.). Journal contents, on the one hand, and scientific activities by various countries, on the other, may be compared, ranked, or analyzed separately. Journal contents are grouped by 27 major thematic areas including 313 specific subject categories and citation data is drawn from over 21,500 titles put out by over 5000 international publishers. Country performance metrics cover up to 239 countries around the world. Data downloaded from the Elsevier's SCImagojr database in August 12, 2016 from: http://www.scimagojr.com/countryrank.php.
2. See also annex Tables A.4.7a and A.4.7b, with figures from this SCImagojr database. The former table includes information on research publications by all countries for which data is available, and the latter including information on research publications by all the Arab countries.
3. The curve representing growth of worldwide scientific research publications may be described by the quadratic equation: $y=7.4867x^2-29874x+3.10^7$. Within the period 1996–2011, this polynomial yields a coefficient of determination $R^2$ of around 0.99.
4. That fewer articles appear to have been published in 2015 than in 2014 may be an artifact of data collection and logging methods, rather than any sudden reduction in the number of publications. As indicated in a variety of references, research publications generally double every 9 years or so, and the only reason for supposing that some reduction might occur in the number of publications might be that researchers are opting for disseminating their research output through alternative modes, both traditional within other journals and nonconventional using blogs and other means electronic means. However, while such trends may have become discernible in some disciplines, it is unlikely to have assumed the large proportions shown by this figure.
5. Evolution of research publications by all Arab countries during the period 1996–2014 conforms to a pattern that is closely described by the polynomial: $y=0.0059x^3-34.925x^2+69441x-5\times 1 0^7$, with a value of 0.998 for the coefficient of determination $R^2$.
6. Growth in the number of publications produced by the Arab countries and worldwide during the period 2002–14 obeys a simple linear relationship of the form $y=ax+b$. The constant term, $b$ in the equation describing growth of publications by the Arab countries equals 4.66, while that for worldwide publications equals 169.19, with $R^2$. Values equal at 0.996 and 0.938, respectively, for research publications worldwide and in the Arab countries, indicating a high degree of reliability for this linear relationship.
7. This comparison is based on data from the aforementioned Elsevier database.
8. These figures ought to be looked at while bearing in mind the fact that the combined populations of the Arab countries, as well as their combined annual nominal GDPs, generally amounted to nearly twice those possessed by Iran and Turkey together, throughout the period in question.
9. During the years 2007–05, the rate of publications of Sudan, Syria, Palestine, Yemen, and Bahrain, closely followed a linear pattern where ($y$) the number of publications in a given year could be calculated to a high degree of reliability using the linear equation: $y=a{\cdot}x+b$; where $x$ is the year in question, while $a$ and $b$ are coefficients characteristic of the country's publications activity. The value of coefficients $a$ for Sudan, Syria, Palestine, Yemen, and Bahrain are 69.0, 59.5, 35.6, 41.5, and 16.3, respectively. The rate of growth of scientific output by Libya followed a rather erratic pattern during this same period, so it may not be adequately represented by the same linear equation.
10. In effect, Iran, Israel, and Turkey are by far the most prolific producers of research publications throughout the Middle East. Israel's research output vies for first and second place with Turkey and Iran in many research disciplines.
11. "Global Research Report, Exploring the Changing Landscape of Arabian, Persian and Turkish Research", J. Adams et al.; February 2011. Thomson Reuters Science. ISBN: 1-904,431-27-5. To refer to this report as TRWK 2011.

12. These two countries dominated publications by the rest of the Arab countries, which combined increased their annual contributions to worldwide publications from 0.81% in 1996 up to 2.19% in 2015; by a factor of around 2.7.
13. An article published by Nature Index states that "Saudi Arabia has quickly risen to become one of the most important players in scientific research in the Middle East, thanks to generous funding from the government and an ambitious plan to shift the country from an oil-based economy to a knowledge economy. Over the past decade, the kingdom has overhauled its research institutes, created top-tier universities and invested in one of the largest overseas education scholarships to meet its target." Posting titled "Saudi Arabia." Published by Nature Index. Downloaded in June 2016 from: http://www.natureindex.com/country-outputs/Saudi%20Arabia.
14. It is undoubtable that the 2003 war, preceded by the Iran–Iraq war and the invasion of Kuwait invasion, and the political upheaval since that war did have negative impact upon university life, including scientific research and related publications.
15. Choice of the year 2014 in holding this comparison—rather than the year 2015 for which data is in fact available from the Elsevier's SCImagojr database—was made since data for the latter year may not be complete for all countries as indicated by the dip taken by the number of total annual publications for the year 2015 relative to previous years.
16. Based on information from the Elsevier database. Downloaded in October 2016 from: http://www.scimagojr.com/countryrank.php?year=1996.
17. See endnote 16.
18. The SCImago database compiles the number of citations to papers published within a given year made throughout the period covered by its data, in the present case 1996 till 2015.
19. Both sets of averages follow a curve that is closely approximated by a simple quadratic polynomial: $y=ax^2+bx+c$; with values for the coefficients for world and Arab country averages, respectively at $a=-0.0888$, $-0.0684$, $b=+3555.27$, $+273.91$, and $c=-355217$, $-274005$. These coefficients yield $R^2$ values equal to 0.988 and 0.972 for World and Arab averages, respectively.
20. The rate of rising self-citations relative to all citations in the former countries, over the years 2008–14 is around 1.15, in small excess to the worldwide rate, of around 0.8, over the same period. A related observation is the rise in the percentage of self-citations within publications produced by the Arab countries from 0.17 in 1996, to around 1.2% in 2015. See annex Table A.4.4.
21. The plot based on data from the Elsevier database on both self-citations and all citations linked to research publications by 200 countries around the world is best described by the equation ($y=279.55e^{-9.975x}$) where $y$ is the country rank and $x$ is the ratio of self-citations to all citations for all countries. Considering the wide variations among the countries included in the value of the coefficient of determination ($R^2$) value is significant at around 0.64.
22. More precisely, the H-index denotes the number of published articles (H) that received at least "H" citations within a given period. The H-index is an author-level measure designed to gauge both the productivity and citation impact of the publications of a scientist or scholar. The index is based on the set of the scientist's most cited papers and the number of citations they received in other publications. This index is also referred to as the Hirsch index.
23. Rankings among all countries of the world based on H-index values, as reported by SCImago Journal and Country Rank. Retrieved March 2016. Values of the H index given between brackets are taken from tables published by the INSEAD report "Global Innovation Index 2015," quoting SCImago. (2007). SJR—SCImago Journal & Country Rank as their source with data retrieved in February, 2015. The INSEAD report "Global Innovation Index 2015." Downloaded in May 2016 from: https://www.globalinnovationindex.org/userfiles/file/reportpdf/GII-2015-v5.pdf.
24. Country rankings according to the total number of research publications monitored by SCImago Journal and Country Rank. See endnote 16. Retrieved in March 2016.
25. The highest increase in the value of the H-index among all the Arab countries was made by Saudi Arabia, which enhanced its H-index rating by 51 points from 144, as extracted from data for the period 1996–2013. However, corresponding gains by Israel and Turkey were 80 and 59.
26. Israel was not considered among countries whose publications and citation statistics were addressed in this report.

27. See figure (5), page 7 of the report (TR 2011).
28. See endnote 11.
29. That is roughly during the second, relative to the first, half of the first decade of the century.
30. That is the Arab countries of the Middle East in addition to Iran and Turkey. Hence, this statement may not be taken to reflect the situation in the Arab countries of the Middle East as such.
31. "Industry-sponsored clinical drug trials in Egypt: ETHICAL QUESTIONS IN A CHALLENGING CONTEXT JUNE 2016." A joint study by Berne Declaration, Centre for Research on Multinational Corporations (SOMO), Wemos Foundation, Egyptian Initiative for Personal Rights and Shamseya for Innovative Community Healthcare Solutions. Published by Raphaël de Riedmatten.
32. See endnote 31.
33. Mission statement by CentreWatch's. Downloaded December 2016 from: http://www.centerwatch.com/.
34. Differences in total articles counts and FC between this table and the annex Table A.4.9 are due to the fact that data entered in these two tables belong to different time periods. Figures in the present table refer to the period indicated in its title, while those in Table A.4.9 refer to the entire calendar year 2015.
35. Nature Index data is essentially intended to be used only as one of a number of metrics required for assessing research performance and institutional ranking. In other words, NI may be taken as a useful sampling rather than overall ranking tool.
36. Several other Arab countries did publish a smaller number of articles in the journals covered by Nature Research. Notably, Kuwait, Oman, and Palestine published nine, eight, and four papers, respectively, while Bahrain, Libya, Mauritania, Syria, and Yemen published one article each during 2015.
37. Including Saudi Arabia within this chart would have impacted clarity and hence readability of information on Arab countries which contributed far smaller numbers of papers.
38. In this sense it is noteworthy that Iraq and Kuwait, two Arab countries with substantial hydrocarbon resources and no dearth of environmental problems, produced no publications at all within environmental and earth sciences at all during the period covered by data on which this chart rests.
39. These countries are Algeria, Egypt, Iraq, Lebanon, Jordan, Qatar, Morocco, and the United Arab Emirates.
40. Significance of this ratio will have to be further substantiated. For our present purposes, however, it is taken as an indication of the extent to which researchers in a given country are in fact capable to publish articles authored by them both solely or contribute to articles authored jointly with researchers from other countries.
41. Thus, respondents in a recent UNESCO survey of science, technology, and innovation in Iraq indicated that at least 116 scientific journals were published in Iraq monthly, quarterly or annually. The same questionnaire also indicated that some Iraqi researchers were also regular contributors to academic reviews published in neighboring countries such as Egypt and Jordan.
42. SJR is the SCImago journal rank indicator. It is designed to measure the journal's impact, expressing the number of weighted citations received in the selected year to documents published in the three previous years.
43. This rating will need to be looked into, considering that the H-index for journals published by Egypt is less than that for Bahrain and only a little higher than that assigned to Lebanon's journals.
44. For the Bulletin of Faculty of Pharmacy, Source Normalized Impact per Paper (SNIP) is quoted at 0.991, while the SCImago Journal's Rank (SJR) is 0.291.
45. For the Journal of the Association of Arab Universities for Basic and Applied Sciences, Source Normalized Impact per Paper (SNIP) is stated as 0.857. The SCImago Journal Rank (SJR) is estimated at 0.278.
46. "Research Performance Indicators; The 3rd Report;" King Abdulaziz City for Science and Technology and Reuters Thomson, 2013. Downloaded in February 2017 from: http://publications.kacst.edu.sa/SystemFiles/Books_Pdf/PDF_635391133237974950.pdf.

47. Indeed, cooperation with scientists from the United States appears to dominate cooperative scientific research activity undertaken by all other industrialized nations around the world, resulting in publications covered by Nature Research.
48. Twelve out of the fourteen remaining countries, i.e., excluding Mauritania, Yemen, and Algeria.
49. However, figures presented in this table were computed for the period March 31, 2015 to April 1, 2016, rather than January 1, 2015 till December 31, 2015.
50. Computed on the basis of WFC.
51. A WFC is a version of a FC in which articles from specialist astronomy and astrophysics journals have been down-weighted. The weighting is undertaken since these journals encompass a much larger proportion of the total publication output than any other field covered by the Nature Index, thereby preventing undue emphasis on publications in these fields. To compute WFC the FC from astronomy and astrophysics journals is multiplied by a factor of 0.2. This factor is to an approximation of the level to which astronomy and astrophysics articles are overrepresented compared to the total publication output of other fields covered by Nature Index.
52. Include all the Arab countries listed in table (Percentages 2) with the exception of Qatar.
53. "Science, Development and Sovereignty in the Arab World;" A. B. Zahlan, New York: Palgrave Macmillan, 2012, and "Research, Innovation, Entrepreneurship and the Rentier Culture in the Arab Countries;" O. Bizri, in "The Real Issues of the Middle East and the Arab Spring;" Ed. T. Andersson and A. Djeflat, Springer, New York 2013.
54. The corresponding number of patents granted before 2002 for five of the Arab countries' closest neighbors is estimated at over 9000 , or 0.35% of the world's total with most of these patents granted to citizens and residents of Israel.
55. It goes without saying that world averages take into account patenting activity in numerous developing as well as developed countries.
56. This may reflect approaches adopted by USPTO as well as enhanced inventive and subsequent patenting activity worldwide as well as in the region.
57. Patent classes addressed in this section are those for which reliable statistical data is available from the database created by the Organisation for Economic Cooperation and Development (OECD).
58. An interesting aspect of this database is its adoption of a FC of patents in case a patent application is initiated by inventors from more than one country. This should ensure that respective contributions by each country are taken into account while multiple counting of such patents is avoided. Thus, supposing a patent was co-invented by one German, one Swede, and two Swiss residents will be counted as one-fourth of a patent for Germany, one-fourth of a patent for Sweden, and half a patent for Switzerland.
59. Patent applications filed under the Patent Cooperation Treaty (PCT) allow applicants international protection for their inventions. PCT is supposed to assist patent Offices with their patent granting decisions and to facilitate public access to technical information relating to those inventions. By filing an international patent application under the PCT, applicants can simultaneously seek protection for an invention in over 150 signatory countries, as indicated in a pronouncement by the World Intellectual Protection Organisation (WIPO). Based on a pronouncement of the World Intellectual Protection Organisation. Pronouncement by the World Intellectual Protection Organisation. Posted in September 2016 and downloaded in November of the same year from: http://www.wipo.int/pct/en/pct_contracting_states.html.
60. Percentages included in the pie charts associated with the present paragraphs are computed in relation to the total number of patents indicated in relevant chart titles.
61. That is with the exception of the classes of "Fixed construction," "Chemistry and metallurgy," "Environment-related technologies," "Physics," "Nanotechnology," and "Mechanical engineering."
62. A preliminary review of Saudi-held patents indicates a predominance of oil, water, and health technologies.

Chapter 5

# Innovation and Entrepreneurship in the Arab Countries

**Chapter Outline**



## 5.1 INTRODUCTION

A variety of technology-based innovations gained recognition over the past two decades for their direct and indirect contributions to economic growth and social development. As a result, acquiring the capacity to innovate now occupies a central position in national and regional policies in the developing as well


**Science, Technology, Innovation, and Development in the Arab Countries**
**http://dx.doi.org/10.1016/B978-0-12-812577-9.00005-2**

as the developed economies, in general, and as bases for novel enterprise and job creation, in particular.

As underlined by reports issued by several international development agencies, based on experience gained during the past two decades, innovation is not just about isolated institutions adopting high-technology products and processes. Rather, it is more to do with the utilization of such products and processes to generate benefits for all segments of society. Additionally, attaining optimal benefits out of innovation policies and initiatives requires that:

- Capacity for innovation be embedded as early and as widely as possible within processes aimed at socioeconomic development, emphasizing education and vocational training;[1]
- Resilient partnerships are established between a variety of national institutions with a view to:
  - fully adapting innovations from external sources to local priorities and recipient ecologies;
  - coming out with fresh home-grown innovations in cases where foreign innovations are inadequate, which often requires adaptive as well as original R&D efforts.

The above partnerships are particularly important to ensure that innovative inputs are ultimately responsive to national priorities and local ecologies, which must often be based on specific R&D capabilities and extension services.

Despite the fact that the following sections embrace current indicators designed to measure and rank innovative capacity, a good deal of caution is needed in interpreting the values of innovation indicators and the rankings they are based upon. This is particularly essential with regard to the implications that these values and rankings hold with regard to social development for the following two closely related reasons. Both, however, are with important implications for innovation policy making and their subsequent implementation strategies. The first relates to the very notion of innovation and the fact that it is often dealt with as an outcome that may indeed find shape in a novel product or process, while several factors would favor regarding it as a process instead. In other words, viewing innovation as delimited and embodied within specific products and processes would diminish the scope of innovation policy directives, aimed at devising implementation measures, fine-tuning practical interventions, forging a variety of partnerships, adopting conducive management, regulatory, and legal arrangements, as well as drafting resource provision strategies.

The second concerns the manner in which current schemes designed to measure innovation tend to neglect its diverse implications for social development, in favor of more immediate commercial returns. In other words, while somewhat effective in gauging capacity for innovative, current indicators used to measure innovation are extremely biased toward the business sector in a manner that severely limits contributions to inclusive sustainable development and poverty reduction. On these issues, most of the principal and composite indicators used by international institutions for measuring innovation are generally quite deficient.

It is with the above-mentioned reservations that the following sections attempt an overview of the Arab countries' performance as measured by indicators designed to evaluate capacity for innovation, technological readiness, and doing business.

With regard to technological readiness, or the ability for taking up and implementing new technologies, relevant sections will show that while some progress has been made by several Arab countries in acquiring further capabilities, essentially in relation to ICT products, a great deal is still needed on many scores; essentially, in terms of the availability of latest technologies in a variety of manufacturing applications, firm-level technology absorption, and wider dissemination of technologies acquired on the basis of FDI activities.

Sections designed to evaluate entrepreneurship in the Arab countries will take a look at their global rankings in relation to a commonly used composite indicator. Evidence will be presented supporting the notion that most Arab countries will need to formulate far more effective policies as well as design-improved implementation modalities to create conducive business ecologies.

In all of the above, efforts will be made to review performance of the Arab countries across time, over the past 5–6 years, and attempt to compare it with that of their closest neighbors.

## 5.2 MEASURING INNOVATION

The key role that innovation can play as a driver of, mainly economic, development has led to the adoption of a variety of indicators to measure the capacity for acquiring and implementing innovation in a wide range of sectors and application areas. International organizations, such as the World Economic Forum (WEF), the Institut Européen d'Administration des Affaires (INSEAD), or the European Institute of Business Administration, and the World Bank, have, over the past decade or so, developed composite indicators that enable evaluation of innovation policies and tracking progress in the implementation of related initiatives. However, it is also essential to recognize the shortcomings that some of these indicators suffer. There is, firstly, the fact that several indicators rely upon opinion surveys, which while unavoidable, considering the fact that many innovation related issues may not be directly measurable, are certain to include a degree of subjectivity into evaluating scores and subsequent rankings. Secondly, and perhaps more importantly, most composite indicators that have been developed thus far emphasize economic- and business-related outcomes and exclude those related to social and cultural implications. Thus, comparable values of certain innovation indicators for two or more countries may not be taken to mean that socioeconomic implications of their innovative capacities are in fact comparable. This is particularly since different national innovation ecologies as well as socioeconomic environments respond differently even when impacted by comparable innovative stimuli.

In short, even the most sophisticated indicators that have so far been developed ought to be used with care and may not be considered perfect tools for tracking innovation, with its impact on a variety of socioeconomic issues. This is especially the case when such indicators are used in comparative analyses across national boundaries, e.g., to compare progress attained by a certain country with regard to the number of patents produced or the size of foreign direct investment attracted. An underlying assumption in undertaking such comparisons is often that all other factors, often strongly related to the issue at hand, remain similar or comparable, which is very seldom, if at all, the case.

Two prime examples of composite indicators in current use for measuring national capacity for innovation are the Global Innovation Index (GII) and relevant pillars, subindicators, or indices used computing the Global Competitiveness Index (GCI).[2]

The following section will first take a look at the Arab countries' scores and world standing as measured by the former, compound GII indicator, while later sections are dedicated to an overview of scores and rankings attained by the Arab countries in relation to the World Economic Forum's GCI and its subindicators of indices.

Both GII and GCI indicators utilize objective as well as subjective data in arriving at scores for their composite indicators and subsequent rankings of countries around the world. Objective data essentially relate to relatively straightforward and measurable attributes of specific aspects of the innovation system in question, e.g., R&D spending, enrolment in various levels of education, Internet use. Subjective data, on the other hand, are derived from opinion surveys directed at business people and government officials.

The following paragraphs provide a summary of the principal components of the two principal indicators in common use for evaluating innovation capacity and related global rankings, namely the GII and selected pillars of the GCI.

### 5.2.1 The Global Innovation Index

GII is an annual ranking of countries around the world in terms of their capacity for, and successful implementation of, innovation. The GII rankings are published annually in partnership between a number of organizations, including INSEAD, Cornell University, and the World Intellectual Property Organization (WTO). GII rankings are based on both subjective opinion surveys targeting business leaders and specialists, as well as objective data, concerning for example the acquisition and dissemination of ICTs, derived from sources, including the International Telecommunication Union (ITU), the World Bank, and the World Economic Forum. The GII ranking for a given country is computed on the basis of a simple average of scores attained by that country on two composite indices, referred to as the Innovation Input Index and Innovation Output Index. The values of both are worked out on the basis of constituent pillars, which are in effect composite indexes in their own right, since they too are made up of constituent indicators. Scores for the composite indices and their constituent pillars are calculated by a weighted average method.

The first GII ranking was produced in 2007. It combines 79 simple indicators that describe the state of national innovation ecologies, taking into account both input and output aspects of their operations. On the other hand, the annual Global Competitiveness Report (GCR) evaluates a set of 12 compound or composite indicators, or pillars, on the basis of a large number of socioeconomic criteria, both directly and indirectly impact competitiveness of a relatively large number of countries around the world including several Arab countries.[3] The basic structure of the GII schemes is summarized within the adjoining Frame 5.1.

## 5.2.2 The Global Competitiveness Report

The GCR is published by the World Economic Forum (WEF), which is a Swiss nonprofit foundation, recognized as an international body with a mission aimed at engaging leaders and policy makers from various spheres including politics, academia, and society at large with view to improving global, regional, socioeconomic development outcomes. Since 2004, the GCR has ranked the world's countries on the basis of its GCI, which is based on more than 110 variables, only one-third of which are based on numerical parameters whose values are available from sources such as the United Nations and the World Bank, while two-thirds are based on executive opinion surveys. Variables used in computing GCI are organized into 12 pillars, representing important determinants of economic competitiveness. In their turn, these pillars are dependent upon a number of constituent subindices or indicators. Of particular interest to the following

**Frame 5.1 Essential Structure of the Global Innovation Index**

**Innovation Input Subindex**

- Institutions: Political environment, Regulatory environment, Business environment.
- Human capital and research: Education, Tertiary education, Research and development.
- Infrastructure: ICTs, General infrastructure, Ecological sustainability.
- Market sophistication: Credit, Investment, Trade and competition.
- Business sophistication: Knowledge workers, Innovation.

**Innovation Output Subindex**

- Knowledge and technology outputs: Knowledge creation, Knowledge impact, Knowledge diffusion.
- Creative outputs: Intangible assets, Creative goods and services, Online creativity.

sections are two pillars, namely, the Technological Readiness Pillar (TRI) and the Innovation Pillar as well as constituent indicators or subindices.[4]

A cursory review of pillars and subindices presented in this table as well as the above basic structure of the GII ranking scheme clearly indicate almost total exclusion of innovation's pillars and indicators designed to measure innovation's social and cultural implications. The only two issues that relate to social development in the above GII scheme concern labor and education with related indices in fact designed to measure benefits that labor relations and educational systems hold for national competitiveness and economic gains rather than wider social benefits.

## 5.2.3 The Global Innovation Index; Arab Countries' Scores and Rankings

### 5.2.3.1 GII Scores and Rankings

Table 5.1 provides a list of 14 Arab countries for which GII values were available in the GII 2016 Report, together with their GII values, their world and regional rankings, as well as their rankings within the income group they belong to, according to the above report.[5] Data for five other countries in the same region, namely Cyprus, Iran, Israel, and Turkey are included for comparison.

As this table indicates, three countries, namely the United Arab Emirates, Saudi Arabia, and Qatar, rank within the first 50 countries worldwide, while nearly half the Arab countries listed in this table rank within the lower half of the world's rankings. Recent scores for Syria, Libya, as well as many other Arab countries are not included in the current issue of the GII Report. However, based on previous issues of this report, scores attained by Syria and Libya fall below the median value for the entire group of Arab countries, indicating that if these two countries were to be included within the above comparison, then at least 16 out of all 22 Arab countries would fall within the lower half world rankings, i.e., below a ranking of around 70.

It is also notable that both Kuwait and Algeria appear to score well below fellow members of their income groups, namely the high and upper-middle income groups, respectively, while Morocco ranks close to the top of the lower-middle income group to which it belongs.

Table 5.1 also includes values of the innovation efficiency ratio (IER), designed to evaluate the effectiveness of national innovation systems.[6] In effect, this is the ratio of the output subindex score over the input subindex score. Several of the Arab countries for which data is available managed to attain efficiency ratios close to the median value for this ratio, which is calculated as 0.71. Such is the case with Kuwait, Lebanon, Saudi Arabia, Tunisia, and Morocco, for instance. However, it is to be remembered that relatively high values of this ratio might also be attained as a result of low inputs toward a national innovation system with middling output levels. The reverse is naturally true with countries attaining low IER values due to high inputs versus smaller outputs. Notably, Algeria, the United Arab Emirates, and Sudan attained low values for the IER.

**TABLE 5.1** Global Innovation Index; Scores and Rankings, and Efficiency Ratio for Fourteen Arab Countries and Five of Their Closest Neighbors (2016)

| Country | GII Score | World Rank | Income Group[a] | Rank (Income Group) | Rank Within Region[b] | Innovation Efficiency Ratio (IER)[c] | EIR Rank |
|---|---|---|---|---|---|---|---|
| **Scores and Rankings for 14 Arab Countries** | | | | | | | |
| UAE | 39.35 | 41 | HI | 38 | 3 | 0.44 | 117 |
| Saudi Arabia | 37.75 | 49 | HI | 42 | 5 | 0.61 | 85 |
| Qatar | 37.47 | 50 | HI | 43 | 6 | 0.56 | 97 |
| Bahrain | 35.48 | 57 | HI | 44 | 7 | 0.58 | 92 |
| Kuwait | 33.61 | 67 | HI | 46 | 10 | 0.73 | 42 |
| Lebanon | 32.70 | 70 | UM | 18 | 11 | 0.73 | 41 |
| Morocco | 32.26 | 71 | LM | 7 | 12 | 0.66 | 64 |
| Oman | 32.21 | 72 | HI | 47 | 13 | 0.53 | 103 |
| Tunisia | 30.55 | 77 | UM | 22 | 14 | 0.60 | 86 |
| Jordan | 30.04 | 82 | UM | 25 | 15 | 0.67 | 58 |
| Egypt | 25.96 | 107 | LM | 20 | 17 | 0.63 | 74 |
| Algeria | 24.46 | 113 | UM | 34 | 18 | 0.49 | 111 |

*Continued*

**TABLE 5.1** Global Innovation Index; Scores and Rankings, and Efficiency Ratio for Fourteen Arab Countries and Five of Their Closest Neighbors (2016)—cont'd

| Country | GII Score | World Rank | Income Group[a] | Rank (Income Group) | Rank Within Region[b] | Innovation Efficiency Ratio (IER)[c] | EIR Rank |
|---|---|---|---|---|---|---|---|
| Yemen | 14.55 | 128 | LM | 29 | 19 | 0.34 | 126 |
| Sudan[d] | 14.95 | 141 | LM | 34 | 32 | 0.37 | 136 |
| **Scores and Rankings for Five Neighboring Countries** | | | | | | | |
| Cyprus | 46.34 | 31 | HI | 30 | 2 | 0.79 | 26 |
| Iran | 30.52 | 78 | UM | 23 | 3 | 0.71 | 51 |
| Israel | 52.28 | 21 | HI | 21 | 1 | 0.81 | 23 |
| Malta | 50.44 | 26 | HI | 26 | 16 | 0.98 | 2 |
| Turkey | 39.03 | 42 | UM | 4 | 4 | 0.84 | 13 |

[a]*Income groups are based on the World Bank's 2013 classification: LI, low income; LM, lower-middle income; UM, upper-middle income; and HI, high income.*
[b]*Regional rankings refer to geographic locations based on UN classification, where all countries listed in this table belong to the region designated "Northern Africa and Western Asia region," with the exception of the Sudan, which belongs to Sub-Saharan Africa, and Iran, which belongs to Central and Southern Asia.*
[c]*The median value for this ratio for all countries for which data is available is 0.71.*
[d]*Data for Sudan is taken from the GII 2015 Report.*
From The "Global Innovation Index 2016; Winning with global innovation;" Soumitra Dutta, Bruno Lanvin, and Sacha Wunsch-Vincent Editors. Published by Johnson Cornell University, INSEAD and WIPO downloaded in March 2017 from: http://www.wipo.int/edocs/pubdocs/en/wipo_pub_gii_2016.pdf.

At any rate, with the exception of Kuwait and Lebanon, values of the IER for the Arab countries included in Table 5.1 fall below the median estimated for all countries included in the GII 2016 report. The IER values for Egypt, Jordan, Morocco, and Saudi Arabia are close at 0.63, 0.67, 0.66, and 0.61, respectively. On the other hand, IER values for the rest of the Arab countries fall within a rather wide range that extends from 0.34 for Yemen to 0.58 for Bahrain. IER values for the Arab countries are also lower than those estimated for their closest neighbors, with the possible exception of Iran whose IER at 0.71, is comparable to those estimated for Kuwait and Lebanon, at 0.73. IER values for Cyprus, Israel, Malta, and Turkey, however, are considerably higher at 0.79, 0.81, 0.98, and 0.84, respectively.

Fig. 5.1 presents a plot of the values of the GII versus per capita gross domestic income computed in purchasing power parity United States dollars for 121 countries around the world as well as 14 Arab countries for which data is available from the GII 2015 Report. This figure clearly shows that almost all the Arab countries underperform with regard to innovation as measured by this index and that the extent to which they underperform increases as their GDP per capita goes up. Thus, while Tunisia, Jordan and Morocco, Sudan, and Algeria fall close to the global trend line, albeit with low values for their respective GII scores, GII values for all the GCC countries fall considerably below that line. In fact Kuwait's GII value fall around 30 points below the trend line.[7] Tellingly, Qatar could not be included in this figure, since its extremely high GDP per capita falls far out of the range adopted for a convenient presentation.[8] As

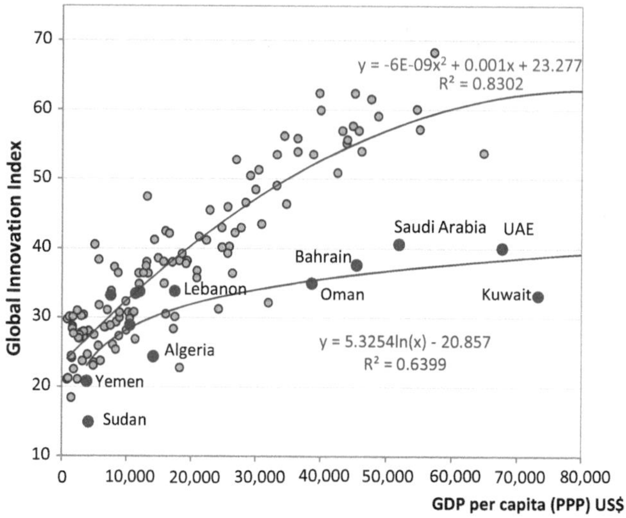


**FIGURE 5.1** Global Innovation Index Values versus GDP per capita in purchasing power parity US$ for 121 countries around the world (*blue circles* (light gray in print versions)) and 14 Arab countries (*red circles* (dark gray in print versions)).

succeeding sections in this chapter will show, the adoption and dissemination of ICT infrastructures and devices do not seem to have translated into improved overall scores on this global index.

## 5.2.4 Scores and Rankings Based on the Global Competitiveness Index

The Arab countries' world rankings based on the GCI, as well as rankings based on the values of its subindices, provide insight on a range of capabilities that ultimately determine a country's competitiveness, including innovative abilities as well as access to, and adaptation of, new technology inputs. An attempt at exploring the extent to which rankings of the innovation index correlate with rankings on other GCI indices is presented in Frame 5.2, demonstrating very strong correlation between innovation, higher education, and training among others, while extremely weak correlation is shown between the innovation index and indices such as market size and macroeconomic environment. One of the policy implications that might be construed from the information presented in this frame concerns the need to devote special attention to improved

**Frame 5.2 Correlating the Innovation Subindex to Other GCI Subindices**

Correlating rankings attained according to the innovation subindex used in determining the overall GCI for the Arab countries and rankings obtained on the bases of other GCI subindices, shown in Table A.5.1, presents a number of interesting observations.

Table 5.2 presents values of $R^2$ or coefficient of determination obtained for linear correlations between the rankings of the innovation index and the remaining 11 subindexes used in obtaining GCI rankings.

As this table indicates, rankings for the innovation subindex are very strongly correlated with rankings based on four other subindices, namely those concerning:

- higher education and training,
- goods market efficiency,
- business sophistication, and
- labor market efficiency.

On the other hand, innovation subindex rankings appear less strongly correlated to three other rankings, namely those based upon:

- financial market development,
- technological readiness,
- health and primary education.

Rankings based on the innovation subindex do not appear to correlate at all with rankings based on the macroeconomic environment, nor market size, while only moderate correlation is apparent between the innovation subindex rankings and rankings based on subindices that address institutional and infrastructural arrangements.

**TABLE 5.2 Coefficient of Determination ($R^2$) Values Correlating the Values of the Innovation Index With Those for 11 Other GCI Indices**

| | |
|---|---|
| Higher Education and Training | 0.95 |
| Goods market efficiency | 0.88 |
| Business sophistication | 0.85 |
| Labor market efficiency | 0.82 |
| Financial market development | 0.78 |
| Technological readiness | 0.72 |
| Health & Primary Education | 0.67 |
| Institutions | 0.58 |
| Infrastructure | 0.55 |
| Macroeconomic environment | 0.08 |
| Market size | 0.01 |

performance in areas, such as higher education and training, business sophistication, goods, and labor market efficiencies to improve the national innovation climate.

Annex Table A.5.1 presents the GCI rankings published by the World Economic Forum's 2016–17 report for 14 Arab countries as well as five of their closest neighbors together with their rankings on the above subindices. The following paragraphs provide a brief discussion of these rankings.

Inspection of the rankings of the Arab countries for which data is presented in Table A.5.1 with regard to their rankings relative to the composite indicators that determine the GCI indicates that the United Arab Emirates, Qatar, Saudi Arabia, Bahrain, and Oman come on top, based on their average rankings on all subindices. These countries are followed by Jordan which comes before their fellow GCC member, Kuwait.

Algeria, Egypt, and Yemen rank among the last three Arab countries with respect to almost all subindices, with exceptions such as Algeria's rank as fifth among all thirteen Arab countries and with regard to its rankings as fourth and fifth with regard to its market size and macroeconomic environment. Clearly, in Algeria, Egypt, and Yemen, radical overhaul of policies is needed to arrive at improved rankings on almost all subindices in order to bring their competitiveness, particularly, with regard to aspects based on innovation and technological readiness up to reasonable levels.

Jordan, Morocco, Lebanon, and Tunisia occupy intermediate positions between the above two groups of countries, with Lebanon attaining 12th position among the rest of the Arab countries for which information is available with regard to subindices that evaluate quality of institutions, infrastructural arrangements, and macroeconomic environment. Should Lebanon perform better with regard to these subindices as well as enhance its market size, on whose subindex it ranks 11th among the rest of the Arab countries, it could readily rank among the top 4–5 Arab countries in terms of overall global competitiveness. Similarly, Jordan's position could also witness enormous GCI improvements should it improve its rankings with regard to subindices that evaluate its macroeconomic environment, market size as well as its healthcare and primary education services.

Tunisia's rankings are generally disappointing and might be explained by recent political strife that engulfed the country. As Table A.5.1 indicates, Tunisia rankings are particularly low with regard to the efficiency of its goods and labor markets as well as the state of development of its financial markets and business sophistication.

In summary, all of the Arab countries would need to enact measures to drastically enhance their performance on all attributes measured by the GCI subindices. In particular, Morocco, Tunisia, Algeria, Egypt, and Yemen would benefit from efforts designed to enhance their rankings across all subindices, since they generally score lower than half the world's countries.

In effect, the Arab countries would need to dedicate special efforts toward improving their performance on several fronts, with particular regard to their labor market efficiency, macroeconomic environment, higher education and training, innovation capabilities, financial markets, and technological readiness. This is particularly evident when their performance, as evaluated by their world rankings relative to subindices concerning these domains, is compared to that of their close five neighbors listed in Table A.5.1. This comparison indicates that on average only the United Arab Emirates and Qatar ranked higher than all their closest neighbors on all subindices taken together, while all Arab countries included in this table, excluding the United Arab Emirates, Qatar, and Saudi Arabia ranked below Malta, on average.

Two of the composite indicators that go into determining the value of the GCI are of major concern for the current discussion. They are namely the "Technological Readiness Index (TRI)," and "Composite Innovation Index (CII);" also referred to as the 9th and 12th pillars of the GCI, respectively.

The CII is included as a factor in evaluating a country's competitiveness on account of the direct impact that innovation has on its ability to generate economic value by merely integrating and adapting available technology inputs. The TRI, on the other hand, is designed to measure a country's ability to access and utilize new technologies, with special emphasis on information and communications technology (ICT). Values estimated for each of these two composite indices, or pillars as they are often referred to, are derived from both statistical data as well as opinion surveys undertaken within the country in question.

The following sections will examine the scores and world rankings of the Arab countries with regard to the above two composite indices and constituent subindices as well as another indicator introduced by the World Bank with the aim of measuring the ease of doing business (EoDB) in countries around the world. Considering scores and rankings of the Arab countries with regard to this latter indicator in addition to the innovation and technological readiness indicators should provide a more comprehensive view of their present abilities and potential for adopting innovations as well as their potential for utilizing such abilities in attaining socioeconomic gains.

### 5.2.5 Arab Countries' Scores and Rankings on the Composite Innovation Index and Its Component Indices

As shown in annex Table A.5.2, subindices used in evaluating the CII, or pillar, include "capacity for innovation," "quality of scientific research institutions," "company spending on R&D," "university–industry collaboration in R&D," "government procurement of advanced technology products," "availability of scientists and engineers," and "PCT patent applications."[9] However, with the exception of the latter subindex, designed to measure national patenting performance,[10] all the other subindices utilized in computing national scores for the innovation pillar are based on responses to surveys conducted within national business communities.

The following paragraphs will attempt a comparative overview of scores and rankings attained by Arab countries on the CII and its constituent subindices as presented in the GCRs for the years 2012–13 and 2016–17.

Fig. 5.2 presents an overview of scores for several Arab countries on the CII, as well as their average value. Scores by the Arab countries included in this figure range between 2.2 for Yemen and 4.9 for the United Arab Emirates.

Fig. 5.2 indicates that the average value of this index in 2016 stood at around 3.3. However, inclusion of innovation scores for the eight Arab countries for which data is not available for 2016, including Iraq, Libya, and Syria,[11] would undoubtedly bring this average down quite considerably, well below the world's median and average values for scores by all 138 countries included in the GCR 2016–17, which are estimated as 3.32 and 3.57, respectively. Incidentally, CII scores achieved by five of the Arab countries' closest neighbors, namely Cyprus, Iran, Israel, Malta, and Turkey, were estimated at 3.21, 3.19, 5.73, 3.71, and 3.31, respectively.

### 5.2.6 Changes in CII Scores and Rankings

On average, scores achieved by the various Arab countries for which data was available went down by around 1% only between 2012 and 2016. However, as indicated in Fig. 5.3, this rather trivial drop masks considerable declines as well as gains in CII scores exhibited by individual Arab countries. Thus, scores attained by Mauritania, Tunisia, Saudi Arabia in 2012 went down by around 19%, 17%, 8% by the year 2016, while scores attained by several other Arab countries, including Lebanon, Bahrain, Yemen,[12] Jordan, the

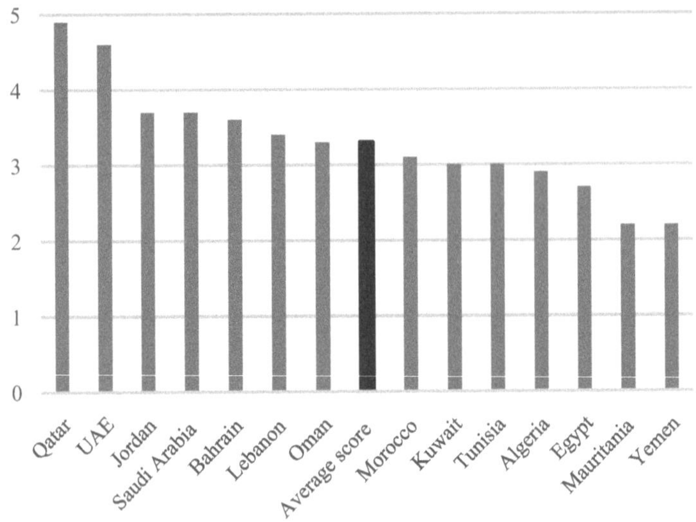


**FIGURE 5.2** Scores of 14 Arab countries on the Composite Innovation Index. *The Global Competitiveness Report 2016–17.*

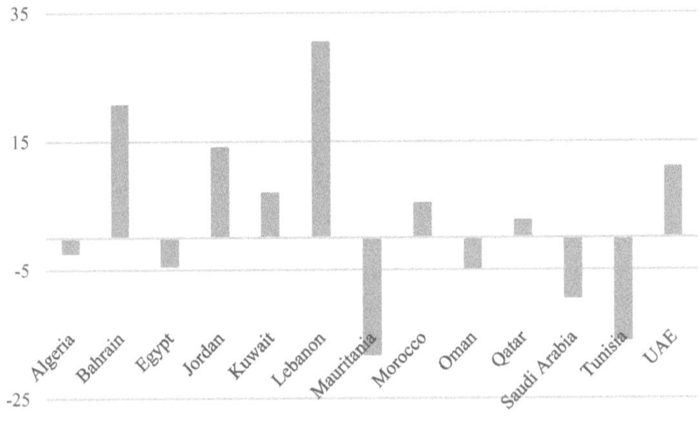


**FIGURE 5.3** Average change in scores of Innovation Subindices between 2012 and 2016.

United Arab Emirates went up considerably, by up to 26% in the case of Lebanon and 16% for Bahrain, as indicated in Table A.5.1. Scores attained by several other countries changed by between 3% and 7%, with Egypt and Algeria losing, and Kuwait and Morocco gaining grounds, compared to their previous scores.

The situation depicted by the CII changes presented above is mirrored by average changes incurred in its subindices, particularly if the values of the index that measures PCT patent applications are excluded from calculating the average of percentage changes across the remaining seven subindices.

Looking at gains and losses sustained on average by all the Arab countries with regard to various subindices, it would seem that the highest average decline took place with regard to the Arab governments' procurement of advanced technologies. Thus, the average value of this subindex for all the 14 Arab countries listed in Table A.5.1 dropped by just over 6%, while declines in the value of this subindex dropped far more acutely, by around 33%–9% for seven Arab countries, including Tunisia (32.5), Mauritania (29.4), Morocco (16.7), Oman (15.9), Saudi Arabia (10.4), Bahrain (9.1), and Qatar (8.6).

Three other subindices of the CII deserve a closer look, in connection to overall changes incurred as well as individual contributions by the various Arab countries to these changes. These are namely the quality of research institutions, university–industry collaboration in R&D, and availability of scientists and engineers.

According to the subindex designed to estimate research institutions, the rather small average decline, by 2.3%, obscures considerable declines for several countries, including Mauritania (22.2), Tunisia (20.0), Algeria (17.1), Saudi Arabia (11.1), Egypt (10.3), and Oman (8.3), while even larger gains appear to have been made mainly by Lebanon (40.0), Bahrain (19.4), and Jordan (14.3). Losses in scores for this subindex registered by other Arab countries, ranging between 3% and 5%, go toward an overall loss in quality of research institutions by the above-mentioned figure of 2.3%.

Considering scores on the subindex designed to assess university–industry collaboration in R&D, the largest losses in scores for this index were recorded for Algeria (28.9), Mauritania (26.9), Saudi Arabia (22.2), Tunisia (18.9), Kuwait (13.3), and Egypt (11.1). Considerable gains were made, on the other hand, by both Bahrain (23.3) and Jordan (15.2).

Mauritania, followed by Tunisia led declines in scores for the subindex designed to evaluate the availability of scientists and engineers with relapses of 34.3% and 20.4%, respectively. While Morocco seems to have sustained a smaller loss in its score on this subindex, of around 11%, Lebanon and the United Arab Emirates increased their scores by around 11% and 8%, respectively.[13]

Concerning scores attained by the Arab countries in relation to the number of PCT patent applications, which is incidentally the only subindex based on numerical records, as opposed to opinion surveys, upon which the remaining subindices are based, both very large increases and much smaller lapses are recorded for the Arab countries on which data were available in the WEF reports. Qatar falls within the first category, with an increase of over 1630% in the number of patent applications in 2016 compared to 2012. Large gains by around 250%, 182%, and 114% are also recorded for Tunisia, Saudi Arabia, and Morocco, respectively. On the other hand, smaller percentage declines in the number of patent applications were recorded for Algeria (82) and Kuwait (25).

As indicated in Table A.5.3, average rankings of the Arab countries based on their CII scores in 2016 witnessed limited changes from the positions they held in 2012. Thus, their average rankings with regard to most subindices

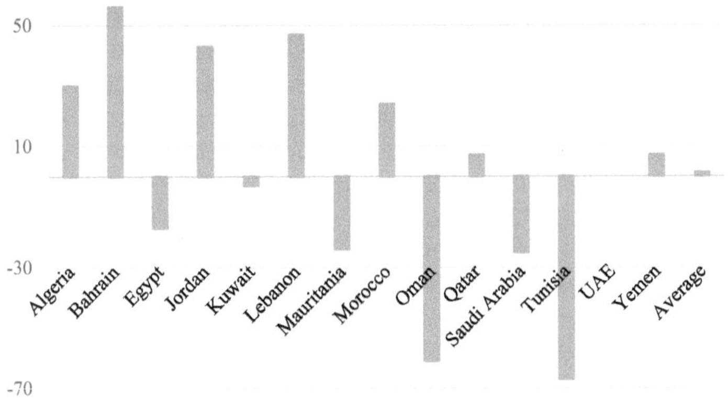


**FIGURE 5.4** Changes in Arab countries' rankings between 2012 and 2016, in relation to the subindex Company R&D Spending.

generally went up or down by between 1 and 2 ranks. With the largest setback experienced in relation to average rankings for the subindex "availability of scientists and engineers," which slipped by three ranks in 2016, compared to 2012. Substantial gains were made, on the other hand, in the average ranking with regard to university–industry collaboration, which went up by 12 positions. The following two points would need to be emphasized, however:

- No data is available for Libya and Syria on the values of the Innovation Index nor any of its constituent subindices, from the Global Competitiveness Report 2016–17. Thus, comparison of average rankings for the Arab countries as reported by the World Economic Forum's GCRs 2012–13 and 2016–17 is limited to countries for which such data is in fact available, namely Algeria, Bahrain, Egypt, Jordan, Kuwait, Lebanon, Mauritania, Morocco, Oman, Qatar, Saudi Arabia, Tunisia, the United Arab Emirates, and Yemen.[14]
- The fact that average rankings showed limited changes masked both dramatic ranking declines and gains by individual countries. Thus, as indicated by Fig. 5.4, which illustrates changes in rankings experiences with regard to the subindex designed to measure R&D spending by the private sector, Algeria, Bahrain, Lebanon, Jordan, and Morocco climbed up 30, 56, 43, 47 and 24 ranks between 2012 and 2016, while ranks attained by Egypt, Mauritania, Oman, Saudi Arabia, and Tunisia slipped down by 17, 24, 61, 25, and 67 points by 2016.

Algeria made overall gains in its CII ranking due to enhanced performance on five of the seven subindices that go into determining the value of this index, namely its capacity for innovation, quality of its research institutions, R&D spending by its companies, university–industry collaboration as well as its government procurement of advanced technologies. As a result, its CII world ranking went up by 29 points from its 2012 rank of 141.

Bahrain achieved gains in its rankings across all CII subindices, most notably on account of its enhanced rankings on three subindices, namely overall capacity for innovation, firms' contributions to R&D spending, university–industry collaboration as well as the availability of scientists and engineers. World ranks regarding these four subindices went up by 52, 56, 71, and 20 points, respectively between 2012 and 2016.

Jordan's enhanced CII rank appears to have been due to improved rankings associated with four subindices, namely quality of research institutions, firms' R&D spending, collaboration between its universities and industrial firms as well as government procurement of advanced technologies. The highest gains were made in relation to firms' R&D spending and university–industry collaboration, with Jordan's world rankings going up by 43 and 55 points, respectively.

Lebanon achieved the largest gains in rankings with substantial gains in all subindices, least of all in relation to the availability of scientists and engineers.

The GCC countries sustained a significant reduction in their CII rankings. Thus, while rankings for Bahrain and the United Arab Emirates went up by 27 and 3 points, respectively, those of Oman, Saudi Arabia, and Kuwait slipped back by 29, 13, and 2 points, between 2012 and 2016. Two countries among the GCC members, namely Kuwait and Oman ranked below Jordan with regard to their CII scores.

Subindices responsible for the above declines in GCC rankings appear to be those designed to measure the quality of research institutions and contributions by firms to R&D, with declines exceeding 20 ranks. Most of the losses in GCC rankings were due to overall unsatisfactory performance by both Oman and Saudi Arabia. Based on average rankings of all seven subindices used in computing the CII, these two countries' average rankings went down by around 21 and 18 positions, respectively, while the average rankings on the same subindices went up by around 33 for Bahrain and by around 5 ranks for Kuwait, Qatar, and the United Arab Emirates.

## 5.2.7 Scores and Rankings on the Technological Readiness Index and Its Component Indices

Fig. 5.5 presents a view of scores obtained by Arab countries on the basis of the composite TRI pillar, as presented in the GCR 2016–17. The average value of scores for the countries included in this figure is estimated at 4.01, which falls below the average and median scores for all 138 countries included in this report, estimated at 4.07 and 4.18. As observed in the case of the similar Fig. 5.5 for scores attained by the Arab countries on the TRI, should scores for Arab countries, such as Libya and Syria, be included then this average would most certainly fall to even lower levels below the world's average.[15] It may also be noted that TRI scores achieved by four of the Arab countries' closest neighbors, namely Cyprus, Israel, Malta, and Turkey, estimated at 4.63, 5.76, 5.8, and 4.16, respectively, fall above the average of all scores attained by the Arab countries included in Fig. 5.5.

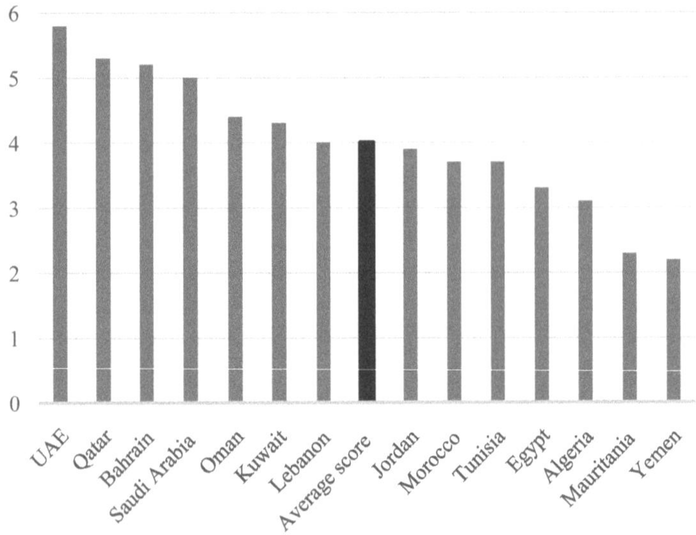


**FIGURE 5.5** Scores of 14 Arab countries on the Technological Readiness Index. *The Global Competitiveness Report 2016–17.*

## 5.2.8 Changes in TRI Scores and Rankings

Annex Tables A.5.4 and A.5.5 present data on the scores and rankings attained by 14 Arab countries on the "TRI" as well as its subindices, as reported by GCRs for the years 2011–12 and 2016–17.[16]

The TRI is computed by WEF reports on the basis of scores attained with regard to two groups of subindices. The first group is designed to evaluate the ability of a given country to adopt and utilize technologies and includes the following subindices: "availability of latest technologies," "firm-level technology absorption," as well as "FDI and technology transfer," while the second addresses: the extent of internet use, the quality of internet services, as indicated by available bandwidth, fixed as well as mobile telephone subscriptions.[17] The first group of subindices is based on opinion surveys conducted within the business sector, while, in the second group data derived from international sources of ICT statistics, is utilized in evaluating subindices.

With the exception of Mauritania, Tunisia, and Yemen, TRI scores computed for the Arab countries witnessed considerable increases in percentage terms, by 2016 from the values they had in 2012. However, closer examination of scores attained by all the Arab countries shows that improvements shown by their TRI scores are essentially due to their acquisition of better ICT capabilities. Thus, the percentage differences in average scores attained by the Arab countries combined, with regard to the subindices associated with availability of latest technologies, technology absorption, FDI, and technology transfer went down by between 8% and 10% between 2012 and 2016. On the other hand, almost all the Arab countries on which data was available in both the 2012 and 2016 WEF

Reports showed remarkable improvements in scores relating to ICT-related subindices, namely, the number of Internet users, broadband Internet subscriptions, and available Internet bandwidth. Yet some of the Arab countries did not do as well as others even with regard to these subindices. Thus, broadband Internet subscriptions per 100 of their population appear to have digressed considerably in both Kuwait and Tunisia, by around 18% and 7%, while in Egypt Internet bandwidth per capita went down, but only slightly, by 0.4%. This is most probably due to the fact that bandwidth has not kept up with population growth.

With respect to world rankings of the Arab countries in relation to TRI and its subindices, data included in Table A.5.5 indicate that collectively, limited progress was made in building technological readiness as defined by constituents of this GCI pillar. Indeed some countries, such as Tunisia, Jordan, and Morocco saw their world ranking on the Technology Readiness pillar slip back by 22, 16, and 15, positions, respectively, between 2012 and 2016. Some of the Arab countries, such as Lebanon, Algeria, and the United Arab Emirates, however, did make some gains in their world rankings, with Lebanon climbing 17 positions, and both Algeria and the United Arab Emirates gaining 12 positions upward on the world ranking in 2016.

While only a small loss in average rankings was sustained by the Arab countries on the TRI, significant losses and lesser gains on this pillar's constituent subindices. Thus, collectively, the Arab countries included in this table lost an average of over 12 and 9 ranks, respectively, in relation to subindices that address firm level technology absorption, on the one hand, and FDI and technology transfer, on the other. Losses of around 3 ranks were also sustained by the same Arab countries in relation to subindices that evaluate availability of latest technologies and available Internet bandwidth. On the other hand, an improvement of 9 ranks was made in the collective average ranking of the Arab countries linked to use of the Internet by their populations.

The most drastic losses in individual rankings of the Arab countries with regard to subindices of the TRI appear to have been sustained by Tunisia, followed by Egypt, Oman, and Saudi Arabia. Thus, in relation to rankings linked to the subindex designed to estimate availability of latest technologies, nearly all countries suffered losses in rankings, of between 1 and 21 positions, while only three countries, namely Morocco, the United Arab Emirates, and Qatar climbed up by 15, 14, and 10 positions, respectively. Losses in Arab country rankings due to reduced ability for technology absorption by firms, averaged 12 positions, with world rankings of Tunisia, Egypt, and Saudi Arabia, slipping back by 56, 43, and 21 positions. Similar losses are observed for rankings on FDI and technology transfer, with Tunisia, Saudi Arabia, and Bahrain registering losses in rankings by 56, 28, and 15 positions.

Fig. 5.6 presents a view of the values of three ICT-related indices, for 14 Arab countries, indicating the superior position occupied by the GCC countries in comparison to the rest of the Arab countries, with the notable exception of Lebanon's position in relation to broadband Internet subscriptions

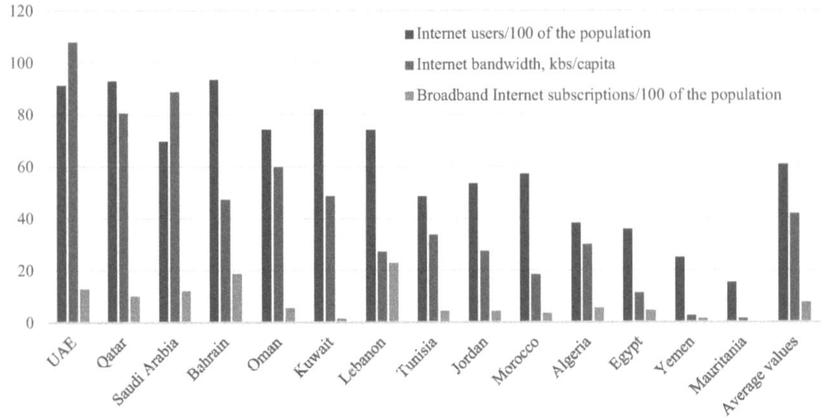


**FIGURE 5.6** Values of ICT-related subindices of the Technological Readiness Index in 2016. *The Global Competitiveness Report 2016–17.*

per 100 of the population, which appears to exceed that of any of the GCC countries. This figure also shows that eight countries, including Lebanon, fall below the average value of Internet bandwidth for the entire group of countries included in this figure, while at least seven countries, including Egypt, Tunisia, Morocco, and Jordan fall below the average value of the index designed to measure broadband subscriptions.

With regard to rankings on subindices concerned with ICT capabilities, considerable gains were observed for all the Arab countries with regard to Internet users, with the notable exceptions of Morocco, Tunisia, Egypt, and Jordan, whose world rankings slipped back by 18, 13, 10, and 5 positions, respectively. On the other hand, available Internet bandwidth, as measured in units of kbs/capita, registered declines averaging 3 ranks across all the Arab countries included in both the 2011–12 and the 2016–17 GCI Reports, with seven countries including Egypt, Yemen, Mauritania, Qatar, Tunisia, Morocco, and Bahrain registering rank declines between 19 positions for Egypt and 8 for Bahrain. Lebanon appears to be the only country which made clear gains on all ICT-related subindices, climbing up in its world ranking by 41, 33, and 16 positions, with regard to subindices that measure Internet use, broadband subscriptions, and available bandwidth per capita, respectively.

### 5.2.9 Innovation and Technological Readiness in the Arab Countries; Concluding Remarks

Comparison of changes measured by the TRI to those attained with regard to the CII indicates that while scores on both indices seem to correlate reasonably well, a small number of Arab countries, namely the United Arab Emirates, Lebanon, Kuwait, Qatar, and Jordan may be counted as having made significant gains on both the CII and TRI fronts, during the 5-year period 2012–16, while

Tunisia and Mauritania showed significant declines in both their technological readiness and composite innovation indices.

Saudi Arabia and Bahrain, which were able to advance their TRI scores by 16% between 2012 and 2016, lost around 8% and 23%, respectively, compared to their previous CII scores by the year 2016.

Closer inspection of cases where enhancing technological readiness did not lead to similar progress along the innovation front would reinforce the notion that advances made in acquisition of improved ICT infrastructures may not automatically lead to innovative prowess. Clearly further measures are needed before such advances are translated into effective capacity for innovation.

As Table A.5.1 clearly indicates, CII and TRI scores and rankings of the Arab countries with the exception of a few of their GCC members leave a good deal to be desired. Thus, with particular regard to technological readiness, only the United Arab Emirates ranked above all of their closest neighbors, while all the remaining 12 the Arab countries listed in table ranked below Malta and 9 below Cyprus. With respect to rankings on the innovation subindex Israel's ranking exceeds all the Arab countries while the rankings of 10 of these countries fall below that of Malta.

It may also be pointed out that Arab countries which performed better than the rest on the TRI and CII scales, such as the United Arab Emirates, Qatar, Oman, and even Tunisia, were generally ones with rather small populations. In effect, their scores and rankings may be linked to special circumstances that could prove difficult to emulate by the more populous Arab countries for reasons including, among others, resource availability as well as absence of relevant educational and training facilities.

## 5.2.10 Role of Information and Communications Technology

The importance of access to ICTs to enhancing scientific research, innovation, and entrepreneurship is now widely recognized. Back in the 1970s and 1980s, personal computers were initially welcomed by the Arab countries as a miracle tool. However, due to inadequate user training, on the one hand, and limited efforts at Arabizing key applications they ended up in offices all over the region, as status symbols and exalted typewriters. This state of affairs has now changed considerably, mainly due to growth in Internet utilization over the past two decades. Thus, the number of Internet users in the Arab countries increased from around 2.5 million in 2000 up to over 170 million in 2017. Annual growth rates in late-comer countries such as Algeria, Morocco, Sudan, Syria, and Yemen often exceeded 100%. As shown in annex Table A.5.7, average penetration of the Internet varies considerably from one country to another, reaching maximum values in excess of 90% in Arab countries of the Gulf region, such as Qatar, the United Arab Emirates, and Bahrain. In the GCC countries Internet users add up to 39 million, or about 23% of all Internet users in the Arab countries.[18]

The status of the Arabic language on the Internet is of considerable importance with particular regard to applications in e-government, e-commerce, e-learning, and e-health. Use of the Arabic language grew by around 6800% over the period 2000–17. This growth rate is the largest among all top 10 languages on the Internet, exceeding by more than twofold growth of the Russian and Indonesian-Malaysian languages and by more than threefold that of the Chinese, all of which languages grew by more than 2000% during 2000–17.[19] However, this large increase in the use of the Internet in the Arab countries does not seem to have resulted in a corresponding increase in most of the above-mentioned applications, essentially due to regulations, concerning various e-applications lagging considerably behind growth in Internet utilization as well as the Arabic language's presence on the Internet.

Available evidence based upon Web interrogation of search engines in the Arabic language reveals preoccupation with religious, cultural, and recreational issues, with a much smaller number of sites as well as visitors viewing content dedicated to socioeconomic development needs, including education, health care, the environment, and promotion of entrepreneurship.[20] Clearly, upgrading Arabic digital content would benefit greatly from enlightened national and regional policies, aimed at enhancing infrastructures, further reducing access costs and promoting content that is relevant to socioeconomic development, armed with suitable incentives. It is only then that the Arab countries' capacity for using the Internet as a tool in the assimilation, adaptation, dissemination, and, indeed, creation of knowledge, may become tangible.

Recent advances in Arabic translation, at least to and from the English language, promise corresponding expansion in the use of the Internet as well as a variety of e-applications, particularly concerning distance learning and publishing. For similar advances to take place in e-government and e-commerce applications, however, obstacles rest with government regulations lagging behind many other countries around the world.

## 5.3 CLIMATE FOR BUSINESS IN THE ARAB COUNTRIES

The Arab region's strategic position, at the crossroad trading routes between three major continents, was crucial in shaping a renowned entrepreneurial tradition among its people. Entrepreneurship in the region suffered numerous setbacks over the past centuries as various major geopolitical transformations with colonial nations invading the globe and technological transformations transformed various aspects of economic activity.

More recently, rules and regulations imposed by successive regimes in several Arab countries played an important part in attenuating the entrepreneurial spirit of the region's population. Thus, the latter half of the twentieth century saw the implementation of regulatory regimes in Arab countries such as Iraq,

Syria, and Egypt, that effectively stifled entrepreneurship. The overbearing role of the state in the Arab countries,[21] in all affairs concerning production and services, continues while untold benefits are bestowed upon select businesses, mostly linked to ruling families, powerful autocrats, and high-ranking army officers.

Among other factors, limited capacity for offering state employment to growing youthful populations in the Arab countries has prompted moves by Arab governments to run enterprise incubation schemes, some with focus on novel technologies, including software development. However, most of the incubators that have been established possess limited facilities and could certainly enjoy much greater support, particularly in terms of funding. In essence, as the following paragraphs will attempt to show, that the overall business climate of Arab countries, with the exception of selected GCC members, remains less than conducive to business enterprise creation and growth.

### 5.3.1 Ease of Doing Business Rankings of the Arab Countries

Table A.5.6 presents rankings attained by 20 Arab countries on the basis of a composite indicator that measures the ease of doing business (EoDB), as well as the rankings assigned on a set of its subindices for the years 2011 and 2017.[22,23] Fig. 5.7 presents a summary view of the rankings of 20 Arab countries for which data was accessible in both 2011 and 2017.

Data presented in this annex table as well as the adjoining figure indicate that with regard to their EoDB rankings in 2011, the Arab countries could be grouped within three categories. The first might include all GCC countries, as well as Tunisia, with world rankings between 11 for Saudi Arabia and 74 for Kuwait. Next comes a group of countries including Egypt, Yemen, Jordan, Lebanon, and Morocco, with world rankings between 94 for Egypt and 114 for Morocco. The third group of countries, including Palestine, Algeria, Syria, and Iraq possessed rankings between 135 for Palestine and 166 for Iraq.

Several remarks might be made concerning the above groups as well as the relative rankings of their members. Thus, in 2011, Saudi Arabia, ranking 11th in terms of the EoDB, and followed at some distance by Bahrain and the United Arab Emirates, stood out among the Arab countries included in Table A.5.6. Indeed, in 2011, Saudi Arabia possessed a world rank ahead of several developed countries with regard to a variety of indicators including property registration, partly explaining its ranking in 11th position worldwide. However, by 2017 the United Arab Emirates took the lead, ranking 26th worldwide, while Saudi Arabia came 94th, preceded by three other GCC countries, namely Bahrain, Oman, and Qatar, in 63rd, 66th, and 83rd ranks worldwide.

Closer inspection of Table A.5.6 indicates that the United Arab Emirates' improved ranking in 2017 is due to its scores on six of the subindices used for ranking countries in that year, namely "Dealing with construction permits,"

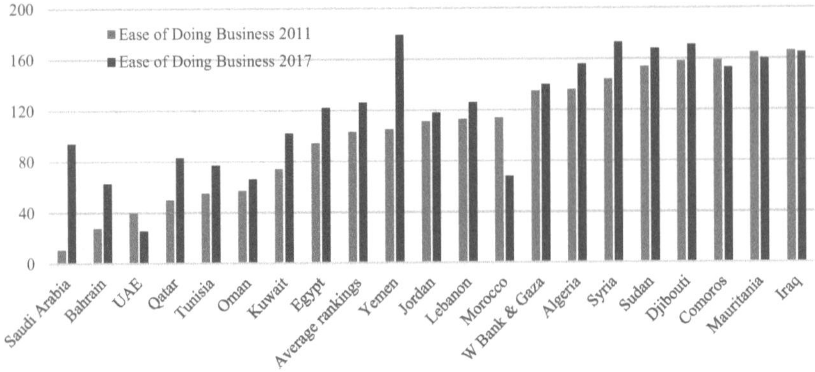


**FIGURE 5.7** Changes in "Ease of Doing Business" for Arab countries in 2011 and 2016. *The Global Competitiveness Report 2016–17.*

"Registering property," "Paying taxes," "Protecting minority investors," "Getting electricity,"[24] and "Enforcing contracts." On the other hand, the rankings attained by the GCC countries, Bahrain, Oman, and Qatar, in 2017, constitute declines with respect to their former status in 2011. Thus, while the GCC countries ranked considerably higher than the rest of the Arab countries, between 26th and 102nd positions worldwide, with the exception of the United Arab Emirates, all remaining five GCC countries, suffered considerable setbacks, with Saudi Arabia, Bahrain, Qatar, Kuwait, and Oman slipping back by 83, 35, 33, 28, and 9 positions. These setbacks may partly be due to the fact that other countries made greater improvements in their rankings. However, they are more likely to be due to setbacks on issues, such as targeting regulations linked to starting and closing businesses, credit availability, and protecting investors. Thus, average rankings linked to these issues are estimated at between 94 and 113, while those concerning other subindicators used in working out the EoDB rankings fall between 16 and 55.[25]

A notable feature of the more recent EoDB rankings is the fact that both Morocco and Tunisia ranked among the GCC countries as top performers among the Arab countries. This is essentially due to the former country's recent reforms allowing it to improve its position, by 46 points, up to 68th rank, while Tunisia suffered a decline of 22 points slipping down to its current 77th rank, but still placing it among the eight top performers, ahead of Qatar, Saudi Arabia, and Kuwait. Closer inspection of ranks allocated to EoDB subindices indicates that, in Morocco's case, it was with regard to protecting investors, paying taxes, enforcing contracts, and registering property that retarded its ranking in 2011. Morocco was able to improve its rankings on all of these subindices as well as three of the remaining subindices, namely, "Starting a business," "Trading across borders," and "Dealing with a construction permit," pulling up into 68th position by 2017. However, its ranking deteriorated with regard to the subindex designed to measure ease of getting credit.

Coming 113th in the 2011 EoDB rankings, Lebanon's position was quite disappointing, considering the country's traditional position as haven for free enterprise in the region and its urgent need to overcome a variety of socioeconomic hurdles. As Table A.5.6 indicates, Lebanon's unexpectedly low EoDB ranking in 2011 is primarily due to difficulties that businesses seem to find in obtaining construction permits as well as protecting investors, enforcing contracts, and declaring a business closed. To complicate matters, a review of rankings attained by Lebanon on the subindices included in Table A.5.6, seems to indicate that little progress was made in resolving obstacles in the above domains since Lebanon's world rank went down further to 126th position in 2017.

At any rate, Lebanon is joined by two other countries of the Arab Levant, namely Jordan and Palestine, which sustained smaller declines in their rankings than some of their neighbors; as Lebanon's rank slipped by 13 positions, while Jordan and Palestine were set back by 7 and 5 ranks, to occupy 118th and 140th positions, respectively.

At 94, Egypt's ranking in 2011 was already somewhat low and certainly insufficient for tackling its unsatisfactory unemployment figures and living standards. Declines in this country's performance on all but one subindex, namely that concerning getting a construction permit, pushed it down by 28 ranks to 122nd position worldwide. The largest setback appears to have been with regard to the subindex designed to measure ease of trading across boundaries, on which Egypt's ranking went from 21st in 2011 to 168th in 2017. Rankings witnessed considerable declines on two other subindices, namely those intended to evaluate protection of minority investors and paying taxes, with Egypt's rankings declining between 2011 and 2017 by 40 and 25, positions, respectively.

The dismal rankings attained by Algeria, Syria, and Iraq, even back in 2011, are noteworthy. The average of these countries' rankings across all subindicators fell below those of Yemen, and Palestine, both of which have always suffered severe socioeconomic hardships over the years. Thus, Algeria, despite the relative political stability it enjoyed over several years past and its dire need for creating employment opportunities for its large youthful population, is in need of drastically improving its ranking on almost all subindicators that go in determining its overall EoDB ranking. With the exception of subindicators that measure legal and regulatory instruments designed to protect investors and procedures required for closing a business, Algeria's 2011 rankings for all other subindices fall between 113 and 168, with seven of these subindicators below 120, granting it 136th rank on the overall EoDB Index.

In 2017, however, things became even worse with an overall EoDB rank of 156 worldwide and its rankings on six of the subindices falling between 175th and 142nd positions worldwide. In fact, improvements seem to have been made by Algeria between 2011 and 2017, including gains of 3, 8, 13, 25, and 36 points in its ranking on the subindices developed to measure ease of registering property, starting a business, paying taxes, enforcing contracts, and dealing with construction permits, respectively. However, these improvements could not compensate for losses sustained on three other subindices, namely those designed to measure ease of getting credit, trading across borders, and the degree of protection provided to minor investors, estimated at 37, 54, and 99 positions, respectively.

Iraq and Syria, on the other hand, are two countries whose entrepreneurial traditions and possession of reasonable physical infrastructures as well as a sizeable stock of human capital should qualify them for much higher rankings. While Iraq's 166 ranking might be at least partly explained away by political upheaval, it had been going through over the past three decades or more that of Syria might be less readily reconciled with the fact that it had enjoyed political calm as well as relatively stable economic conditions for many years past. At any rate, both Iraq and Syria ranked far below Turkey, whose EoDB ranking was estimated at 65 in 2011, and considerably lower than that of Iran which ranked 129th. Needless to say, the rankings achieved by both Iraq and Syria remain rather dismal with the former country having improved its status by one rank between 2011 and 2017 and the latter country's rank diving to 173rd position worldwide.

The position attained by Yemen with regard to the EoDB indicator in 2011 and its subsequent decline is also noteworthy. Thus, in that year Yemen achieved a ranking of 105th worldwide on the composite E0DB indicator, ahead of several other countries, including Lebanon, Morocco, and Jordan. Indeed, had Yemen been able to improve its position on some subindicators, such as "Getting credit," "Paying taxes," and "Protecting investors," it would have scored far higher than it eventually did. However, its rank by 2017 had fallen down to 179, inevitably due to the armed conflict that has raged through most of the country's territories, particularly, its capital, several of its main cities and towns as well as its land and sea trading outlets.

### 5.3.2 Ease of Doing Business in the Arab Countries Compared to Their Closest Neighbors

The Arab countries' five closest neighbors occupy rankings that are superior to many Arab countries, despite the declines in rankings that some of them sustained between 2011 and 2017. In particular, Cyprus followed by Israel occupied second and third ranks on a regional list of EoDB rankings, following the United Arab Emirates. On the same list, Turkey comes seventh immediately behind Morocco, while Malta comes eighth just ahead of Tunisia. Iran, on the other hand, ranks 14th ahead of several Arab countries, including Egypt and Lebanon.

A review of changes that took place with regard to doing business in countries within the immediate vicinity of several Arab countries, namely in Cyprus, Iran, Israel, Malta, and Turkey, reveals that with the exception of Iran all of the Arab countries' neighbors witnessed limited losses in their EoDB rankings. The largest decline by 23 ranks was sustained by Israel, apparently due to greater difficulties met in getting credit and trading across borders. Thus, Israel's world ranking in relation to subindices designed to measure these two attributes, declined by around 40 and 50 points between 2011 and 2017. In Turkey's case, declining ranks on subindices designed to evaluate procedures for starting a business, registering property, getting credit, and paying taxes were partly offset by gains in dealing with construction permits and protection of minor investors, resulting in a drop in its EoDB ranking by 4 points. A somewhat similar situation was faced by Cyprus, where considerable declines in rankings linked to starting a business, dealing with construction permits and trading across borders were compensated in past by improved protection on investors, resulting in an overall drop of 8 points on this country's EoDB world ranking.

Iran, on the other hand, made large gains in rankings linked to dealing with construction permits, registering property, offsetting considerable losses on other attributes, including procedures involved in starting a business and trading across borders. As a result, its world EoDB ranking moved up by 9 points.

To summarize, most countries in the region saw their EoDB rankings decline between 2011 and 2017. Arab countries that saw the most serious declines included Saudi Arabia, Yemen, Bahrain, Qatar, Syria, Egypt, Kuwait, Tunisia, Algeria, and Sudan; witnessing declines of between 83 ranks in Saudi Arabia and 14 ranks in Sudan.

Countries that witnessed acute political upheaval and armed conflict, such as Syria, Iraq, Libya, and Yemen ranked 173rd, 165th, 188th, and 179th, respectively, among the lowest ranking 20 countries in the world. Based on available records two of these countries, namely Syria and Yemen suffered the most drastic declines in their EoDB rankings, dropping by 29 and 74 positions, from their previous rankings in 2011. These two countries are now joined by a group of other countries with historically poor performance according to the same indicator, namely Djibouti, Sudan, and Somalia, which ranked in 171st, 168th, and 190th positions, respectively.

That Israel, Cyprus, and Turkey also witnessed declines, albeit less drastic, in their world rankings than those seen by their Arab neighbors, might be taken as an indication of global factors, such as the economic recession. However, the fact that oil-producing Arab countries directly involved in armed conflicts and political turmoil head the list with EoDB rankings of Saudi Arabia, Yemen, Bahrain, Qatar, and Syria declining by 83, 74, 35, 33, and 29 positions, would point elsewhere.

On the other hand, the fact that other countries, such as Morocco, the United Arab Emirates, and even Iran managed to climb up by 46, 14, and 9 ranks, despite global and regional conditions, must be credited to policies implemented by these countries with reference to a variety of regulatory arrangements that seem to vary from one country to another, with common factors including simplifying regulations concerning starting a business, paying taxes, and trading across borders.

An overview of the state of doing business in the Arab countries would indicate that they generally ranked lowest with regard to getting credit and in relation to starting a business enterprise, both in 2011 and in 2017. This is essentially due to the complexity of procedures and time required for obtaining bank loans, venture capital, and other sources of enterprise funding. Enforcing contracts came next as an obstacle in 2011 but was replaced by "Protecting investors" in 2017 as the penultimate rank for the Arab countries taken together.

In 2011, the average ranking based on the "Starting a business" subindex for all 20 Arab countries listed in Table A.5.6 came third from the bottom of the list for all average rankings. It was replaced, quite understandably, by trading across borders as the third worst ranking in 2017. Still, "Starting a business" retained a pretty low ranking in 2017, in fourth, rather than third lowest average ranking in 2011, which must be linked to the complex bureaucratic procedures and time taken to get a business enterprise going.

On the other hand, the Arab countries' average world ranking is rather high in relation to paying taxes. This is probably due to the fact that many GCC countries require minimal taxes and even exempt a variety of enterprise operations from taxes altogether.

## 5.4 CONCLUDING REMARKS

Based on the scores and world rankings discussed above, the Arab countries will all need to radically revise a range of policies and measures directed at building capacity for innovation and facilitating the creation and efficient functioning of business enterprises.

In particular, the Arab countries would do well to emphasize measures aimed at improving the quality of education and contents of curricula at all levels, closer cooperation between universities and business enterprises in the private as well as the public sector, higher levels of support for R&D activities across the board. With regard to their business ecologies, the main requirements are for simplification of legislative and regulatory arrangements throughout, but particularly with regard to access to credit, enforcing contracts, and starting and folding up business enterprises.

However, while reforms aimed at improving their scores and rankings linked to the above composite indicators and their subindices should in principle provide certain economic rewards, they may not be expected to deliver the wide range of urgently needed social outcomes, in terms of creating jobs and reducing poverty. This is essentially due to the fact that compound indicators and their constituent indices are intrinsically biased toward immediate commercial rather than overall socioeconomic outcomes. Assuming that socioeconomic benefits once achieved would translate into widespread social development has been proved time and again in the Arab countries and worldwide. Indeed, given predominant governance models in the region, improved performance in line with the narrow scope set for some of the above indicators might well improve a country's position on global or regional innovation and competitiveness ranking schemes while greater damage is sustained in other domains.

With particular regard to tangible outcomes of innovation, indicators used in evaluating national performance, as well as in the design of national visions and innovation policies are generally concerned with supply-side conditions, activities, capabilities, and interactions. Additionally, both justifications and instruments for innovation policies have been dominated by concerns exclusive to developed country requirements and ecologies.[26] Both factors have severely limited innovation policies' potential to deliver tangible and lasting goals on a variety of fronts, particularly in the case of developing countries enduring difficult conditions, such as witnessed by many Arab countries at present.

Far greater attention will need to be paid to demand side issues with developing country ecologies at the center of attention for innovation to deliver as well as it might toward socioeconomic goals. In the final analysis, nothing short of the widest reform of socioeconomic development policies at all levels would suffice for bringing full benefits of innovation and entrepreneurship to the Arab countries.

It is also well to emphasize that the future of building capacity/momentum for innovation has to be firmly established on the basis of fresh partnerships, among a wider variety of institutions, including local community bodies and civil society institutions as well as concerned government agencies, targeting problem areas as well as the uptake of promising technologies with specific development needs in sight.

Special attention will be needed in support of capacity building for innovation in stressed national and local economies where balance must be struck in allocating resources, whether for infrastructure building and renovation, or in education and vocational training, between ameliorating inadequate and failed governance systems and building novel structures and related support systems. Striking such a balance should constitute basis for surveys and further research, without which the outcomes of future capacity building efforts would certainly be ephemeral.

# ANNEX TABLES

**TABLE A.5.1** Rankings of Arab Countries and Five of Their Closest Neighbors According to the

| | Global Competitiveness Index | Pillars | | | | |
|---|---|---|---|---|---|---|
| | | Institutions | Infrastructure | Macroeconomic Environment | Health and Primary Education | Higher Education and Training |
| **Rankings for Arab Countries** | | | | | | |
| Algeria | 87 | 99 | 100 | 63 | 73 | 96 |
| Bahrain | 48 | 25 | 32 | 113 | 34 | 44 |
| Egypt | 115 | 87 | 96 | 134 | 89 | 112 |
| Jordan | 63 | 34 | 56 | 118 | 80 | 51 |
| Kuwait | 38 | 59 | 52 | 6 | 76 | 94 |
| Lebanon | 101 | 119 | 117 | 136 | 52 | 66 |
| Mauritania | 137 | 135 | 129 | 106 | 130 | 138 |
| Morocco | 70 | 50 | 58 | 49 | 77 | 104 |
| Oman | 66 | 28 | 38 | 81 | 69 | 85 |
| Qatar | 18 | 10 | 18 | 2 | 27 | 30 |
| Saudi Arabia | 29 | 24 | 31 | 68 | 51 | 46 |
| Tunisia | 95 | 78 | 83 | 99 | 59 | 93 |
| UAE | 16 | 7 | 4 | 38 | 40 | 34 |
| Yemen | 138 | 137 | 136 | 138 | 117 | 136 |
| **Rankings for Five of the Arab Countries' Closest Neighbors** | | | | | | |
| Cyprus | 83 | 62 | 50 | 115 | 36 | 55 |
| Iran | 76 | 90 | 59 | 72 | 49 | 60 |
| Israel | 24 | 31 | 28 | 48 | 28 | 24 |
| Malta | 40 | 38 | 40 | 21 | 18 | 38 |
| Turkey | 55 | 74 | 48 | 54 | 79 | 50 |

From World Economic Forum's The GCR 2016–17. Downloaded in March 2017 from: http://www3.weforum.org/docs/GCR2016-2017/05FullReport/TheGlobalCompetitivenessReport2016-2017_FINAL.pdf.

Global Competitiveness Index and Its Pillars (2016–17)

| Goods Market Efficiency | Labor Market Efficiency | Financial Market Development | Technological Readiness | Market Size | Business Sophistication | Innovation |
|---|---|---|---|---|---|---|
| | | | | | | |
| 133 | 132 | 132 | 108 | 36 | 121 | 112 |
| 22 | 35 | 43 | 37 | 92 | 33 | 45 |
| 112 | 135 | 111 | 99 | 25 | 85 | 122 |
| 43 | 87 | 68 | 75 | 75 | 36 | 40 |
| 85 | 115 | 65 | 60 | 51 | 61 | 110 |
| 55 | 104 | 69 | 72 | 76 | 50 | 58 |
| 136 | 131 | 137 | 133 | 128 | 138 | 137 |
| 64 | 124 | 83 | 81 | 55 | 76 | 96 |
| 51 | 82 | 55 | 57 | 68 | 66 | 76 |
| 7 | 17 | 21 | 33 | 50 | 18 | 18 |
| 41 | 65 | 47 | 41 | 14 | 31 | 42 |
| 113 | 133 | 119 | 80 | 69 | 101 | 104 |
| 3 | 11 | 28 | 18 | 27 | 13 | 25 |
| 131 | 137 | 138 | 136 | 89 | 129 | 138 |
| | | | | | | |
| 33 | 48 | 20 | 51 | 114 | 62 | 85 |
| 111 | 134 | 131 | 97 | 19 | 109 | 89 |
| 32 | 21 | 19 | 22 | 57 | 21 | 2 |
| 30 | 41 | 41 | 20 | 126 | 40 | 41 |
| 52 | 126 | 82 | 67 | 17 | 65 | 71 |

**TABLE A.5.2** Scores of 14 Arab Countries on the Composite Innovation Index and

| | Composite Innovation Index[c] | | Percentage Difference Between 2012 and 2016 | Capacity for Innovation[c] | | Quality of Research Institutes[c] | |
|---|---|---|---|---|---|---|---|
| | 2012–13 | 2016–17 | | 2012–13 | 2016–17 | 2012–13 | 2016–17 |
| Algeria | 3.0 | 2.9 | −3.3 | 2.9 | 3.7 | 4.1 | 3.4 |
| Bahrain | 3.1 | 3.6 | 16.1 | 2.5 | 4.1 | 3.1 | 3.7 |
| Egypt | 2.8 | 2.7 | −3.6 | 3.0 | 3.1 | 2.9 | 2.6 |
| Jordan | 3.3 | 3.7 | 12.1 | 3.3 | 4.4 | 3.5 | 4.0 |
| Kuwait | 2.8 | 3.0 | 7.1 | 2.6 | 3.8 | 3.2 | 3.3 |
| Lebanon | 2.7 | 3.4 | 25.9 | 2.6 | 4.6 | 2.5 | 3.5 |
| Libya | 2.5 | – | – | 2.5 | – | 2.7 | – |
| Mauritania | 2.7 | 2.2 | −18.5 | 2.7 | 2.1 | 2.7 | 2.1 |
| Morocco | 3.0 | 3.1 | 3.3 | 2.6 | 3.8 | 3.2 | 3.1 |
| Oman | 3.4 | 3.3 | −2.9 | 3.4 | 3.8 | 3.5 | 3.2 |
| Qatar | 4.7 | 4.9 | 4.3 | 4.6 | 5.2 | 5.8 | 5.6 |
| Saudi Arabia | 4.0 | 3.7 | −7.5 | 3.9 | 4.1 | 4.5 | 4.0 |
| Syria | 2.5 | – | – | 2.1 | – | 2.6 | – |
| Tunisia[b] | 3.6 | 3.0 | −16.7 | 3.4 | 3.8 | 4.0 | 3.2 |
| UAE | 4.2 | 4.6 | 9.5 | 4.0 | 5.4 | 4.6 | 4.9 |
| Yemen | 1.9 | 2.2 | 15.8 | 1.9 | 3.1 | 1.9 | 1.8 |
| Average scores[a] | 3.34 | 3.31 | −0.9 | 3.10 | 3.93 | 3.54 | 3.46 |
| Differences in average rankings | −0.3 | | – | 0.83 | | −0.08 | |

[a]*Average scores based on data from both the 2012–13 and 2016–17 GCRs, excluding scores for both Libya and Syria, for which no data was available from the 2016–17 WEF Report.*
[b]*Data included for Tunisia are taken from The World Economic Forum's GCR 2011–12.*
[c]*Survey questions requested responses on a scale of 1–7, where an answer of 1 corresponds to the worst, while 7 relates to the best possible outcome. Country scores are reported with a precision of one decimal point. However, exact figures are used to determine rankings.*
From World Economic Forum's GCRs 2012–13. Downloaded in March 2017 from http://www3.weforum.org/docs/WEF_GlobalCompetitivenessReport_2012-13.pdf; World Economic Forum's The GCR 2016–17. Downloaded in March 2017 from: http://www3.weforum.org/docs/GCR2016-2017/05FullReport/TheGlobalCompetitivenessReport2016-2017_FINAL.pdf.

## Component Indices (2012–13 and 2016–17)

| Company R&D Spending[c] | | University–Industry Collaboration[c] | | Government Procurement of Advanced Technologies[c] | | Availability of Scientists & Engineers[c] | | PCT Patent Applications | | Average Percentage Differences Between Scores |
|---|---|---|---|---|---|---|---|---|---|---|
| 2012–13 | 2016–17 | 2012–13 | 2016–17 | 2012–13 | 2016–17 | 2012–13 | 2016–17 | 2012–13 | 2016–17 | |
| 2.9 | 2.8 | 3.8 | 2.7 | 2.6 | 2.9 | 3.9 | 3.8 | 1.1 | 0.2 | 4.3 |
| 2.7 | 3.4 | 3.0 | 3.7 | 4.4 | 4.0 | 4.2 | 4.4 | 2.1 | 3.1 | −3.3 |
| 2.6 | 2.4 | 2.7 | 2.4 | 3.3 | 3.2 | 4.2 | 4.3 | 0.6 | 0.8 | −13.2 |
| 2.8 | 3.5 | 3.3 | 3.8 | 3.6 | 3.6 | 5.2 | 5.1 | 0.5 | 0.8 | −2.2 |
| 2.7 | 2.8 | 3.0 | 2.6 | 2.9 | 3.1 | 3.9 | 3.7 | 0.4 | 0.3 | 2.2 |
| 2.6 | 3.2 | 3.3 | 3.6 | 2.3 | 2.9 | 4.5 | 5.0 | 1.1 | 1.5 | 11.2 |
| 2.2 | – | 2.5 | – | 3.0 | – | 3.4 | – | 0.5 | – | – |
| 2.6 | 1.9 | 2.6 | 3.3 | 3.4 | 2.4 | 3.5 | 2.3 | 0.0 | 0.0 | −1.5 |
| 2.6 | 3.0 | 3.0 | 3.1 | 3.6 | 3.0 | 4.5 | 4.0 | 0.7 | 1.5 | −0.7 |
| 3.4 | 2.9 | 3.8 | 3.6 | 4.4 | 3.7 | 3.7 | 3.8 | 0.4 | 0.4 | −8.0 |
| 4.6 | 5.2 | 5.4 | 5.2 | 5.8 | 5.3 | 5.3 | 5.5 | 1.3 | 22.5 | −0.3 |
| 4.2 | 3.5 | 4.5 | 3.5 | 4.8 | 4.3 | 4.8 | 4.7 | 2.2 | 6.2 | −6.3 |
| 2.1 | – | 2.4 | – | 3.0 | – | 4.1 | – | 0.0 | – | – |
| 3.4 | 2.9 | 3.7 | 3.0 | 4.0 | 2.7 | 5.4 | 4.3 | 0.2 | 0.7 | −28.2 |
| 4.2 | 4.5 | 4.6 | 4.5 | 5.0 | 5.6 | 5.0 | 5.4 | 4.5 | 6.8 | 3.8 |
| 1.7 | 2.2 | 1.9 | 1.9 | 2.1 | 2.3 | 2.7 | 2.8 | 0.0 | 0.0 | −0.5 |
| 3.07 | 3.16 | 3.47 | 3.35 | 3.73 | 3.50 | 4.34 | 4.22 | 1.08 | 3.20 | −3.1 |
| 0.09 | | −0.12 | | −0.23 | | −0.12 | | 2.12 | | – |

**TABLE A.5.3** Rankings of 16 Arab Countries on the Composite Innovation Index and

| | Composite Innovation Index | | Capacity for Innovation | | Quality of Research Institutes | | Company R&D Spending | |
|---|---|---|---|---|---|---|---|---|
| | 2012–13 | 2016–17 | 2012–13 | 2016–17 | 2012–13 | 2016–17 | 2012–13 | 2016–17 |
| Algeria | 141 | 112 | 143 | 112 | 141 | 99 | 143 | 113 |
| Bahrain | 72 | 45 | 117 | 65 | 107 | 75 | 113 | 57 |
| Egypt | 109 | 122 | 80 | 135 | 114 | 128 | 116 | 133 |
| Jordan | 57 | 40 | 57 | 43 | 77 | 57 | 95 | 52 |
| Kuwait | 108 | 110 | 113 | 93 | 103 | 104 | 112 | 115 |
| Lebanon | 119 | 58 | 114 | 38 | 129 | 91 | 122 | 75 |
| Libya | 129 | – | 123 | – | 122 | – | 138 | – |
| Mauritania | 121 | 137 | 108 | 138 | 121 | 137 | 114 | 138 |
| Morocco | 97 | 96 | 115 | 94 | 104 | 112 | 119 | 95 |
| Oman | 47 | 76 | 50 | 97 | 74 | 105 | 45 | 106 |
| Qatar | 19 | 18 | 18 | 19 | 5 | 14 | 18 | 11 |
| Saudi Arabia | 29 | 42 | 29 | 72 | 37 | 58 | 23 | 48 |
| Syria[b] | 125 | – | 134 | – | 126 | – | 136 | – |
| Tunisia[b] | 37 | 104 | 44 | 99 | 52 | 111 | 42 | 109 |
| UAE | 28 | 25 | 27 | 15 | 35 | 27 | 22 | 22 |
| Yemen | 144 | 138 | 142 | 136 | 143 | 138 | 144 | 137 |
| Average rank[a] | 80.6 | 80.2 | 82.6 | 82.6 | 88.7 | 89.7 | 87.8 | 86.5 |

[a]*Average rankings based on data from both the 2012–13 and 2016–17 GCRs, exclude rankings for both Libya and Syria.*
[b]*Data included for both Syria and Tunisia are taken from The World Economic Forum's GCR 2011–12.*
[c]*The GCI Report 2016–17 ranked 138 countries, while the 2012–13 report ranked 144 countries.*
From World Economic Forum's GCRs 2012–13. Downloaded in March 2017 from: http://www3.weforum.org/docs/WEF_GlobalCompetitivenessReport_2012-13.pdf;
World Economic Forum's The GCR 2016–17. Downloaded in March 2017 from: http://www3.weforum.org/docs/GCR2016-2017/05FullReport/TheGlobalCompetitivenessReport2016-2017_FINAL.pdf.

**Component Indices (2012–13 and 2016–17)[c]**

| University–Industry Collaboration | | Government Procurement of Advanced Technologies | | Availability of Scientists and Engineers | | PCT Patent Applications | | Average of Differences Between Rankings |
|---|---|---|---|---|---|---|---|---|
| 2012–13 | 2016–17 | 2012–13 | 2016–17 | 2012–13 | 2016–17 | 2012–13 | 2016–17 | |
| 144 | 120 | 142 | 105 | 72 | 81 | 91 | 94 | 22.6 |
| 115 | 44 | 20 | 18 | 62 | 42 | 52 | 55 | 32.1 |
| 128 | 137 | 95 | 72 | 61 | 46 | 73 | 74 | −8.9 |
| 93 | 38 | 69 | 43 | 11 | 13 | 74 | 72 | 21.9 |
| 120 | 125 | 123 | 75 | 77 | 96 | 82 | 87 | 4.1 |
| 95 | 50 | 141 | 101 | 37 | 18 | 67 | 67 | 40.8 |
| 133 | – | 118 | – | 118 | – | 75 | – | – |
| 129 | 81 | 82 | 133 | 114 | 138 | 119 | 121 | −14.4 |
| 116 | 100 | 72 | 86 | 38 | 67 | 71 | 66 | 2.0 |
| 54 | 53 | 17 | 35 | 93 | 79 | 80 | 82 | −21.6 |
| 9 | 10 | 1 | 2 | 8 | 4 | 62 | 28 | 4.3 |
| 31 | 56 | 5 | 13 | 26 | 33 | 49 | 48 | −17.6 |
| 133 | – | 116 | – | 63 | – | 90 | – | – |
| 58 | 107 | 42 | 116 | 8 | 48 | 73 | 76 | −51.8 |
| 26 | 25 | 3 | 1 | 15 | 5 | 45 | 46 | 4.4 |
| 143 | 138 | 143 | 135 | 140 | 132 | 106 | 121 | 3.1 |
| 89.7 | 77.4 | 68.2 | 66.8 | 54.4 | 57.3 | 74.6 | 71.4 | −0.6 |

**TABLE A.5.4 Scores of 14 Arab Countries on Technological Readiness and Its Component**

| | Technological Readiness[b] | | Availability of Latest Technologies[b] | | Firm Level Technology Absorption[b] | | FDI and Technology Transfer[b] | |
|---|---|---|---|---|---|---|---|---|
| | 2011–12 | 2016–17 | 2011–12 | 2016–17 | 2011–12 | 2016–17 | 2011–12 | 2016–17 |
| Algeria | 2.8 | 3.1 | 4.0 | 3.7 | 3.7 | 3.6 | 3.8 | 3.6 |
| Bahrain | 4.5 | 5.2 | 6.1 | 5.6 | 5.7 | 5.1 | 5.3 | 4.9 |
| Egypt | 3.3 | 3.3 | 4.3 | 3.9 | 4.7 | 3.8 | 4.7 | 4.4 |
| Jordan | 3.8 | 3.9 | 5.5 | 5.4 | 5.4 | 5.0 | 4.9 | 4.6 |
| Kuwait | 3.7 | 4.3 | 5.2 | 5.0 | 5.4 | 4.4 | 3.4 | 3.6 |
| Lebanon | 3.4 | 4.0 | 4.8 | 4.3 | 4.8 | 4.5 | 4.0 | 3.5 |
| Libya | 3.1 | – | 3.9 | – | 4.3 | – | 3.6 | – |
| Mauritania | 2.5 | 2.3 | 4.0 | 3.3 | 3.6 | 3.4 | 3.2 | 2.4 |
| Morocco | 3.7 | 3.7 | 5.1 | 5.0 | 4.7 | 4.6 | 4.9 | 4.5 |
| Oman | 4.1 | 4.4 | 5.5 | 4.9 | 5.2 | 4.7 | 4.9 | 4.1 |
| Qatar | 4.7 | 5.3 | 6.0 | 6.0 | 6.0 | 5.7 | 6.1 | 5.4 |
| Saudi Arabia | 4.3 | 5.0 | 5.8 | 5.4 | 5.7 | 4.9 | 5.5 | 4.8 |
| Syria | 3.1 | – | 4.1 | – | 4.8 | – | 4.1 | – |
| Tunisia | 3.8 | 3.7 | 5.3 | 4.6 | 5.1 | 4.1 | 5.2 | 4.3 |
| UAE | 4.9 | 5.8 | 6.1 | 6.2 | 5.9 | 5.8 | 5.5 | 5.7 |
| Yemen | 2.4 | 2.2 | 3.5 | 2.9 | 4.0 | 3.3 | 2.7 | 2.8 |
| Average scores[a] | 3.7 | 4.0 | 5.1 | 4.7 | 5.0 | 4.5 | 4.6 | 4.2 |
| Difference in average scores | 0.31 | | –0.36 | | –0.50 | | -0.39 | |

[a]*Average rankings based on data from both the 2012–13 and 2016–17 GCRs, exclude rankings for both Libya and Syria.*
[b]*Survey questions requested responses on a scale of 1–7, where an answer of 1 corresponds to the worst, while 7 relates to the best possible outcome. Country scores are reported with a precision of one decimal point. However, exact figures are used to determine rankings.*
From World Economic Forum's The GCR 2016–17. Downloaded in March 2017 from: http://www3.weforum.org/docs/GCR2016-2017/05FullReport/TheGlobalCompetitivenessReport2016-2017_FINAL.pdf; The World Economic Forum's GCRs 2011–12. Downloaded in March 2017 from: http://www3.weforum.org/docs/WEF_GlobalCompetitivenessReport_2011-12.pdf.

Indices (2011–12 and 2016–17)

| Differences Between Average Values of TR Subindices 1–3 in 2012 and 2016 | Internet Users/100 of the Population | | Broadband Internet Subscriptions /100 of the Population | | Internet Bandwidth, kbs/capita | | Differences Between Average Values of TR Subindices 4–6 in 2012 and 2016 |
|---|---|---|---|---|---|---|---|
| | 2011–12 | 2016–17 | 2011–12 | 2016–17 | 2011–12 | 2016–17 | |
| −0.33 | 12.5 | 38.2 | 2.5 | 5.6 | 1.0 | 30.1 | 19.3 |
| −0.50 | 55.0 | 93.5 | 12.2 | 18.6 | 7.9 | 47.2 | 28.07 |
| −0.72 | 26.7 | 35.9 | 1.8 | 4.5 | 18 | 11.3 | 1.73 |
| −0.54 | 38.0 | 53.4 | 3.2 | 4.2 | 2.4 | 27.5 | 13.83 |
| −0.34 | 38.3 | 82.1 | 1.7 | 1.4 | 3.7 | 48.6 | 29.27 |
| −0.46 | 31.0 | 74.0 | 4.7 | 22.8 | 0.6 | 27.3 | 29.47 |
| – | 17.0 | – | 1.1 | – | 1.6 | – | – |
| −0.75 | 3.0 | 15.2 | 0.2 | 0.2 | 0.1 | 1.5 | 4.53 |
| −0.45 | 49.0 | 57.1 | 1.6 | 3.4 | 2.3 | 18.3 | 8.63 |
| −0.68 | 62.6 | 74.2 | 1.9 | 5.6 | 3.1 | 59.8 | 24.00 |
| −0.43 | 69.0 | 92.9 | 9.2 | 10.1 | 13.9 | 80.6 | 30.50 |
| −0.64 | 41.0 | 69.6 | 5.5 | 12.0 | 11.6 | 88.7 | 37.40 |
| – | 20.7 | – | 0.3 | – | 0.3 | – | – |
| −1.03 | 36.8 | 48.5 | 4.6 | 4.3 | 4.9 | 33.8 | 13.43 |
| 0.04 | 78.0 | 91.2 | 10.5 | 12.8 | 26.6 | 107.9 | 32.27 |
| −0.60 | 10.9 | 25.1 | 0.3 | 1.5 | 0.1 | 2.5 | 5.93 |
| – | 39.4 | 60.8 | 4.3 | 7.6 | 6.9 | 41.8 | – |
| −0.53 | 21.36 | | 3.36 | | 34.92 | | 19.88 |

**TABLE A.5.5** Rankings of 14 Arab Countries on Technological Readiness and Its Component

| | Technological Readiness | | Availability of Latest Technologies | | Firm Level Technology Absorption | |
|---|---|---|---|---|---|---|
| | 2011–12 | 2016–17 | 2011–12 | 2016–17 | 2011–12 | 2016–17 |
| Algeria | 120 | 108 | 122 | 125 | 134 | 128 |
| Bahrain | 39 | 37 | 22 | 36 | 20 | 36 |
| Egypt | 95 | 99 | 110 | 117 | 78 | 121 |
| Jordan | 59 | 75 | 42 | 43 | 37 | 38 |
| Kuwait | 65 | 60 | 58 | 55 | 39 | 82 |
| Lebanon | 89 | 72 | 79 | 91 | 68 | 68 |
| Libya[b] | 110 | – | 125 | – | 108 | – |
| Mauritania | 137 | 133 | 124 | 132 | 140 | 134 |
| Morocco | 66 | 81 | 65 | 50 | 74 | 63 |
| Oman | 51 | 57 | 45 | 66 | 49 | 57 |
| Qatar | 33 | 33 | 31 | 21 | 7 | 11 |
| Saudi Arabia | 43 | 41 | 36 | 42 | 23 | 44 |
| Syria | 105 | – | 119 | – | 70 | – |
| Tunisia | 58 | 80 | 57 | 76 | 50 | 106 |
| UAE | 30 | 18 | 25 | 11 | 16 | 7 |
| Yemen | 139 | 136 | 139 | 136 | 123 | 136 |
| Average rank[c] | 73.1 | 73.6 | 68.2 | 71.5 | 61.3 | 73.6 |

[a]*The GCI Report 2016–17 ranked 138 countries, while the 2011–12 report ranked 142 countries.*
[b]*Figures quoted for Libya are taken from the GCR 2011–12.*
[c]*Average rankings based on data from both the 2012–13 and 2016–17 GCRs, exclude rankings for both Libya and Syria.*
From World Economic Forum's The GCR 2016–17. Downloaded in March 2017 from: http://www3.weforum.org/docs/GCR2016-2017/05FullReport/TheGlobalCompetitivenessReport2016-2017_FINAL.pdf; The World Economic Forum's GCRs 2011–12. Downloaded in March 2017 from: http://www3.weforum.org/docs/WEF_GlobalCompetitivenessReport_2011-12.pdf.

Indices (2011–12 and 2016–17)[a]

| FDI and Technology Transfer | | Internet Users/100 of the Population | | Broadband Internet Subscriptions/100 of the Population | | Internet Bandwidth, kbs/capita | | Average of Differences |
|---|---|---|---|---|---|---|---|---|
| 2011–12 | 2016–17 | 2011–12 | 2016–17 | 2011–12 | 2016–17 | 2011–12 | 2016–17 | |
| 123 | 121 | 103 | 95 | 87 | 84 | 90 | 80 | 4.3 |
| 18 | 33 | 41 | 5 | 45 | 48 | 57 | 65 | −3.3 |
| 67 | 71 | 86 | 96 | 91 | 87 | 86 | 105 | −13.2 |
| 51 | 47 | 67 | 72 | 80 | 89 | 80 | 81 | −2.2 |
| 135 | 116 | 66 | 27 | 93 | 105 | 71 | 64 | 2.2 |
| 112 | 123 | 79 | 38 | 73 | 40 | 98 | 82 | 11.2 |
| 136 | – | 103 | – | 100 | – | 82 | – | – |
| 140 | 138 | 132 | 125 | 118 | 121 | 122 | 135 | −1.5 |
| 54 | 58 | 49 | 67 | 95 | 93 | 82 | 92 | −0.7 |
| 46 | 85 | 36 | 37 | 89 | 83 | 73 | 58 | −8.0 |
| 2 | 7 | 29 | 8 | 56 | 69 | 39 | 50 | −0.3 |
| 8 | 36 | 59 | 49 | 67 | 63 | 45 | 42 | −6.3 |
| 106 | – | 95 | – | 111 | – | 110 | – | – |
| 25 | 81 | 69 | 82 | 74 | 88 | 66 | 77 | −28.2 |
| 10 | 3 | 19 | 11 | 49 | 60 | 28 | 32 | 3.8 |
| 142 | 135 | 111 | 103 | 110 | 104 | 117 | 131 | −0.5 |
| 66.6 | 75.3 | 67.6 | 58.2 | 80.5 | 81.0 | 75.3 | 78.1 | – |

**TABLE A.5.6** Ease of Doing Business Rankings for Arab and Five of Their Closest Neighboring

| | Ease of Doing Business | | Starting a Business | | Dealing With Construction Permit | | Registering Property | |
|---|---|---|---|---|---|---|---|---|
| | 2011 | 2017 | 2011 | 2017 | 2011 | 2017 | 2011 | 2017 |
| Algeria | 136 | 156 | 150 | 142 | 113 | 77 | 165 | 162 |
| Bahrain | 28 | 63 | 78 | 73 | 17 | 19 | 29 | 25 |
| Comoros | 159 | 153 | 168 | 161 | 68 | 92 | 99 | 90 |
| Djibouti | 158 | 171 | 175 | 172 | 125 | 120 | 140 | 168 |
| Egypt | 94 | 122 | 18 | 39 | 154 | 64 | 93 | 109 |
| Iraq | 166 | 165 | 174 | 164 | 102 | 104 | 96 | 115 |
| Jordan | 111 | 118 | 127 | 106 | 92 | 109 | 106 | 96 |
| Kuwait | 74 | 102 | 141 | 173 | 91 | 144 | 90 | 67 |
| Lebanon | 113 | 126 | 103 | 139 | 142 | 135 | 111 | 103 |
| Libya | – | 188 | – | 163 | – | 187 | – | 187 |
| Mauritania | 165 | 160 | 152 | 80 | 153 | 118 | 73 | 102 |
| Morocco | 114 | 68 | 82 | 40 | 98 | 18 | 124 | 87 |
| Oman | 57 | 66 | 76 | 32 | 70 | 52 | 21 | 35 |
| Qatar | 50 | 83 | 111 | 91 | 30 | 21 | 58 | 26 |
| Saudi Arabia | 11 | 94 | 13 | 147 | 14 | 15 | 1 | 32 |
| Somalia | – | 190 | – | 184 | – | 187 | – | 148 |
| Sudan | 154 | 168 | 121 | 156 | 139 | 145 | 40 | 89 |
| Syria | 144 | 173 | 134 | 136 | 134 | 187 | 80 | 154 |
| Tunisia | 55 | 77 | 48 | 103 | 106 | 59 | 64 | 92 |
| UAE | 40 | 26 | 46 | 53 | 26 | 4 | 4 | 11 |
| West Bank & Gaza | 135 | 140 | 173 | 169 | 157 | 157 | 76 | 93 |
| Yemen | 105 | 179 | 57 | 161 | 50 | 94 | 53 | 83 |
| Average rankings | 102.8 | 126.0 | 106.8 | 121.5 | 93.7 | 95.1 | 76.0 | 93.7 |
| Difference between averages | −23.2 | | −14.7 | | −1.3 | | −17.7 | |

| Countries[a] | | | | | | | | | |
|---|---|---|---|---|---|---|---|---|---|
| Getting Credit | | Protecting Investors | | Paying Taxes | | Trading Across Borders | | Enforcing Contracts | |
| 2011 | 2017 | 2011 | 2017 | 2011 | 2017 | 2011 | 2017 | 2011 | 2017 |
| 138 | 175 | 74 | 173 | 168 | 155 | 124 | 178 | 127 | 102 |
| 89 | 101 | 59 | 106 | 14 | 4 | 33 | 82 | 117 | 110 |
| 168 | 118 | 132 | 145 | 96 | 168 | 135 | 107 | 152 | 179 |
| 176 | 181 | 179 | 178 | 60 | 106 | 38 | 155 | 160 | 184 |
| 72 | 82 | 74 | 114 | 136 | 162 | 21 | 168 | 143 | 162 |
| 168 | 181 | 120 | 123 | 54 | 52 | 179 | 179 | 141 | 138 |
| 128 | 185 | 120 | 165 | 29 | 79 | 77 | 50 | 129 | 124 |
| 89 | 118 | 28 | 81 | 9 | 6 | 113 | 157 | 114 | 66 |
| 89 | 118 | 93 | 145 | 36 | 67 | 95 | 134 | 122 | 127 |
| – | 185 | – | 185 | – | 121 | – | 114 | – | 143 |
| 152 | 157 | 147 | 123 | 172 | 188 | 163 | 137 | 83 | 83 |
| 89 | 101 | 154 | 87 | 124 | 41 | 80 | 63 | 106 | 57 |
| 128 | 133 | 93 | 118 | 8 | 12 | 88 | 67 | 104 | 60 |
| 138 | 139 | 93 | 183 | 2 | 1 | 46 | 128 | 95 | 120 |
| 46 | 82 | 16 | 63 | 6 | 69 | 18 | 158 | 40 | 105 |
| – | 185 | – | 190 | – | 190 | – | 156 | – | 109 |
| 138 | 170 | 154 | 187 | 94 | 141 | 143 | 184 | 146 | 147 |
| 168 | 170 | 109 | 87 | 110 | 81 | 120 | 176 | 176 | 159 |
| 89 | 101 | 74 | 118 | 58 | 106 | 30 | 92 | 78 | 76 |
| 72 | 101 | 120 | 9 | 5 | 1 | 3 | 85 | 134 | 25 |
| 168 | 118 | 44 | 158 | 28 | 101 | 111 | 99 | 93 | 122 |
| 152 | 185 | 132 | 132 | 146 | 92 | 123 | 189 | 34 | 142 |
| 122.4 | 139.9 | 100.1 | 129.8 | 67.3 | 87.3 | 86.6 | 129.7 | 114.5 | 115.0 |
| −17.4 | | −29.7 | | −20.0 | | −43.1 | | −0.6 | |

*Continued*

**TABLE A.5.6** Ease of Doing Business Rankings for Arab and Five of Their Closest

| | Ease of Doing Business | | Starting a Business | | Dealing With Construction Permit | | Registering Property | |
|---|---|---|---|---|---|---|---|---|
| | 2011 | 2017 | 2011 | 2017 | 2011 | 2017 | 2011 | 2017 |
| **Data on Two Close Neighbors for Comparison** | | | | | | | | |
| Cyprus | 37 | 45 | 26 | 53 | 75 | 125 | 66 | 91 |
| Iran | 129 | 120 | 42 | 102 | 143 | 27 | 156 | 86 |
| Israel | 29 | 52 | 36 | 41 | 121 | 71 | 147 | 126 |
| Malta | – | 76 | – | 132 | – | 82 | – | 147 |
| Turkey | 65 | 69 | 63 | 79 | 137 | 102 | 38 | 54 |

[a]*TRI subindices in this table exclude "Getting electricity," "Closing a business," and "Resolving insolvency" due to mismatches between subindices used in determining TRI scores and rankings in 2011 and 2017.*
From World Bank and International Financial Corporation's Doing Business Statistics; Downloaded September 2011 from: http://www.doingbusiness.org/rankings; World Bank's Report "Doing Business 2017; Equal Opportunity for All." 14th Edition. Downloaded in March 2017 from: http://www.doingbusiness.org/~/media/WBG/DoingBusiness/Documents/Annual-Reports/English/DB17-Report.pdf.

Neighboring Countries[a]—cont'd

| Getting Credit | | Protecting Investors | | Paying Taxes | | Trading Across Borders | | Enforcing Contracts | |
|---|---|---|---|---|---|---|---|---|---|
| 2011 | 2017 | 2011 | 2017 | 2011 | 2017 | 2011 | 2017 | 2011 | 2017 |
| | | | | | | | | | |
| 72 | 62 | 93 | 27 | 32 | 34 | 19 | 45 | 104 | 109 |
| 89 | 101 | 167 | 165 | 115 | 100 | 131 | 170 | 49 | 70 |
| 6 | 44 | 5 | 9 | 82 | 96 | 10 | 59 | 96 | 89 |
| – | 139 | – | 32 | – | 33 | – | 40 | – | 58 |
| 72 | 82 | 59 | 22 | 75 | 128 | 76 | 70 | 26 | 33 |

**TABLE A.5.7** Internet Users and Internet Penetration in the Arab Countries and Five of Their Closest Neighbors

| | Users, in December 2000 | Internet Usage, March 31, 2017 (Millions) | % Population (Penetration) |
|---|---|---|---|
| Algeria | 50 | 18.85 | 45.2 |
| Bahrain | 40 | 1.279 | 90.1 |
| Comoros | 1.5 | 0.06 | 7.3 |
| Djibouti | 1.4 | 0.15 | 16.5 |
| Egypt | 450 | 34.80 | 36.5 |
| Iraq | 13 | 14.00 | 36.2 |
| Jordan | 127 | 5.70 | 72.4 |
| Kuwait | 150 | 3.20 | 78.1 |
| Lebanon | 300 | 4.577 | 75.8 |
| Libya | 10 | 2.80 | 43.7 |
| Mauritania | 5 | 0.71 | 16.7 |
| Morocco | 100 | 20.21 | 57.3 |
| Oman | 90 | 3.310 | 69.8 |
| Palestine | 35 | 3.008 | 44.7 |
| Qatar | 30 | 2.200 | 94.1 |
| Saudi Arabia | 200 | 20.814 | 63.6 |
| Somalia | 0.2 | 0.66 | 5.8 |
| Sudan | 30 | 10.89 | 25.8 |
| Syria | 30 | 5.502 | 29.1 |
| Tunisia | 100 | 5.80 | 50.5 |
| UAE | 735 | 8.515 | 90.6 |
| Yemen | 15 | 6.773 | 24.1 |
| Cyprus | – | 0.845 | 71.1 |
| Iran | 250,000 | 56.700 | 70.0 |
| Malta | – | 0.334 | 79.4 |
| Israel | 1,270,000 | 6.351 | 76.3 |
| Turkey | – | 46.283 | 59.6 |

From Internet World Statistics. Adapted from a table downloaded in June 2017 from: http://www.internetworldstats.com/stats.htm; World Economic Forum's The GCR 2016–17. Downloaded in March 2017 from: http://www3.weforum.org/docs/GCR2016-2017/05FullReport/TheGlobalCompetitivenessReport2016-2017_FINAL.pdf.

**TABLE A.5.8** Top 10 Languages Used on the Web[a] (Number of Internet Users by Language)

| Languages | Internet Users by Language (Millions) | Internet Penetration (% Age of Population) | Users Growth in Internet (%) Between 2000 and 2017 | Internet Users% of World Total (Participation) | Populations Speaking Language (2017 Estimate) (Millions) |
|---|---|---|---|---|---|
| English | 952.1 | 66.3 | 576.4 | 25.5 | 1434.9 |
| Chinese | 763.3 | 53.5 | 2262.8 | 20.4 | 1425.4 |
| Spanish | 293.8 | 57.6 | 1516.1 | 7.9 | 510.4 |
| Arabic | 173.5 | 42.5 | 6805.9 | 4.6 | 408.2 |
| Portuguese | 155.0 | 55.1 | 1946.4 | 4.1 | 281.6 |
| Indonesian/ Malaysian | 154.7 | 52.4 | 2599.7 | 4.1 | 295.1 |
| Japanese | 118.5 | 94.0 | 151.6 | 3.2 | 126.0 |
| Russian | 104.6 | 72.9 | 3272.7 | 2.8 | 143.4 |
| French | 100.6 | 24.8 | 738.3 | 2.7 | 405.6 |
| German | 84.0 | 88.4 | 205.1 | 2.2 | 94.9 |
| World's top 10 languages | 2900.0 | 56.6 | 877.2 | 77.5 | 5125.7 |
| Rest of the languages | 839.7 | 35.1 | 1207.5 | 22.5 | 2393.3 |
| Totals | 3739.7 | 49.7 | 936.0 | 100.0 | 7519.0 |

[a]*The above figures are March 31, 2017 estimates.*
From Internet World Statistics. Adapted from a table downloaded in June 2017 from: http://www.internetworldstats.com/stats.htm.

## ENDNOTES

1. See for example "Innovation for Development;" A discussion of the issues and an overview of work of the OECD directorate for science, technology and innovation, May 2012. Downloaded in April 2017 from: http://www.oecd.org/innovation/inno/50586251.pdf.
2. Details of pillars, subindicators, or indices used for computing the Global Competitiveness Index were obtained from World Economic Forum's Global Competitiveness Report 2016–17. Downloaded in March 2017 from: http://www3.weforum.org/docs/GCR2016-2017/05FullReport/TheGlobalCompetitivenessReport2016-2017_FINAL.pdf.
3. Smaller Arab countries, such as Comoros, Djibouti, and Somalia often go unmentioned in this report. Additionally, countries for which data is unavailable due to armed conflict are excluded. Thus Syria and Libya were not reported upon in the WEF GCI report for the year 2016–17.
4. "The Global Innovation Index 2015; Effective Innovation Policies for Development." Downloaded in March 2017 from: http://www.wipo.int/edocs/pubdocs/en/wipo_gii_2015.pdf.
5. The GII Report 2015 covers 141 countries and economies around the world. No information is provided by the report on several countries suffering armed conflict such as Syria, Iraq, and Libya. Information is also lacking with regard to the smaller economies worldwide, including, among the Arab League member countries, Somalia, Djibouti, and Comoros.
6. The Innovation Efficiency Ratio is the ratio of scores attained on a subindex measuring an innovation system's output over that achieved for the subindex score measuring the system's input. It assesses the effectiveness of innovation systems and policies. While a high IER score must be considered an indication of a well-functioning innovation system, it is to be remembered that relatively high IER scores might also be attained as a result of excessively low inputs, while outputs might be mediocre.
7. A similar view is obtained when a plot of the World Bank's innovation index was plotted against the GDP per capita values using 2011 values for both the innovation index and per capita GDP.
8. The pattern exhibited in Fig. 5.1 is consistent with previous attempts at correlating the Arab countries' capacity for innovation with their per capita gross domestic products. See for example "The Real Issues of the Middle East and the Arab Spring; Addressing Research, Innovation and Entrepreneurship;" T Andersson and A Djeflat, (Eds.) Chapter 8, titled "Research, Innovation, Entrepreneurship and the Rentier Culture in the Arab Countries." Omar Bizri. Figure 8.8, on page 216 in this book, attempted a correlation of 2011 figures for the World Bank's Knowledge Assessment Methodology's Innovation Index belonging to nearly 140 countries worldwide, including several Arab countries with per capita GDP data from the same year. Almost identical trend lines to those presented in figure (GII), above, resulted with R2 values close to those exhibited in Fig. 5.1; 0.675 for the Arab countries' trend line and 0.868 for the correlation concerning all other countries around the world.
9. Criticism is often voiced with regard to the survey approach in determining a sizeable proportion of scores ascribed to various subindices that go into evaluating pillar scores and ultimate country rankings. While such criticism is often justified, it does seem to ignore the fact that most decisions with economic and social implications are based on judgments by business people as well as other decision makers. Thus, while evaluations of subindices are based on verifiable and objective statistical data, the viewpoints of business people do carry significant weight and might be used, together with other factors, in judging current and future paths taken by national economies, with regard to building capacity for innovation and technological readiness, among other national development goals.
10. The patents subindex is based on the number of PCT patents filed per million of a country's population.
11. The score for both Libya and Syria reported by the Global Competitiveness Report 2011–12 and the Global Competitiveness Report 2011–12 were 2.5. Since the scores of both countries could not have improved in the intervening years, even the inclusion of their previous values would bring the Arab countries' average below its current value in 2016.

12. Gains achieved by Yemen on the value of its Composite Innovation Index as well as five related indices are rather difficult to explain given the current armed conflict that raged since 2013. High scores for some of the innovation indices are particularly incomprehensible, e.g., in reference to the 63.2% rise in this country's capacity for innovation and the 29.4% rise in company contributions to R&D, when available information, admittedly based on press reports, indicates that the Yemeni economy, has suffered an enormous downturn as a result of political turmoil and military operations. Further suspicions are raised with regard to Yemen's score on company contributions to R&D by the fact that no change is reported with regard to university–industry collaboration in R&D between 2012 and 2016.
13. It is tempting to link this increase in Lebanon's score on the availability of scientists and engineers to observations by media reports indicating that refugees from the fighting in Syria included a considerable proportion of scientists and engineers.
14. Additionally, Tunisia's ranking used in computing the change in its ranking corresponding to this subindex were taken from WEF data published for 2011.
15. Figures for Syria and Libya were not included in the Global Competitiveness Report 2016–17. In the 2011–12 issue of this report, Syria's technological readiness score was estimated as 3.1 and the 2012–13 issue of the same report reported Libya's score also as 3.1. Clearly, the scores of both countries could not have improved in the intervening years. Even the inclusion of their previous values would considerably depress the Arab countries' average below its value in 2016.
16. On account of armed conflict and security situations suffered by countries such as Iraq, Libya, and Syria. Additionally, the GCI reports on which the table is based does not provide data on three additional countries, namely Comoros, Djibouti, and Somalia.
17. More recent editions of the GCI Report have also included mobile broadband subscriptions.
18. This, despite the fact that inhabitants of the Arab countries in the Gulf region make up around 13% of the Arab countries' total populations. A likely explanation of this anomaly is that many Gulf country inhabitants are expatriate workers and enterprises with high rates of Internet access and utilization.
19. Incidentally, growth of Russian, Indonesian-Malaysian, and Chinese approximated 3273%, 2600%, and 2260%, respectively, during the same period.
20. "Third Report on Cultural Development;" Arab Thought Foundation. ISBN: 978-9953-0-1963-5, published 2010.
21. Lebanon still stands as a singular exception to this trend, while the public sector was far more pervasive in many other Arab countries during that period.
22. To rank countries around the world in terms of ease of doing business, the following subindicators were developed by and used by the International Financial Corporation (IFC) and the World Bank (WB): Starting a Business, Dealing with Construction Permits, Registering Property, Getting Credit, Protecting Investors, Paying Taxes, Trading cross Borders, Enforcing Contracts and Closing a Business. Available data for the EoDB indicator in 2011 covered only 16 Arab countries, while available values of this indicator in 2017 cover all 22 Arab countries, with additional subindicators dedicated to concerning ease of getting electricity.
23. EoDB statistics are published by the World Bank and International Financial Corporation. Statistics published by the World Bank and International Financial Corporation. Accessed in January 2017 at: http://www.doingbusiness.org/rankings.
24. This particular subindex is not included in Table A.5.6 since it was introduced in the 2017 EoDB Report with no equivalent in 2011.
25. Strictly speaking comparisons of EoDB rankings published in 2017 to those of 2011 may be somewhat flawed since an additional subindex was added in 2017. Nevertheless, such comparisons might be instructive, particularly with regard to rankings linked to subindices of the composite EoDB indicator that witnessed no changes in scope or definition between 2011 and 2017. Such comparisons might in fact help indicate points of strength and weakness in the Arab country's business ecologies, due to recent developments as well as measures taken by Arab governments with positive or negative impact upon their business environment.

26. "Local Needs, Global Challenges: The Meaning of Demand-Side Policies for Innovation and Development;" Jakob Edler, Manchester Institute of Innovation Research, Alliance Manchester Business School, University of Manchester. Published as part of the GII 2016 Report. Downloaded in March 2017 from: http://www.wipo.int/edocs/pubdocs/en/wipo_pub_gii_2016.pdf.

Chapter 6

# Arab Science, Technology, and Innovation Systems: Challenges, Policy Regimes, and Future Directions

## Chapter Outline




Science, Technology, Innovation, and Development in the Arab Countries
http://dx.doi.org/10.1016/B978-0-12-812577-9.00006-4

## 6.1 INTRODUCTION

That the Arab countries have consistently underperformed in many areas of science, technology and innovation (STI) is widely acknowledged and patently demonstrated by figures concerning both input and output parameters. Previous chapters repeatedly stressed this notion as they addressed various issues pertaining to some of the main aspects of STI systems and their activities in the Arab countries, including governance, policies, institutional settings, human and financial resources allocated, and output produced, as embodied in scientific papers and patents registered by the Arab countries during the past few years.

Through a synthesis of findings underlined in previous chapters with regard to the above cornerstones of the Arab countries' STI systems, the first part of this chapter will underline principal challenges these systems face in relation to governance, education, and due to poverty and inequality.

Secondly, an overall characterization is presented of attempts by several Arab countries at the formulation of long-term national visions and STI policies, often with the aim of transforming into knowledge economies, together with consideration of issues related to funding STI activities. The third part of the chapter is dedicated to possible approaches that hold promise for a better future of Arab STI systems and, by association their contributions to inclusive sustainable developments in the Arab countries.

It is essential to stress at the very outset that the dynamic nature of the challenges faced by the Arab countries in relation to building their STI capabilities, in addition to global and regional factors, will necessitate equally dynamic responses. Some of which may already be evolving as these lines are being typed.

It may not be taken for granted, however, that all such responses would be beneficial in setting up robust STI capabilities. Rather it is more likely that at least some of these responses might retard work that has already been carried out in the past along such lines. Such, for instance, is the case where regional tensions coupled to global developments and power games seem to have succeeded in diverting enormous resources toward armament acquisitions and a considerable buildup in the region's defense industries.

Furthermore, it is also essential to point out that targeting human development may not always go hand in hand with commercial implementation of STI outcomes, often touted as a primary indicator of ultimately successful STI systems. Finding formulae that allow commercial success to feed into sustainable human development and vice versa is neither an easy nor a simple task. To the contrary, it is quite a complex undertaking calling for astute governance, in general, and STI initiatives, in particular, aimed at:

- nurturing forward-looking STI systems;
- creation of equitable and efficient educational systems;

- effective R&D activity aimed at tackling national development priorities;
- eradicating poverty and reducing income and wealth inequalities.

This, of course, is not to diminish the importance of engaging in certain areas of fundamental or basic research, with no apparent application, however. Thus, fundamental research holds the following extremely important, albeit indirect, practical benefits that materialize, in

- enhancing the quality of education, in general, and tertiary education, in particular;
- creating expertise in resolving applied research problems; and
- providing the scientific community in the country in question with improved credentials enabling cooperative ventures and partnerships, which, if well managed, may bring eventual benefits for all concerned.

## 6.2 CHALLENGES FACING STI CAPACITY BUILDING IN THE ARAB COUNTRIES

The following sections in this chapter will attempt to show that a great deal of the challenges faced, and continue to hinder Arab development in general and STI capacity building in particular, are linked to governance practices deeply rooted in rentierism, leading to attendant corruption and cronyism.

Thus, while governance in many of the Arab countries must be credited with making measured advances on certain fronts, they must also be held accountable for stubborn impacts upon various spheres of socioeconomic activity. In general, the lackluster performance of Arab STI systems is inextricably linked to inadequate educational services, poverty, and inequality that succeeded in creating the most violent brands of fundamental extremism witnessed by humanity since Europe's middle ages and crusader invasions of the Middle East.

### 6.2.1 Governance, Rentierism, Corruption, and Cronyism

A question that is likely to be posed by some readers might run: Why does a book on science, technology, and innovation in the Arab countries need to even mention issues such as rentier economics, cronyism, corruption, tribalism, as well as fundamental extremism in Islam and associated sectarianism? The answer to such a question lies in the history as well as the web of relationships than determine this region's socioeconomic status, of which acquiring and implementing science, technology, and innovation are but one manifestation. In other words, the root causes of the Arab countries' underperformance in various spheres of science, technology, and innovation are inadequate governance practices, not only in terms of how STI systems are run but in relation to how entire Arab economies are modeled and managed.

Lest this read like a political manifesto, however, reference is made in the next section to recent studies linking entrenched rentier approaches to corruption and cronyism, responsible for severely hindering socioeconomic development in, for example, Egypt, a country that is not devoid of natural resources and whose population are no less creative and enterprising than any other around the world.

The following few paragraphs will attempt a brief explanation of the historical background to the strong hold that rentierism has had, and continues to have, on political and socioeconomic development in the Arab region.

#### *6.2.1.1 Rentierism in the Region's History*

The roots of rentierism date back far beyond the present era of hydrocarbon-generated wealth. In parts of the Arab world that managed to remain beyond the direct influence of empires that ruled its major cities, rentier tendencies prospered with tribes that raided neighboring territories with their chieftains distributing loot among their followers, naturally favoring their closest relatives. Sectarianism came into the picture as means of justifying raids and associated looting. However, sectarianism was not always essential as justification for plunder, as tribes of the same faith, sect, and political allegiance also managed to find reasons for raiding and looting one another.

All the Arab countries were occupied directly, or indirectly, through treaties with local rulers by the Ottomans, since the early 15th century. The period of Ottoman rule only reinforced and deepened rentier tendencies, as the High Port sought to dominate the vast expanse of its dominion through regional and local chieftains and overlords, often feudal lords in their own right to start with and sometimes selected on ethnic, religious, and sectarian grounds, while creating new overlords as necessity called. Rent collected by overlords were more or less faithfully transmitted to the High Port, with portions of the rent kept as the rightful share of the local rulers and used to buy loyalties as well as ensure a perpetual hold on power, for themselves and their offspring.

To this extent, the situation might be described as merely medieval with a good chance that it would disappear as countries under Ottoman rule gained their independence, with the help of European colonialists. However, history was not to be so kind, as colonial powers, which held virtually all countries of the region under mandate or direct colonial rule also found it convenient to hand power upon their departure over to prominent tribal leaders in the Arab Gulf region and went on to do business whenever convenient with putsch leaders who assumed power in other countries of the Levant and North Africa. The plot thickened considerably as untold hydrocarbon wealth was discovered in the region.

With only very few exceptions, power held by colonial powers was handed over in almost all the Arab countries, either to prominent sheikhs, feudal lords, or heads of semifeudal families. And, in all but the kingdoms and emirates of the Arab world, coup d'états ensued, resulting in the installment of military officers who managed the countries they ruled as fiefdoms benefiting few at the expense of many.

What the onset of the region's hydrocarbon era managed to achieve on this score was merely to reestablish and reinforce ages-old rentierism on a new footing. In essence, revenue was now being generated by international firms contracted to extract underground wealth, transport, and sell it to the highest bidders, handing part of the proceeds to rulers of the land, who then distributed it in such a manner that guaranteed continuation of this exceedingly convenient status quo.

Multinationals in liaison with governments of their headquarters, on the one hand, and ostensibly acting as agents for governments in the region did little to change rules of the old game. Indeed, the old rules were further reinforced particularly since only limited or no popular representation was allowed and free speech had always been viewed as hostile to national interests.

Rentierism comes at a rather heavy price, exceeding mere economic damage. Thus, alliances wrought between government authorities with sectarian and tribal roots, and Islamic groups, both outright fundamentalist as well as diluted versions, were essentially born out of an urgent need for added legitimacy, on the part of the authorities and the quest for more adherents to their cause by the religious groups. Alliances of this nature have had an immense, mostly negative impact over the past decades on socioeconomic development, affecting women's participation in public life and even extensively influencing teaching curricula in a manner that retards skill building in a range of subjects.

### *6.2.1.2 Cronyism, Corruption, and STI Capacity Building*

Several Arab countries are afflicted by relatively high rates of corruption compared to other countries around the world. A report issued in 2010 by Transparency International ranked 12 Arab countries between 91 and 178 out of 178 countries worldwide in terms of perception of corruption.[1] Thirteen Arab countries scored lower than the median and average scores for all countries of the world. Six Arab countries, namely Libya, Yemen, Comoros, Sudan, Iraq, and Somalia had scores among the 10 lowest worldwide.

As demonstrated in a 2015 paper on cronyism in Egypt by Diwan et al.,[2] privileges accorded to firms on the basis of political connections reduce the chances for entry by more efficient operators and diminish possibilities of creative destruction, thereby also reducing the chances of introducing innovation. Since more efficient operators are in many instances those with better capabilities for utilizing various STI inputs, it stands to reason that cronyism may eventually retard STI capacity building. See Frame 6.1.

Aghion has also indicated[3] that as cronyism ensures "fewer threats from frontier entrants, incumbents have fewer incentives to invest in innovation."

The impact of alliances between political figures and selected firms goes beyond mere bribes and kickbacks with influences that can eventually impact the entire national economy and the paths taken for social development. Thus,

**Frame 6.1 The Case of Cronyism in Egypt**

In their analysis of Egypt's large corporations Chekir and Diwan[1] focus on the magnitude, mechanisms, and effectiveness of their political connections.[4] The main findings of this paper are summarized by the following three points:

1. While connected firms (CFs) enjoyed access to valuable privileges, they did not perform as well as the nonconnected firms (NCFs).
2. Important constraints to economic growth in Egypt have been political rather than economic as such.
3. Through shifting political constraints, the uprising witnessed by Egypt could produce important impacts on the country's future growth and development. However, this is predicated on a more open political system that would eliminate or constrain the need for granting privileges to business insiders and foster a more effective use of capital as well as more competition, thereby enhancing growth and generating wider employment opportunities.

A database of publicly traded Egyptian firms was used in the analysis undertaken by the authors, who incidentally indicate that some of the results they arrived at could have been reached before the 2011 uprisings. However, based on information revealed by the market following the uprising allowed them to verify the information they had, concerning the nature of connections between the firms they looked at and politically influential persons in government. Thus, while the identity of Egyptian businessmen closely connected to government figures was well known to the Egyptian public and to financial markets before the 2011 uprisings, the trials of previously connected businessmen, and falls in the market value of their businesses, made the claims about their preferential treatment more credible.

The study focused on the efficiency of investments and the manner in which capital was being misallocated, in the sense that it was used to support less efficient CFs rather than NCFs where it would have produced higher economic returns. Thus, CFs borrowed a larger share of national savings not due to the fact that they were more efficient, but on the strength of state guarantees that rendered lending them more attractive to banks despite their low efficiency. As the market fell in early 2011, the value of state guarantees virtually disappeared as well as a portion of their value linked to their market domination.

Had credit been allocated to NCFs instead, an estimated 3.5% return per year on investment would have been obtained, creating more jobs and attendant welfare. While such returns are significant, they are admittedly small since they represent a once-off effect only. Taking into account accumulative effects would most likely have revealed a host of economic, political, and social dynamics with far larger net effects of the political connections on growth. It is such dynamics that ought to be carefully investigated in future research.

on the economic front, exclusionary mechanisms imposed by such alliances reduce entry by new actors, which are more likely to be innovative while the privileges that CFs enjoy often ensure that the chances of creative destruction are minimal. Thus, in the sample studied by Diwan and colleagues, NCFs essentially did not grow in the last decade preceding the uprising, thereby creating limited competition for incumbent firms. Additionally, with the reduced threat of new entrants, incumbents had fewer incentives to invest in innovation and push efficiency frontiers.

Clearly, conclusions such as those reached by Diwan et al.[5] would require further substantiation, possibly based on enterprise surveys within and across selected sectors enabling better understanding of the dynamic impact of cronyism on innovation by both FCs and NCFs. For instance, case studies could be undertaken comparing the performance of various sectors under cronyism, using an especially constructed index that helps divulge the extent of cronyism in various sectors and segments. Thus, the cement industry in Egypt, unlike steel, was able to maintain a reasonable level of competitiveness, while both served the same sector, namely construction, and both were capital intensive. Hence, an evaluation of the manner with which they handled innovative inputs as well as political connections might help reveal the dynamics and impacts of political connections on competitiveness, and possibly even that associated with innovative inputs, within essentially the same sociopolitical ecology.

The unwillingness and inability of autocratic regimes to truly liberalize policies ended up generating social discontent, in countries such as Syria, Tunisia, and Egypt, thereby subjecting the corporate sector, including connected firms, to greater risks. Thus, the decade preceding the uprising is reported as witnessing private investment in Egypt struggling to stay above 10% of GDP, and capital investment estimated at over $5 billion/year,[6] indicating that risks were perceived to be generally high for all firms, whether connected to political power or not.

Cronyism had both direct and indirect impact on productivity and innovation following the uprisings in all the Arab countries, since it went hand in hand with repressive practices against various opposition groups alternating with attempts at their appeasement. While such tactics allowed the regimes to survive longer, they eventually led to (1) creation of a climate of uncertainty and (2) enhancing the level of popular frustrations and factional tensions, possibly due to heightened perceptions of inequality. Thus, the eventual collapse of government authority, coupled to the fact that opposition was essentially disorganized, with unclear agenda, resulted in several years of sharp stagnation on many fronts, including competitiveness and innovation.

More insidious and long-lasting implication of cronyism in light of the above factors are the risks of falling into a situation in which political tensions feeding on inadequate economic performance render transition to a more stable political order even harder, even encouraging extreme fundamentalist tendencies as has demonstrably been the case in almost all the Arab countries.

Stating that there is nothing inherently detrimental about close state-business relations, and citing the case of South Korean Chaebols to illustrate how acceptable performance was possible even when state-business relations were characterized by cronyism, Diwan et al. indicate that while it is still possible to advance growth, profitability, and innovation, despite such contraventions,[7,8] however, due to their political connections and the fact that they were perceived as less risky by likely debtors, CFs in Egypt appeared more capable of attracting capital, and possibly certain associated innovations as well, since politicians seemed to care more about the favors they garnered than about economic growth,[9] a situation that could only arise due to the dichotomy between market valuation of firms and politicians' prerogatives.

An emerging body of literature has taken up mechanisms that enable firms to enhance their profits based on the political connections they were able to establish. Several factors come into play. Thus, it is often possible for such firms to enhance their profits through preferential access to government contracts[10,11] as well as gain protection through regulatory arrangements. Additionally, as firms grow on the basis of their preferential status, they would gain greater market power, providing more tax revenues and thereby gaining even more preferential treatment.

Diwan et al. have pointed out that privileges allocated to CFs in Egypt, included:

- privileged access to energy subsidies;
- cheap land for development into housing and tourism projects;
- operating in sectors protected by trade barriers.

Clearly, being able to operate within such a preferential environment would not induce the management of CFs to innovate beyond what is absolutely necessary to continue their operations within the overall setting they function in.

Other benefits alleged by Diwan and colleagues would be difficult to quantify, including lax application of competition rules, early access to sensitive policy decisions, and even influence on policy making in areas that impact their business interests.[13]

An interesting aspect concerning the performance of CFs is that while they did obtain privileges that should have boosted their performance, there were other factors that acted in the opposite direction, depressing their profitability. Thus, apart from reduced profits due to their need to pay back their patrons in various ways, CFs in Egypt were often urged by the government to grow much faster than was practicable in return for the valuable privileges that they enjoyed, resulting in enhanced costs and reduced profitability. The results were especially detrimental in cases where CFs were originally "directed" to operate and grow in sectors with low profitability to start with.

Another unquantifiable aspect of CF operations is that when politicians are more concerned with their share of the cake and continued loyalty by CFs, it is often that they would settle for management with mediocre skills, further reducing their profitability.

In short, regimes that emerged following economic liberalization in the Arab countries during the past three decades were merely Egyptian, Syrian, Tunisian, and Jordanian versions of "crony capitalism." Within such regimes, competition was stifled and business people with close connections to the regime were able to obtain most if not all major contracts, and bend and even break planning regulations and a variety of legal constraints whenever they found it useful and convenient. In return for favors extended by well-situated government officials, businesses that won contracts had to accept a good deal of both reasonable and unreasonable demands by these same officials as well as other better situated officials, who might be able to hinder their future operations. While steps to discipline client businessmen were attempted at times, as political conditions deteriorated in the late 2000s, many of the businessmen and associated government officials prospered.

To conclude this section, with corruption having become more widespread than ever and with tales of politicians and businesses implicated in corrupt practices becoming regular news items throughout the planet, it is helpful to attempt an answer to the question: What makes corruption even more harmful in a developing country situation?

In effect, the following two reasons may be given as to why corruption is rendered particularly pernicious, with more deleterious effects in the developing countries:

- Due to lower resource levels depriving wider segments of the population of resources that might not be compensated by other means;
- Since media, and even the judiciary, in the developing countries tend to be closely controlled by governments which generally harbor corrupt officials and their cronies, it is far more difficult to divulge cases of corruption, and even more difficult to prosecute corrupt individuals. As a result corrupt practices are far more difficult to contain and continue to consume increasing resources that might otherwise be available to fuel growth and infrastructure building as well as provide much needed resources for social services.

Policies designed to stem out corruption and cronyism are an essential requirement for future development. At least initial support in implementing such policies is offered by the adoption of STI inputs designed to improve access to information on a wide range of development projects and processes. Reference is made here to the initiative by the Iraqi government, referred to in Section 3.2.1.2, aimed at the introduction of e-government capabilities for a number of causes, chief among which is combatting corruption. Whether the Iraqi government's efforts will succeed in their quest is very much an open question. However, the manner in which its initiative is implemented and allowed to evolve is worth a good deal of attention.

It is also well to bear in mind that the war against corruption will need to use strategies of constant escalation, adopting more intensive measures with time to deal with emerging sophistication of corrupt practices. In the long term, durable outcomes may be guaranteed only through initiatives addressing root causes on a broad front, targeting a variety of domains, including the educational and cultural domains.

#### 6.2.1.3 STI Capacity Building in the Shadow of Rentier Economics

Rentierism has influenced STI capacity building in a multitude of ways. By way of an illustration, few motives exist for an official or a group thereof, in charge of water desalination in a given Arab country, for example, to establish local capabilities for building such facilities, given that most essential technologies may be commercially accessible from a variety of sources. While, steeped in rentier practices, it makes more sense, for such an official or group of officials, to simply contract the erection, and often running and maintenance, of such a facility to an established foreign contractor. As a result, this official or group of officials would be able to make a variety of gains, as rent on their positions as decision makers and purse holders, extending from financial gains, in the form of kickbacks, to the opportunity of appointing relatives and associates, who may not necessarily be well qualified, in favorable positions in cooperation with the contracted firm.

A claim often made in open defense of such practices is that local capabilities are not up to scratch, and lacking in expert human resources to handle likely problems, which may very well be the case; however, attempts are never made to ensure that local capabilities are brought up to scratch and visited with sufficient means to ensure sufficient dynamism and expertise, in countries known to depend now and in the foreseeable future on water desalination.

While water desalination is used in the above example, similar situations arise with regard to a multitude of domains, including renewable energy plants, petrochemicals processing facilities, and other areas of industrial activity, where it would be logical to build national and regional capacity if only as a means of bringing down costs. Incidentally, the above arguments must in no way be taken as seeking to cut links with outside sources of technology, commercial or otherwise. On the contrary, such links must persist and might even grow in volume. However, in the absence of policies and practical steps aimed at national and regional capacity building, such practices are often wasteful and counterproductive since they:

- are inherently more costly;
- result in creating a vicious cycle with ever-increasing dependence on commercial sources of technology and related know-how from outside the country/region;
- hinder real growth of national/regional STI capabilities;
- provide benefits for a meager proportion of officials, thereby enhancing inequality.

In short, the manner with which STI inputs are acquired, and later disseminated, adapted, and utilized, may not be readily divorced from the economic climate in which such activities take place and the manner in which business affairs are conducted, often very much influenced by inherited power structures. This is particularly the case in the Arab countries in which rentierism is well entrenched, and in which legal and regulatory arrangements provide various means for manipulation, essentially by politicians and community leaders for a variety of purposes.

### 6.2.2 Education in the Arab Countries: Achievements and Enduring Challenges

STI capacity building is not something to be targeted for its own sake. Rather, it is to be essentially conceived as an essential means for attaining inclusive and sustainable development goals with a country's population as prime beneficiaries. Hence, attempts at enhancing STI capacity invariably necessitate focus on human capital building. And, benefits are likely to be realized within various spheres of development, enabling society to meet numerous challenges in health, security, environment, and social welfare, only where education has succeeded in molding the expectations and responsibilities of individuals and communities.

In effect, education plays an indisputable role in attaining economic growth and social development. The fact that the Arab countries are home to the world's most youthful populations renders education of even greater importance for the Arab countries. Thus, in 2015, over 130 million inhabitants of the Arab countries, or around 33.2% of their total population, were below the age of 14. Around 109 million inhabitants of only eight of the Arab countries, namely Algeria, Egypt, Iraq, Morocco, Saudi Arabia, Sudan, Syria, and Yemen, fell within this age group. The resource-endowed Arab countries of the Gulf, often referred to as the Gulf Cooperation Council (GCC) countries possess only around 10% of this total.[12] Yet, more than half the population of the largest GCC country, Saudi Arabia, are below 25 years of age.[13] On the other hand, several Arab countries, some with severely limited resources, such as Comoros, Palestine, Somalia, and Yemen have populations with 40%–47% within this age group.

Increasing demand coupled to budget constraints and inadequate management have led to severe weaknesses in many, if not all Arab education systems, with several indications to inefficient allocation of resources in public education as key deficiency throughout. Thus, the largest portion of educational budgets is used to cover operating expenditure, wages, maintenance, and infrastructure repair, with little left to finance curriculum development and innovative approaches. This has resulted in reduced quality of teaching and overall efficiency and productivity of Arab educational systems. In Algeria, for example a mere 12% of school children entering primary education graduate from higher education. Regional disparities with regard to access to education and levels of educational achievement are likely to increase if public subsidies continue to be on the low side and inadequately distributed.

While problems are evident in all stages of education, they are most acute in higher education, where the State bears most of the financial burdens and where innovative curricula and teaching methods are still lagging behind neighboring countries as well as many developed and some developing countries. This has invariably resulted, throughout the Arab world, in an oversupply of graduates with qualifications ill-suited to undertake innovative tasks in modern businesses, and even in government institutions intent upon adopting innovative practices, rare as they are in the first place.

Vocational training systems are often managed inefficiently with little regard to actual needs and near-future expectations for key economic sectors. They are often run as basic and residential training facilities for pupils who underperformed in their basic or secondary education, rather than agents and enablers of technical and economic change. Training centers run by government institutions, in countries such as Syria and Iraq, for example, imparted skills that did not correspond to modern requirements of their economies. While, on the other hand, private sector institutions that provide vocational training confined mainly to desk jobs in accountancy, typing, and rudimentary IT-related skills operated unsupervised, placing financial gains above quality of training.

Apart from the trauma suffered by education in countries such as Yemen, Syria, Libya, and Iraq, some of the main reasons why education is in need of special and urgent attention in the Arab countries include:

- obsolete curricula and teaching methods retaining their hold in both public and private educational institutions, with content that has little or no relevance to national socioeconomic development and market needs;
- mediocre management with little, if any, attention to quality considerations and limited scope for reinvention and constant improvement;
- political and religious indoctrination that has succeeded only in nurturing extremism and restricting human rights across the board;
- inadequate funding and infrastructural arrangements across all stages of the educational process, leading to acute shortages of new technology inputs, including computers, Internet access, and smart boards.

This is not to say that the past decades were devoid of some positive developments. Thus, progress was indeed made on several fronts, resulting in enhanced literacy as well as wider enrolment in various stages of education for both males and females across the Arab countries. In particular, the adult literacy rates for males and females throughout the Arab countries increased from around 55% and 25%, in 1980–82 and 62%, respectively, in 2005. Gross male and female enrolment ratios in all levels and types of secondary schooling also increased from a median of about 57% and 38%, respectively, to 70% and 65%, respectively, by 2006. Considerable gains were also made in tertiary enrolment.

Nevertheless, close to 60 million of Arab populations, two-thirds of whom are women, are illiterate, and many millions of school-age children are out of school, not only in countries that failed to tackle the literacy challenge, but also due to armed conflict and political turmoil in Iraq, Libya, Syria, and Yemen.

In addition to the above challenges, crucial issues relating to the quality of education in many of the Arab countries must be considered and remedied.

Failing this, pronouncements made by high officials in several Arab countries with a view to setting up national knowledge economies and information societies may not progress beyond mere daydreams, or the establishment of isolated enclaves for select populations and business enterprises.[14]

The need for radical overhaul of educational policies, with inclusive and sustainable socioeconomic development as principal goals has never been more urgent. Novel STI inputs should facilitate such endeavors.

#### 6.2.2.1 *Spending on Education by the Arab Countries*

Table 6.1 presents information on government expenditure on preuniversity education as a percentage of GDP in the Arab countries and five of their closest

**TABLE 6.1** Expenditure on Education in the Arab Countries

| **North Africa** | | **Arabian Peninsula** | |
|---|---|---|---|
| Algeria[a] | 4.3 | Bahrain | 2.6[g] |
| Libya[c] | 2.3 | Kuwait[d] | 3.8 |
| Mauritania | 2.9 | Oman[e] | 5.0 |
| Morocco[e] | 5.3 | Qatar | 3.5[h] |
| Tunisia | 6.3[g] | Saudi Arabia | 5.1[a] |
| **NA average** | **4.25** | **GCC average** | **4.0** |
| **East Africa** | | Yemen | 4.6[a] |
| Comoros | 5.1[h] | | |
| Djibouti | 4.5[b] | | |
| **EA average** | **4.8** | | |
| **The Levant** | | **Arab countries' closest neighbors** | |
| Jordan | 4.9 | Cyprus | 6.4 |
| Lebanon | 2.6[c] | Iran | 2.9 |
| Palestine | 1.3[f] | Israel[f] | 5.9 |
| Syria | 5.1 | Malta | 8.3 |
| **Levant's Average** | **3.5** | Turkey | 4.8 |
| **Nile Valley** | | **Averages** | |
| Egypt | 3.8[a] | **Arab countries** | **3.96** |
| Sudan | 2.2[e] | **Closest neighbors** | **5.66** |

*1. No figure was available for Iraq, The United Arab Emirates and Somalia.*
*2. Superscripts indicate the year when data was produced: a. 2008, b. 2010, c. 1999, d. 2006, e. 2009, f. 2015, g. 2012, h.2014.*
From The UNESCO Institute of Statistics. Downloaded in December 2016 from: http://data.uis.unesco.org/.

neighbors. As indicated elsewhere, data concerning expenditure on education in the Arab countries is not consistently available and several figures presented in Table 6.1 date quite a few years back. Indeed, figures for Jordan and Libya date as far back as 1999. Secondly, it may be remarked that no figures were found at all for three countries, namely Iraq, Somalia, and the United Arab Emirates. At any rate, it may be readily seen that levels of government expenditure on education vary widely from one Arab country to another, within a range that includes 1.3% of GDP in Palestine, on one end, and 6.3% for Tunisia on the other, with an overall Arab country average of a little less than 4%. However, at an average of about 5.7% of GDP, expenditure on education by the Arab countries' five closest neighbors is much higher: by more than 40% than the Arab average.

It may be further remarked that Algeria, Morocco, and Tunisia were ahead of all other Arab countries, devoting an average of 5.3% of their GDP to preuniversity education. Yemen and the smaller African Arab countries of Comoros, Somalia, and Mauritania come next, dedicating 4.6% of their GDP, followed by the GCC countries with the exception of the United Arab Emirates which dedicate 4% of their GDP to education. The Levant countries Jordan, Lebanon, Syria, and Palestine devote an average of 3.5%, while Egypt and the Sudan come last with only 3% of their GDP earmarked in support of preuniversity education.

### 6.2.2.2 Enrolment in Primary and Secondary Education

Primary level enrolment rates witnessed considerable improvements during the past few decades in all of the Arab countries. Fig. 6.1 presents trends recorded for the growth since 1970 of primary enrolment rates for both boys and girls in the Arab countries taken together, countries of the Middle East and North Africa (MENA), which include many of the Arab countries, as well as the world

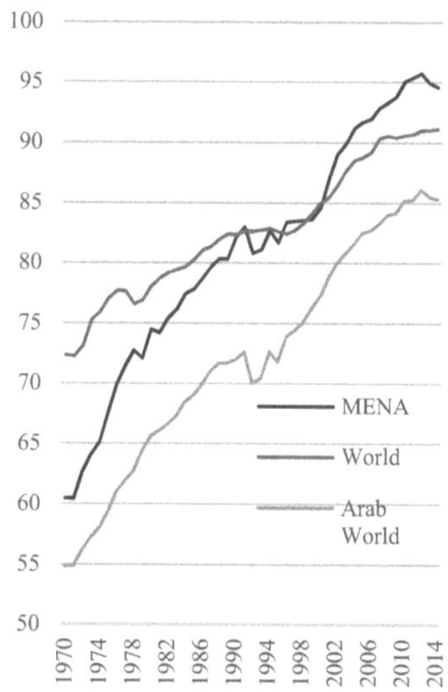


**FIGURE 6.1** Primary enrolment rates 1970–2014; both sexes (%).

at large. Fig. 6.1 clearly indicates that despite considerable growth in their primary enrolment, the Arab countries still lag behind growth witnessed by the entire world as well as the MENA countries. Fig. 6.1 also indicates that growth of primary enrolment in both the Arab and MENA countries started to stagnate toward the start of the present decade, while growth continued worldwide, albeit at a lower rate, which might be expected from experience in similar situations as saturation is approached.

With the aim of examining quantitative as well as qualitative aspects related to basic education in the Arab countries, the Arab Knowledge Report for the year 2009 indicated that on the quantitative side, statistics on net primary school enrolment rates showed that only four Arab countries, presumably, Libya, Bahrain, Tunisia, and Algeria, approached percentages in excess of 95%. Eight more countries had enrolment rates between 80% and 94%, while six others, namely, Djibouti, Mauritania, Oman, Palestine, Yemen, and Saudi Arabia, fell below this range. Notably, only two countries, Bahrain and Tunisia, appeared to have attained saturation for female enrolment. The situation in 2014 was far more rosy. Thus, a more detailed look at primary enrolment in the Arab countries for the year 2014 or the most recent year for which data is available in seven Arab countries, namely Algeria, Bahrain, Egypt, Kuwait, Libya, Morocco, and Tunisia, claimed primary enrolment at or higher than 98%, for both sexes. On the other hand, six Arab countries, comprising all of the above with the exception on Libya, achieved similar goals, i.e., enrolment beyond 98%, for female primary enrolment.

Indeed, it is comforting to see that the gender parity index continued to improve throughout the period 1970–2014 and that its growth did not seem to suffer during the tumultuous period that witnessed stagnations and declines in several Arab countries. It should be noted, however, that several other Arab countries still performed below world and Arab country averages where primary enrolment is concerned. In particular, it is rather disconcerting to see that Lebanon, a country that formerly led many of its neighbors in the establishment of educational institutions at all levels, with primary enrolment figures for both sexes and for girls at 89% and 86%, respectively, below the world's corresponding averages. Its neighbor, Syria, is in a much worse state with a primary enrolment percentage for both sexes in 2014 at around 71%, following a 4-year period, 2006–2009, during which primary enrolment stood at or around 98.2%.

Fig. 6.2 presents an overall view of the growth in gross secondary enrolment in the Arab countries compared to the worldwide trend. Fig. 6.2 clearly indicates that gross secondary enrolment for both sexes in the Arab countries approached and even exceeded the world's average between the years 1990–1991, before retreating significantly around 2005 and considerably by 2011. Several reasons might be put forward for advances and relapses of this nature. Some of the reasons might be linked to political upheaval and civil strife that engulfed several countries in the region over the past decades as well as the rise and spread of fundamentalist tendencies, taking out girls from the educational

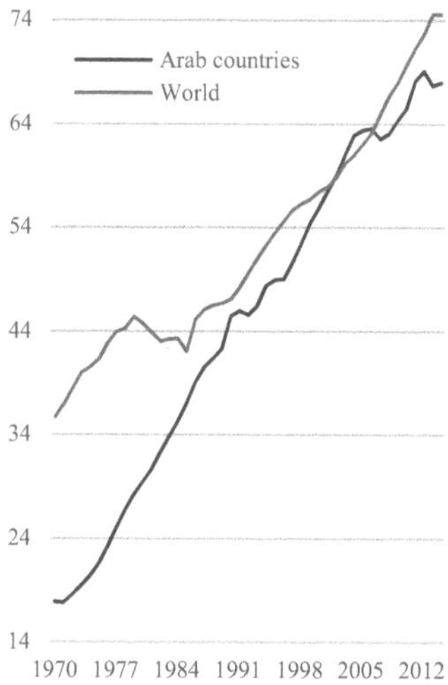


**FIGURE 6.2** Gross secondary enrolment ratio, females (%).

system altogether. It is doubtless, on the other hand, that proactive policies coupled to civic action at the local community level would have retained such advances and drastically reduced subsequent reversals.

Nevertheless, the situation now is such that the gross secondary enrolment in the Arab countries is lower by over 5% than the world's average. Additionally, countries that contribute to this situation are not merely the small Arab countries such as Somalia, Djibouti, Mauritania, and Comoros, with chronically limited income levels. Rather, it is countries that made important advances in the past and enjoyed considerable resources that need to take blame.

### *6.2.2.3 Evaluating Education in the Arab Countries*

Several schemes are now available for the evaluation of the performance of national systems of education across the world. The following section presents the results of two evaluation exercises. The first is based on an index devised specifically for the purpose by the United Nations Development Program (UNDP) and the Mohammed bin Rashid Foundation (MBRF) in the United Arab Emirates, while the second, which was conducted by the Organisation for Economic Cooperation and Development (OECD) adopts a direct approach based upon standardized tests conducted within a selection of schools in the target countries. The latter scheme is usually referred to as the PISA scheme denoting its full title: "Programme for International School Assessment."

It goes without saying that the results of evaluation exercises may not be taken at face value. Thus, indexes derived from data on enrolment and spending, as well as the results of standardized tests would need to be interpreted with a good deal of care and attention. Nevertheless, evaluations such as the ones discussed below do provide essential pointers which may be supplemented by further research and in-depth analysis to enable formulation of viable education policies.

#### 6.2.2.3.1 Assessment of Preuniversity Education by UNDP/MBRF

Table 6.2 presents values of the UNDP/MBRF index devised with a view to evaluating preuniversity education in the Arab countries. According to the UNDP/MBRF scheme, an assessment of preuniversity education in the Arab countries seems to place these countries within four somewhat overlapping groups. On top of the list with values for the preuniversity indicator ranging from 60 to 70, there are five GCC countries, namely the United Arab Emirates, Saudi Arabia, Bahrain, Qatar, and Kuwait. This group is followed by seven Arab countries, including Oman, Lebanon, Algeria, Egypt, Jordan, Morocco, and Palestine, with scores between 50 and 60 for the same indicator. Iraq, Tunisia, Syria, Sudan, and Yemen come next with values for the indicator between 40 and 50. Trailing the above three groups of countries are Djibouti, Libya, Mauritania, and Somalia with scores ranging between 20 and 40. That resource rich countries occupy top ranks according to this indicator comes as little surprise. However, scores accorded to countries such as Tunisia and Libya may be questionable.[15]

**TABLE 6.2** UNDP-MBRF Index of Preuniversity Education in the Arab Countries

| | | | |
|---|---|---|---|
| Algeria | 56.0 | Morocco | 50.4 |
| Bahrain | 66.5 | Oman | 57.4 |
| Comoros | 28.8 | Palestine | 52.1 |
| Djibouti | 35.9 | Qatar | 66.2 |
| Egypt | 55.4 | Saudi Arabia | 67.4 |
| Iraq | 48.5 | Somalia | 27.5 |
| Jordan | 55.3 | Sudan | 41.3 |
| Kuwait | 63.0 | Syria | 41.4 |
| Lebanon | 56.3 | Tunisia | 43.1 |
| Libya | 26.5 | UAE | 68.6 |
| Mauritania | 21.2 | Yemen | 40.4 |

From Arab Knowledge Index. Report by the United Nations Development Program (UNDP) and Mohamed bin Rashid Al Maktoum Foundation (MBRF). Downloaded in December 2016 from: http://www.knowledge4all.com/uploads/files/AKI2015/PDFAr/AKI2015_Full_Ar.pdf.

At any rate, values of this index vary widely between a minimum of 21% for Mauritania and nearly 69% in the case of the United Arab Emirates. Average and median values for this indicator are estimated at 48.6 and 51.3, respectively. The six GCC countries come at the forefront with an average value for this indicator at nearly 65, whereas seven Arab countries, namely Lebanon, Algeria, Morocco, Egypt, Jordan, Palestine, and Iraq possess values of around 53. Four other Arab countries, namely Tunisia, Syria, Sudan, and Yemen possess indexes valued between 40.4 and 43.1. This relatively poor performance by Tunisia and Syria might well be linked to the political upheaval and violence they have been witness to over the past 5–6 years. Libya, Comoros, Djibouti, Mauritania, and Somalia trail all the rest with an average value for the preuniversity education index of around 28. Here too, the low value, of around 27, assigned to Libya, may be linked to the current situation in that country.

### 6.2.2.3.2 Results of the PISA Evaluation Exercise

Table 6.3 presents the results of the evaluation conducted by the Programme for International Students Assessment (PISA) in science, mathematics, and reading for six Arab countries that took part in the 2015 exercise, namely Algeria, Jordan, Lebanon, Qatar, Tunisia, and the United Arab Emirates, as well as four of their closest neighboring countries, namely, Cyprus, Israel, Malta, and Turkey.[16] This table also includes:

- trends in PISA scores in all three subjects for countries that took part in the PISA exercise over the past 3 years;
- information on the share of top performers in at least one subject, i.e., those achieving levels 5 or 6, as well as the share of low achievers in all three subjects, i.e., those achieving levels 1 or 2 in all three subjects.

As readily indicated by Table 6.3, mean scores attained by the above-mentioned Arab countries in the subject categories of science, mathematics, and reading, fell considerably below the OECD average[17] as well as average scores achieved by their closest neighbors, i.e., Cyprus, Malta, Turkey, and Israel. With averages of mean scores by the Arab countries falling markedly below those for their neighbors, by between 12 and 17 points, and for all the OECD countries, by 23–29 points.

Entries in the second and third columns of Table 6.3 indicate that the United Arab Emirates and Qatar ranked highest among the six Arab countries that took part in the PISA 2015 exercise, while Tunisia and Algeria ranked lowest. Still, schools tested even in the former two countries attained scores 60–70 points lower than the average OECD country scores, while Algeria, Lebanon, and Tunisia, on the lower end of the scale attained scores around 110–120 points lower than the OECD average. It may also be noted that the United Arab Emirates exhibited a downward trend in terms of mean PISA scores attained 3 years ago by 12 points while Qatar gained 21 points compared to the mean score it attained 3 years ago. Trends observed for mean PISA scores are repeated—more or less faithfully—by scores attained for individual skills in science, mathematics, and reading.

**TABLE 6.3** A Snapshot of Arab Students' Performance in Science, Reading, and Mathematics Compared to OECD and Neighboring Country Averages (2015)

| | Science | | | Reading | | Mathematics | | Science, Reading, and Mathematics | |
|---|---|---|---|---|---|---|---|---|---|
| | Mean Score in PISA 2015 | | Average of Three-Year Trend | Mean Score in PISA 2015 | Average of Three-Year Trend | Mean Score in PISA 2015 | Average of Three-Year Trend | Share of Top Performers in At Least One Subject (Level 5 or 6) | Share of Low Achievers in All Three Subjects (Below Level 2) |
| | Mean Score in PISA 2015 | Rank | Score dif. | Mean | Score dif. | Mean | Score dif. | % | % |
| **OECD average** | **493** | – | **–1** | **493** | **–1** | **490** | **–1** | **15.3** | **13.0** |
| **Scores and Rankings for Arab Countries Taking Part in PISA 2015** | | | | | | | | | |
| UAE | 437 | 47 | –12 | 434 | –8 | 427 | –7 | 5.8 | 31.3 |
| Qatar | 418 | 57 | 21 | 402 | 15 | 402 | 26 | 3.4 | 42.0 |
| Jordan | 409 | 62 | –5 | 408 | 2 | 380 | –1 | 0.6 | 35.7 |
| Lebanon | 386 | 66 | – | 347 | – | 396 | – | 2.5 | 50.7 |
| Tunisia | 386 | 67 | 0 | 361 | –21 | 367 | 4 | 0.6 | 57.3 |

| | | | | | | | | | |
|---|---|---|---|---|---|---|---|---|---|
| Algeria | 376 | 70 | – | 350 | – | 360 | – | 0.1 | 61.1 |
| Averages | 402 | – | 1.0 | 383.7 | –3.0 | 388.7 | 5.5 | 2.2 | 46.4 |
| **Scores and Rankings for Four Closest Neighbors** | | | | | | | | | |
| Cyprus | 433 | 50 | –5 | 443 | –6 | 437 | –3 | 5.6 | 26.1 |
| Malta | 465 | 41 | 2 | 447 | 3 | 479 | 9 | 15.3 | 21.9 |
| Israel | 467 | 40 | 5 | 479 | 2 | 470 | 10 | 13.9 | 20.2 |
| Turkey | 425 | 53 | 2 | 428 | –18 | 420 | 2 | 1.6 | 31.2 |
| Averages | 447.5 | – | 1 | 449.3 | 4.8 | 451.5 | 4.5 | 9.1 | 24.9 |

**A country's rank is the mean score for science among 71 countries that participated in the PISA exercise in 2015.*
From OECD Report in 2016 under the title: "PISA Results in Focus, 2015." Downloaded in November 2016 from: https://www.oecd.org/pisa/pisa-2015-results-in-focus.pdf.

However, since performance by students and schools is never uniform, it is of interest to note the differences between the shares of top performers and low achievers in the Arab countries to corresponding average values for their OECD and neighboring country counterparts. In essence, differences between the percentages of top performers and low achievers indicate the ability of curricula and teaching methods adopted in schooling systems to bring the skills of their student population up to comparable levels.

As indicated in the penultimate column of Table 6.3, which presents the percentages of top performers in at least one subject, and the last column with percentages of low achievers in all three subjects, results attained by these two percentages in the Arab countries are far more diverse than those for the OECD countries on the one hand, as well as the Arab countries' closest neighbors, on the other. Thus, the percentages of top and low performers tended to assume closer values with regard to the OECD average as well as the Arab countries' closest neighbors, while they exhibited wide disparity in the case of the Arab group. Taken as a group, the average percentage of low achievers in the Arab countries amounted to 46.4% while top performers were a mere 2.2%. Corresponding average percentages for the OECD countries were estimated at 13% and 15.3%, respectively, and for neighboring countries at 24.9% and 9.1%, respectively. Wide variations appear to exist between these two indicators among the Arab countries with the greatest divergence between these two groups pertaining in the case of Algeria where the ratio of these two percentages exceeded 600-fold, going down to around 100-fold in Tunisia and has its smallest value, of around six, in the United Arab Emirates. The above performance disparities highlight the need for radical revisions and reforms of educational resources and methods adopted by various schools and with regard to various groups of students, to ultimately impart comparable capabilities with which to embark upon higher studies and later life.

### 6.2.2.3.3 Information and Communications Technologies in Education

A review of relatively recent literature reveals that the use of technology in general is still wanting. Thus, a UNESCO report issued in 2013 with the aim of evaluating access to electricity and telephone services in five Arab countries, namely Egypt, Jordan, Oman, Palestine, and Qatar, indicated that around 30% of schools in Egypt and between 20% and 27% of schools in Jordan lacked access to telephone communications facilities.[18]

Nevertheless, while even rudimentary technologies, such as electricity and telephone communications facilities may be lacking in a significant proportion of Arab schools, it is still essential to consider as far as possible the status of ICTs at schools in the Arab countries. Table 6.4 presents information on access to ICTs by schools in 14 Arab countries as well as 5 of their closest neighbors, namely, Cyprus, Iran, Israel, Malta, and Turkey. Table 6.4, based on data presented in the 2016–2017 Global Competitiveness Report, includes both scores

**TABLE 6.4** Internet Access in Schools in 14 Arab Countries and Five of Their Closest Neighbors

| | Score | World Rank |
|---|---|---|
| **All Arab Countries** | | |
| Algeria | 3.1 | 124 |
| Bahrain | 5.0 | 39 |
| Egypt | 2.6 | 133 |
| Jordan | 4.5 | 55 |
| Kuwait | 3.8 | 91 |
| Lebanon | 3.9 | 86 |
| Mauritania | 2.2 | 135 |
| Morocco | 3.6 | 109 |
| Oman | 4.0 | 84 |
| Qatar | 5.9 | 11 |
| Saudi Arabia | 4.3 | 65 |
| Tunisia | 3.5 | 112 |
| UAE | 6.1 | 4 |
| Yemen | 1.7 | 137 |
| **Five Closest Neighbors** | | |
| Cyprus | 4.4 | 62 |
| Iran | 3.5 | 113 |
| Malta | 5.6 | 19 |
| Israel | 5.3 | 27 |
| Turkey | 4.1 | 79 |

and rankings, given to various countries as a result of opinion surveys that covered 138 countries worldwide. Table 6.4 clearly indicates an inadequate state of affairs with around six Arab countries ranking below 100, with at least three countries, namely Algeria, Egypt, and Mauritania achieving close to bottom rankings at 124, 133, and 135. The eight Arab countries not included might be expected to have scored similarly. That only three Gulf Arab countries, namely the United Arab Emirates, Bahrain, and Qatar, ranked among the top third countries of the world covered by the data is also significant. The state of affairs is also disappointing in relation to Tunisia, Morocco, and Lebanon, with scores well below median ranking, at 112, 109, and 86.

### *6.2.2.4 Tertiary Education in the Arab Countries*

Higher education in the Arab World is no recent phenomenon. Rather, it is deeply rooted in various urban centers of the Arab Middle East and North Africa. Before the spread of Islam, cities such as Antioch and Alexandria hosted centers of learning and higher studies with a decidedly cosmopolitan outlook. Following the region's Islamization, local madrasas mostly linked to mosques disseminated higher learning with notable examples in cities such as Baghdad in the Levant, as well as Cairo, Qairawan, and Fez in North Africa. See Frame 6.2.

European and American missionaries are credited with founding many schools and institutions of higher learning in the Middle East and North Africa, especially during the latter half of the 19th century.

**Frame 6.2 Prominent Examples of Higher Education Institutions in the Arab Countries[19]**

As indicated in the table below, institutions of higher learning in the Arab countries date back centuries ago. Many were funded from Islamic Waqf or charitable endowment.

**Examples of Early Institutions of Higher Learning in the Arab Region**

| Institution | Date established (approximate) | Country |
|---|---|---|
| Al Mustansiriyah | 984 | Iraq |
| Beit al Hikma | 790 | Iraq |
| Al Azhar | 970 | Egypt |
| Al Zaitouna | 734 | Tunisia |
| The Qarawiyyin | 859 | Morocco |

While, admittedly, confined to the elite and leaning toward religion and jurisprudence in the past, some of the earlier institutions of higher learning in the Arab world, for example, Beit Al Hikma, which was established toward the end of the 8th century in Baghdad by the Abbasid Caliph Harun al-Rashid, taught sciences including astronomy, physics, mathematics, medicine, chemistry, and geography.

Institutions of higher learning in the Arab countries originated intellectual movements such as humanism and scholasticism, taken up by Western institutions after the 12th century.

In 1720, Sultan Ahmed III sent delegations of scholars to Europe with a view to promoting translation of Western scientific literature. He was later emulated by Egypt's Mohamed Ali (1805–49), when dozens of modern institutions of higher learning were established on the European model.

The number of Arab universities went up from 233 in 2003 to 286 in 2006, of which 57% were state institutions. In 2012, the number of universities rose to around 500.

Several universities were established by government as well as foreign missionary societies during the latter decades of the 19th and around the first half of the 20th centuries. These universities were forerunners for modern institutions of higher education all over the Arab countries. Lebanon, in particular, was the first Arab country to host institutions of higher education established by both Catholic and Protestant missionaries. On the other hand, Egypt may be credited with establishing the first of government institutions of higher learning in the region.

Tertiary education occupies a special position with its more immediate impact on STI capacity as well as growth and socioeconomic development. Particular attention is justified to tertiary students pursuing their studies abroad as they often act as agents for the transfer of knowledge back to their home countries, with possibilities for dissemination of a variety of notions and methodologies among wider segments of their home populations.

The following paragraphs will take a look at statistical data concerning growth of tertiary enrolment in the Arab countries during the period 1999–2014. Comparisons will be attempted to five of the Arab countries' closest neighbors as well as other groups of countries across the world. The essential objective of this overview is to identify past trends and to predict, as far as possible, future evolution of tertiary enrolment in the Arab countries.

Based on data published by the UNESCO Institute of Statistics (UIS), Fig. 6.3 compares gross tertiary enrolment in 20 Arab countries in 1999 to that attained by 2014, or the most recent date for which reliable data is available, indicating, for the most part, considerable gains over intervening years. Fig. 6.3 shows that apart from Djibouti and Comoros, whose tertiary enrolment figures were

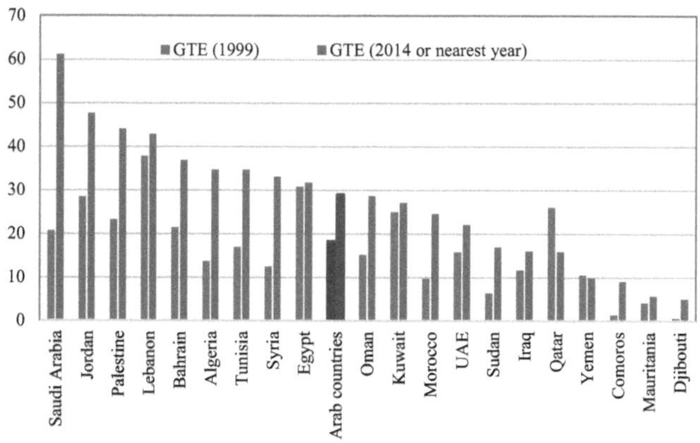


**FIGURE 6.3** Gross tertiary Enrolment (GTE) in 15 Arab countries in 1999 and 2014 or nearest year.

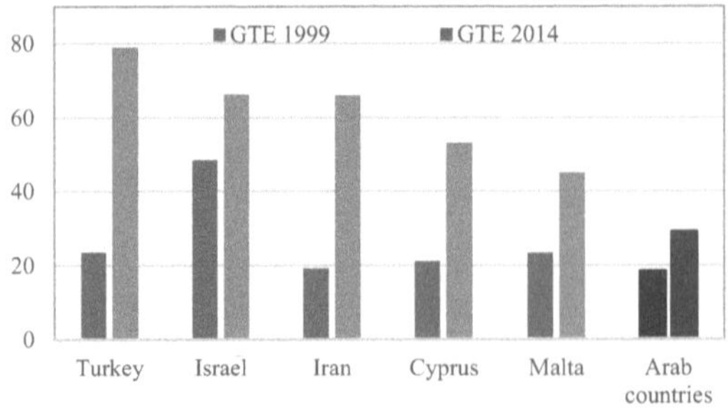


**FIGURE 6.4** Average gross tertiary enrolment figures for the Arab countries in 1999 and 2014 or nearest year (%).

miniscule in 1999, Arab countries which made the most notable gains increased their tertiary enrolment several times over since the start of the century to reach the present status they now occupy. Values for gross tertiary enrolment increased in all the Arab countries with a minimum of 17% for Sudan to 61% in Saudi Arabia.

Fig. 6.3 also indicates that several other countries such as Egypt, Kuwait, and Lebanon made minimal gains over the period 1999–2014. Average performance of the Arab countries taken together is also shown to be rather mediocre. Qatar and Yemen stand out as having had negative overall growth in tertiary enrolment, with the percentage attained by Qatar in 2014 considerably lower than that it had in 1999.

Fig. 6.4 compares past and current tertiary enrolment figures in the Arab countries taken together as well as gains made during the period 1999–2014, to corresponding figures for five of their closest neighbors. The comparison clearly does not favor the Arab countries. Thus, taken together, the Arab countries performed considerably worse than their five closest neighbors, in terms of the average percentage of those enrolled in tertiary programs and made smaller gains over the period 1999–2014 relative to their initial tertiary enrolment figures compared to their closest neighbors with the exception of Israel, which may be argued as having attained rates of tertiary enrolment superior to any other country in the region, back in 1999.[20] Gains made by Turkey and Iran stand out. Thus, gross tertiary enrolment in Turkey and Iran was reported as close to 80% and 66%, respectively, in 2014, a 3.4- and 3.5-fold increase on their values in 1999. In fact, these two countries stand ahead of all other countries in the region, with Turkey's enrolment figures now exceeding the corresponding percentage for Israel, and with Iran at par with it, at 66 as its percentage gross tertiary enrolment in 2014.

Incidentally, while gross tertiary enrolment in the Arab countries was almost equal to the world's average in 1999, it declined to around 5% below this average in 2014, reaching levels only very slightly higher than the average for all developing countries and over 10% lower than the group of upper-middle income countries to which several Arab countries belong.

Figs. 6.5–6.7 provide some further insight into the state of university education in the Arab countries. Thus, Fig. 6.5 indicates that Arab countries with smaller populations, such as Palestine, Jordan, Lebanon, and Libya stand ahead of the rest in terms of the percentage of university students out of their entire populations. Figs. 6.6 and 6.7, on the other hand, reflect standards and quality of university education, with Fig. 6.6 indicating exceedingly high rates of student enrolment per institution in Egypt, Algeria, Mauritania, and Libya, compared to other Arab countries, including Tunisia, Oman, and Lebanon, for example. Thus, the average number of students enrolled in a given university, both government or private, in Egypt was estimated at around 80,000 in 2011, while the corresponding average in Lebanon, for example, was around 10,000.

Fig. 6.7, which presents average ratios of students to staff at universities in the Arab countries, indicates a disproportionately high ratio of students to staff in countries such as Algeria, Sudan, and Egypt, compared to countries such as Tunisia, Lebanon, Qatar, Iraq, and Bahrain. The average for all the Arab

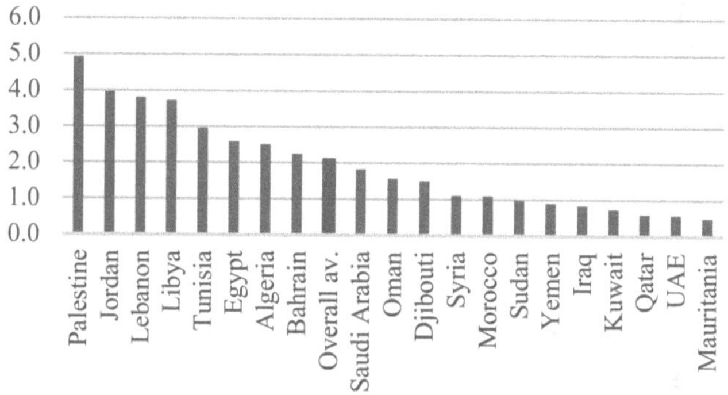


**FIGURE 6.5** Percentage of university students out of Arab country populations in 2011 (%).

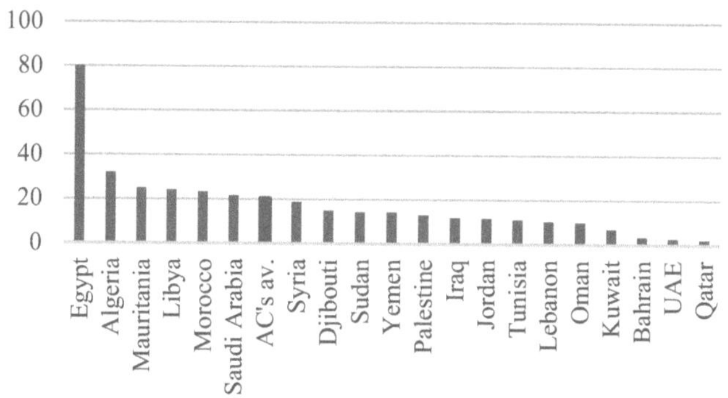


**FIGURE 6.6** Average number of students enrolled in a given university, both government or private, in 2011 (thousands).

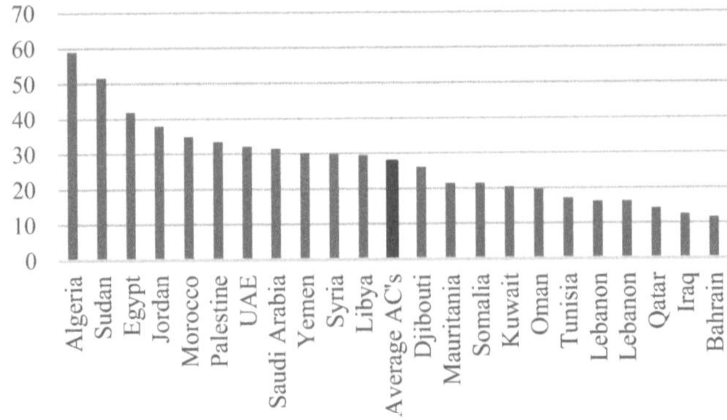


**FIGURE 6.7** Average ratios of students to staff at universities in the Arab countries (2011 statistics).

countries falls close to 30 students per staff member, which is around double the recommended average.

#### 6.2.2.4.1 Expenditure on Higher Education by the Arab Countries

In countries around the world, government spending on university education provides essential means for enhancing uniform development of human capital skills. With limited information concerning spending by nonprofit and for-profit institutions in the Arab countries, government expenditure on higher education, as reported by national statistical authorities and ministries of higher education to the United Nations Education, Science and Culture Organization (UNESCO) remains the most reliable measure of financial resources dedicated to higher education. Based on this measure it would seem that Tunisia heads the list of Arab countries, as well as their closest neighbors, dedicating 1.8% and 5.6% of its GDP and total government expenditure, respectively, to this stage of the educational process. Average percentages for all the Arab countries fall at around 0.98% and 3.2%, respectively, while among the Arab countries' neighbors the highest percentage of spending relative to GDP is made by Malta and Turkey at around 1.6 each, and by Iran, at 5.5 relative to total government expenditure. Interestingly, at 0.7% of GDP Lebanon spends less than the Arab countries' average and allocates only 2.5% of all government spending toward higher education, compared to the Arab countries' average of 3.2%.

#### 6.2.2.4.2 World Rankings of Arab Universities

Several schemes have recently come to the fore with a view to classifying universities worldwide, including several Arab universities. According to one of the most widely acknowledged schemes, namely the Times Higher Education (THE) classification, only one Saudi University, namely the King Abdulaziz

University was listed among the first 251–300 leading universities across the world. Four other universities, namely the American University of Beirut, King Fahd University of Petroleum and Minerals, King Saud University, and United Arab Emirates University were listed among the leading 501–600 universities. Nine universities fell into a third group of the leading 601–800 universities. Egypt and Jordan are represented within this group, with three universities from the former and two from the latter country. Five more Arab universities fall within this group, namely Morocco's University of Marrakesh, Doha's Qatar University, the United Arab Emirates' American University of Sharjah, and Oman's Sultan Qaboos University. Thus, in all only 14 Arab universities out of 41 universities in the region were among the leading 800 universities across the world. A feature that is worth noting in statistics presented in connection with these universities is the preponderance of foreigners within their student populations, compared to other universities in the region.

With regard to gender distribution, it is comforting to see that females are well represented as students in most Arab universities listed among the top 800 universities worldwide. Indeed, their percentage in at least 7 of the 14 universities in the list exceeds that of their male counterparts, outnumbering them in at least two universities, namely the United Arab Emirates University and Qatar University, by factors of 3.3 and 2.6, respectively. Notably, no female students appear to have registered at the King Fahd University for Petroleum and Minerals, indicating that fewer female students in Saudi Arabia take up science and engineering specializations linked to the country's abundant mineral resources.

Student to staff ratios at the Arab universities in the above list range between 6.8 students to staff members, for King Fahd University for Petroleum and Minerals and 46.3 for Cairo University, averaging overall at around 18. The fact that this ratio falls below a benchmark of 10 students per staff member in at least 5 universities, and below a benchmark of 20 students per staff member at 9 out of the 14 universities is notable. On the other hand, the Universities of Cairo as well as Morocco's University of Marrakech Cadi Ayyad seem to possess exceedingly high student to staff ratios at 46.3 and 44.1, respectively, while the University of Alexandria attained an interim position with a ratio of around 27.

Table 6.5 presents scores given with respect to individual criteria used in ranking Arab universities among the 800 top universities across the world. With regard to international outlook—an attribute presumably linked to the range of nationalities represented by the student body and faculty staff as well as linkages to foreign institutions of higher education—all GCC universities listed among top 800 universities worldwide score considerably higher than the rest. The average score for the seven GCC universities in the list is around 88, in contrast to 50 for the remaining Arab countries. Universities in Egypt rank lowest on this attribute with an average score of around 32.

Saudi universities take the lead in relation to income from industry, with an average score of about 72 for this attribute, as opposed to an overall average of

**TABLE 6.5** Values Given to Individual Criteria Used in Ranking Arab Universities Among the 800 Top Universities Across the World

| THE Ranking | University | Teaching | International Outlook | Industry Income | Research | Citations | Percentage of Foreign Students | Ranking in Asia |
|---|---|---|---|---|---|---|---|---|
| 251–300 | King Abdulaziz University | 23.9 | 93.0 | 73.1 | 11.8 | 76.6 | 21 | 26 |
| 501–600 | American University of Beirut | 27.7 | 93.0 | – | 11.2 | 31.9 | 23 | 84 |
| 501–600 | King Fahd University of Petroleum and Minerals | 25.4 | 84.5 | 81.2 | 16.1 | 24.4 | 21 | 66 |
| 501–600 | King Saud University | 20.0 | 75.5 | 57.4 | 17.5 | 22.3 | 7 | 96 |
| 501–600 | United Arab Emirates University | 18.8 | 95.0 | 30.1 | 14.8 | 21.7 | 28 | 101–110 |
| 601–800 | Alexandria University | 20.3 | 33.7 | 29.7 | 8.2 | 14.1 | 84 | – |

| | | | | | | | | |
|---|---|---|---|---|---|---|---|---|
| 601–800 | American University of Sharjah | 12.4 | 95.6 | 33.3 | 10.6 | 13.3 | 3 | 161–170 |
| 601–800 | Cairo University | 18.9 | 31.0 | 30.0 | 11.1 | 11.9 | 2 | – |
| 601–800 | University of Jordan | 19.0 | 55.7 | 28.8 | 7.7 | 7.8 | 17 | 191–200 |
| 601–800 | Jordan University of Science and Technology | 16.4 | 62.2 | 31.4 | 6.2 | 16.4 | 20 | 181–190 |
| 601–800 | University of Marrakech Cadi Ayyad | 17.4 | 42.7 | 29.2 | 6.3 | 12.1 | 1 | – |
| 601–800 | Qatar University | 9.9 | 99.9 | 55.2 | 10.3 | 22.3 | 43 | 121–130 |
| 601–800 | Suez Canal University | 22.6 | 31.6 | – | 5.9 | 16.0 | 2 | – |
| 601–800 | Sultan Qaboos University | 21.3 | 71.5 | 37.6 | 10.6 | 16.0 | 3 | 131–140 |

From THE rankings from: https://www.timeshighereducation.com/world-university-rankings/2017/world-ranking#!/page/0/length/25/sort_by/rank/sort_order/asc/cols/stats.

43 for all other universities included in Table 6.4. Universities in Jordan, Egypt, the United Arab Emirates, and Morocco scored considerably lower on this attribute, averaging around 30.

Scores attained by all universities in the table with regard to research activity are somewhat similar with an overall average of 11. Universities from Saudi Arabia and the United Arab Emirates stand ahead of the rest, with an average score of around 18, 16, and 15, respectively, for King Saud, King Fahd, and the United Arab Emirates universities. The average score for the rest of the universities in Table 6.5 stands at 9, with some universities, notably from Jordan and Egypt and Morocco, scoring at 6.

The view that King Abdulaziz University attained the ranking it did on account of citations attributed to its faculty and affiliates is indicated in this table, with the score for its citations approaching 77, in contrast with scores for the remaining universities ranging between about 8 and 32, and a corresponding average score of around 17. This needs to be viewed in relation to the score achieved by this university with regard to research activity, which is inferior to the other two Saudi universities included in this table, namely KFUPM and KSU. Scores achieved by the universities with regard to teaching quality vary between around 10 for the University of Qatar and 25.4 for King Fahd University of Petroleum and Minerals, with an overall average score of 20. The discrepancy between the citation and research scores attained by King Abdulaziz University might well be linked to its controversial bid, referred to in Chapter 4, toward enlisting highly cited specialists in various fields of science and technology as affiliates.

#### 6.2.2.4.3 Students on Study Missions Abroad

Information concerning Arab students undertaking tertiary studies abroad are rather scarce and generally unreliable. A notable exception is provided by Saudi Arabia whose Statistical Year Book presents ample information on Saudi students undertaking study missions abroad, including other Arab countries. The total number of both male and female students pursuing their studies abroad in 2015 amounted to 183,532. Around 26% of these students were females. Their distribution across levels of study is presented by the pie chart in Fig. 6.8,[21] which indicates that only a small percentage, around 13%, of students were registered in PhD programs. On the other hand, the majority of Saudi students studying abroad were pursuing Bachelors and Masters courses, in proportions estimated at 58% and 26% of the total, respectively.

Official records on studentships granted to over 177,400 Saudis on study missions abroad in 2015 indicate that around 57% of this total were meant to seek qualifications in engineering, health, and the physical and agricultural sciences, 41% in education, arts, humanities, and finance and business studies, and 2% in unspecified specializations.

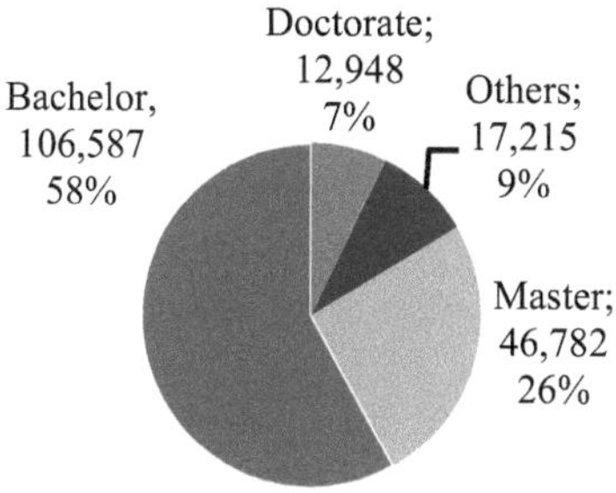


**FIGURE 6.8** Distribution of all Saudi students on study missions abroad in 2015 across levels of study.

### *6.2.2.5 Concluding Remarks*

While notable in some respects, achievements attained over the past few decades by educational systems in most of the Arab countries evidently proved inadequate for meeting challenges that faced their societies and seem even less capable of meeting challenges of the future. Major indicators used in evaluating achievements in basic and higher education point to the Arab systems' underperforming on several scores. Several aspects of both basic and higher education require urgent attention. Of particular concern in the former is the inability of educational methods to even out skills among student populations as measured by common criteria. With regard to higher education, but still in relation to providing quality higher education to the largest possible segments of Arab youth, two points must be kept in mind, the first is the low number of entrants into the tertiary phase. Thus, enrolment in this phase is considerably lower in the Arab countries than in other developing counterparts and fall far short of corresponding rates in neighboring countries. Secondly, the rising number of private and for-profit higher education institutions, most of which operate without adequate government supervision, could also add further disparities to existing imbalances.

Piecemeal reforms are likely to be of little value when radical improvements in levels of funding, curricular contents, teaching methods, and quality monitoring practices are all urgently needed.

The impact that political upheaval and violent conflict that heralded the present decade on education in several Arab countries is plain for all to see. Less obvious are reasons for the lackluster performance of the education sector in countries that have remained free from violent events and political upheaval while enjoying relatively high levels of national income. Complacency in this regard is all the more difficult to understand when lost opportunities due to inaction and inadequate policies are taken into account.

With regard to tertiary students undertaking their studies abroad, the extent to which a given country constitutes a source for tertiary-level students travelling to study abroad is linked to a variety of factors, including national policies designed to promote higher education. Such policies are expected to influence the capacity and quality of available institutions providing tertiary courses at home as well as initiatives aimed at supporting study missions abroad. Issues to be considered in both respects include enhanced resources, allowing various government departments, enterprises, and even families, to support tertiary-level students abroad.

Violent conflict, in countries such as Lebanon, during the 1970s and 1980s, Iraq since the early 1990s, and more recently, Libya, Syria, and Yemen, have come to play a more important part. Families in all of these countries, as well as in Palestine, sent their teenage children abroad on study missions in a bid to secure a better future for them as well as take them out of harm's way. Naturally, only families with sufficient financial resources and often with relatives abroad would manage to do so. While reliable statistics on those who eventually returned home having attained higher qualifications are unavailable, it is highly likely that a substantial proportion chose to remain abroad.

What may be stated with confidence nevertheless, is that during the past few decades, and even more so during the past few years, several countries in the Middle East and North Africa exported an increasing proportion of their talented youthful populations abroad to study and work when they were most needed at home. The cost of bringing up and educating regiments of talented youth only to leave their home countries and join the ranks of their well-educated counterparts in countries of Europe and North America must surely count as reverse aid, from the South to the North.

## 6.2.3 Unemployment, Poverty, Inequality, and Labor Productivity in the Arab Countries

Challenges faced by STI capacity building as well as overall socioeconomic development assume even greater significance in the Arab countries on account of their specific demographics, imposing higher dependency ratios than many other parts of the world, while a variety of regional and international development contribute to rising expectations by the region's largely youthful populations.

### *6.2.3.1 Unemployment in the Arab Countries*

Estimates of unemployment, poverty, and inequality in the Arab countries were referred to whenever available, in Chapter 2. In general, information on such issues in most Arab countries is either lacking or outdated. Prime reasons for this situation include lack of transparency and a general aversion to research in the social sciences. However, as indicated in various sections of Chapter 3, apart from countries that are presently going through severe political strife and military conflict, Egypt, Sudan, Mauritania, Morocco, and Tunisia appear to be facing severe unemployment, with particular regard to their youthful populations, for which unemployment ratios of the order of 30%–40% are a common occurrence.

### *6.2.3.2 Poverty in the Arab Countries*

Figures on poverty quoted by international reports on the Arab countries rely on official sources, which tend to present a far rosier picture of reality. However, based on figures quoted in the UNDP 2015 Human Development Report, it is estimated that at least 67 million of the combined populations of the 12 Arab countries for which poverty data was available[23] lived below national poverty lines and that at least 14.4 million live on under $1.25/day.

#### 6.2.3.2.1 Arab Populations in "Multidimensional Poverty"

The concept of multidimensional poverty was introduced by UNDP to reflect the fact that it is not only those living in abject poverty who cease to contribute to and benefit from socioeconomic developments but also a sizeable proportion of the less endowed population of a given country, living on limited resources will suffer lack of, or limited access to a variety of services, including STI inputs that go into health services, education, and living standards, hence making limited contributions to sustainable development, in general. See Frame 6.3.

Fig. 6.9 provides a view of the percentage of the MPI attributed to deprivations in education, health, and essential components of proper standards of living in the Arab countries for which data is available. Factors that determine access to education, health, as well as living standards, which include ready

**Frame 6.3 The United Nations Development Program's Multidimensional Poverty Index**

The UNDP Multidimensional Poverty Index (MPI),[22] was devised to highlight the plight of populations who, while not suffering extreme poverty in the sense of having incomes below the national poverty line, may still be deprived of access to educational and health services, be undernourished, or lack access to safe drinking water. This broader concept of poverty is reflected as a weighted average of 10 indicators. MPIs are classified within three categories, namely Health, including child mortality and nutrition, Education, including years of schooling and school attendance, and Living Standards, including availability of cooking fuel, water, sanitation, electricity, floor conditions, and assets.

UNDP considers a person as suffering multidimensional poverty when deprived in relation to at least a third of the above indicators, with each indicator having a defined deprivation level. Estimates issued in 2015 suggested that about 1.5 billion persons worldwide live in multidimensional poverty.

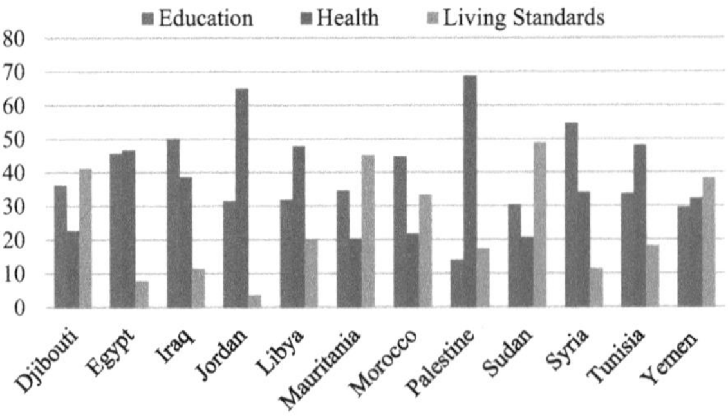


**FIGURE 6.9** Contributions of deprivation to overall poverty in Arab countries for which data is available.

access to clean water and modern fuels are, naturally very closely related to dissemination of STI inputs.

Fig. 6.9[24] clearly indicates that with respect to most Arab countries for which multidimensional poverty data are available, contributions by living standards to the MPI is limited, as compared to those due to education and healthcare. Exceptions to this trend include Yemen, Djibouti, Mauritania, and Sudan, in which inadequate living standards contribute more to the MPI than both education and healthcare.

It is also noteworthy that Egypt, Iraq, Morocco, and Syria stand out in terms of the size of the contribution of deprivation attributed to education among the Arab countries for which data is available, whereas Palestine, Jordan, Egypt, and Tunisia take lead positions with regard to the contribution of deprivation to the Index due to healthcare.

With regard to multidimensional poverty, it is estimated on the basis of available figures that around 45.6 million of the populations of the Arab countries for which data is available live in multidimensional poverty. This is almost one-seventh to one-eighth of the entire population of all Arab countries.

### *6.2.3.3 Inequality in the Arab Countries*

The impact of inequality on human development and education is particularly important. The 2010 Human Development Report[25] listed human development index (HDI) and Education Index (EI) values, adjusted by taking income inequality into account. Figures for several Arab countries, including all the six GCC states, are absent.[26] Fig. 6.10 presents a summary view of percentage change in the original HDI and EI values as a result of the corrections introduced. A disquieting, albeit predictable, observation is that the values of both indices suffered the largest corrections, i.e., they are depressed most, for the poorer Arab countries.[27] It is also noteworthy that the values of the education index for Egypt and Tunisia suffered considerable reductions, approaching 45% for the former and 35% for the latter when corrected for income inequality.

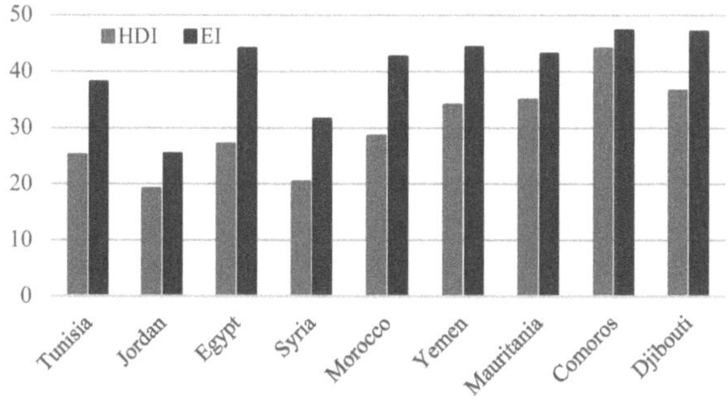


**FIGURE 6.10** Percentage corrections for income inequality of human development index (HDI) and education index (EI) values for Arab countries.

Limited information is available on income as well as wealth inequality in the Arab countries, in fact customary references, such as the World Bank's World Development Indicators databases and the UNDP's Human Development Report, do not include any figures concerning inequality in several Arab countries particularly those in the Gulf region. Judging by available figures from the above-mentioned database, values of the Gini index for the 10 Arab countries for which figures were quoted for or around the year 2013,[28] ranged between around 64 in Comoros and 32 in Egypt, averaging at 39.7. The view is rather similar with regard to data obtained from a UNDP database with figures dated in 2013.[29] Here again Comoros and Egypt occupy both extremes at around 64 and 31, respectively, with an average for the 11 countries for which figures were available at 38.8. A rather similar situation is reported by both sources concerning inequality in the Arab countries' closest neighbors with Gini index values for Iran, Israel, and Turkey in possession of Gini index values at around 38, 39, and 43, respectively.

However, the "Global Wealth Databook 2014," issued by the Credit Suisse Research Institute indicates figures of the Gini index presumably taking into account wealth as well as income inequality points to a much higher degree of inequality within the Arab countries as well as their closest neighbors. Thus, according to this report, values of the Gini index for the Arab countries ranged between 64.2 in Sudan and 85.5 in Lebanon, with an average value for the 19 Arab countries for which data was available,[30] of nearly 73.

Promoting social cohesion and welfare, reducing inequality, and alleviating poverty may not be taken simply as humanitarian gestures. Thus, accumulating evidence from a variety of sources seems to indicate that inequality fosters poverty and that both inequality and poverty hamper economic growth in addition to setting back social development.

Results of an analysis issued in 2014 by the Organisation for Economic Cooperation and Development (OECD)[31] suggested that income inequality induced statistically significant negative impact on medium-term growth. This OECD analysis also indicates that rising inequality by 3 Gini points, which corresponds to the average increase in inequality recorded for the OECD countries over the 1990s and the first decade of the century, appears to have reduced economic growth by 0.35% point per year for the previous 25 years.[32,33]

While many of the national development policies and vision documents discussed in Chapter 3 do acknowledge the need to create employment opportunities, it is very seldom that issues of poverty and inequality are addressed directly. Indeed, most innovation and enterprise promotion documents appear to favor more advantaged segments of society leaving populations in rural and marginalized areas to their own devices.

### 6.2.3.4 *Arab Labor's Productivity*

In addition to the state of unemployment, poverty, inequality, but still closely linked to inadequate education and training as well as availability of STI inputs and economic diversification in the Arab countries, are trends linked to labor productivity. Fig. 6.11 presents a view of estimated and forecast labor productivity published by the International Labour Organisation for the Arab countries, the world as well as three of the Arab countries' closest neighbors, namely Iran, Israel, and Turkey. Fig. 6.11 indicates that labor productivity in the Arab countries expressed as the contribution per worker to GDP, assumed a very small increase in 2010–2011 before declining markedly between 2011 and 2016. Fig. 6.11 also indicates that according to the ILO forecasts, while labor productivity worldwide as well as in the Arab countries' neighborhood has been constantly

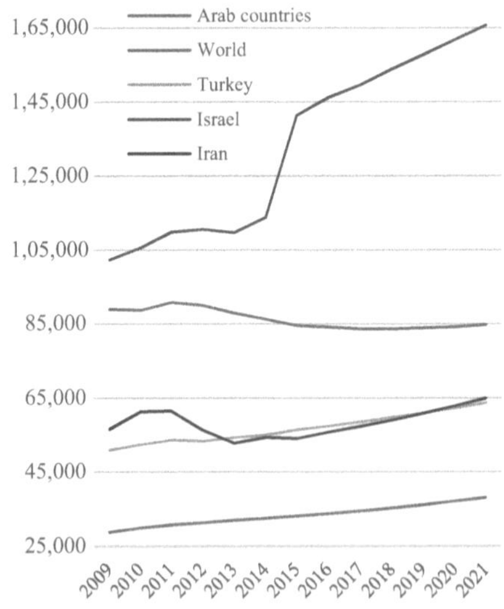


**FIGURE 6.11** Estimated and forecast productivity growth in the Arab countries and the world (GDP constant 2011 international $ in PPP).

rising and is expected to rise further till the end of the forecast period in 2021, was expected to remain flat or rise only marginally in the Arab countries by comparison.

## 6.3 STI CAPACITY BUILDING IN THE ARAB COUNTRIES; POLICIES, RESOURCES, AND INSTITUTIONS

As indicated in Chapters 2 and 3, all countries of the region adopted measures aimed at building their own STI systems. The following sections are intended for a brief discussion of the main characteristics of recent vision and STI policy documents, resource allocations in support of their implementation, as well as novel institutional forms established in this regard.

### 6.3.1 National Vision and STI Policy Documents

The past few years witnessed the launch of several policy and long-range vision documents as well as the customary national plans by most Arab countries. Many of these documents accentuated moves toward knowledge economies and information societies. Issues such as economic growth and diversification were clearly prominent in documents produced by the Arab countries in the Gulf region, while even documents authored by countries presently undergoing distressful conditions, such as Iraq and Syria, underscored as late as 2014 and 2015, the need to adopt STI-based approaches to their future economic development.

#### *6.3.1.1 National Vision Documents*

Only a few Arab countries, mainly GCC members have had the opportunity to produce full-fledged national vision documents charting intended paths for their future development. Most vision documents emphasize the need to attain concrete objectives with regard to customary issues including:

- economic diversification;
- reduced dependence upon foreign labor and enhanced employment opportunities for nationals.

All vision documents display preoccupation with the financial sector and investments as means for national wealth creation. Enhancing the quality of education is also a common concern. Some, such as Qatar's document, however, address sectoral issues in some detail.

A limited number of vision documents present numerical benchmarks. An outstanding exception in this regard is Saudi Arabia's Vision 2030 document, very briefly discussed in the following paragraphs, with particular reference to some of its declared numerical goals.

Saudi Arabia's Vision 2030 document received a good deal of attention on account of the importance of this country's future for the region and on the international arena. Chapter 2 included a brief discussion of this Vision's salient features, indicating its rather cursory treatment of STI-related goals. Thus, while long-term objectives sought by the Vision, in many spheres are linked directly and indirectly to STI capacity, the vision document does not address head-on the need to access, adapt and internalize such capacity within respective institutions. Additionally, the Vision's outward orientation in stipulating specific goals to be attained, by stating for instance that the country would enhance its international ranking on the "Logistics Performance Index" from 49 to 25 with or having at least five Saudi universities among the top 200 highest ranking universities, seems to sidestep the need to prioritize specific national challenges, rather than world rankings.

Additionally statements of some numerical goals do not appear to be sufficiently inclusive. Thus, in the education sector, the Vision document's stated goal is to train at least half a million employees through distant learning, while no indications are given of similar goals targeting other segments of the population. Similarly, numerical goals concerning ICT development as well as industry appear contrived in reference to international standing rather than directed at specific challenges at home.

Other features that characterize the Saudi Vision 2030 document include its limited specificity with regard to broader industrial and technological development, while singling out emphasis on localizing a considerable percentage of the country's defense spending within its borders. Its overall leaning toward financial investment as a principal means of national wealth creation is another noticeable feature.

#### *6.3.1.2 STI Policies*

While some of the Arab countries attempted to formulate national science and technology policies back in the 1960s and 1970s, few were in fact officially adopted and almost none were fully implemented. In many instances, most benefits gained were due to groups of national scientists getting together to discuss related issues and establish contacts with international experts, engaged mostly by United Nations organizations to guide policy formulation exercises, providing their organization's perspectives and preferred methodologies. Results of such exercises were mostly indirect and informal. Nevertheless, it may be claimed that they reflected positively on later institutional orientations and measures.

Innovation and related entrepreneurship activities were introduced into policy regimes by several Arab countries toward the end of the 1990s and early 2000s, with the formulation of full-fledged STI policies attempted by several Arab countries within the first few years of the century.

As indicated, in Chapter 3, STI policies, along with other government policies in many Arab countries, were seldom formulated on the basis of in-depth analysis of salient socioeconomic realities and future ambitions, then widely debated before being translated into cost-effective programs, with benchmarks and monitoring modalities. To the contrary, STI policies may often be based on disparate even contradictory decrees and official statements and it is only rare that serious research findings are employed in drafting STI policies.

The claim that the series of actions undertaken by a given ministry or department entrusted with STI governance constitute a virtual policy is simply untenable. In effect, no matter how well conceived and implemented such actions are, they may never be a viable substitute for a comprehensive national policy that has been collectively crafted, publicly debated, and then incorporated within national plans and implementation strategies.

The need for well-designed policies and implementation strategies should in no way be taken as an objection to a healthy measure of dynamism in policy and strategy regimes, allowing revisions, as and when needed, in view of changing STI ecologies, national priorities, and results attained through past endeavors. Indeed, such revisions should be carried out routinely. However, changes must be introduced (1) with due regard to widely agreed objectives, (2) on the basis of detained examination of outcomes already achieved, and (3) with view to fresh and well-defined goals and evaluation criteria.

### 6.3.1.2.1 STI Policies and the Quest for Knowledge Economies

STI policies and knowledge economy initiatives attracted considerable attention within Arab government circles during the past two decades. In particular, the notion that knowledge-intensive industrial and services activities had become the core of economic growth and social development and that economic activity throughout the world had now embarked on an entirely new era dominated by knowledge-based activities gained many adherents throughout the Arab countries.

Thus, several Arab countries, unaffected by events that swept the region have announced national plans for establishing knowledge economies and information societies, by the end of next decade. This will doubtless confront their socioeconomic development with immense challenges that require considerable resources and planning. Access to, and utilization of, modern media and institutional networks is likely to be the least of their troubles. More cumbersome and time-consuming will be goals such as human capital development, establishing quality standards, implementing long-distance learning and life-long educational and training programs, as well as modern legal and regulatory systems.

While developments over the past few years do not auger well for the immediate and short-term futures, it would be a grave mistake to abandon such initiatives. On the other hand, some of the notions at the core of bids toward realizing knowledge economies in the countries of the region will need to take into account much wider applications of knowledge, particularly of the scientific and technological varieties than is varieties held essential.

Thus, accepting the idea that some modern economies are in some sense more knowledge-intensive does not necessarily mean that only a restricted number of industrial and service sectors are the standard bearers of the new knowledge economy. Rather, it is essential to realize that knowledge bases of mature industries are also immensely deep and complex, as well as institutionally distributed, in a manner that must be recognized by policy makers.[34]

A more nuanced interpretation of the terms "knowledge economy" and "knowledge society" has been put forward taking into account "distributed knowledge bases," with diffuse systemic, institutional, and even geographical locations. Thus, knowledge with essential implications for many key activities may be distributed among a variety of institutions and knowledge domains. The implications for policy making of such a realization boils down to the need for better understanding of embodied and disembodied knowledge flows between components of "distributed knowledge bases," including institutions, groups of professionals, even within local communities of practitioners as well as qualified experts.

This is especially important for the developing economies of the region's growth trajectories do rest as much or even more on such sectors as engineering, food, wood products, vehicles, and so on, as they do on radically new sectors such as ICTs and biotechnologies. At the same time, innovations in all these latter sectors may well contribute to improved productivity and profitability in the more established sectors. While resting upon extensive knowledge bases, growth in the less fashionable "low-tech" sectors is certainly innovation based, leading to intermittent changes of varying degrees. The lesson in all this for future policymakers is that growth must be targeted not just within newly created sectors but also through constant technological upgrading of sectors that are already well established leading to transformations in their productivity and profitability.

However, for such transformative capabilities to be attained, there is need for complex innovation systems capable of creating, inserting, and disseminating advanced knowledge, sometimes linked to fundamental fields of research.

Thus, perhaps mostly by virtue of their maturity, "low-tech" sectors—such as food production, printing, publishing, even certain petrochemicals—intensively utilize scientific knowledge that is far more widely distributed across various fields of knowledge and institutions than the new high-tech sectors. Additionally, they may also be associated with a range of services, with significant but indirect science and technology inputs. In many instances, these science and technology inputs are supported by links to universities, research centers that are seldom explored in current literature addressing knowledge economies, that constituted the main sources for current STI and knowledge economy visions authored for the Arab countries.

In short, it is essentially futile to brand "low-tech" industries as knowledge nonintensive and both policymakers and practitioners should gain much greater awareness of their importance for balanced economic growth and social development. Plans aimed at creating knowledge economies must take their requirements into account, not simply as mature bearers and users of knowledge but also as drivers of further scientific and technological change.

On the one hand, there is also need to recognize that the more mature "low-tech" industries may well be averse to taking up novel science and technology inputs due to costs involved in both tangible and intangible assets they may have accumulated over many years. However, such may not be the case in many of the Arab countries, considering that while many may have taken up several traditional "low-tech" centers, most would not be so capital intensive as to impede their uptake of novel knowledge inputs at various levels, even when significant capital investments might be involved. At any rate, the above would necessitate the need to analyze knowledge development and industrial innovation decisions in such industries, asking questions such as:

- Under what conditions might it be possible for a firm engaged in food products, for example, to muster the resources to invest in the complex physical and intangible assets essential for knowledge-intensive approaches to production?
- What research and development, even basic research activity, might be necessary for rendering such transformations more facile and profitable?

It should be borne in mind that such questions are not meant as issues for corporate strategy designers, rather they ought to be answered on a case-by-case basis through effective interaction between a variety of agents including STI experts, economists, and industry specialists.

The above issues possess profound policy implications. And while policy-makers in many parts of the world, including the Arab countries, seem mesmerized by biotech and nanotechnologies, parallel, if not equal, attention might be allocated to areas of knowledge that may produce disruptive but profitable change across mature industries. This may not create a source of much worry, when resources are available for tackling knowledge needs of both the "low-tech" as well as high-tech industries. However, such is not the case in many of the Arab countries suffering severe constraints in terms of human capital as well as funding.

Clearly, the above may not be understood in preference for leaning toward knowledge creation activities and institutions that favor the needs of "low-tech" as opposed to high-tech industries. On the contrary, it must be taken to acknowledge that growth rests as much on technological upgrading of the former industries as well as creation of high technology sectors and that asymmetries in handling various sectors of Arab economies must be resolved in favor of more even-handed approaches regarding knowledge creation and utilization.

In addition to the need for producers of new knowledge inputs, countries attempting transition to knowledge- and innovation-based economies require firms operating within their boundaries to design and develop cutting-edge products and processes to maintain a competitive edge, thereby moving toward higher value-added activities. Such transitions also require an environment that is conducive to innovative activity and is nurtured by both the public and the private sectors. To create such an environment, there is need to provide sufficient R&D investment, by the private as well as the public sectors. See Frame 6.4.

**Frame 6.4 Enhancing R&D Allocations in the Arab Countries**

Rentier practices that invariably favor contracting foreign firms for expertise as well as off-the-shelf STI products and processes are mostly to blame for the regions' limited R&D capabilities. However, some of the blame must also be shared by the limited role played by scientists and engineers in influencing decision makers through open dialog as well as lobbying by various means. A commonly quoted excuse for their reticence is that most scientists may not possess political and public relations skills that enable them to engage in securing funding for their research institutions and projects and in promoting collaboration schemes with public, private, and civil society institutions. Hence, imparting such skills should become part of their training. At the same time steps should be taken to allow young researchers better visibility within the scientific and educational community and in society at large through taking part in debates and discussions about present and future research, and overall development plans would undoubtedly increase chances of higher funding levels. One thing is certain, no such funding would come about while researchers remain confined within their laboratories, coming out only when they achieve a breakthrough of one kind or another.

Participating in conferences and scientific meetings, engaging in small budget research projects and disseminating their own research outcomes should lead to greater opportunities for funding. Then there is a strong need for making scientists aware of the numerous international and regional, as well as the national opportunities that they might access.

However, while it is essential for the research community to press for more funding, researchers will also need to firstly demonstrate benefits that their research outcomes carry for society; secondly they should make the most of the available resources as well as seek to diversify sources of funding. To justify enhanced government expenditure, researchers must highlight society's essential needs for research-based knowledge as means for future development. This would merit specific research programs undertaken at the initiative of educational and development oriented institutions, with funding from public, private, and philanthropic institutions.

Other crucial requirement for creating such an environment include high-quality scientific research institutions, adequately equipped for application of research results, adaptation of acquired technologies to local needs and conditions, as well as the ability to come up with entirely new technologies. Extensive collaboration in research and technological developments between universities and industry is a key factor to achieve meaningful goals in all the above.

## 6.3.2 Resource Allocations for STI Capacity Building

Information on STI resource allocations in general is extremely scarce. Hence, emphasis in the following paragraphs will be placed on human capital and funding allocations to R&D, for which some figures are available, although outdated and in some instances inaccurate.

In essence, both in terms of human capital and regarding R&D expenditure as a percentage of GDP, the Arab countries now lag behind many developed, and even some developing countries. Expenditure by the Arab countries, recently averaged at less than 0.2% of GDP, a percentage that is inferior, not only in comparison to developed countries but also in relation to the world's average as well as expenditure by several developing countries.

With regard to human capital dedicated to R&D activity, the number of full-time equivalent researchers (FTER) per million inhabitants ranged between a minimum of 49.5 in Iraq and a maximum of 661 in Morocco, while neighboring countries such as Cyprus, Iran, and Turkey had 752, 750, and 803 full-time equivalent researchers per million inhabitants. Significantly, R&D expenditure and FTER figures were available for only 10 and 7 Arab countries, respectively, out of all 22 Arab countries. Additionally, salaries of academics in several Arab universities are often linked to their teaching load, with evidence pointing to even established and senior staff members at universities in some Arab countries struggling to enhance their teaching load to supplement their income, thereby reducing the time they might be able to commit to actually conducting their own research or supervising research activity by junior staff. The academic life of a young researcher is often occupied with lecturing, marking test papers, invigilating, and laboratory demonstrating, leaving them little time to contribute to research activity, in order to improve their knowledge and academic standing.

Whichever indicators are used to measure the level of national R&D expenditure, it is patently clear that funding allocated to R&D activities at universities, ministries, and research centers in the Arab countries is in need of massive enhancements through well-planned measures. In addition to R&D spending in the Arab countries being generally inferior to world averages, it is also likely that disparate norms are used in working out national R&D expenditures, leading to distorted figures, with a variety of unrelated expenses considered as R&D expenditures.

Steps must also be taken to improve R&D institutions' capacity to make optimal use of their resources and improve demand for their output by both the public and private sectors.

Limited information is available concerning spending on ventures designed to promote innovation, whether in enterprises and in order to enhance national capabilities in e-learning, e-government, e-health, and e-commerce. Nevertheless, figures presented in various parts of Chapter 3 clearly indicate that allocations toward innovation are generally small and particularly out of proportion to actual needs.

Most of the science and technology policy documents produced by Arab countries over the past two decades pledged increased resource levels for R&D activity. It is essentially in only a few countries, including Tunisia and Morocco that such pledges have been met, and then only in part. At any rate, it is unlikely that such increases, once realized, would lead to immediate improvement in overall R&D performance. This is essentially due to the time needed for establishing new research facilities, train essential and auxiliary cadres, as well as reach an agreement with likely end-users on priority issues and collaboration modalities.

Time will also elapse before improved quality of R&D activity attracts the attention of likely partners, both regional and international, inviting research networking and collaboration, thereby launching a benign cycle, leading to lasting improvements in R&D volume and quality. In practical terms, this means that at least a decade may be expected to pass before research in the Arab countries taking measures to radically enhance their research spending would achieve research intensity, productivity levels, and ensuing impact that have already been realized by their close neighbors, Iran and Turkey, let alone leading R&D performers around the world.

At the end of the day, merely increasing budget allocations for R&D may never be sufficient, of its own accord, for producing the desired outcomes. A host of interventions would have to be launched in unison. Improved management practices, selection of research priorities that are closely linked to sustainable development objectives, strengthened links between R&D institutions and various end-user sectors, as well as institution of effective monitoring and evaluation measures are some of the main issues that require urgent attention. In most instances there will be a need to formulate fresh R&D policies or at least introduce drastic changes into existing policy schemes.

## 6.3.3 Novel STI Institutional Forms

Several Arab countries established research centers as well as auxiliary science and technology facilities, such as testing, calibration, and quality assurance facilities earlier in the 1960s and 1970s. The rise of novel institutional forms combining scientific research and technology development with enterprise and industrial clusters commenced only toward the end of the 1990s, in emulation of widely reported successes by the Silicon Valley in the United States, Sophia Antipolis in France, and in the Cambridge Science Park in the United Kingdom, established in the 1970s and 1980s.

The following paragraphs will very briefly outline efforts made in the region to establish novel institutional forms.

### *6.3.3.1 Technology and Innovation Parks*

Reference was made in several parts of Chapter 3 to novel institutional entities established by countries such as Egypt and Saudi Arabia with a view to enhancing their STI capabilities, including technology and innovation parks as well as a variety of incubation schemes dedicated to promoting innovation and innovation-based enterprises. A comment that may safely be made with regard to all novel institutions established during the past few years is their overwhelming preoccupation with almost pure commercial outcomes to the neglect of socioeconomic objectives that may not produce immediate commercial returns, while however, possessing enormous other benefits that should eventually translate into future economic growth.

Novel institutional schemes in the Arab countries, whether associated to university institutions or otherwise, have mostly targeted ICTs and related applications. Additionally, several such schemes have been largely taken over by branches of multinationals seeking to expand their businesses in the region. While detailed impact assessments are overdue, it may safely be assumed that these schemes may well have had positive impacts, albeit confined to the institution's clients and their businesses, with particular accent on ICTs.

Extending the domain of novel institutional entities into other branches of technology-based enterprises, while still awarding ICTs priority status, is a pressing need. Initiatives in this direction should ensure injection of novel STI inputs into various manufacturing sectors, including mature industries, agricultural production, as well as services sectors, including education and transportation, which are in dire need for enhanced productivity through novel technology inputs in almost all the Arab countries.

### *6.3.3.2 Enterprise Incubation Schemes*

Almost all Arab countries, including some that suffered political turmoil and armed conflict over the past few years had established enterprise incubation schemes, mainly on public university premises with the aim of giving graduates and staff opportunities for launching enterprises based on research outcomes and other novel technology inputs.

A brief review of such facilities reveals the need for securing a fuller range of preincubation services, to help rapid takeoff by technology-based enterprises. Similarly, establishing close links between national university incubation schemes and those run by European and other overseas universities as well as international incubators associations should facilitate acquisition of best practices and enhance the productivity of national incubation schemes.

As indicated in Chapter 3, incubation schemes in most Arab countries are allocated low levels of funding. Additionally, only a small number of universities in each one of these countries possess incubation schemes and associated facilities. Both issues would have to be taken up in fresh policy formulation efforts.

Equally important is the danger of commercialization tendencies at the core of incubation activities sidelining and even derailing urgently needed incubation ventures aimed at disseminating specific technologies and associated services within marginalized and impoverished rural and urban communities. Examples of such schemes might include dissemination of solar power systems in rural schools and small farm houses, development of online and other distant educational schemes. While such efforts do not need to be nonprofit ventures, they might attract limited interest on account of their lower profit margins and long-term returns. At any rate, it is essential that such projects are properly valorized and subsidized by concerned authorities, with specific modalities adopted to ensure that such incubation ventures are allocated priority as and when needed.

### 6.3.4 Concluding Remarks

Chapters 2 and 3 included summary accounts of several national development policies and national vision documents authored by concerned institutions in the Arab countries during the past decade or so. Drawbacks suffered by concerned institutions in many such documents are due to their adoption of statements concerning the intention to create knowledge economies and societies while essential conditions for such quests, as embodied in educational systems, scientific research, technological development, and innovation in public and private enterprises, are not tackled with the seriousness they merit. Additionally, lack of sectoral development policies deprived STI policies of much needed focus, resulting in diluted STI support and feedback into crucial economic development activities.

Problems that some of these documents suffer often appear to be due to their adoption of one-size-fits-all STI policy templates provided by international development agencies, with limited due serious consideration of socioeconomic requirements and conditions within which such policies would be implemented.

As evidenced by successful experiences across the globe, optimal outcomes are contingent upon coordinated action on several fronts, with STI and associated knowledge capabilities close to the center, but never impeding initiatives seeking sustainable development on the periphery. Additionally while certain goals may be achieved through astute technology acquisition, other, even more important, goals in education, enterprise, and job creation will remain elusive in the absence of overall development policies targeting wide-reaching political and economic development with emphasis on eroding and ultimately dismantling deep-rooted rentier practices and attendant relations.

In other words, knowledge society initiatives are no panacea for all development problems, and action is needed on several fronts, with well-researched and result-oriented STI policies, knowledge society initiatives, as well as commensurate resource allocations playing a central role in achieving many socioeconomic development goals at acceptable financial and societal costs.

To many, the present state of STI capabilities in countries, such as Egypt and Lebanon, both of which took relatively early steps toward setting up widely recognized higher education and some research facilities, is clear confirmation of retrograde trends engulfing the entire region for several decades now. Thus, Egypt which took several steps during the late 19th and early 20th century to build higher education and research facilities was ranked 128th worldwide and 12th among 13 Arab countries included in the survey conducted by the World Economic Forum's Global Competitiveness Report for 2016–2017.[35] Lebanon, one of the first countries in the region to host the precursor of the American University of Beirut, in 1866, and to formulate a research policy back in 1960, was classified on the basis of the quality of its research institutions in 91st position among 140 countries evaluated by the same report.

## 6.4 THE ROAD AHEAD: FUTURE DIRECTIONS FOR STI-BASED DEVELOPMENT

Events of the past few years have clearly shown that the way ahead toward sound and inclusive socioeconomic development for the Arab countries is extremely arduous. Historical and cultural factors conspiring with economic, demographic, and resource-related considerations might see to it that it remain so for several years to come.

With the future of coming generations in and around the Arab region at stake, a most urgent need arises for alternative paths to those that led the region into its present quagmire. As many would certainly agree, solutions will not be simple, least of all when unidimensional and with linear trajectories. The current state of the globe's economy, environment, and political climate will see to it that this would certainly not be the case.

While hopes must ultimately rest with the endurance and resilience of human beings in general, action must be taken to avert many more repetitions of the kind of crises recently witnessed by several countries of the region.

Action on many fronts, whether social, economic, or cultural stands to receive enormous contributions from science, technology, and innovation inputs. However, resorting solely to earlier models and agencies for future action will only guarantee much of the same unfolding in the not-too-distant future.

Daily events, whether in the region, and around the world clearly indicate that new actors and new modalities are called upon with instruments of science, technology, and innovation acting as enablers and impact multipliers on many fronts.

### 6.4.1 New Roles for Traditional and Novel Actors

The following paragraphs constitute a summary review of proposals intended to guide progress toward improved STI capacity with the ultimate goal of inclusive and sustainable development. All that is included in these paragraphs will need to be looked into and carefully customized to specific conditions that present themselves in the various Arab countries, with a view to their inclusion into future national as well as regional and local community agenda.

#### *6.4.1.1 Government and Public Sector Institutions*

Roles for governments and public sector institutions in STI capacity building have so far been tilted toward funding the education sector, limited research within selected institutions of higher education, as well as a small collection of research facilities, in many cases dedicated principally to agricultural development. With only a few exceptions, little attention has been paid by governments and public institutions to:

- mobilizing greater demand, particularly for scientific research and technology development and adaptation;
- support for innovation, particularly within small and micro enterprises, beyond a few showcases that appear to be designed to benefit the elite, rather than the poor and marginalized, segments of society.

Special government programs may be designed along lines taken up by developed countries to generate demand for research that is then transformed into viable businesses through public procurement initiatives.[36]

### 6.4.1.2 The Private Sector

Private sector institutions can aid the development of STI capacity with particular regard to approaching knowledge economies by improving access to finance. While some programs have in fact been designed for such purposes in several countries, particularly in the Arab countries of the Gulf region, resources allocated have generally fallen far short of needs and the fact that these programs are confined to national boundaries, rather adopt regional or subregional perspectives, considerably limit their effectiveness and reach. Clearly, appropriate regulatory frameworks are of extreme importance, particularly where regional and subregional STI collaboration is concerned. However, the fact that several financial as well as other private institutions are active in several Arab countries should facilitate the establishment and enforcement of such regulations.

### 6.4.1.3 Civil Society Organizations in STI Capacity Building

Movements that spread across parts of the Arab world in 2011 may be read as an expression of frustration, by the younger generations in particular, with lacking accountability and transparency as well as widespread corruption and cronyism in the Arab countries. Despite significant gains in countries such as Tunisia, as well as precautionary measures taken by other countries, still untouched by protest movements, several other Arab countries still face pressing problems—civil unrest, unemployment, and dysfunctional institutions due to continuing armed conflict.

While civil society organizations may not be counted upon to correct all of the above, their role may be invaluable in arriving at wider participation in tackling many of the problems Arab societies are facing, particularly in areas where governments are yet unable to function, such as the case in many parts of Iraq and Syria.

Precautionary measures will certainly have to be taken to ensure that strategies devised by civil society organizations and the links they establish with foreign governments and international organizations are not viewed as separatist movements or initiatives seeking to fragment national entities. On the contrary, for civil society organizations to succeed, it is essential for them to be widely acknowledged as bridging modalities seeking to engage a wider diversity of stakeholders and beneficiaries, including government departments wherever they are capable of functioning.

Several areas may be earmarked for effective contributions by civil society organizations. However, considering the constraints they face in many parts of the Arab world, the following domains might be awarded priority:

- Delivery of food aid and food processing equipment, enabling communities to utilize and market available agricultural resources;
- Basic education and vocational training, with particular accent on training girls and young women with family responsibilities;

- Essential health services and delivery of medicines and medical supplies;
- Microcredit schemes enabling aspiring entrepreneurs to launch their own small and microbusinesses;
- Setting up water treatment and sewage facilities.

Action on only some of the above priorities is already being taken by civil society organizations in refugee camps in, for example, Jordan, Lebanon, and Turkey. And while not directly aimed at the delivery and dissemination of STI inputs as such, all of the above tasks involve contributions to STI capacity building at community levels, setting the stage for further STI-based development as normalcy returns.

Under more peaceful conditions, experiences gathered through interventions in the above domains should foster the emergence of widespread collaboration with state institutions and with greater emphasis on actual implementation of novel technologies for socioeconomic development.

Information gathered so far concerning community relief in Arab countries afflicted by violence, points to limited participation by scientists and engineers,[37] with professional contributors to interventions such as the above made by volunteer medical personnel. While this may be partly understandable in the case of university scientists and engineers actively pursuing teaching and research duties at universities and research centers, a good deal might be gained through undergraduate and graduate research projects, supervised by established scientists and engineers to design and deliver STI-based interventions to impoverished and marginalized communities. Contributions by active as well as retired Arab scientists and technologists abroad should also play an important part in improving conditions within such communities.

Clearly, for action along the above lines to succeed, there is need for universities and other educational and vocational training establishments to adopt conducive policies, possibly in cooperation with international agencies with good track records and solid field experience. There is scope here as well for action by associations of scientists and engineers, particularly in liaison with private sector enterprises with a developed sense of social responsibility.

In support of the above-mentioned role by civil society organizations, concrete contributions must be made by scientists and engineers toward national and community development. Contributions along these lines have traditionally been channeled through formal routes, mediated by both public and private institutions, which have traditionally bypassed the most needy communities for a variety of reasons. Clearly, contributions by scientists and engineers would often have to be compensated although recent events have widely demonstrated willingness on the part of some to donate their expertise and services without pay. At any rate, any expenses must also be taken up by fresh policy measures aimed at equitable and sustainable development, by governments as well as donor and civil society organizations.

Finally, civil society interventions aimed at STI promotion might well be expected to meet their most fertile grounds within marginalized and impoverished rural and urban local communities. Judging by experience in the Indian subcontinent, alliances between civil society and business promotion agencies should provide added impetus to inclusive development in such communities.

#### *6.4.1.4 International Organizations*

Two main issues are raised in criticism of campaigns launched by international organizations, such as the World Bank, among others with the aim of STI promotion. The first concerns the need for approaches to delve much deeper into the particularities of the country they are supposed to help rather than adopt one-size-fits-all approaches. The second concerns entry points chosen when launching such campaigns. Thus, several examples may be put forward to illustrate how the choice of a particular institution or ministry resulted in rivalries among concerned national institutions rather than consolidation of their efforts.

In general, international, and indeed regional, organizations ought to orient their STI campaigns, including those aimed at creating their own knowledge economies and information societies as means of unifying efforts on the part of as many concerned organizations as possible, rather than single out just one lead national institution as the sole operator. Incidents in which the latter approach was utilized have often led to less than optimal outcomes for a variety of reasons, particularly in the absence of an overriding vision spelled out by well-articulated implementation strategies.

In fact, a good deal of the successes achieved by STI institutions taking part in collaborative research may be ascribed to the fact that the programs and constituent projects they engaged in included inherent components that demanded networking and collaboration allowing all participants to reap concrete benefits.

At any rate, several examples may be put forward to illustrate the need for inculcating an institutional culture that favors STI collaboration including a disposition toward networking. Thus, the experience of the consecutive Framework programs appears to demonstrate the importance of adopting projects that promote the creation of networks, of even three to four parties, to begin with, but with the possibility of expanding, through attracting more funding and wider interest in participation by qualified STI teams.

Additionally, for optimal results, STI collaboration will need to be based on well-formulated national policies, with clearly stated collaboration modalities, transparent funding, management procedures, accounting methods, and well-designed evaluation modalities.

### 6.4.2 Building New Institutional Forms

All Arab countries must be credited with establishing novel STI institutions to cater to scientific research and innovation. Total reliance on overseas think tanks has been the rule, rather than the exception whenever it was necessary to formulate STI policies or create novel STI institutions. While consultations with overseas experts and specializations may be an absolute necessity at all times, creating national and regional capacity for doing policy research, monitoring STI developments through specialized observatories, as well as optimizing technology transfer through national and regional sectors would go a long way toward making optimal use of all expertise, home-grown as well as overseas in origin.

### 6.4.2.1 STI Policy Research Units

While various new institutional forms have in fact been created by the Arab countries to promote innovation and research, none appears to have established permanent units to undertake policy research with the objectives of streamlining STI policy formulation, examining the needs of particular sectors and application areas as well as evaluating outcomes and drawing lessons for future interventions. Establishing such units, even in virtual space, should transform many aspects of STI policy formulation; through providing timely analysis of the state-of-the-art of technologies required for priority applications and exploring particular STI needs of priority sectors.

Given the state of flux which characterizes many aspects of development in the region, there are strong reasons for extending the scope of foresight exercises undertaken by STI policy units to include STI inputs required for socioeconomic development and environmental protection. The results of such extended foresight exercises should result in improved funding and human capital allocations to R&D activities as well as actual planning of subsequent action concerning STI interventions, including the locating of, and partnering with, regional and international actors.

No better example of the benefit that such exercises might have is possible than the case of drought and soil salinity that hit Syria's north eastern regions driving many inhabitants into abject poverty and providing extremists with fertile recruitment grounds. Had such exercises been carried out in the late 2000s, then large segments of the population might have been spared the turmoil and human misery that they eventually had to endure. Extending the scope of STI foresight exercises into domains that include social, demographic, and healthcare issues should also help exploration of STI trends in combatting ailments such as the spread of liver diseases in Egypt and diabetes in the Gulf countries. Devoid of such capabilities, many Arab countries are likely to fall prey to costly ad hoc interventions.

### 6.4.2.2 STI Observatories

One of the main shortcomings of past practices in policy formulation and implementation in the Arab countries has been the lack of reliable data upon which policy directives and subsequent strategies and programs must surely rest. While some of the Arab countries have announced plans for setting up observatories to monitor scientific and technological development at the behest of some United Nations organizations, it is unlikely that any of these observatories has in fact attained planned staff recruitment targets and adequate operational maturity. Apart from observatories especially dedicated to monitoring science and technology developments, most of the Arab countries now possess central government statistical offices, some of which publish Statistical Year Books and special reports some of which touch upon educational as well as national science and technology-related activities.[38]

Policies formulated to facilitate STI capacity building often suffer limited access to hard data on crucial issues, which may be one of the reasons why many such policies are often quite short on details of achievements made in the past and goals that need to be targeted on a multitude of issues. Uncertainty, on the part of policymakers, concerning resource allocations and shifting priorities might offer additional explanations for this lack of specificity.

At any rate, it is essential to reiterate that an essential condition for future progress based on STI inputs, must rest upon unfettered access to data crucial for STI policy formulation, monitoring, and outcome evaluations.

### 6.4.2.3 *Technology Transfer Centers*

As several past surveys and studies have indicated,[39] neither academic and research communities nor industrial enterprises, were actually in a position to develop new, competitive technologies or products in response to local needs or to compete on the regional and international markets. Nearly all such technology inputs were imported, while capabilities for unbundling and reproducing imported technologies remain underdeveloped. In effect, few or no bridging mechanisms link academic and research communities to production and services sectors.

Technology transfer centers (TTCs) are valued as tools for bridging the technology gap suffered by production and services sectors in the developing countries. Through such centers, it is possible to disseminate valuable product and process technologies.

The ultimate responsibility for running TTCs, for networking and monitoring their activities would best be assigned to the national agency or consortium of agencies charged with implementing the country's innovation policy.

The need for TTCs to engage in training activities besides merely providing technological expertise has been demonstrated in various parts of the world, in Egypt, for example, where TTCs have gained an important role in offering demand-oriented support for local industry. Engaging in pilot projects and trainings in innovation management as well as technology adaptation and development would constitute valuable offerings by TTCs.

In summary, by addressing issues on the supply and demand sides as well as processes involved in implementing fresh technological knowledge, TTCs could substantially improve innovation capabilities of Arab enterprises, allowing for greater competitiveness on the regional as well as international markets.

Additionally, while setting up TTCs might be considered costly in terms of human capital development as well as hardware acquisition, their positive impacts on the national innovation system are expected to more than compensate such investments.

#### 6.4.2.4 Manufacturing and Technology Clusters

No traces of a cluster policy may be detected in most of the Arab countries. However, it is essential to acknowledge the presence of promising sectors that might gainfully adopt clustering approaches as well as concrete possibilities for the formulation and implementation of a clear cluster policy with moderate efforts and investment levels. Analyses of strengths and weaknesses as well as the future opportunities should be undertaken in formulating such a policy. Quick wins might be attained by targeting specific manufacturing sectors with collaborative ventures within selected competitive sectors, e.g., in specialty textiles, traditional food products, furniture, and copperware items. However, rather than merely resuscitating artisanal crafts, such clusters would need to adopt a much wider outlook with regard to its range of products, processes, and markets.[40] Targeting existing industrial zones with cluster initiatives should help accelerate moves toward agglomeration of similar manufacturing and services activities.

The expected socioeconomic impact of industrial clusters in the Arab countries is expected to be considerable in many ways. In particular, should clusters be established on sound economic and environmental grounds, then, they might be expected to make effective contributions to sustainable development, with rather limited investments. Success factors that need to be guaranteed include adequate preliminary analysis of pertinent conditions as well as management practices that motivate cooperation among cluster members and their surroundings, including local authorities and civil society institutions. Effective links to local and nearby universities and other educational establishments should go a long way toward problem solving as well as promoting new ideas.

The cluster approach could be adopted in support of efforts aimed at reviving impoverished and marginalized local communities and regions. Such efforts would, however, need to be based upon in-depth analysis of economic, environmental factors, as well as the innovation potential of targeted regions, in addition to traditional supply and demand factors.

### 6.4.3 Alternative Means of Funding STI Activity

With regard to STI funding, a considerable gap exists between programs and budgets available, on the one hand, and actual needs, on the other. Additionally, there is need for designing program that matches different competencies from academia and industry.

To start with, government institutions would certainly have to enhance their own STI allocations, particularly emphasizing spending on R&D priorities as well as enterprise incubation schemes. At the same time, however, alternative means for increasing overall STI allocations must be considered. Access to various charitable and religious endowments normally dedicated by individuals and institutions in the countries of the region for aiding the impoverished and often used for providing food, clothing, and, less often, educational grants for poorer populations. In several countries of the region, these endowments are collected by a special ministry and used almost entirely for the upkeep of religious institutions and monuments as well as institutions of religious education. Using at least a small portion of such endowments, which incidentally amount to enormous sums in countries such as Egypt for example, in support of STI activities, in general, and R&D on pressing health and environmental issues, in particular, should be permissible on both ethical and practical grounds, because it would produce clear benefits for the same populations that the initial endowments were intended for in the first place.

Private R&D institutions are virtually nonexistent in many parts of the Arab world. Very few foundations have recently been set up, in selected countries, including Lebanon, Saudi Arabia, and the United Arab Emirates with sociopolitical and cultural studies as essential domains of activity.[41] New legislative and framework conditions are needed for extending the scope of activity of such institutions and for creating fresh facilities to cater to studies and actual experimental research in the physical, natural, and health sciences, ensuring their subsequent sustainability. Nevertheless, it should be possible and indeed beneficial for federations of chambers of industry, in particular, to embark on programs aimed at setting up auxiliary STI facilities providing precompetitive technical advice relating to an assortment of issues including product quality, environmental compatibility, and health-related matters as prelude to setting up R&D facilities targeting specific priority areas.

### 6.4.3.1 Making Use of International Research Funding

A recent evaluation of the EU seventh Framework cooperation program (FP7) reveals that the level of funding that EU funding that is dedicated to non-EU partners, essentially all Mediterranean countries is ~10% of total allocations to FP7. Nevertheless, it is well to remember that several European countries, including France, Italy, Spain, Germany, and even the United Kingdom, but perhaps to a lesser extent, actively fund bilateral programs with countries in the Middle-East and the North Africa, particularly in the Maghreb Arab countries.

To make optimal use of such funding sources, as well as funding that might become available through bilateral collaboration among the Arab countries themselves, or through bilateral arrangements with other countries in the region, including Turkey and Iran, there is need for the formulation of sound policies with clear objectives and follow-up modalities.

Further exploration of the above means for enhancing R&D spending as well as additional funding avenues, would need to be undertaken essentially by members of the scientific community aided, as needed by development, legal and financial experts.

### 6.4.4 Promoting Innovation and Entrepreneurship in the Arab Countries

Global economic conditions during the last 3 years of past decade as well as economic policies adopted by many Arab governments due to lower oil prices, led to uncertainties within many spheres of economic activity. While enterprise incubation, technology parks, and various other initiatives aimed at promoting innovation may have prospered in some countries, the above developments had the effect of driving more graduates to seek secure employment within the public sector. The following few paragraphs outline recommendations with regard to promoting innovation and entrepreneurship in the Arab countries.

#### *6.4.4.1 Raising Awareness of Innovation and Entrepreneurship*

Terms such as "innovation," "innovation policy," and "innovation management" do not seem to have the same meaning and relevance to all stakeholders in many parts of the Arab world. In fact, even some decision makers appear to consider R&D and innovation synonymous or very closely linked to one another. This raises the need for activities designed to improve understanding of innovation and the role played by technology in creating, disseminating, and implementing innovations. Steps proposed by several national innovation policies include making optimal use of universities to promote innovation as well as entrepreneurship. Additionally, awareness campaigns, competitions, and training courses sponsored along these lines by professional associations and industry and trade federations and implemented by their communities of experts should reap short-term benefits, with relatively high leverage and manageable costs.

#### *6.4.4.2 Nurturing a National Innovation Culture*

Innovation friendly regulations are but one component of an innovation culture. Other prerequisites include a series of networking activities created and maintained to transfer research outcomes and all kinds of innovative ideas into commercial, as well as noncommercial applications targeting particular development objectives.

Promoting a nourishing culture that favors innovative practices across the board are essentially tasks for many actors including public as well as private sector institutions in addition to civil society. Advocacy, coupled to concrete measures including training activities and support for specific innovative ventures on the part of all the above actors, is needed. Civil society institutions, industry and trade federations as well as professional associations in particular, could play an important part in accelerating moves toward a durable and dynamic innovation culture.

Programs designed for awarding grants and prizes on competitive basis to innovators should also help create and maintain an innovation-friendly culture. While returns based on success in the marketplace might be considered as sufficient compensation, publicizing successful innovations aimed towards sustainable development objectives could also go a long way toward promoting a conducive culture.

In other words, there is also need to advertise and celebrate innovations with impact on development, but with limited or no immediate financial returns. To attain tangible results along this path, it would be essential to secure government intervention aided by national stakeholders concerned with promoting innovation for sustainable development. At any rate, to discover, support, and reward innovators must be an on-going task for a variety of public and private institutional actors. By highlighting examples of successful innovations, the national innovation culture is revitalized and a benign cycle is set with positive impacts leading to further actors taking innovative paths, creating demand for further innovative activity and stimulating increased output.

Other avenues toward creating and sustaining an innovation culture might include:

- Entrepreneurial conferences, in close cooperation with the universities, in various regions and local communities;
- Dissemination of success stories, both local and international;
- Courses to impart entrepreneurial skills at selected university departments and other institutes of higher education;
- Funding program for entrepreneurs.

### *6.4.4.3 Roles for Universities in Promoting Innovation and Entrepreneurship*

The university landscape in many Arab countries is sufficiently well developed for it to support initiatives designed to promote innovation and entrepreneurship within a relatively broad spectrum of disciplines and sectors. Recently established private universities could become leading actors on the innovation/entrepreneurship landscape. At any rate, both public and private universities would need to match their educational curricula to the development of socially responsible, as well as economically viable, innovation and entrepreneurship cultures, with emphasis on selected production and services sectors identified as proponents and home for a wide range of innovations. With this in mind, business development and associated risk management practices should become integral parts of educational curricula linked to such sectors.

In combination with preincubation and incubation schemes, briefly discussed below, changes in educational curricula should foster the emergence of new ways of learning modalities, combining research, education, and business creation. Universities would in effect become partners in promoting innovative enterprises that are conducive to inclusive and sustainable socioeconomic development, at least partially absorbing initial start-up costs as well as providing infrastructure support. Course projects might include the design, planning, and even experimental implementation of early-stage innovative enterprises.

#### *6.4.4.4 Innovation Friendly Regulations*

While policies devoted to promoting innovation have recently been formulated in several Arab countries, concrete regulatory arrangements designed expressly in support of innovation are either absent or embryonic in many parts of the Arab world.

Prompted by international organizations, regulations in several Arab countries have tended to overemphasize directives linked to intellectual property rights (IPRs). However, IPRs are essentially only one segment of the mosaic required for launching effective innovation systems. Since in addition to protecting innovators, whether at home or abroad, there is need for reforms in numerous areas before viable innovation systems may prosper. There is a strong need, for example, to establish regulatory arrangements to:

- channel funding toward essential and promising domains of innovative activity;
- adopt measures limiting associated risks and promoting their management;
- ensure that implementation of regulations proceeds in a socially responsible, equitable, and transparent fashion, while still creating economic gains.

While it is essentially a task for the government to formulate and implement national regulations to tackle the above issues, it would be imperative for a variety of nongovernmental organizations to take part in proposing and drafting relevant regulations. And, in addition to concerned public sector institutions, it would be essential for the banking sector, industry, and trade federations as well as civil society institutions concerned with job creation and poverty reduction, to contribute. Such contributions are urgently needed to avoid situations in which guidance by international organizations and donor agencies is implemented with little regard to specific national and sectoral particularities.

#### *6.4.4.5 Funding Innovation*

Funding innovation in most Arab countries is distributed among several institutions, rendering well-designed policy directives and national initiatives complex exercises. Deficiencies suffered by funding innovation in the Arab countries are linked to the following two, somewhat related attributes:

- relatively small allocations in view of evident needs;
- reticence by various actors, but principally governments, to underwrite costs in relation to innovations targeting purely social returns, with limited financial returns, but likely enormous long-term benefits for overall development.

In most cases, there will be need for a high-level body to coordinate funding decisions, particularly for specific cross-cutting innovation schemes, e.g., in relation to introducing e-government or e-commerce schemes. While several Arab governments did set up coordinating and supervisory bodies to cater for higher education, scientific research, and technology development, innovation as such has remained an orphan topic without a dedicated institution; assigned specific responsibilities for promoting and funding related activities.

Finally, setting up agencies to provide funding for innovations and monitor their implementation may not proceed unless governments decide to spend significant amounts on innovation support programs, through policies targeting the creation and maintenance of well-tuned national innovation system.

## 6.4.5 Cooperation and Networking for STI Capacity Building

Scientific networks in certain disciplines and subdisciplines could be promoted with considerable returns in terms of knowledge creation as well as manpower training. Additional resources would naturally be required to create and further support networking arrangements. However, the volume of such resources should not be in anyway prohibitive. In fact, it could, with astute management, reduce resource requirements to a minimum.

### *6.4.5.1 Tapping the Resources of the Arab Diaspora*

Tough economic conditions coupled to sectarian and political strife turned several countries in the region into net exporters of human capital. Traditionally, Lebanon, Syria, and Palestine topped the list of Arab diaspora initially targeting the Americas as a preferred destination, during the late 19th and early 20th centuries but gradually moving in greater numbers toward countries in Europe and Africa.

This trend returned with added vigor during Lebanon's civil war and several other countries joined in, including some such as Iraq, Egypt, and Sudan, whose populations never harbored émigré traditions before. With so many of its people now away from their homes in neighboring countries as well as overseas, Syria's diaspora is probably exceeded only by Lebanon's with many professionals, including experts in various fields of science, technology, and innovation seeking new lives and careers, principally in Europe and North America.

A recent report estimated that nearly 20 million people of Middle Eastern and North African (MENA) descent, i.e., around 5% of the Arab countries' population, live and work away from their countries of origin, mostly in Europe and the Americas.[42] Estimates that place the number of those who took up careers in science, technology, and management are unreliable. However, their proportion is estimated at one-fourth of this total. At any rate, examples abound of young men and women of Egyptian, Lebanese, Iraqi, Palestinian Moroccan, Algerian, and Tunisian ancestry, who over the past few decades made outstanding careers in various fields of specialization in their adopted countries. The latter three countries are said to have 2.8, 1.7, and 1.2 million citizens in the diaspora, respectively.

On a personal level, members of the diaspora appear quite willing to contribute to the welfare of relatives in their countries of origin. Thus, remittances by MENA emigrants and their descendants sent amounted to around US $53 billion in 2014, with remittances to Lebanon and Jordan making up about 15% of GDP in each country. On the other hand, little has been done by governments to tap the rich network of expertise embodied in the collective Arab diaspora.

Results of a survey conducted among members of the MENA diaspora for this report regarding contributions to development of their countries of birth[43] indicated that "mentoring" was one of the most common forms of ongoing involvement by the survey's respondents with home populations. However, the survey also underlined the need to overcome mistrust on the part of diaspora communities of MENA governments, to which "elite diaspora" members seemed to be particularly susceptible.[44] Such mistrust may be understood in the context that it was originally their governments' policies that forced them to leave their home countries for greener pastures. At any rate, Arab émigrés emphasized the need for their respective governments to facilitate their engagement as partners, capable of contributing to such development.

Annual conferences are held, for example by Lebanon's Ministry of Foreign and Affairs and Immigrants, for émigré community members, including business people as well as scientists and technologists. Official as well as unofficial networks of scientists and engineers were launched by Syria, Lebanon, Palestine, and Iraq over the past two decades or so. While detailed analysis of the impact achieved by such networks is yet to be undertaken, outward signs indicate that limited substantive outcomes have in fact been achieved to date. An explanation often put forward by members of émigré communities is that conditions in their home countries were still far from conducive for launching networking programs with the intention of promoting collaboration between members of endogenous STI community and their counterparts among the country's diaspora. Indeed, to many intents and purposes, a good deal of the networking has merely served to find positions for more qualified young people to escape conditions in their home countries for better opportunities in academia and business. At any rate, indications are that involvement by Arab diaspora in building STI capabilities in their countries of origin remains a low-keyed affair compared to what is undertaken by several other countries with sizeable expatriate STI communities, such as China and India. Reasons for this are many, but may be linked primarily to meager resources allocated by the concerned government institutions for undertaking serious collaborative ventures.

Turning this vicious cycle into a benign one is a task for both government authorities, civil society institutions, as well as professional associations at home and abroad, in countries where members of the Arab diaspora live and work. Chances for positive developments, at least in areas linked to new technology-based industries, are presented by the need for European and North American enterprises to access relatively low-cost but expert labor, to face competition by China and other emerging industrial nations. An example of this trend is provided by recent investments by aircraft manufacturers Bombardier of Canada, aimed at producing aircraft parts in Morocco, which followed an earlier decision by Boeing in 2001, that was initiated by an expatriate Moroccan working for Boeing, with the Moroccan government moving in to facilitate both ventures.[45]

However, as illustrated by the above example from Morocco, little would have been achieved without the government adopting a proactive position, with measures designed to encourage contributions by their diaspora to socioeconomic development in their countries of birth.

Countries such as Lebanon, with a diaspora numbering millions around the world, some of whom have achieved high status in their adopted countries stand to reap immense benefits should its government adopt conducive measures. Such measures would need to address both sustaining links to members of its diaspora, e.g., through giving Lebanese mothers married to non-Lebanese the right to pass their citizenship and the right to vote on to their children, on the one hand, as well as adopt an open attitude to their likely contributions without prejudice based on sectarian or political considerations. Naturally, there is always the need to ensure that investments along the lines hosted by Morocco hold concrete benefits for the entire country and are socially responsible as well as economically viable.

In conclusion, a great deal may be achieved should future STI policies, and revisions thereof place particular emphasis on mobilizing the Arab diaspora to promote economic development, regional integration, and entrepreneurship. To achieve such ends, governments must facilitate the role that their émigré communities could play through encouraging direct investment they might mediate, as well as the transfer of knowledge, and professional networking capabilities they are capable of bringing about. Additionally, given governments' legendary lack of agility, greater emphasis is needed on the roles of civil society, professional associations, as well as industry and trade federations in realizing concreted contributions by the Arab diaspora's to evolving STI systems at home.

#### *6.4.5.2 Novel Avenues for International Collaboration*

European Union collaborative research programs offer a useful example for international STI cooperation. EU funding is essentially project-oriented and organized on the basis of competitive schemes. A research institution taking part in a given research project would receive funding in addition to regular provisions made available from the national budget.

Given the political situation that now prevails in the region, and how things might be expected to evolve over the coming few years, prospects for formal multilateral cooperation in research might be limited to a small number of the Arab Gulf countries, with few prospects for wider regional collaboration for the time being. On the other hand, it is far more likely that bilateral schemes would show positive outcomes in the short run. Nevertheless, it would be essential to undertake radical revisions of collaboration modalities of such schemes if they are to proliferate and constitute adequate bases for wider multilateral collaboration. At any rate, it is essential to realize that expanding existing or future bilateral cooperative arrangements may not be looked at as an alternative to multilateral cooperation. Rather, it would be quite useful to continue and even further develop existing bilateral arrangements while cultivating wider regional collaboration.

### *6.4.5.3 Collaboration Among the Arab Countries*

That collaboration in scientific research and technological development among the Arab countries is rather limited should come as no surprise, since regional collaboration on almost all other fronts, economic, political, and educational is minimal. On the other hand, it would be true to state that collaboration in higher education and research, both formal and informal, is probably in a better state than collaboration in many other areas of likely joint endeavors. Thus, mostly informal collaboration in higher education takes place, through recruitment of Arab nationals by university institutions in countries other than those of their birth.[46] The Arab Gulf countries count as prominent beneficiaries of such collaboration in substantive terms. With their tradition as recipients of many Arab and overseas experts in a wide range of subjects, they could well qualify as home to actual and virtual networks in a variety of STI domains.

More formal collaboration is also on the rise with regard to joint research publications in international journals as well as through cooperative in agricultural research, mediated by organizations such as ACSAD and ICARDA.

Several Arab organizations established over the past few decades with the intention of promoting science, technology, and innovation should facilitate Arab research collaboration. Notable among these organizations are the Kuwait Foundation for the Advancement of Science (KFAS) and the Mohammed bin Rashid Foundation. With the exception of attending joint regional and international workshops and issuing reports, dedicated to STI affairs, few if any collaborative STI programs appear to be undertaken by the above institutions. The situation might be somewhat different in the case of new institutions such as the Qatar Foundation (QF) and Al Masdar City for Science and Technology. Both institutions report support for STI activities, however, it may be too early to pass judgment on their roles as hubs for STI collaboration, given that they have been active on this front for a brief period of time.

A major obstacle to making better use of collaborative research funding is the lack of expertise in such arrangements. In other words, more than lack of funding, it is the inability to manage collaborative research and innovation activities that hinders effective research cooperation. On the other hand, participation in collaborative research within international networks without requisite consolidation strategies would most certainly result in diluting even the most useful outcomes. Clearly, consolidation strategies would need to be based on well-founded national and institutional policies as well as reflect national and regional priorities.

## 6.5 CONCLUDING REMARKS

Collectively, the Arab countries possess many assets that should enable them to meet many of the challenges they face along with many other developing countries across the world. If well managed, their geographic location and mineral endowments, on the one hand, and the fact that they possess one of the largest youthful populations in the world, on the other, hold an enormous promise for a bright future.

However, immense efforts will be needed to address a range of shortcomings, in governance and socioeconomic policies. Reforming educational and STI policies and institutions and enhancing their resource allocations may not be relied upon to rid the Arab countries of all the above shortcomings, they will certainly, however, create conditions for tackling them through gradual and peaceful means. Equitable, inclusive, and sustainable socioeconomic development may not be realized otherwise.

Populations of the Arab countries are anxious to see not only economic growth and better employment opportunities for the skilled among advantaged segments of society but also to experience tangible improvements in basic services, with special emphasis upon improving the quality of the country's education and vocational training as well as healthcare systems. With regard to the latter point, it is essential that improved services are extended to all rural areas, to help reduce disparities and inequalities among social groups and particularly those affecting girls and women. Carefully designed and targeted STI inputs should certainly help people in the region to improve their access to basic services and enhance their future prospects.

The road toward a future in which acquiring and utilizing STI knowledge is both a central goal as well as the source of guidance and a tool for inclusive and sustainable socioeconomic development is littered with many threats and challenges, but the rewards are well worth every sacrifice. Additionally, for future visions and policy regimes to succeed in building effective and durable STI capacity and attain worthwhile development objectives, there will be a strong need to discard entrenched rentier systems throughout the Arab countries. Obstacles facing this goal must be acknowledged upfront, as well as tackled through action on several fronts.

Successful schemes in all fields of human endeavor must be based on a clear vision, comprehensive policies, carefully designed strategies, adequate logistics, capable institutions, and qualified individuals. Education, scientific research, technological development, and associated innovations are no exception. It is in this sense that science, technology, and innovation must act, as both catalysts, and primary objectives for a much better future in the months and years ahead.

## ENDNOTES

1. Transparency International's Report for 2010. Downloaded in June 2017 from: https://www.transparency.org/cpi2010/in_detail#.
2. "Crony capitalism in Egypt." Journal of Globalization and Development (2015). Chekir, Hamouda, and Ishac Diwan. Downloaded in June 2016 from: scholar.harvard.edu/idiwan/publications/crony-capitalism-egypt-1JGD 2014.

3. "The Effects of Entry on Incumbent Innovation and Productivity;" Philippe Aghion, Richard Blundell, Rachel Griffith, Peter Howitt, and Susanne Prantl. Downloaded in June 2016 from: http://personalpages.manchester.ac.uk/staff/rachel.griffith/PublishedPapers/AghionBlundellGriffithHowittPrantl2009.pdf.
4. Large corporations in Egypt include several enterprises that are controlled by the country's armed forces.
5. See endnote 2.
6. "Illicit Financial Flows from Developing Countries: 2000–2009 Update with a Focus on Asia;" D. Kar and K. Curcio, Global Financial Integrity, January 2011. Downloaded in June 2016 from: http://www.gfintegrity.org/storage/gfip/documents/reports/IFF2010/gfi_iff_update_report-web.pdf.
7. See endnote 2.
8. The Journal of Finance, vol. LII, No. 2, June 1997. Downloaded in June 2016 from: http://scholar.harvard.edu/files/shleifer/files/surveycorpgov.pdf.
9. See endnote 2.
10. "Political Connections and the Allocation of Procurement Contracts;" Eitan Goldma, Jörg Rocholl, and Jongil So, December, 2010. Downloaded in June 2016 from: http://idei.fr/sites/default/files/medias/doc/conf/icied/papers_2011/goldman.pdf.
11. "Politically Connected Boards of Directors and The Allocation of Procurement Contracts;" Eitan Goldman, Jörg Rocholl and Jongil So. Published by Review of Finance, Oxford Journals 2013. Downloaded in June 2016 from: https://rof.oxfordjournals.org/content/early/2013/01/10/rof.rfs039.full.
12. Available population statistics for the GCC countries do not differentiate nationals from expatriates. Hence, the above figures may be biased toward higher age groups, with the median age of GCC populations being actually lower, and the proportion of the population below the age of 14 higher, than actually reported.
13. Official version of Saudi Arabia's Vision 2030 document. Downloaded in January 2017 from: http://vision2030.gov.sa/en.
14. The corresponding percentage exceeds 80% in industrialized developed nations and the countries of Central Asia.
15. The UNDP/MBRF report does not in fact discuss the results of its evaluation schemes. Rather, it merely defines its methodology and lists the results of its implementation.
16. The PISA survey for 2015 focused on science, with reading, mathematics, and collaborative problem solving as ancillary areas for evaluation. Approximately 540,000 students completed the assessment in 2015, representing about 29 million 15-year-olds in the schools of the 72 participating countries and economies. See also "PISA Results in Focus, 2015." Downloaded in November 2016 from: https://www.oecd.org/pisa/pisa-2015-results-in-focus.pdf. Available under the Creative Commons Attribution-Non Commercial-ShareAlike 3.0 IGO (CC BY-NC-SA 3.0 IGO).
17. In this connection, it is well to keep in mind that although the OECD countries comprise some of the richer countries in the world, they also include quite a few other countries in which per capita incomes are comparable to those enjoyed by some of the more resource-rich Arab countries.
18. "Information and Communication Technology in Education in Five Arab countries;" Information Paper by UNESCO; April 2013. http://uis.unesco.org/sites/default/files/documents/information-and-communication-technology-ict-in-education-in-five-arab-states-a-comparative-analysis-of-ict-integration-and-e-readiness-in-schools-en_0.pdf.
19. "Scientific research and higher education in the Arab World;" Presentation by Dr. Sultan T. Abu-Orabi Secretary General, Association of Arab Universities.
20. Comparing the Arab countries as a group to individual countries in their neighborhood might be called into question on a point of methodology, since resources at their command vary widely. However, such comparisons might be justified on socioeconomic and cultural similarities.
21. See endnote 13.

22. The Human Development Report issued on annual basis by UNDP has reported MPI estimates since 2010.
23. These countries are Djibouti, Egypt, Iraq, Jordan, Libya, Mauritania, Morocco, Palestine, Sudan, Syria, Tunisia, and Yemen.
24. Fig. 6.8 is based on data from the UNDP's Human Development Report 2015.
25. The Human Development Report 2010. Published by the United Nations Development Program (UNDP). See: http://hdr.undp.org/en/reports/global/hdr2010/chapters/.
26. From an original set of around 160 countries, inequality data on some 30 countries is lacking. Eight of the group of countries for which data are lacking are Arab countries.
27. A strong negative correlation may be demonstrated between the magnitude of corrections for income inequality and the original HDI and EI values. That is to say, smaller HDI or EI values are associated with larger relative corrections. This is clearly an outcome income inequality gaining grounds as HDI and EI values decrease. The trend applies for the entire set of data on all countries of the world for which statistical information is available.
28. These countries included Algeria, Comoros, Djibouti, Egypt, Jordan, Mauritania, Palestine, Sudan, Tunisia, and Yemen.
29. UNDP Website accessed in June 2017 at: http://hdr.undp.org/en/content/income-gini-coefficient.
30. That is all the LAS member countries with the exception for Iraq, Palestine, and Somalia, for which no data was available from the report by Credit Suisse.
31. "Focus on Inequality and Growth; Does income inequality hurt economic growth?" Publication issued by the Directorate for Employment, Labour and Social Affairs, OECD. December 2014. https://www.oecd.org/social/Focus-Inequality-and-Growth-2014.pdf.
32. Rising inequality is estimated by this publication as having reduced growth in half of the OECD countries by more than 4% points, over two decades, while greater equality before the global financial crisis in 2007–2008 helped increase GDP per capita in countries such as Spain.
33. That the richest in several countries across the world are accumulating wealth at an ever-increasing rate is highlighted by Thomas Piketty in his "Capital in the 21st Century;" necessitating policy options that ensure a fairer distribution of resources to promote inclusive growth.
34. This notion is highlighted, for example by a UNU/Intech Discussion Paper published in 2002 under the title "What is the 'Knowledge Economy'? Knowledge Intensity and Distributed Knowledge Bases." by K Smith, Downloaded in April 2017 from: http://www.intech.unu.edu/publications/discussion-papers/2002-6.pdf.
35. The World Economic Forum's Global Competitiveness Report 2016–2017.
36. The example of the Small Business Innovation Research (SBIR) program of the United States is often quoted in this regard. The SBIR program apparently allocates 2.5% of the total research budgets of all federal agencies with budgets approaching $100 million to contracts or grants designed to benefit small business enterprises.
37. Personal communication, through activities undertaken by the Social Support Society at Syrian refugee camps in the Beqa' valley in Lebanon.
38. See endnote 13.
39. See for example the study performed by the Deutsche Gesellschaft für Internationale Zusammenarbeit (GTZ) on behalf of the Federal Ministry of Economic Cooperation and Development. Listed under endnote 106 in Section 3.2.5.5.
40. The danger that such clusters would become too engrossed with artisanal activities targeting local markets with traditional processes would have to be resisted in favor of innovative action seeking the incorporation of novel inputs such as renewable energy, environmentally sound manufacturing.
41. Incidentally, the foundations established in both Gulf countries may not be described as purely private since they were set up by top members of ruling families in these two countries.

42. The governments of Morocco, Tunisia, and Algeria estimate the sizes of their diaspora at around 2.8, 1.2, and 1.7 million, respectively. Both Morocco and Tunisia have set up government offices to cater to issues posed by their diaspora communities. Both governments also allow dual citizenship and overseas voting. "Arab World Diaspora's Strong Attachment to Home could Play Role in Regional Development;" World Bank document, downloaded in May 2017 from: http://www.worldbank.org/en/news/feature/2017/03/28/arab-world-diaspora-s-strong-attachment-to-home-could-play-role-in-regional-development.
43. "Arab World Diaspora's Strong Attachment to Home could Play Role in Regional Development;" World Bank document, downloaded in May 2017 from: http://www.worldbank.org/en/news/feature/2017/03/28/arab-world-diaspora-s-strong-attachment-to-home-could-play-role-in-regional-development.
44. See endnote 43.
45. See endnote 43.
46. Thus, many Syrian, Egyptian, Lebanese, Palestinian, and Jordanian professors and researchers are employed by universities and research centers in the Arab Gulf countries.

# Index

*Note*: Page numbers followed by "f" indicate figures, "t" indicate tables and "b" indicate boxes.

## H

## J



## K

## L

## M

## N

## O



## P

## S

## V



## W

## Y



## Z

Kolibri 360 GmbH
Ferdinand-Jühlke-Straße 7
99095 Erfurt, Deutschland
produktsicherheit@kolibri360.de